Wireless Security Architecture

Wireless Security Architecture

Designing and Maintaining Secure Wireless for Enterprise

Jennifer (JJ) Minella

WILEY

Published by John Wiley & Sons, Inc., Hoboken, New Jersey.

Published simultaneously in Canada and the United Kingdom.

978-1-119-88305-0
978-1-119-88306-7 (ebk.)
978-1-119-88307-4 (ebk.)

Library of Congress Control Number: 2021952991

Cover images: © Leigh Prather/Adobe Stock Photos
Cover design: Wiley

SKY10033345_022622

This book is dedicated to my parents: to my Dad, who taught me to "follow the packet" and to my spunky Mom who we're told was the first woman in the world with a WLAN certification.

—Jennifer Minella

About the Author

Having begun working in the tech industry at the tender age of 14, **Jennifer (JJ) Minella** has spent the last 20 years helping security-conscious organizations solve network security challenges in meaningful ways.

Throughout that time, Jennifer has led strategic research and consulting for hundreds of organizations ranging from government, education, healthcare, and mid-market up to Fortune-5. This book is based on experience accumulated through the thousands of consulting and implementation engagements across these diverse environments. Jennifer holds, or has held, numerous technical certifications over the years ranging from switching and routing to firewalls, remote access, NAC, Wi-Fi, and IoT wireless, along with CISSP and other security certifications.

She's also a published author, internationally recognized speaker on network security and identity, analyst with IANS, member of Forbes Technology Council, and former chairperson of $(ISC)^2$. Combining master-level network architecture skills with CISO and compliance experience, she continues her mission of aligning security and networking as Founder and Principal Advisor of Viszen Security.

Jennifer is also known for introducing mindfulness-based leadership to individuals and organizations in infosec. And aside from meditation and security, she's a competitive powerlifter and dancer. It's also been reported that she loves Figment, the Imagination Dragon. For more information visit her company at `www.viszensecurity.com` or award-winning blog at `www.securityuncorked.com`.

About the Technical Editor

Stephen Orr is a twenty-three-year veteran of Cisco and serves as a Distinguished Architect for Cisco's Global Systems Engineering Organization. Stephen currently participates in multiple industry working groups such as the Wi-Fi Alliance where he is currently serving as the Chair for the Security Marketing and Security Technical Task Groups focusing on WPA3 as well as one of the Vice-Chairs for the IEEE 802.11TGbh Working Group (Operation with Randomized and Changing MAC Addresses).

Acknowledgments

They always say it takes a village to raise a child, and the creation of this book is no exception. Having worked in many areas of technology and information security for the last 20 years, the one thing I'm acutely aware of is exactly how much I *don't* know. It's a fool's errand to be the master of all technology in today's world. And while this book was based on my experiences with clients over the past twenty years, the depth of value in this content would not be made possible without the support of the communities.

First, Stephen Orr deserves a special callout; not only did he do exceptional work with the foreword and technical editing; he went above and beyond, helping me research and dig into complex topics, often sifting through lengthy documents to help aggregate data to be included here. As a Distinguished Cisco Engineer and leader of the Wi-Fi Alliance Security Task Group, I knew Stephen would ensure this text was accurate and inclusive and offer balance in many ways including supplementing my knowledge gaps.

We had a bit of a relay with some of the technical editing and I'm deeply appreciative of Adrian Sanabria, Mick Douglass, and Matthew Gast for their contributions throughout the book and for helping us stay on schedule. Their insights have been invaluable.

When certain topics of this book extended beyond my reach, the infosec and Wi-Fi communities were there. Instead of making educated guesses on the operation of certain technology, I solicited feedback from subject matter experts in addition to our dedicated technical editors. With that, additional thanks are sent to Peter Mackenzie, Troy Martin, Andrew von Nagy, Keith Parsons, Jonathan Davis, Sam Clements, and the lovely Tauni Odia. Additional thanks go to my infosec family, especially Chuck Kesler, Ryan Linn, my fellow IANS faculty, and friends at SecureIdeas.

This book wouldn't be possible without an introduction from none other than David Coleman to Wiley, and special thanks goes to Jim Minatel, Pete Gaughan, Tom Cirtin, and the editors at Wiley who indulged my completely ludicrous timeline to release this book.

Last but not least, special thanks are owed to my "work family" including Julie Allen, other colleagues from my twenty years with Carolina Advanced Digital, Inc., and amazing clients over the years who have trusted their infrastructures to me and let me into their lives and datacenters. You're the reason this book exists.

Contents at a Glance

Contents

Foreword

With all the innovations and alphabet soup of the different 802.11 standards, Wi-Fi security is one of those functions at the core of all modern protocol development. We are in an age of ensuring that we protect not only the information in transit, but also the identity of the user and device, and we need to balance this with protecting the access to personal and corporate resources. Modern-day security development and deployment for Wi-Fi is a delicate balancing act where all these nuances must be considered. If you make security too difficult to use/deploy, people will bypass it or turn it off; if it's not secure enough, you will surely make the headlines.

Full disclosure, Wi-Fi security is a topic near and dear, so I am slightly biased to its importance. My journey started by asking a simple question: how is Wi-Fi Protected Access different than WEP and could this be deployed for use in government agencies? Years later this continued interest in Wi-Fi security led to an opportunity to participate in the Wi-Fi Alliance Security Working Groups, where I was introduced to a talented group of individuals focused on advancing wireless security capabilities. Fast forward 10+ years and this group has addressed vulnerabilities in Wi-Fi Certified WPA2, launched Wi-Fi Certified WPA3 (multiple releases), addressed the challenges of Open networks with Wi-Fi Certified Enhanced Open, and finally, set a new bar for wireless security by eliminating WPA2 and Open networks in new MAC/PHY bands starting with 6 GHz. Introducing new security requirements is hard and not always fast. Status quo is easy and change is difficult specifically with security, but sometimes ripping the bandage off may need to be done because the major headlines get made when there is an issue with Wi-Fi security.

When asked to write the foreword for JJ's book on Wireless Security Architecture, I needed to think about how to frame the journey readers will take because the context of the title is important. The topic isn't securing wireless—it's wireless security architectures, which is an overarching defense-in-depth approach to creating an architecture that meets your business objectives, starting at the wireless access layer with security at the forefront and continuing that into the enterprise network. Most don't look past the 802.11 layer when we talk about Wi-Fi security, however 802.11 is now the dominant access layer and care must be taken on its integration into your enterprise.

Entire books have been written on the intricacies, nuances, and in-depth protocol discussions of 802.11 security. They have done so from the point of view of the interpretation of the standards bodies (IEEE 802.11) or the certification organizations such as the Wi-Fi Alliance, but not from the perspective of the CXO, network operator, network architect, or the user. Where this book is different is that it is all about context. JJ takes a unique approach to making wireless security relevant and peels back the complexities to show how security at the wireless access layer should be integrated as part of your overall architecture design and strategy, as opposed to bolted on as an independent afterthought. Additionally, the discussions and insights on how compliance will cause decisions to be made and impact architecture designs are invaluable for those dealing with customers that must meet requirements set forth by PCI, HIPAA, NIST, etc.

I hope you enjoy reading the book as much as I did, and I will leave everyone with a few final thoughts.

Security is a continual process; requirements and designs must always be reevaluated. What was good three years ago may not provide the same security levels that you need today; "good enough" security is never "good enough."

Raise the bar. Security can and should be at the forefront of the discussions with your vendors, consultants, and architects during the acquisition process, not after. Certifications are critical to the security conversation—not only do they define interoperability, but they ensure conformance. You shouldn't fall back to compromising your network because the latest security standards are not supported (remember you will make the news, not them).

Ensure your use cases are driving development by asking questions, contributing to forums, and getting involved.

Stephen Orr

Preface

This book was envisioned with the goal of empowering a broad category of network and security professionals to design and maintain secure wireless networks in enterprise environments.

It focuses on the most current and relevant details and offers appropriate background information and explanations that will persist for years, even as technology evolves. This book teaches network, wireless, and security architects "how to fish" and make ongoing decisions about how best to secure networks.

Wi-Fi-connected devices alone have tripled in the past six years, surpassing 20 billion in 2021. That growth, coupled with the projected 75 billion connected IoT devices by 2025 means network and security teams will be faced with new challenges securely connecting all of these "things" and protecting enterprise assets from the ones that introduce risk. With wireless connections growing exponentially, security threats increasing, ransomware rampant, and new initiatives like zero trust strategies and digital transformation, yesterday's technologies and techniques aren't sufficient to secure tomorrow's enterprise. And in fact, they're not even adequate to protect us today.

Plus, these new initiatives necessitate tighter integration between network and security disciplines. Cross-functional teams mean greater communication and knowledge sharing, and this book bridges the divide between risk management and network architecture.

This book merges the concepts of enterprise security architecture with wireless networking and teaches professionals how to design, implement, and maintain secure enterprise wireless networks. It covers everything a technology professional needs to know to make sure the organization's wireless matches the organization's risk models and compliance requirements.

Who This Book Is For

The reader is assumed to be an IT or infosec professional with basic networking knowledge (in line with Network+ or Cisco CCNA). Wi-Fi experience, while helpful, is not required. There are portions of the book written for technical and non-technical leadership, summarizing more complex technical recommendations into business requirements.

For organization size, this book is appropriate for security-conscious organizations of all sizes and industries from small schools to universities, healthcare systems, state and local government, commercial enterprise, and even federal agencies. The only requirement being that this content is most relevant for environments using fully managed switches.

Wireless, and especially Wi-Fi, touches so many aspects of the enterprise network, from wired infrastructure to authentication servers, endpoint management, and security monitoring. This book is for you if you are:

- Network, wireless, and security architects responsible for designing secure network and wireless infrastructures. You're the primary audience for the book.
- IT professionals, network engineers, and system admins interested in learning more about securing wireless. Even if your role isn't architecting, you'll learn a lot about supporting and maintaining secure wireless systems, devices, and users.
- Technical leaders with strategic responsibility for network security and security controls compliance. While you may prefer to skip some of the more technical minutia, this book provides valuable insight into what's possible, what's practical, and the complexity of meeting stringent security postures.
- Non-technical executive leadership and boards with oversight of risk management. Only portions of this book are appropriate for non-technical audiences, but they provide valuable actionable business-level guidance.

While the vast majority of my own experience has been with clients in the United States, this book addresses the variations in policies and compliance requirements across the globe, and the standards and technologies used are applicable in all parts of the world, with slight variations in implementation such as the RF frequencies, which are localized.

I truly hope this book will spark conversations in the networking and infosec communities, and that each professional will take aspects of this content, make it their own, and further expand the reach of not only the book, but of their own experiences and contributions.

As always, we have done our best to be as complete and accurate as possible, but alas we are human, and errors are to be expected. Although I had amazing technical editors, I accept full responsibility for any errors or omissions, and graciously appreciate any feedback including corrections. If you come across an error, please email `wileysupport@wiley.com` with the subject line "Possible Book Errata Submission." For other feedback you can contact me on LinkedIn or at `info@viszensecurity.com`.

Distinctive Features

This book separates itself from others in many ways. While other books, authors, and publishers of network security topics add valuable content and context, this book offers these unique features:

- This book is vendor-neutral, but references vendor-specific features and documentation where applicable to ensure guidance is actionable. Many vendors call the same feature by different names, and that's addressed throughout the book.
- Advice throughout this book is based on hundreds of real-world implementations, not academic or theoretical research.
- This book addresses current and, at times, relevant future technologies, and anticipated changes.
- While other books focus on technical standards and specifications, this book provides an applied how-to approach to use that knowledge in practice.
- Unlike many training programs and books, this book seamlessly merges security, risk, and compliance considerations with network architecture at each step.
- This book prepares professionals for successfully executing their work, not merely for passing a certification test.
- The content is presented in a casual and conversational tone, making it accessible to a wide audience by not relying on niche terminology and acronyms.

Introduction

In business environments, wireless and especially Wi-Fi networks are configured and maintained by a breadth of technology professionals—from Wi-Fi specialists to the sole IT professional left to juggle everything from networking to managing endpoints, applications, and servers. This book brings deep technical details to the seasoned wireless professional and summarizes best practices in easy-to-follow advice for those wearing many hats.

After fifteen years of stale Wi-Fi security suites and a limited focus on IoT security, the world of wireless is finally putting the spotlight on security. But designing secure wireless networks isn't nearly as straightforward as it seems. The newest WPA3 security suite greatly enhances security, but also introduces complexity as organizations move from legacy security to the latest standards.

This book reframes and redefines architecting secure wireless, opposing outdated guidance in favor of more robust security practices meant to address today's and tomorrow's evolving wireless networks. Its contents walk professionals through the decision-making steps of architecting secure networks, starting from risk and compliance considerations to detailed technical configurations. Along the way, it offers practical guidance, best practices, and specific recommendations for a variety of environments, vendor implementations, and security needs.

Overview of the Book and Technology

Securing wireless networks today requires a new way of thinking and a new set of tools. The best tool in the architect's toolbox is knowledge, and that's what this book delivers.

Along with recommendations for designing with best practices, *Wireless Security Architecture* also offers deep technical journeys into areas of encryption, authentication, authorization, segmentation, certificates, roaming, hardening, and more. Each chapter offers practical guidance along with technical details, empowering professionals to make informed decisions about how best to secure their environments.

One unique aspect of this book is that the full work is presented in a conversational tone, eliminating the rigidity of academic wording that can be a barrier to easy comprehension. Also, all chapters have been written by a single author, and common connected topics are woven and cross-referenced throughout.

This book addresses the full breadth of enterprise wireless products, with a strong focus on Wi-Fi. In it, architecture and security design considerations are offered for:

- Controller-managed Wi-Fi
- Cloud-managed Wi-Fi
- Autonomously managed Wi-Fi
- Small, medium, and large environments
- Specific manufacturers including Cisco, Aruba, Juniper Mist, and Extreme, among others
- Specific industries such as healthcare, government, and education
- Non-traditional endpoints and IoT devices

How This Book Is Organized

Wireless Security Architecture is sure to become the definitive guide for designing and maintaining secure wireless networks in any size organization. With content for wireless specialists and technology generalists alike, this book covers deep technical topics with appropriate introductory concepts that will allow non-wireless IT professionals to learn and follow along. And while it includes foundational knowledge, extraneous historical details such as the history of WEP have been deliberately omitted to keep the reader's focus on current technologies.

To remain vendor-neutral, the language used in this book is based on natural language. Where appropriate, in the more technical areas of the book, current vendor terminology and features are called out, allowing readers to easily find and further research topics within the context of their environment. Some vendors' configuration guides exceed 2,000 pages for a single product, and most enterprise Wi-Fi deployments incorporate several integrated solutions for management and monitoring, easily bringing the total to more than 5,000 pages of documentation—which often don't include details on security hardening.

At times, the terminology used may be purposefully deviant from a particular vendor feature name to avoid confusion. As one example, Cisco has a feature called Infrastructure MFP, which is easily confused with (but very different from) the IEEE 802.11w standard for Management Frame Protection (MFP), also known as Protected Management Frames (PMF) by the Wi-Fi Alliance. To avoid confusion, PMF is used when referring to 802.11w.

You may also notice some acronyms are intentionally repeated along with their full text, breaking traditional editorial conventions. Just within wireless, acronyms can be frequently re-used: "PSK" for example may mean pre-shared key but can also refer to phase-shift keying. Introducing security and IoT disciplines only complicates matters further: in this book "SOC" could mean security operations center, or system on a chip. To avoid confusion and make the text more accessible to a broad audience, these are often spelled out repeatedly.

The following is a brief summary of each chapter's content. In many cases, the chapters build on one another, adding technical context at each step. Expect to see topics repeated as the book progresses but presented in evolving context along the way. To help readers that may be skipping to specific portions of the book, cross references are included.

Part I, "Technical Foundations," introduces technical foundations for the reader and encompasses Chapters 1–4.

Chapter 1, "Introduction to Concepts and Relationships," is a level-set to get diverse professionals on the same page with foundational concepts of information security, high-level technical concepts that impact security, and an overview of wireless technologies and architectures. This chapter sets the tone by defining identity, authentication, and offering an entry to cryptography concepts.

The first technical content, Chapter 2, "Understanding Technical Elements," provides the underpinning of all content that follows with a deeper dive into data paths, segmentation methodologies, and the first dive into security profiles for Wi-Fi including the new WPA3 security suite.

Chapter 3, "Understanding Authentication and Authorization," is filled with every detail a professional ever wanted to know about authentication and authorization, including 802.1X, EAP, RADIUS, certificates, and MAC-based authentications.

Rounding out Part I, Chapter 4, "Understanding Domain and Wi-Fi Design Impacts," explains the symbiotic relationship of secure wireless architecture to network design elements and RF planning, with a strong focus on secure roaming protocols.

In Part II, "Putting It All Together," the reader is taken on the journey of planning the network and security architecture based on technical concepts from Part I.

Chapter 5, "Planning and Design for Secure Wireless," walks the reader through the author's own design methodology for planning secure wireless. It includes pointed questions to ask during scoping, and several planning templates and worksheets for the reader to use or modify—including complex policy matrices as well as simplified planners.

Hardening the infrastructure is the focus of Chapter 6, "Hardening the Wireless Infrastructure," with extensive guidance, tiered recommendations, and relevant vendor-specific sidebars.

Part III, "Ongoing Maintenance and Beyond," picks up where planning and design of Parts I and II left off.

Chapter 7, "Monitoring and Maintenance of Wireless Networks," addresses monitoring and maintenance of wireless networks including pen testing and audits, along with ongoing management with WIPS, and specific recommendations for logging, alerting, and reporting. As an added bonus, troubleshooting tips are included here as well.

Chapter 8, "Emergent Trends and Non-Wi-Fi Wireless," segues into the less evergreen topics, covering the more variable technologies of IoT and emergent trends such as remote workforces, BYOD, and zero trust.

The appendices present four topic areas, starting with Appendix A, "Notes on Configuring 802.1X with Microsoft NPS." Appendix B, "Additional Resources," offers hints on navigating IETF and IEEE documents, and Appendix C, "Sample Architectures," includes the much-requested examples of secure wireless architectures. A few niche topics are covered in Appendix D, "Parting Thoughts and Call to Action."

Why Read This Book

This book delivers insightful knowledge based on hundreds of real-world implementations and aggregates data and recommendations from thousands of pages of standards, vendor documents, and best practices white papers.

Offering a blend of relevant technical details alongside summarized best practices, the book offers advice within the context of a flexible framework that allows network and security architects to adapt and layer concepts as needed to meet their needs.

Whether you're a Wi-Fi professional, network admin, security architect, or anything in between—this book will be your go-to resource for planning and maintaining secure wireless networks.

What's on the Website

In addition to the material provided in the book, additional and updated supplemental materials such as downloadable planning templates and cheat sheets can be found online at the books' website `www.securityuncorked.com/books`.

The contents of Chapter 5's section "Notes for Technical and Executive Leadership" are also made available to download and share. Our hope is this messaging will empower you to further negotiate within your organization and create allies with non-technical stakeholders.

Congratulations

On behalf of the author, technical editors, and the Wiley team, we appreciate the opportunity to share this body of work with the technology community at large, and hope you find it a useful tool for helping to make our world a safer place.

Oh, and one last note. We may be technologists, but we're all still just human. Don't be afraid to ask questions, ever. If this book teaches you nothing else, it should demonstrate the complexity and breadth of information technology. Asking what an acronym means, how a technology works, or how something is done when you don't know is the only way to learn, grow, and share.

Part

I

Technical Foundations

In This Part

Technical Foundations

In This Chapter

CHAPTER

1

Introduction to Concepts and Relationships

Before we dive in to the how-to, I want to introduce some concepts at a high level and explain the relationships among them so there's at least a baseline context moving through the subsequent chapters.

The overview here will look at the roles and responsibilities of the various technical teams involved in wireless architecture and security, then touch on basic security concepts that will be referenced throughout this book, and after that will cover some key foundational wireless concepts.

Since this book is written for a variety of technology professionals, some of you will have deep knowledge in one or more areas, and others may be looking to fill some gaps. The following sections serve to get us all on the same level of understanding and taxonomy, so we have a common starting point for the architecture conversations that follow.

NOTE While this chapter introduces concepts that will be referenced in the context of wireless security, the terms and constructs here transcend wireless or even network security topics. I've purposefully avoided confounding textbook-style definitions and tried to maintain a focus on the application of the definitions within the scope of this book's content.

First, we'll look at the roles and responsibilities of various people and teams within the organization, how they may relate to your wireless security architecture, and what to do if that person or role is nonexistent in your organization. After that is an introduction to basic security concepts relevant to wireless security and an explanation of each specifically in the context of wireless security (vs. diving into lengthy definitions). Finally, we'll cover a few foundational wireless concepts that will impact your architecture, since we're not assuming everyone reading this book has deep (or possibly any) enterprise wireless experience.

This chapter covers an introduction to concepts and relationships organized in three parts:

- Roles and Responsibilities
- Security Concepts
- Wireless Concepts

Roles and Responsibilities

Organizations of different sizes, industries, and security profiles will have different types of teams and resources. This discussion of roles and responsibilities will help outline some common titles and roles within an organization to provide context of how these different people and teams interact throughout the secure wireless planning and implementation. Also, this section should help you identify areas and roles for which your organization may not have specific people assigned. Whether officially or unofficially, where there is a role absent in the organization, there's an opportunity (and expectation) for you or your team to fill those gaps (yourself or with help from other teams or third parties).

As an example, in its purest form some might say this book is written for people with the title of "enterprise security architect." However, many organizations don't have a single person or team dedicated to this task, in which case you as a network or wireless architect have an opportunity to use the guidance here to fill some of that void. With each section there's an explanation of how to engage with the various roles, and some tips and tricks for navigating the organization.

Network and Wireless Architects

We'll start with covering the roles and responsibilities of network and wireless architects since it's likely most readers will identify most closely with this area of expertise.

Depending on the size of the organization, the wired and wireless teams may be the same, or may be different groups or people. For purposes of brevity, "network and wireless architects" will be discussed as simply "network architects" for this section. Aside from LAN networking and Wi-Fi professionals, organizations are likely to have specialists for other networking specialties such as datacenter, WAN, and cloud—all of which we'll consider lumped into the networking category here.

The role of the network architect is to design the systems, interactions, and integrations that provide the appropriate connectivity, availability, quality of service, and security for an organization's networked communications. Typically, the network architect will participate in or define requirements for configurations and specify or recommend products to meet the objectives.

The network architect and (if existent) the enterprise security architect (or teams) should work together to define the wireless architectures covered in this book and specifically to ensure the security controls (including policies/procedures, configurations, and monitoring) are in line with the organization's risk management strategy. In the absence of the enterprise security architect, the majority of work described here will fall to the network architect, with input from other security resources.

SIZING IT UP: IDENTIFYING WHO HAS WHAT ROLE

In large organizations, once a system is in place the organization may rely on different teams with operations, engineering, and system admin roles to manage the daily operations and maintenance of the systems that were defined and designed by the architect. Of course, in small organizations it is quite likely the network architect, network admin, wireless professional, and help desk escalation contact may be one person or a small team.

Security, Risk, and Compliance Roles

Discussing the roles of the various security, risk, and compliance resources gets a bit dodgy since only the most mature organizations will have these roles well defined. Having said that, almost every organization should have someone responsible for making decisions about the organization's risk tolerance. Risk tolerance may not be well defined and may be more qualitative than quantitative, but there will be a risk manager, a Chief Information Security Officer (CISO), a compliance officer, a board of directors, and if nothing else—an owner who cares and makes decisions about risk tolerance.

Risk and Compliance Roles

In many organizations, and specifically any organization that falls under industry or governmental regulation, there will be a risk and/or compliance officer. This person or team is responsible for ensuring the organization adheres to any regulatory requirements. They'll be accountable for audits, documenting policies and processes, and ensuring compliance.

For example, in healthcare there will always be someone responsible for managing compliance with the Health Insurance Portability and Accountability Act (HIPAA). In regulated power and utilities organizations, someone is responsible for North American Electric Reliability Corporation (NERC) Critical Infrastructure Protection (CIP) compliance. Organizations processing certain volumes of credit card transactions will have someone responsible for Payment Card Industry Data Security Standard (PCI DSS) compliance. There are similar risk and compliance requirements covering IT systems for manufacturing, pharmaceuticals, and any organization working with the US federal government, such as requirements for the Defense Federal Acquisition Regulation Supplement (DFARS) and Cybersecurity Maturity Model Certification (CMMC), and the list goes on. Some of these regulations will require the organization to designate a person responsible for that compliance program, while others may simply require specific reporting and auditing.

As it relates to our wireless security architecture, the risk and compliance requirements will be a primary driver in dictating certain documented policies, product configurations, monitoring, and mitigation of risks. Note that in many industries, the risk and compliance officer likely resides outside of the IT organization and therefore may not be our first "hop" toward getting risk information. If there's a CISO or similar title or office, that person or group will be intimately familiar with the organization's compliance requirements and will have the added benefit of some familiarity with information technology. We'll talk about the CISO role next since that's a likely starting point for risk conversations.

Chief Information Security Officer Roles

The role of the CISO may be covered in a different title. First let's set the expectation that the reporting structure of a CISO role and the expertise of that professional will vary wildly from organization to organization. While in many instances, the CISO or Chief Security Office (CSO) will report to a Chief Information Officer (CIO) or Chief Technology Officer (CTO), sometimes the CISO may report directly to the Chief Executive Officer (CEO), other executive leadership, or even to a board of directors. In other cases, the CISO role may even be a dual responsibility of a CIO, CTO, or other technical leadership.

To add to that complexity, the role of the CISO is not always defined consistently across organizations, and therefore the expectations of that person's skills and experience may vary greatly. Some CISOs will have a technical background, while others a governance, risk, and compliance (GRC) background. I've worked with clients whose CISO was a web developer two years prior, and I've worked with clients whose CISO ran a full security practice with exceptional risk management maturity.

This is worth noting because if we're relying on the CISO (or similar security professional) to help us in our endeavor, we need to understand what to ask for, and what to expect in terms of a response.

If there is a CISO (or similar role) in your organization, start there. Ask about the organization's risk framework, and specifically what policies and compliance requirements you need to consider as it relates to network security architecture and wireless connectivity. We'll cover this more in depth throughout the topics under "Security Concepts for Wireless Architecture" later in this chapter, but for context, the answers will likely include requirements for specific encryption, device inventory and management, hardening the systems, and segmentation, among other things.

If there isn't a CISO available to you, or if the CISO is not equipped with the answers to your questions, the risk and compliance person should be your next stop for answers. If that's how you get your answers, you may have a bit more work to do—something we'll cover in the section "Considering Compliance and Regulatory Requirements."

Security Operations and Analyst Roles

Depending on the organization, security operations and the roles of technical security professionals can range from security analysts in mature security operations centers (SOC) to the person designated to monitor firewall logs. Because of that, there will be expected inconsistencies in how we interact with security teams while planning our architecture.

In general, the security operations and analyst teams are responsible for managing the security tools, and more specifically, they're responsible for monitoring the infrastructure for security including vulnerability management as well as incident detection and response. These teams have tools and workflows that ingest logs and configuration files, help identify indicators of compromise, prioritize patching based on risk, and generally monitor for and investigate potential incidents. If your organization has a SOC team, they'd be the ones managing the logging and event correlation, with tools such as security information and event management (SIEM), security orchestration, automation, and response (SOAR), and any detection and response tools such as endpoint detection and response (EDR) or the newer extended detection and response (XDR) platforms, which encompass more than just endpoints. The SOC and analyst teams would also be responsible for forensics as part of incident response handling.

For your secure wireless architecture, you'll want to talk to the security team to understand what they need to monitor the wireless infrastructure, how they're connecting to and monitoring key components such as endpoints and wireless APs, controllers, and your wireless-specific security tools such as wireless intrusion prevention systems (WIPS). This team is also involved in determining how to identify incidents, how to manage threats/vulnerabilities, defining what constitutes an incident, and how to respond to and mitigate any attacks. The SOC team, likely responsible for vulnerability management, should also communicate with you about patching the wireless infrastructure to address security vulnerabilities as they're discovered.

SIZING IT UP: YOUR ROLE WITH AND WITHOUT A SOC

As noted, the capabilities and responsibilities of the security operations and analyst teams can vary greatly. Large, highly regulated, or very mature organizations will most likely have a full SOC practice either internally or with a managed service platform such as a managed detection and response (MDR) provider. Due to regulatory requirements, these organizations likely have well-defined practices for everything security-related including patching and vulnerability management as well as security monitoring and incident response. These teams will be a great resource to the network architect in planning security for wireless.

However, many organizations don't have experienced security analysts, SOC infrastructures, or well-defined security processes. In many cases, the only technical resource with a security title may be a firewall or other system administrator. If that's the case, you'll have a bit of extra work to do to ensure your architecture has robust security throughout the system's life cycle. You may get a bit of relief and help if you have a mature network operations center (NOC) team at your disposal, which we'll cover next.

Identity and Access Management Roles

Identity and access management (IAM) teams dole out the access rights for users and digital identities such as non-person entities (NPEs), controlling access to data, applications, and other resources.

The IAM group has principal responsibility for account and identity life cycles including provisioning, authorization, maintenance, governance, and deprovisioning. As such, the group will most likely be responsible for access policies and processes around users and endpoints accessing the network.

It's also possible a different team may be responsible for subsets of users or endpoints—such as in healthcare where clinical engineering groups have ownership of the biomedical device life cycle, or digital transformation teams that may be responsible for managing Internet of Things (IoT) devices. It's common also for organizations with operational technology (OT) teams for that group to have express ownership of those systems, sometimes completely outside the visibility of IT teams.

Operations and Help Desk Roles

If you're reading this book, then you probably already have an intimate level of familiarity with network operations and user support structures. You've probably been a technical escalation point, and you may even have served in on-call rotations for after-hours support. We'll forego the 101-level overview and jump straight to how operations and help desk are related to your wireless security architecture.

Network Operations Teams

Network operations encompasses all the tactical daily care and feeding of the network infrastructure. Depending on the products deployed, the tools, and the team skillsets, they may or may not be monitoring and managing wired and wireless together. This team or toolset will usually manage uptime, configurations, software updates, and basic monitoring of the systems—for example, bandwidth and resource utilization. This is another area in which you likely have expertise directly or indirectly, so we'll skip the formalities beyond that overview.

The network operations team, if one exists aside from you, is a resource for you to collaborate with to ensure the wireless infrastructure is being monitored and managed in a way consistent with your expectations as the architect, with the business's expectations of security, and with the uptime service level agreements (SLAs), if any, set by management.

Some organizations have blurred the lines between network and security operations (NOC and SOC), and in fact in many cases, the NOC team has "SOC-like" responsibilities for some degree of security monitoring and response. They're likely managing some type of logging and may also have SIEM-like tools. If that's the case, then you'll work with the NOC team also for the vulnerability management (patching) and security monitoring.

Help Desk and End-User Support Roles

The help desk and end-user support teams have a thankless and often overlooked area of responsibility but play a critical role in maintaining secure environments.

With any technology, there will be problems, and with wireless now considered a critical business resource, uptime and availability of the system is top priority for organizations. As it relates to your wireless security architecture, you'll want to ensure the procedures for troubleshooting, workarounds, and end-user assistance fall within the accepted security practices you'll be identifying (or defining) in later sections.

External and Third Parties

Another often-overlooked area in security architecture is the role of external and third-party entities. This could entail everyone and everything from the wireless product manufacturer to your integrator or consultant, to the sources of your threat intel feeds.

Technology Manufacturers and Integrators

Between the product manufacturers' field teams (including the field systems engineer you work with) and the manufacturers' technical assistance centers (TACs) or support teams, as well as your integrator—you may have a lot of hands in your wireless network "pot" as it were. These should all be considered as valuable resources during the planning of your secure wireless, and can have input and offer assistance through upgrades, architecture changes, and support escalations. They are also a point of security vulnerability to consider and manage.

With more platforms moving to cloud-managed or cloud-monitored models, understanding who has access to your environment and to what degree is a crucial part of your security architecture. Often manufacturers have built-in backdoors for support or product development teams. You may have temporarily granted access to a support engineer or other field SE helping you with a support case. Identifying, tracking, and monitoring these privileges should be on your to-do list. As part of a longer-term strategy if there is an ongoing need, you may want to consider a secure privileged remote access (PRA) solution. In fact, it's very likely your organization already has a vendor management program or at least a PRA tool in place that you could piggyback on.

Vendor Management and Supply Chain Security Considerations

Your organization may have (or may soon have) more stringent supply chain and vendor management policies that may dictate how and when you're able to interact with and exchange information with vendors, and how you source and procure technology systems, usually including anything that stores or transmits data.

This is an important shift from traditional business operations where IT teams have been permitted to procure items however they wanted, and/or to allow remote access to various partners or TAC engineers. It's valuable to understand the organization's expectations up front and have any formal vetting done before there's an emergency or outage and you're stuck in paperwork hell or red tape in the midst of chaos.

Depending on the size of the organization, the vendor management process can be as simple as a requisition and single form to a full vetting by a third party who evaluates the risk posture of the vendor. Vendor management and supply chain policies are typically communicated down the chain from leadership, and in the absence of this you can proactively reach out to your manager, the CISO, or risk and compliance officer—probably in that order.

During the process of vetting vendor solutions from a technical perspective, there is a great opportunity for a network or security architect to participate.

Their own security architecture should be assessed, either by requesting findings from a reputable audit framework such as System and Organization Controls 2 (SOC 2) or through direct testing.

Security Concepts for Wireless Architecture

This section covers foundational concepts in the domains of both security and wireless networking. Again, the purpose is to get us all to an even playing field with concepts and taxonomy before we dive into the architecture sections.

For the hard-core networking and wireless professionals, I understand some of these topics may seem mundane and not relevant to your objectives, but I promise great results if you stick through this section. This lays the groundwork for communicating everything you do in your architecture in terms of business objectives and risk, which puts you at a great advantage. It also provides a structure for organizing the threats and vulnerabilities in wireless you're already familiar with to a common nomenclature and model understood by other parts of the organization.

For the non-wireless professionals, the wireless architecture concepts covered here will give you greater insight into the parts of today's Wi-Fi solutions and context of how they're related. In later sections, we dive into more technical concepts and then into the architecture design itself.

The following sections in this chapter only serve to offer some preliminary perspectives of the relationships of various elements so you have a rough mental model before we get into the details. In them, we define integrity, availability, and confidentiality and describe them in the context of protecting your wireless infrastructures; then go on to describe how to get started with aligning your architecture to the organization's compliance and security requirements, based on risk tolerance. After that is a short tour through policies, standards, and procedures, followed by segmentation and then authentication concepts and the high-level view of their relationship to wireless security.

Security and IAC Triad in Wireless

Integrity, availability, and confidentiality (IAC) is the holy trinity of all things security. As much as I hate to even beat this drum, the truth is that everything we're going to do in our wireless architecture should (and does) map back to one or more elements of this trio. If you want to dive further into security concepts relevant for system engineering, you'll find resources later in the book. For now, let's just look at how integrity, availability, and confidentiality play a role in our planning; you're going to get my non-textbook definition of these words. See Figure 1.1.

By the way, if you're wondering why this isn't presented as "CIA" instead of "IAC," I invite you to give a security presentation in the presence of government and academia outside the US. I made that mistake only once and decided the IAC acronym would raise fewer eyebrows.

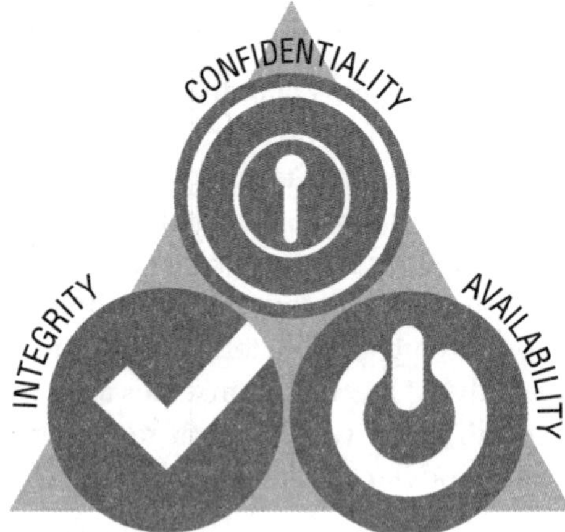

Figure 1.1: Elements of integrity, availability, and confidentiality are pervasive in secure network architectures (Source: https://falcongaze.com/en/pressroom/publications/articles/cia-triad.html)

Integrity in Secure Wireless Architecture

Integrity in security has to do with the validity or trustworthiness of the data. In this case, data may mean the wireless controller configuration, or it may mean the authenticity that the software updates we're getting from the manufacturer aren't tampered with. When it comes to wireless communications, integrity also has to do with assurance that each entity—each user, device, server, or controller—is who it says it is, and hasn't been spoofed or tampered with in any way. It also covers non-repudiation as an offshoot of that.

Some examples of integrity in network architecture include:

- Verifying the integrity of software packages to ensure they were not tampered with, often by comparing hashes or validating certificates
- Authenticating a server to an endpoint or vice versa to prove identity
- Building a trusted infrastructure where only known APs are adopted and provisioned, with the ability to detect rogue APs
- Implementing mechanisms that prevent device spoofing, including MAC spoofing
- Using Protected Management Frames (PMF) between wireless infrastructure and endpoints
- Enforcing change management processes and controlling management access to network devices to prevent tampering or unauthorized changes

For example, when a RADIUS server authenticates itself to a client with a server certificate, that is enforcing integrity. Following on from that the key exchanges and encryption would be an example of confidentiality, which we'll get to shortly. Chapter 3, "Understanding Authentication and Authorization," covers authentication and authorization, including RADIUS in great detail.

Availability in Secure Wireless Architecture

Availability is as straightforward as it sounds—it's making the system or data available to the users or devices that need it. In networking, uptime is a great measurement of availability at the most basic level. As it pertains to our architecture designs, availability also factors into general resiliency, how we manage high availability across the system, and the accessibility of the resources as defined by the business requirements.

Some examples of availability in network architecture include:

- Detection and mitigation of denial-of-service (DoS) attacks over the air
- High availability and failover configurations for controllers
- A proper wireless RF design with appropriate coverage and signal quality
- Overlapping AP coverage and settings for dynamic power and channel settings
- Redundant cable connections and power supplies for supporting wired infrastructure
- Appropriate security and segmentation controls to allow users or devices access

Confidentiality in Secure Wireless Architecture

Confidentiality covers protecting data and systems, ensuring sensitive data remains inaccessible from unauthorized or unintended parties, primarily through the use of encryption and secure mutual authentication. With data privacy laws growing by the day, confidentiality is a central element of many compliance requirements, as you'll see later.

Some examples of confidentiality in network architecture include:

- Secure key exchanges and encryption of client data with 802.1X-secured networks
- Encryption of user traffic between an AP and controller or gateway
- Secure authentication for accessing protected resources
- Use of Protected Management Frames (PMF) in Wi-Fi
- Proper segmentation of networks

As we get into sections mapping compliance and regulatory requirements, you'll see integrity, availability, and confidentiality are common threads that manifest in various requirements for our wireless security architecture.

Using the IAC Triad to Your Advantage

An important thing to note for network and wireless architects is that these three concepts in the IAC triad can be used to your advantage to justify just about anything you need in your infrastructure designs. As you see, even if a need doesn't strictly fall in what you may classify as "security," many requests can be justified in the name of integrity or availability.

I can't emphasize enough the importance the business has placed on wireless connectivity and uptime. Network connectivity is considered a critical business resource just like power and phone services. You can use this to your advantage and justify your requests in these business terms throughout your projects—including requesting additional tools, switching products or manufacturers, budgeting for upgrades, and (as you'll see later) for requesting professional development in the form of training or conferences. Just some of the items that support the security triad include tools used for testing, analysis, and monitoring of the wired and wireless networks.

Now that we have some context of what integrity, availability, and confidentiality mean in the real world, let's look at how we use these concepts to align our wireless architecture to organizational risk.

Aligning Wireless Architecture Security to Organizational Risk

The primary objective of our secure wireless infrastructure is to design our systems in a way that is aligned with the organization's risk tolerance. That's easier said than done, and there are entire books written on this topic, so I'll refrain from a deep dive. The pertinent points here are to identify and understand whether you're designing a wireless infrastructure for a high, medium, or low risk tolerance organization, and then understand how your configuration options and designs play into that model.

Identifying Risk Tolerance

Identifying risk tolerance is another task in which your resources may vary greatly; you may have very specific guidance from your organization about their risk tolerance and a specific risk model to work from, or you may have nothing other than gut feelings and qualitative data. In the absence of specific guidance from your organization about the risk tolerance, you can move along with two inputs you can identify yourself—factors that influence risk tolerance, and a basic risk classification based on that.

Factors Influencing Risk Tolerance

Factors that influence risk tolerance include compliance requirements (covered next), privacy risk, security threats, classification of data or assets (how valuable they are), and corporate culture. Here are some questions to consider:

- Does the organization or portions of the organization fall under compliance regulations such as PCI, GDPR, HIPAA, NER CIP, FedRAMP, etc.?
- In the networked environments, is there personally identifiable information (PII) that requires special protection? Note that PII is defined differently not only by country but even by state within the United States.
- Are there current security threats that are relevant to the scoped environment, target assets, or industry?
- Has the organization classified data in the environment and were there high-value assets identified such as PII or intellectual property (IP)?
- Does the executive leadership or general culture in the organization seem to prioritize security and privacy?

If you answered "yes" to any of these questions, you could safely assume the organization has a low to moderate risk tolerance, which we'll cover next.

Assigning a Risk Tolerance Level

With the preceding information and your knowledge of the environment, you can probably prescribe a risk tolerance level. These levels may apply to the entire organization, or they may apply to specific network segments. When we get into the architecture design, we'll talk about the security profiles for SSIDs in terms of a low, moderate, or high risk tolerance—or conversely a high, medium, or low sensitivity.

High Risk Tolerance High risk tolerance means an organization is willing and able to accept a higher level of risk for more operational flexibility. This is not a highly regulated environment, and the organization doesn't have compliance requirements or contractual relationships that demand strong security controls or processes.

Moderate Risk Tolerance Moderate risk tolerance means an organization has identified assets that need to be protected, likely has some compliance or reporting requirements, but affords some flexibility in trade-off with security controls.

Low Risk Tolerance Low risk tolerance applies to organizations that are in highly regulated industries such as finance, healthcare, utilities, and many federal government agencies. These organizations prioritize security over most everything else, other than requisite business operations. Figure 1.2 demonstrates the inverse relationship of an organization's risk tolerance to its security requirements.

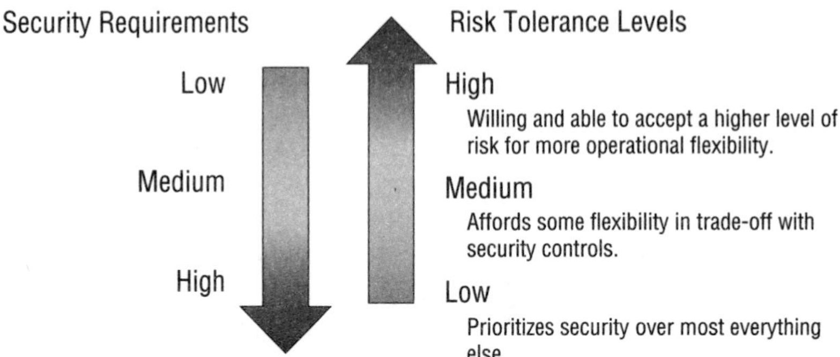

Figure 1.2: As security requirements increase, risk tolerance decreases, and vice versa

As with everything in life, there are trade-offs when it comes to security. As security controls are increased, the related risk level decreases—but on the flip side, a reduction of risk as it relates to information security may cause an increase in opportunity loss and decrease in competitive advantage.

For example, when a hotel has a mobile app for online booking, it has surely increased its risk exposure and introduced more attack surface, but it would be at a huge competitive disadvantage if the only way guests could book was by calling. As a more relevant example to our tasks, it's more secure for an organization to determine it's not going to allow wireless connectivity to any of its networks, but it would again be a huge impediment to innovation and put the organization at a strategic disadvantage.

Along with external considerations, the impact of increased security on internal operations should also be considered. Additional security controls—especially ones that impact end users such as multi-factor authentication (MFA)—bring with them additional operational overhead and a general increase in friction for both users and IT teams.

The take-away is that organizations aren't aiming for zero risk. Instead, their goal is to identify and quantify risk and create an environment that allows the flexibility of competitive operations and digital transformation, while operating within the defined level of tolerated risk.

SIZING IT UP: IT AND ENTERPRISE RISK MANAGEMENT

Large and highly regulated organizations will have robust and mature risk management programs. An enterprise risk management program tracks not only information security and IT risks, but also general business risks such as reputational risks from bad press, and human resource and loss of life risks as well.

Considering Compliance and Regulatory Requirements

The first major undertaking in wireless security architecture will be to identify and then map designs to any compliance requirements that need to be met. If you've followed along in the chapter, you know to hunt down the CISO, risk, or compliance officer to get started. If you don't have any resources to guide you on compliance requirements, don't worry—the architectures and templates provided in this book will make sure you're following best practices.

The bad news is there are myriad compliance regulations and controls frameworks to track against. The good news is, for network security, a lot of the requirements are consistent across the various frameworks. Because there are so many, we're not going to cover them all here. Instead, we'll look at a few key examples of the controls we need to implement, and then in Chapter 5, "Planning and Design for Secure Wireless," you'll get specific guidance with examples that will cover the majority of your use cases and specific recommendations for the various risk tolerance levels.

Compliance Regulations, Frameworks, and Audits

As you're working through the process, there are two terms you'll see throughout the book that might be easily conflated if you don't work in information security: compliance regulations and frameworks.

Cybersecurity *compliance regulations* are the rules or laws from industry or government that prescribe specific (usually auditable) requirements for an organization as it relates to its information security program. Common compliance regulations include programs such as PCI DSS for protecting payment card data, HIPAA for protecting patient health data, Europe's General Data Protection Regulation (GDPR), the California Consumer Protection Act (CCPA), the US government's Cybersecurity Maturity Model Certification (CMMC), and the list goes on. The compliance regulations are requisite and prescriptive to varying degrees.

Many of these regulatory standards are audited—you've probably heard of PCI DSS Qualified Security Assessors (QSAs) that assess and report on the organization's posture against those requirements. And for CMMC there are Certified CMMC Assessors (CCAs). Other regulations, like HIPAA, may not be audited formally, but there could be hefty fines and consequences for an organization that is found to be in violation. Many companies in the US also use SOC 2 Type II audits, which are performed by accredited auditing firms, to show adherence to one or more of Trust Services Principles—Security, Privacy,

Confidentiality, Processing Integrity, and Availability. In the EU and other parts of the world, ISO 27001 and other ISO 27000-series audits may be used to certify their security programs.

Frameworks come in many flavors. There are frameworks for risk management, compliance, and controls as a few examples. Frameworks give us a common language to define and describe security controls and posture. Most of the frameworks we reference in enterprise security architecture are cybersecurity frameworks that describe or rate the maturity of security-related policies, processes, or controls.

We'll touch on policies and processes in a bit. Examples of cybersecurity frameworks include National Institute of Standards and Technology (NIST) Cybersecurity Framework (CSF) as well as its more robust NIST SP 800-series, International Organization for Standardization's ISO 27001 and ISO 27002, Center for Internet Security (CIS) Controls, and HITRUST Cybersecurity Framework (CSF) for healthcare. Many of these compliance regulations will map to one or more frameworks. Some frameworks are also auditable, but most of the time in our role of doing enterprise security architecture, we're simply focusing on a subset of the controls that describe how we configure, manage, and monitor the network system for security.

Cybersecurity frameworks establish the minimum requirements for an organization to:

- Not be considered negligent with reasonable expectations for security and privacy
- Comply with applicable laws, regulations, and contracts
- Implement the proper controls to secure your systems, applications, and processes from reasonable threats

Whether we're talking about compliance or a framework, or both, they're going to dictate certain aspects of our designs, such as:

- The cryptographic strength of algorithms we need to use
- How and when we need to authenticate users or devices
- How and when to segment networks or apply role- or attribute-based access controls (RBAC/ABAC)
- How to harden the management access to infrastructure devices
- How to enforce auditable management access with logging
- Direction on maintaining system inventory including endpoints, infrastructure, and applications
- Policies and processes related to accessing resources and assets
- Policies and processes for vulnerability and change management
- Policies and processes for security monitoring and response

The world of compliance and frameworks is complicated and twisty. If you're working with the CISO or risk and compliance officer, you'll have some great resources already. Otherwise, don't get bogged down in too much detail here and just know that at times people may refer to compliance requirements. For the purposes of this book, the architecture design guidance is mostly aligned with guidance in the various controls frameworks, which can then be mapped to the compliance line items.

The Role of Policies, Standards, and Procedures

As technologists, we tend to throw around the word "policy" loosely when describing the rules of what users and systems are allowed to do within the organization. The more accurate depiction is an interrelated hierarchy, with policies living outside of our purview. When we say "policy" in this context we mean some blend of standards and procedures, as defined next and demonstrated in Figure 1.3. In this hierarchy, there are also guidelines that serve as more of an "FYI" with general statements and recommendations that complement the more formal structure. For our purposes, we're not covering guidelines and we'll instead focus on the hierarchy of policies, standards, and procedures.

NOTE Network and security administrators will of course recognize the phrase "policy" as a common component of access rights such as those in firewalls and access lists. Just to clarify and distinguish the two: here it's meant as organizational policies, not technical control policies.

A NOTE ON POLICIES

This overview is for informational purposes only to help guide conversations you may have with security, risk, and compliance professionals who use this model. In daily use, it's both common and acceptable within the IT community to refer to the collection of formal policies, standards, and procedures as simply "policies."

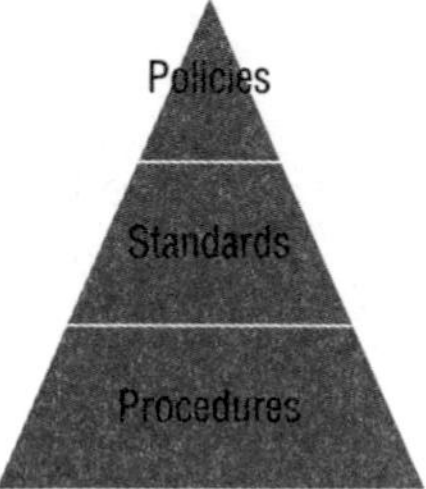

Figure 1.3: In the hierarchy of policies, standards, and procedures, one broad policy may have multiple standards and procedures to meet the policy objective

Policies

Policies are high-level statements from executive leadership, and they ultimately speak to expectations of the security.

Policies:

- Focus on desired results, not on means of implementation
- Are further defined by standards, procedures, and guidelines
- Require compliance by users and outline consequences if they fail to comply
- Change rarely and are meant to be more evergreen than the more detailed standards and procedures

Standards

Standards define a mandatory action designed to support the policy requirements. They're specific to an area of practice, technology, or system but are not as granular as procedures.

Standards:

- Define models or methodologies required to meet policy objectives
- Specify the mechanisms and application of controls without defining the discrete steps
- Can be platform-specific and may include secure baseline configurations
- Will change over time as product features and industry standards evolve

Procedures

Procedures (also called processes) document the operational steps necessary to implement the policy.

Procedures:

- Describe who does what, when, how, and under what conditions
- Include a series of steps and instructions
- Are specific to systems and platforms
- Change frequently as products are updated, techniques are improved, or new information is learned

Example with Wireless Security

Here's an example of the hierarchy using wireless security. The relationship is represented in Figure 1.4, as well.

Policy The organization will allow managed devices to connect to network resources including the wireless network. The managed devices should be permitted to connect only if both the device and the user are authenticated, and the connection must be encrypted.

Standard Internal secured SSIDs will be configured as minimum WPA2-Enterprise with 802.1X authentication with Microsoft PEAP-MSCHAPv2 or EAP-TLS. Managed client machines may be authenticated via the directory structure or machine certificate. Users will be authenticated by Microsoft login credentials. The encryption must be minimum of AES 256.

Procedure Includes a description of who does the work (e.g., an authorized network admin) and the exact steps of what they will do, such as the step-by-step instructions for logging in to the wireless controllers and configuring ACME-Corp SSID to be a 802.1X-secured network, and the steps to configure connection to the authentication server, as well as configuring a policy that requires machine and user authentication, selects the options in the menus for proper supported EAP methods as described in the standard, and the steps and checklist for configuration of any supporting infrastructure, network services, and endpoint configuration.

Hierarchy Item	Scope	Example
Policy	Broad, high-level statement of intent	The organization will allow managed devices to connect to network resources including the wireless network. The managed devices should be permitted the connection only if both the device and the user are authenticated, and the connection should be encrypted.
Standard	Specific requirements within a given technology	Internal secured SSIDs will be configured as minimum WPA2-Enterprise with 802.1X authentication with Microsoft PEAP-MSCHAPV2 or EAP-TLS. Managed client machines may be authenticated via the directory structure or machine certificate. Users will be authenticated by Microsoft login credentials. The encryption must be minimum of AES 256.
Procedure (Process)	Step by step process	Includes a description of who does the work (e.g., an authorized network admin) and the exact steps of what they will do, such as the step-by-step instructions for logging in to the wireless controllers and configuring ACME-Corp SSID to be a 802.1X-secured network, and the steps to configure connection to the authentication server, as well as configuring a policy that requires machine and user authentication, selects the options in the menus for proper supported EAP methods as described in the standard, and the steps and checklist for configuration of any supporting infrastructure, network services, and endpoint configuration.

Figure 1.4: Recap of the intent and relationship of policies, standards, and procedures

Segmentation Concepts

Segmentation plays a critical role in securing networks, including and especially wireless networks. Whether we're talking about standard 802.11 WLANs, private cellular/CBRS, or IoT-based sensor networks, there will be some need for segmentation to divide the environment into logical segments, and/or to control or filter traffic between the wireless and wired environments.

Why and When to Segment Traffic

Traffic is segmented in different ways and for different purposes such as to:

- Secure management and control access from end users
- Separate different classes of networks, defined by sensitivity or asset value
- Isolate and protect legacy endpoints that present security risk
- Prevent disruption from large network broadcast domains
- Protect network segments in scope for compliance requirements (such as PCI DSS)

Methods to Enforce Segmentation

Segmentation can happen in many ways and at many layers of the Open Systems Interconnection (OSI) network stack, from the physical layer (layer 1) up through application layer (layer 7) with "micro" segmentation technologies, and everywhere in between with virtual LANs (VLANs) and access control lists (ACLs) at layers 2 and 3, respectively. And, if you can't segment in one place, your architecture will reveal that gap and help identify another place to do it. As a refresher, the OSI model is shown in Figure 1.5.

For example, most enterprise wireless vendors offer a feature like firewall or ACLs through policies or roles on the wireless system. With that, you could certainly implement segmentation to keep, for example, guest users from accessing the internal network. Depending on the network architecture, you could also enforce segmentation through non-routable VLANs, or routed networks with ACLs in place, or even internal firewalls. These are mechanisms covered in more detail in the coming chapters.

Almost every compliance requirement and cybersecurity framework will reference and mandate segmentation in one way or another. While it doesn't prescribe how you do it, it will outline the requirements for when to segment, and to what degree.

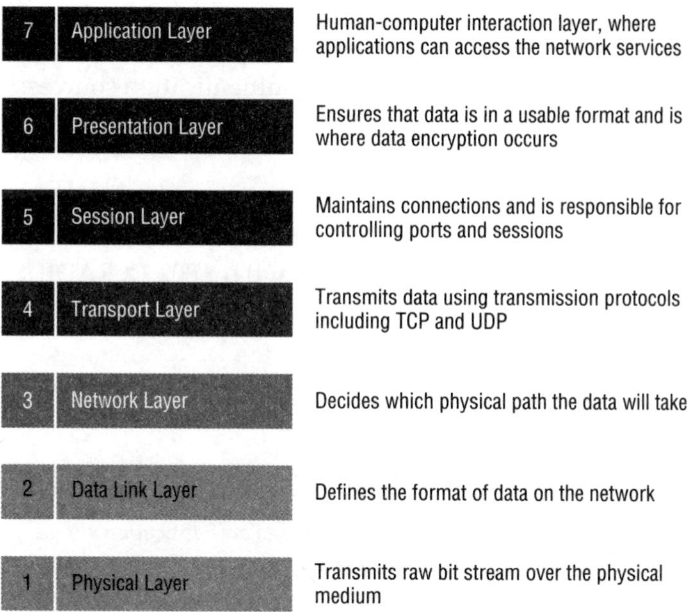

7 Application Layer
Human-computer interaction layer, where applications can access the network services
6 Presentation Layer
Ensures that data is in a usable format and is where data encryption occurs
5 Session Layer
Maintains connections and is responsible for controlling ports and sessions
4 Transport Layer
Transmits data using transmission protocols including TCP and UDP
3 Network Layer
Decides which physical path the data will take
2 Data Link Layer
Defines the format of data on the network
1 Physical Layer
Transmits raw bit stream over the physical medium

Figure 1.5: Overview of the seven layers of the OSI model. Layer 2 is the MAC layer, and layer 3 the IP layer

To recap, segmentation methods most relevant to wireless architecture include:

- Layer 1 physical segmentation (air gap)
- Layer 2 segmentation through non-routable VLANs (on wired infrastructure) or inter-station blocking and other methods (over the air)
- Layer 3–4 segmentation with ACLs (this can be applied on the wired or wireless network segment)
- Layer 3–7 segmentation with software-defined networking or virtualized networking (SDN, VXLAN, IPSec, SSL) (again this can be on wired or wireless segment)
- Layer 7 and network-based microsegmentation in zero trust architectures (covered later)

Authentication Concepts

Authentication is another fundamental security concept that plays a vital role in networking and especially wireless security architecture. Due to the flexible nature of wireless and the ability to broadcast multiple networks at once (vs. most wired deployments), and the more dynamic nature of the connectivity, managing authentication of users and devices in wireless is a big piece of the planning activities.

In wireless architecture, there are five main authentication needs to be considered, and with each need or use case comes different options and considerations for the actual means of authentication and the authentication sources:

- Authentication of users (specifically end users)
- Authentication of devices (wireless endpoints)
- Authentication of administrative users (for system administration)
- Authentication of the servers (for captive portals and/or 802.1X RADIUS)
- Authentication of the wireless infrastructure components (APs to management and vice versa)

We'll spend the most time covering authentication of the endpoints and users.

AUTHENTICATION VS. IDENTIFICATION IN INFOSEC

There's a common misconception about what exactly qualifies as authentication versus identification of a user or device. There's an elegant and simple explanation that states identification is an assertion of an identity, whereas authentication is the act of proving an assertion by verifying that identity.

By this definition, using a device's MAC address to authorize it to a network is considered identification, not authentication. Similarly, joining a network using a pre-shared key (PSK) or passphrase is considered identification (at best) and not authentication. Even in the case of per-device passphrases, there's only an assertion and assumption of the identity and not validation.

This topic gets a little hairy as we proceed into the technical chapters of the book, since by 802.11 standards terminology, the packet-level process of joining the wireless network (regardless of mechanism) involves an exchange labeled "authentication."

To further complicate the matter, there are industry-accepted features such as MAC Authentication Bypass which, by name, indicates that using a MAC address is a form of authentication.

Just keep this in mind as you continue through this and future chapters. The strict protocol definition of authentication may not align with what's accepted in information security and compliance, and you'll need to adjust accordingly.

Authentication of Users

Authentication of users (and devices, which we'll cover next) is a primary use case of authentication in wireless. When we authenticate a user, it may be with a username-password combination or a user-based certificate or token.

Anything short of an identity and a credential (e.g., a username and password, or certificate that serves as both) is not considered true authentication and is more of an identification, at best. This definition is more applicable when we talk about authenticating devices and the use of MAC addresses, but to translate to user concepts, this means having a pre-shared key to an SSID is not considered a form of authentication even if it is, in fact, proving some type of authorization just by the virtue of joining a specific network and being allowed access to resources. When a user keys in a pre-shared key to join a network, there is no guarantee that the intended user is the one who entered it. Even in the case of personal or multiple pre-shared key (MPSK/PPSK) enabled systems, the identity of the user is still assumed and not validated or authenticated.

In Chapter 2, "Understanding Technical Elements," we dive further into authentication and authorization concepts at a technical level, but for now just know that most user-based authentication in wireless is managed with username-password combos such as those stored in Active Directory for internal domain users, and in external directories such as products for portal management of non-domain users and devices.

Authentication of Devices

Authentication of endpoint devices is another primary use case for authenticating in wireless networks. Contrasted with authentication of users, authentication of the device can be a bit trickier because in most cases (service accounts aside), devices won't authenticate with a username-password combo. Instead, Wi-Fi endpoints may authenticate with a device certificate or through a process based on MAC address identity.

Device certificates for endpoints can be managed by embedded Trusted Platform Module (TPM) chips (on supported devices), issued to internal devices through an internal Public Key Infrastructure (PKI) infrastructure, issued by an organization through mobile device management (MDM) tools, and/or pushed from a third-party certificate service such as those managed by SecureW2 for Eduroam (popular in higher education).

The other main way devices are authenticated to the network is based on MAC address. Remember we technically don't define the use of MAC address as authentication, but instead would consider it identification (at best). Having said that, I'll contradict my definition here because there are several protocols based on using the MAC addresses as the devices' identity for "authentication" including MAC-based port security on switches, MAC Authentication Bypass (MAB) as a subset of the IEEE 802.1X standard, and other network access control (NAC) and authentication products that use non-RADIUS (meaning not 802.1X-) based authentication and enforcement. More on that later. MAC addresses are easily spoofed and not immutable as we've seen with the advent of MAC randomization on most endpoints.

Other wireless technologies such as private cellular (CBRS/4G/5G) authenticate devices using subscriber identity module (SIM) identities, tied to embedded TPM chips for certificate-based identity and encryption. There are different types of SIM technologies available in different types of endpoints ranging from traditional laptops and tablets to cell phones and industrial IoT with machine-to-machine communication.

Authentication of Administrative Users

In wireless, as with any network infrastructure, we also authenticate the management and administrative access through protocols such as TACACS+ or RADIUS. Ideally—and specifically to meet most compliance requirements and security best practices—you'll want to ensure all parts of the wireless infrastructure management and monitoring systems use identity-based user access (such as authenticating with domain credentials) and not using shared "admin" or "root" accounts. This applies to the APs (if applicable) as well as the controllers or gateways, cloud management platforms, and any monitoring tools. When we cover hardening the infrastructure later, we'll dive more into this topic with specific recommendations.

Authentication of the Servers (for Captive Portals and/or 802.1X RADIUS)

If your wireless architecture includes an 802.1X-secured network and/or any captive portals, you'll need to plan for server certificates. And really, even outside of those use cases, you should plan for certificates as part of administrative access hardening covered later.

Server certificates in wireless are often overlooked or misunderstood. Even in authentication methods where the user is authenticating with a username-password (as in 802.1X MS-PEAP) or not at all (as in some captive portals) the servers still require certificates.

As with the rest of this section, we'll cover this topic more in depth later. For now, know that you'll need to plan for two types of certificates:

- For 802.1X-secured networks the RADIUS server will need a certificate that is issued by a root CA trusted by the endpoints (e.g., a domain or internal PKI-issued certificate or something you can push with MDM).
- For captive portals, the portal will need a third-party publicly trusted certificate so that unmanaged and guest devices already have and trust the root CA certificate.

Authentication of the Wireless Infrastructure Components

Another often overlooked component of secure wireless architecture is the authentication of the various wireless infrastructure components to one another.

As we'll discuss more in Chapter 6, "Hardening the Wireless Infrastructure," there are several mechanisms by which we can authenticate (and authorize) components to one another—specifically, we'll consider how to whitelist, adopt, or authenticate APs to the management infrastructure, and vice versa.

Authenticating the infrastructure bolsters the overall security posture and adds a layer of integrity to the system.

Cryptography Concepts

Without writing a novel on cryptography, I want to share a few basic terms that will help in your wireless architecture journey. Again, the aim is to describe these concepts as they relate to one another and specifically within the context of wireless security, and therefore I'm avoiding deep dives in this section. There are countless wonderful books and resources on cryptography available should you want to dig deeper.

Cryptographic Keys, Key Exchanges, and Key Rotation

Cryptographic keys are what's used in the cryptographic algorithm (described next) to encrypt the data. Crypto keys have a long history of challenges, starting with, but not limited to, complications with getting truly random numbers to feed the key generation process, securely exchanging the keys, and having an appropriately long key length.

Think about cryptography and key exchanges the way you would consider a virtual private network (VPN) or any type of point-to-point connection—specifically for there to be a point-to-point tunnel, there must be two ends, and they must communicate and negotiate securely. It's the same with exchanging keys over wireless. Similarly, the less a key is used or exposed, the more secure the system is, and so rotating keys is another important aspect of cryptographic operations in wireless security.

In Chapter 2, "Understanding Technical Elements," you'll find the Wi-Fi Protected Access WPA2 vs. WPA3 comparison including the specific cryptographic enhancements in WPA3 that divorce passphrase length (in passphrase-based networks) from the cryptographic key generation, which produces a strong cryptographic key even in the presence of a short passphrase.

Cryptographic Algorithms and Hashes

The cryptographic algorithms are the "formulas" for taking data from plain text to encrypted (and usually also in reverse). They describe the process to use the cryptographic keys and other inputs to secure the data. There are several classes of algorithms such as:

- Symmetric-key algorithms
- Asymmetric-key algorithms
- Hash functions

The high-level view is that symmetric-key algorithms use the same key to encrypt and decrypt data. This is fast and efficient, but if we're exchanging encrypted data, it means two parties (or devices) need the same key for this to work, and getting the keys sent securely is troublesome (without some help).

Asymmetric-key algorithms, on the other hand, are based on public key cryptography where each party (or device) generates a pair of keys (one public/shared, one private/secret) that are mathematically related but unique. These key pairs allow the parties (in our example, Sally and Nigel) to exchange public keys. Sally will encrypt data using Nigel's public key, and he is able to decrypt it with his private key; and vice versa. Historically, the problem with asymmetric key cryptography algorithms such as RSA is that they rely on very large numbers and are therefore computationally intense and slow. While it solves the secure key exchange problem, it's not viable for large volumes of data encryption, and may not be sustainable for smaller and less-capable devices.

Hash functions are one-way cryptographic algorithms used to verify integrity. Comparing a calculated hash value to a known value indicates that the contents of the data or file were not altered in any way. Hash functions are covered more in Chapter 6, "Hardening the Wireless Infrastructure."

Figure 1.6 offers a simplified representation of public key cryptography and its use of public and private keys. In this example Sally and Nigel exchange public keys used to encrypt the traffic and use their own private keys to decrypt.

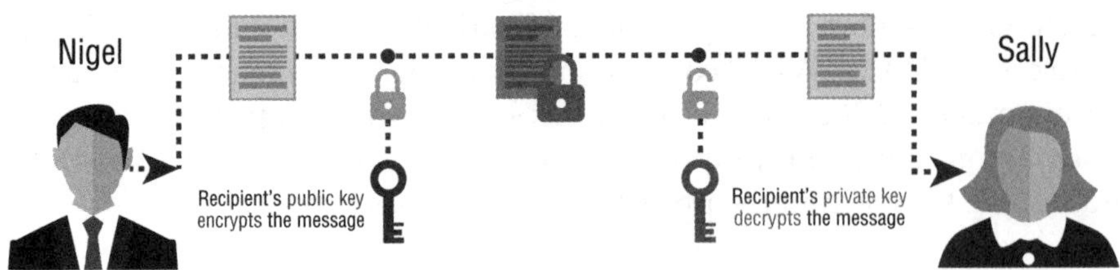

Figure 1.6: A simplified view of public and private key use in public key cryptography

Tying It All Together

All of this comes together in a beautiful symphony of hybrid cryptography. To solve the limitations of the secure key exchanges in symmetric algorithms and the computational overhead of their asymmetric counterparts, we brilliantly use asymmetric cryptography to establish a secure channel, then we switch to symmetric encryption for the data in transit.

What you'll consider in a secure wireless architecture is using 802.1X for secure mutual authentication and key exchanges and the key rotation scheduling, along with the symmetric encryption settings that will describe the algorithm (e.g., AES) and the key length (such as 256). Of course, when we're looking at networks with pre-shared keys the encryption keys are derived from the passphrase that both parties have, using symmetric encryption directly without the secure key exchanges that happen through 802.1X.

Figure 1.7 provides a summary of the pros and cons of symmetric and asymmetric keys. Combined, symmetric and asymmetric keys provide fast and flexible security that leverages the benefits of each model while compensating for the deficiencies.

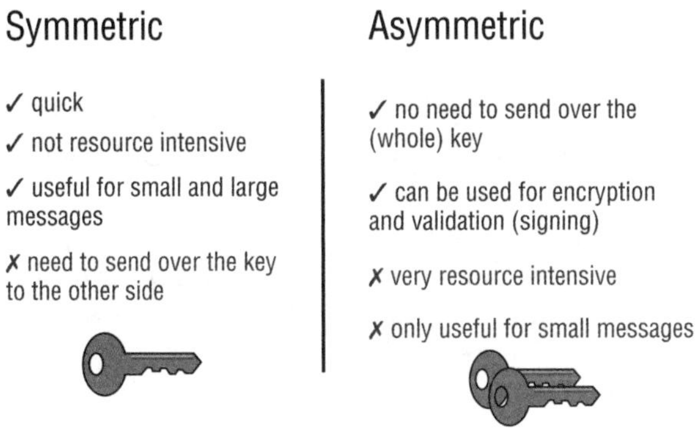

Symmetric
✓ quick
✓ not resource intensive
✓ useful for small and large messages
✗ need to send over the key to the other side
Asymmetric
✓ no need to send over the (whole) key
✓ can be used for encryption and validation (signing)
✗ very resource intensive
✗ only useful for small messages

Figure 1.7: A side-by-side comparison of pros and cons of symmetric and asymmetric keys

As an aside, in WPA3 and future wireless security protocols we'll also see new encryption models such as elliptic curve cryptography (ECC), which removes some of the overhead of traditional asymmetric encryption and allows greater security with shorter key lengths, making it a better choice for lightweight and battery-powered IoT devices.

Since this section is all about understanding the foundational concepts and how they relate, the last note on cryptography is that we'll revisit these key exchanges later as we explore how the various roaming protocols impact the key distributions for clients as they're roaming between APs. Spoiler alert: roaming protocols have a huge impact on key exchanges and therefore the integrity and speed of the wireless access.

Next, we'll switch domains and move from the high-level security concepts to the wireless concepts and components and how they're related to one another, and to your security architecture.

Wireless Concepts for Secure Wireless Architecture

In this section, you'll be introduced to the main wireless protocols and elements and learn how they're related within secure wireless architecture. Remember this entire chapter is an introduction to the concepts and relationships, with the next chapter being dedicated to the more technical deep dives.

In the wireless concepts, we'll look briefly at the various standards and protocols, review SSID security profile settings for authentication and encryption, outline factors related to endpoint features, and cover how the network architecture and distribution impacts wireless security architecture.

Wireless Standards and Protocols

There are thousands and thousands of pages of books and standards documents covering the wireless standards and discrete protocols. If you're interested in a deeper dive than what we cover in this and the next chapter, I invite you to check out the additional recommended resources and trainings found in the section "Training and Other Resources" in Chapter 7, "Monitoring and Maintenance of Wireless Networks."

For our introductory purposes, we're going to cover a few basic areas—a quick overview of wireless standards and technologies applicable here, description of generations of WLAN technologies, and the role of NAC and IEEE 802.1X in wireless.

Wireless Standards and Technologies

Just so we're on the same page, this book is titled *Wireless Security Architecture*, not "Wi-Fi Security Architecture" because we're covering technologies outside of Wi-Fi.

The Institute of Electrical and Electronics Engineers (IEEE) is a global organization that creates the technology standards for virtually all of our networked technologies, including local area networks (LANs)—both wired Ethernet (IEEE 802.3) and WLANs (802.11), among others.

IEEE 802.11 WLANs are the Wi-Fi networks we know and love. It's what (for now) makes up the bulk of our enterprise-connected wireless devices such as laptops, tablets, and many headless devices like printers. But it's not the only wireless standard. In this book, we'll also cover (to a lesser degree) security architecture considerations for private cellular (CBRS/4G/5G), cellular for IoT (LPWANs), and other wireless technologies common in IoT and machine-to-machine deployments such as 802.15 and 802.15.4.

The IEEE 802.15 standard encompasses Bluetooth and Bluetooth Low Energy (BLE) technologies, used in the enterprise for shorter-range, in-room connectivity or in-building connectivity for wearables, smart home technologies, and asset tracking.

The IEEE 802.15.4 standard is the basis for a large selection of wireless technologies described as low-rate wireless personal area networks (LR-WPAN) specifically designed for various IoT applications. Protocols like Zigbee, ISA100.11a, WirelessHART, 6LoWPAN, and Thread all offer ways to connect and secure connected IoT devices in mesh and other topologies. The IEEE, Wi-Fi Alliance, and IETF logos are shown below in Figures 1.8, 1.9, and 1.10.

Figure 1.8: IEEE logo (Source: www.ieee.org)

The Wi-Fi Alliance (WFA) is an organization whose purpose is to test and certify the operations and interoperability of WLAN devices. Among other programs, the Wi-Fi Alliance certifies against the operation of the Wi-Fi Protected Access (WPA) security suite used in WLANs.

You'll see the use of "WLAN" and "Wi-Fi" used interchangeably, but the Wi-Fi Alliance actually owns the branding of Wi-Fi, which is a designation reserved for use with its certifications. To avoid confusion with other wireless technologies, after this section I avoid using "WLAN" on its own since not everyone outside the wireless industry understands that's specifically 802.11; instead, I'll call out 802.11 networks as such or as "Wi-Fi."

Figure 1.9: The Wi-Fi Alliance manages the testing and certification of 802.11 WLAN specifications for functionality and interoperability (Source: www.wi-fi.org)

The Internet Engineering Task Force (IETF) is responsible for most of our Internet protocol standards including RADIUS, EAP, TLS, and even opportunistic wireless encryption (OWE) that you'll learn about in Chapter 2.

Figure 1.10: The IETF creates most of the world's Internet protocols in use today

Aside from these prominent organizations and standards bodies, there are countless others that play a role in wireless architecture and security.

In the cellular world, 3GPP (3rd Generation Partnership Project) participates in standards and interoperability for a multitude of mobile telecommunications technologies, including 3G, 4G, and 5G networks as well as their IoT-capable counterparts.

From an RF and spectrum management perspective there's global oversight by the International Telecommunication Union (ITU-R) (a specialized agency within the United Nations) as well as regional oversight such as the US Federal Communications Commission (FCC).

There are organizations, standards, consortiums, and alliances for every niche technology from 802.11 WLANs to cellular, Bluetooth, private cellular, and more. Additional organizations will be introduced throughout the book.

Generations of 802.11 WLANs

Talking about "IEEE 802.11a, b, g, n, ac" and now "ax" isn't a very friendly way to communicate the different generations of Wi-Fi technologies. To make matters worse, some manufacturers still represent the 5 GHz radio as "802.11a radio" and the 2.4 GHz radio as the "802.11b/g radio" in management consoles. To solve this confusion, a few years ago the industry took a page from the book of cellular and retroactively incremented a generational numbering, as follows, from most recent to older:

- **Wi-Fi 6E:** 6th generation Wi-Fi, 802.11ax in 6 GHz
- **Wi-Fi 6:** 6th generation Wi-Fi, 802.11ax in 2.4 GHz and 5 GHz
- **Wi-Fi 5:** 5th generation Wi-Fi, 802.11ac in 5 GHz
- **Wi-Fi 4:** 4th generation Wi-Fi, 802.11n in 2.4 GHz and 5 GHz

The earlier generations are not officially named in the new scheme, but you get the picture.

Wi-Fi 6E is a new and special implementation of Wi-Fi 6 (802.11ax) extended over the (newly FCC-released) 6 GHz spectrum. We go into Wi-Fi 6E more later.

NAC and IEEE 802.1X in Wireless

The IEEE 802.1X standard was originally developed for use in wired networks—it's 802-dot-1-X not 802-dot-11-X. Instead, the convention is that these are protocols used by wireless.

IEEE 802.1X is a standard for port-based network access control in the 802.1 (LAN/MAN) working group, and although not a wireless-specific standard, it's used in wireless more than in wired port security. When used in wireless, the standard's terminology and functions translate since we still consider each endpoint to have a connection to the "wireless port" just as it does on the wired equivalent.

Implementing 802.1X for secure authentication of users and devices should be part of your scope in designing secure wireless architectures, and it may be the primary objective. 802.1X is considered the "gold standard of Wi-Fi security" when it comes to standards-based protocols, and the security features it brings will map directly to various security and compliance requirements for the organization.

Let me go ahead and reiterate this—if you're designing a secure wireless network, you must master 802.1X. Passphrase-secured networks may get you through in a pinch, but they should not be used in secure enterprise wireless unless absolutely required by a subset of devices. We'll cover which devices and use cases warrant an exception from 802.1X, and how to design mitigating security controls if your environment requires pre-shared key networks.

Related to the 802.1X standard, but fundamentally different in many ways, are network access control (NAC) products. NAC products vary greatly from applications that centralize and boost authentication services and features to very complex integrations that layer device profiling, posturing, and dynamic authorization properties. NAC products may include one or more of these features:

- Device authentication (through RADIUS or MAC address)
- User authentication (through RADIUS or captive portal registration)
- Device profiling (querying or inspection of the device or its traffic to determine what type of device it is, such as an HP printer or an Apple iPad, and what OS version)
- Device posturing (a security assessment of the endpoint against predefined policies for minimum OS or patching levels)
- Dynamic authorization (assignment of a VLAN or ACL based on then-current dynamic properties such as a combination of device posture and logged-on user)

TIP It's safe to make the general statement that all enterprise NAC products (such as Cisco ISE, Aruba's ClearPass, Forescout, and Fortinet FortiNAC) offer all the aforementioned features, but each vendor solution supports each subfeature to varying degrees, and each organization may have chosen to only use a subset of features, such as only using NAC for authentication.

As organizations finalize their zero trust strategies, there may be additional impact on the wireless architecture. While the majority of products marketed as zero trust solutions are designed to support remote users, there are a few zero trust products aimed at network-based *microsegmentation* on the LAN (including the wireless LAN). Zero trust is covered more later, but warrants mention here alongside NAC since the products designed for on-premises enforcement are really just another evolution of NAC. Zero trust is covered more in Chapter 8, "Emergent Trends and Non-Wi-Fi Wireless."

SSID Security Profiles

Now that we have a bit of foundation of the security concepts and Wi-Fi standards, let's look at the various options for configuring security profiles on SSIDs, the authentication and encryption attached to each one, and what may impact your decisions during design.

We can break this down to the three classes of SSID security profiles—open, personal, and enterprise. The next chapter will cover these in more detail and explore differences between generations of security suites (specifically WPA2 vs. WPA3 and open vs. Enhanced Open technology).

Open Wi-Fi Security

Open (and newer Enhanced Open) networks don't require any form of pre-authentication to connect to the wireless network and are most often used for guest access and captive portal experiences. Through captive portals, there can be additional authentication enforced before the user proceeds and is granted access to resources (hotel captive portals requiring a room number and name are a great example), but the term "open" here means the device is able to complete a basic association to the network.

Legacy Open networks have also not provided encryption or data privacy, something addressed in the discussion of Enhanced Open technology in Chapter 3.

TIP In the most technical terms, open security SSIDs perform an "Open System Authentication" function, which is best described as a non-verified or null 802.11 authentication, followed by an 802.11 association to the AP. This is not authentication of the device or user to the network, but instead just an open-door policy with the AP and endpoint that allows the endpoint to associate and connect. For the sanity of the readers who are not wireless professionals, for this chapter's purpose I'm describing "open" as no authentication to avoid confusion.

Personal (Passphrase) Wi-Fi Security

A bit of a misnomer, *personal* designated networks are simply networks that use a passphrase to connect, and are prevalent in just about every enterprise environment, one way or another. As noted earlier, PSK- or passphrase-secured networks should not be your go-to for secure Wi-Fi. We'll start with 802.1X-secured networks and fall back to using passphrase-based networks when there's no other choice, and then we'll secure it properly to reduce risk.

Enterprise (802.1X) Wi-Fi Security

Enterprise-secured networks use IEEE 802.1X with EAP for authentication and key exchange, as described earlier. This is our gold standard for authenticating users and devices on the wireless network, which may entail username-passwords, tokens, certificates, or some combination for multi-factor authentication (MFA). 802.1X requires the wireless infrastructure and the endpoints to support the 802.1X protocol and have some ability to authenticate itself—both servers and endpoints—to one another. It also requires there be an authentication server (specifically a RADIUS server) properly configured.

The table in Figure 1.11 offers a summary of the three classes of SSID security, the relative security level, authentication used, and encryption.

SSID Security Profile and Feature	Open	Personal (Passphrase)	Enterprise (802.1X)
Use case	Guest portals or BYOD registration	IoT and endpoints that don't support 802.1X	Secured enterprise devices
Security level	Low	Low-Medium	High
Authentication	None	Passphrase*	802.1X via username-password or certificates
Encryption	None (unless using Enhanced Open)	Yes, with limited functions	Yes

* Remember that technical authentication processes as described here don't necessarily meet the requirements for authentication in compliance.

Figure 1.11: Summary of three classes of SSID security profiles with use cases and supported security features

Endpoint Devices

So many organizations get caught up with planning the wireless infrastructure that they neglect to consider the endpoints' capabilities. I've seen an entire hospital system send back consumer-grade tablets they planned to deploy to their nursing staff because the devices blue-screened every time they tried to connect to an Enterprise 802.1X-secured network, and the manufacturer had no intention of resolving it.

With security, it takes two to tango, and the quality and capability of the endpoints is just as important as the APs and controllers selected. Here are a few considerations with endpoints, and how they may impact your security planning.

Realizing that it's very likely your team is not responsible for procuring or specifying endpoints, it will be important to communicate these requirements effectively to other teams and to management. Later, we'll outline the complications that arise when different groups in an organization (shadow IT) procure endpoints without proper vetting and review from infrastructure or architecture teams.

INDUSTRY INSIGHT: 802.1X SUPPORT ON ENDPOINTS

Here's an interesting fact. For devices to be certified by the Wi-Fi Alliance, only WPA2-Personal and WPA3-Personal are mandatory. Our beloved WPA2-Enterprise and WPA3-Enterprise security is optional! The result? Not all devices support 802.1X.

All the more reason to validate your security architecture's endpoint requirements before purchasing devices.

Form Factors

Form factors play a role in wireless security architecture because of their mobility and possible limitations of processing resources and battery life.

Encompassing the entirety of an enterprise wireless, form factors may include standard laptops, tablets, smart phones, as well as headless (that is, userless) devices such as printers, plus facilities management and IoT devices such as sensors and biomedical devices.

The form factor dictates how mobile the device is and other limitations. Devices that transmit small packets and have other battery-conservation features will likely be provisioned, secured, and monitored differently due to limited capabilities. Also, smaller and less capable devices don't have the processing and memory resources required for advanced computational functions such as those in certain cryptography processes.

User-based vs. Headless

Whether a device has a user attached, rotating users, or no user attached will also inform the security requirements. A userless or headless device may not have a way to authenticate with a username-password credential set, limiting your options to certificate-based authentication and/or MAC-address–based security.

RF Capabilities

RF capabilities and support for various wireless protocols will determine secondary security considerations, such as the key exchanges in roaming protocols (mentioned earlier and covered later in more depth). If an endpoint isn't capable of supporting the latest wireless technology (regardless of the standard), you'll be forced to design backward-compatible networks for legacy devices which—aside from severely impacting performance and availability—will also mar your security posture.

As you'll begin to see in this book, the newer technologies will require newer security. Once we get to Wi-Fi 6E, WPA3 will be the only security suite supported, and the options to use WPA2 will be gone. There will also be a requirement for Wi-Fi Protected Management Frames (PMF), another protocol the endpoint will have to support in order to enable it fully in the wireless network.

Security Capabilities

Similar to the RF capabilities, obviously the endpoint can't participate in secure architecture if it doesn't support the proper protocols. In addition to the latest RF standards, it's important to ensure the endpoints will be able to support the latest security standards and cipher suites. While most security upgrades can come in the form of software updates, there is certainly a limit to how far you can go, and as your organization may move to more secure wireless, features like embedded TPM chips for secure device authentication and on-board hardware certificates may be a major consideration.

Ownership

Ownership of the device is another primary factor in security architecture for wireless. As I jokingly say, there are "50 shades of guest" and in today's climate of dramatic consumerization and BYOD, CYOD, or COPE (bring your own device; choose your own device; or corporate-owned/personally-enabled) models, understanding ownership, BYOD policies, and legal requirements or liabilities is important when planning the wireless security. Determining how to sift and sort, identify and securely provision and manage personally-owned devices takes planning. Later I'll share a NAC and zero trust policy matrix to help with your planning.

Network Topology and Distribution of Users

The underlying network architecture and connectivity of the wireless infrastructure components describes the distribution of the environment and users. Typical distribution models include traditional campus, remote workers, and remote branch. This section provides an overview of each, considerations, and relationship to the prior elements we covered in security and wireless domains.

Campus Environments

"Campus" here doesn't mean specifically school campus—we use this term to describe environments with centralized network architecture, which could be a school campus but also could be a series of buildings in local government or a commercial organization. In campus deployments, there may be one building or a group of buildings, which will be connected in a LAN topology, locally or with longer-range fiber runs.

In this campus topology, the organization has a strong on-premises presence of network devices, resources, and users. From a wireless perspective, it may have wireless controllers or gateways, on-premises monitoring tools, and the option to connect users with network resources by various data path models (e.g., bridging or tunneling, as described in the next chapter). In this model, network and domain services are typically centralized and within a layer 2 or layer 3 boundary from the users and devices, meaning campus deployments have a lot of options and flexibility of deployment in wireless architecture.

To visualize the connectivity models, Figures 1.12, 1.13, and 1.14 offer a simplified topology of the three models, which we discuss in detail in the next section. The organization depicted in Figure 1.12, is a single campus environment, with all buildings centralized to a local Internet egress. In Figure 1.13, a more complex topology is shown, with a main office and two branch offices—one connected via an existing WAN connection and the other with only an Internet connection. Finally, Figure 1.14 depicts the Internet-only connectivity for remote home users.

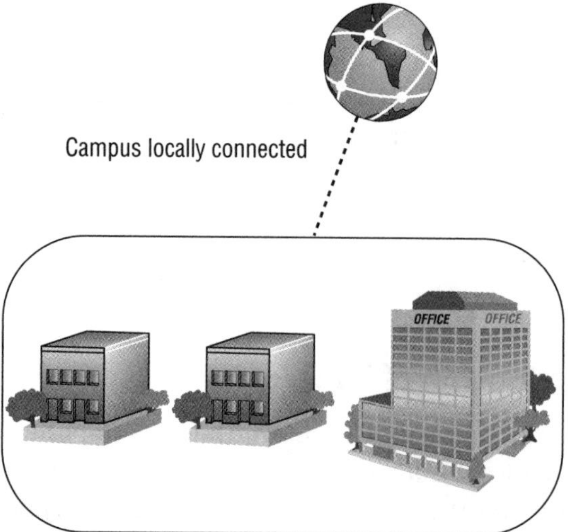

Figure 1.12: In campus topologies, the users, infrastructure, and internal resources are all co-located

Remote Branch Environments

Remote branch environments connect a remote office back to a centralized campus or datacenter, and/or connect a series of branch offices to one another. The former is a hub-and-spoke topology common in retail, hospitality, and consumer-facing environments where each location needs to access shared resources but has little need to communicate between sites. The latter is more of a mesh topology common in connecting peer or distributed offices with shared or interconnected resources.

Branch office sites may service a handful of users and devices, or they may need to provide access to thousands of users and devices. Planning wireless security architecture for branch offices can be taxing, since there are several ways to go about the topology, and different vendors offer different solutions in this area with varying (and frequently changing) capabilities and recommendations.

The following are a few of the ways we may design wireless access and security in remote branch sites.

Extension of Wireless LAN Services over Existing WAN Connections

In this model, the organization will have some type of site-to-site wired network connectivity, possibly through MPLS or via site-to-site IPSec VPN tunnels. In these cases, the organization's LAN is extended via the WAN, and there's generally a main office or main datacenter housing the organization's internal network infrastructure core components—including the wireless controllers or gateways (where applicable), possibly authentication or domain services, and/or other resources. In larger branch offices though, some resources may be local while others are hosted at the main office or datacenter.

The key consideration here is that in planning wireless security architecture, we need to carefully consider the control plane, data plane, domain services, and authentication sources, which may be more distributed than our campus topology.

Connection of Wireless LAN Services over Internet Without WAN

In cases where the branch office doesn't have persistent network connectivity to a main datacenter, or in cases where there is a VPN connection but it's over a high-latency, low-bandwidth link—the wireless architecture will vary yet again. When connecting remote branch offices over the public Internet versus a WAN service or already-established tunnel, we're forced to rely on the wireless infrastructure to create the tunnel.

You have a few options here depending on the size of the site and the connectivity needs. If you're using a cloud-based wireless platform you'll have an obvious advantage, as these solutions will include an option to tunnel or terminate remote and branch APs into the infrastructure over the Internet. Alternatively, there are similar options for Internet-based connectivity for legacy controller architectures, but these may require additional controllers or devices that are separately managed, and therefore, separately secured.

Lastly, in cases where the remote office is connecting users only to Internet-hosted software-as-a-service (SaaS) resources (and therefore doesn't require access to any on-premises datacenter resources), the deployment is drastically simplified because there's no need to connect to remote LAN resources for purposes of wireless infrastructure management or client data pathing. In these simplified deployments, our primary security consideration is ensuring we have full visibility and centralized security policy control of wireless at the remote sites.

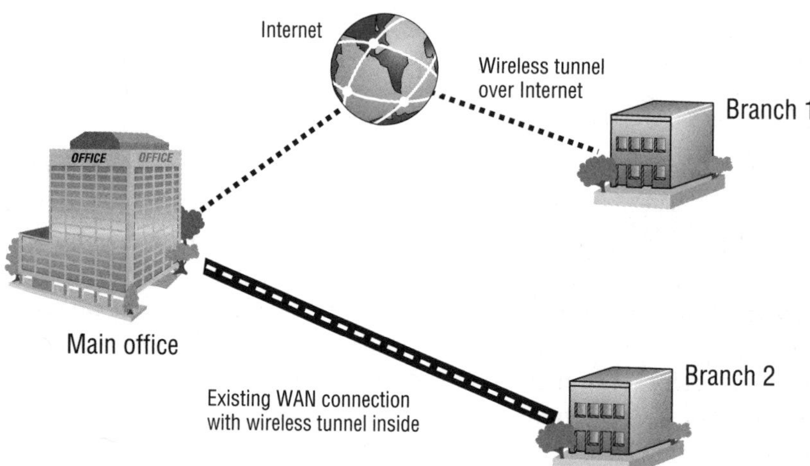

Figure 1.13: With remote branch topologies, satellite locations are connected back to the main office or datacenter via existing WAN connections or using purpose-built tunnels over standard Internet links

Remote Worker Environments

In the remote worker scenario, the goal is to provide secure wireless access to a user, commonly working from a home office but also supporting models of workers traveling in the new "work from anywhere" era. Remote worker and home office solutions vary by manufacturer (and again here, the capabilities change over time). The solutions are most often aimed at providing access to just a single user, but again some of the product features and limitations are very dynamic.

Remote workers connecting to the Wi-Fi infrastructure will connect in one of two ways—with remote AP hardware or with a remote Wi-Fi VPN application.

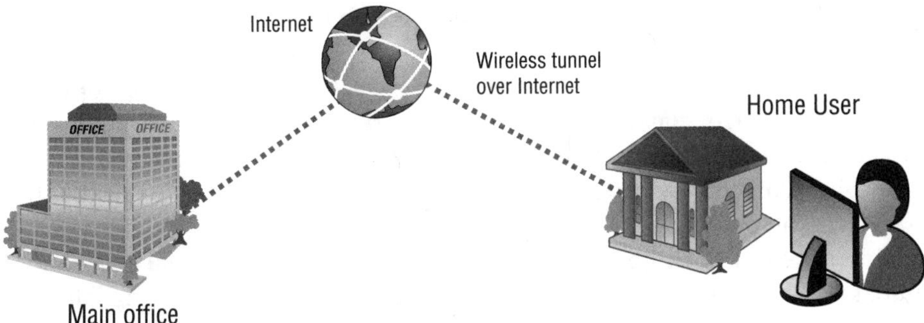

Figure 1.14: In remote worker environments, the user either has a remote AP at the house or relies on Wi-Fi-specific VPN connections

The Issue of Connectivity

Before we compare the two options, it's worth mentioning one shared requirement that makes every remote worker solution a bit complicated—any time a remote user is accessing resources hosted within the organization, it means we must provide some type of connectivity (specifically, over the Internet) back into the organization. Just as with traditional VPNs, we need a secure tunnel, and two ends to accomplish this.

It sounds obvious, but you wouldn't believe how many meetings I've been in where this isn't addressed adequately by the wireless manufacturer sales team up front, and the IT teams are left to figure it out on their own later. This is true for both models described next, with the difference just being how that tunnel over the Internet is created, and what the user-side tunnel termination point is (remote AP hardware or software on an endpoint).

Remote AP Hardware

Remote APs are just what they sound like—they're Wi-Fi access points that can be sent home with a user (usually an employee), and they extend the enterprise-secured wireless networks to the user's home office. They can come in different flavors and models, ranging from compact desktop form factor APs (sometimes with wired ports) to standard model APs just like you'd deploy in the organization.

As described previously, to extend a corporate network and its resources to a user remotely, the remote AP will have to connect back into the organization, and this requires there to be another end back at the organization to terminate the connection—usually a physical or virtual controller or gateway.

This can be more complicated than it seems because it requires deployment of the other end in the enterprise DMZ, firewall policies to allow the specific traffic and connections, and ongoing security monitoring of that external entry point to the network. Depending on the manufacturer and product version, you may also have to pre-provision the remote APs on the network before sending them home with a user.

Other solutions have more simplified *zero touch provisioning* (ZTP) options. Additional security considerations here are privacy and monitoring. Specifically for privacy, there may be certain wirelessly enabled services you'll want to disable on the remote AP (such as deactivating BLE location services or other RTLS tracking on APs used for remote workers in their homes). There are also considerations of how the data is processed and routed, and we'll cover more on that and the implications of full tunnel versus split tunnel routing later.

Lastly if users have an AP at their home, they're not within the standard protection areas of your other monitoring tools, which may encompass monitoring of Internet-bound traffic, and/or monitoring of RF-based attacks that can only be detected with wireless intrusion prevention systems (WIPS).

Remote worker APs can come in many forms, and in fact with most vendors, any model AP can be used as a remote AP. However, there are models designed for this use case that tend to be smaller and have wired ports to support additional home office connectivity. Figure 1.15 shows images of two such offerings, one from Aruba Networks and one from Juniper Mist.

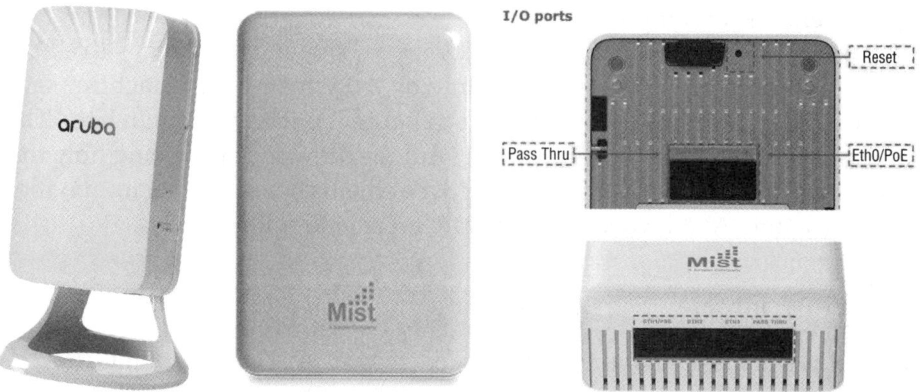

Figure 1.15: Sample remote AP products from Aruba Networks and Juniper Mist

Remote Wi-Fi VPN Client

Several enterprise Wi-Fi vendors will also offer a VPN client-based remote access solution. This option looks and works just like a traditional enterprise client VPN solution, but with the differentiator being that the tunnel is terminated to the enterprise Wi-Fi management instead of the organization's VPN gateway or VPN concentrator.

These clients are typically a hybrid SSL-VPN and IPSec-VPN (like many other traditional VPN remote access products). Just as with remote APs, the Wi-Fi VPN client must have an end termination for the secure tunnel inside the organization, which still means we have the complexity of working out secure connectivity to the virtual or physical controller, gateway, or management system in a DMZ.

Although less commonly used, the remote Wi-Fi VPN client has potential security benefits over the remote AP discussed prior. As noted, if the enterprise's secure Wi-Fi networks are being broadcast in homes and remote locations not logically or physically controlled by the organization, there is an increased risk that the network could be attacked without the organization's knowledge or ability to stop or mitigate it.

Summary

This chapter delivers only the most foundational knowledge around wireless and security domains. Although the content thus far is the least technical of the book, understanding the relationship of network architecture to security architecture is vital for creating a truly secure wireless architecture.

Having an intimate knowledge of the organization's business objectives, risk tolerances, compliance requirements, and security ecosystem is the launch point for your architecture design. Wireless and wired networks have, in large part, been designed and maintained in a silo of operations—a legacy trend I hope will end with the dawn of new security requirements and transformative initiatives such as zero trust.

In the following three chapters, we'll take a deeper technical dive into the components that comprise a secure wireless network including management architectures, segmentation and data paths, and an in-depth look at authentication including 802.1X with Extensible Authentication Protocol (EAP).

As you proceed, keep in mind the concepts and terms introduced here, notably recalling the difference in what qualifies as identification versus authentication for the purposes of information security and compliance.

CHAPTER

2

Understanding Technical Elements

Now that we've covered the foundational concepts of security and wireless domains, we'll take a deeper dive into the specific elements, starting with a more technical explanation of wireless infrastructure and management architecture, their impact on data paths, and how that impacts security architecture.

After that, we get into the more technical nuances of wireless security profiles including WPA2 and WPA3, followed by an intimate look into authentication and authorization schema in Chapter 4. Chapter 5 includes considerations for network domain services for wireless as well as a section on non-security Wi-Fi design elements that impact our security architecture, such as roaming protocols and designs for resiliency.

Understanding Wireless Infrastructure and Operations

The first topics to explore are the options for wireless infrastructure and management, and specifically what that means in terms of the data and control planes. Later in the chapter we'll also look more deeply at the data paths specifically for client traffic, and how that impacts security. For now, you'll just get an architecture overview and general teaser for the data path conversation to come.

The history of wireless architectures has been quite cyclic—just like client applications have moved from mainframes to distributed computing, back to centralized light virtual desktop infrastructure (VDI) technologies—so, too, has wireless followed the cycle of heavy→light→heavy, and hybrid management models.

Cloud-based management introduces a newer model, as does the more distributed gateway solutions the industry is favoring as we centralize management in the cloud for organizations that still benefit from certain functions remaining on premises.

In the following sections, we'll look at cloud-managed Wi-Fi, controller- and gateway-connected Wi-Fi, local cluster managed Wi-Fi, and remote APs. With each is a comparison of the data and control plane models and considerations for how that impacts your security architecture.

Management vs. Control vs. Data Planes

In network architectures, we typically talk about three planes or layers of operations related to controlling the infrastructure's data flow for management and user packets—the management plane, control plane, and data plane.

The management plane controls the infrastructure components. The control plane tells the infrastructure components how to handle the client data. The data plane is comprised of the end-user traffic.

Management Plane

The management plane comprises the communications and functions related to the management, configuration, and (usually) monitoring of the network infrastructure devices.

In the wireless world, it's the management plane that connects APs to either controllers or cloud management, pushes the configurations and software to them, and performs similar primary management functions.

Control Plane

The control plane is the part of the network function that controls how data is forwarded, while the data plane (covered next) is executing the actual movement of the packets or data, based on the decisions of the control plane.

The control plane is responsible for populating routing tables and discovering network topology and adjacency, among other things. Conversely, the data plane uses the framework provided by the control plane to get the packets forwarded. The control plane makes the decisions, and the data plane follows the instructions.

In wireless, the control plane manages the standard layer 2 and routing functions and would also encompass managing wireless-specific operations such as radio management (coordination of power and channel settings), key distributions for secure roaming, handling of multicast protocols, and any wireless-configured policies related to roles, traffic filtering, or ACLs. Figure 2.1 shows the three planes and the interaction in both a wireless controller environment as well as with cloud management. The client traffic in the data plane is localized with the organization's network and includes the endpoint, the AP, and may also include the wireless controller. A controller will fully encompass both the management and control planes. Despite the simplified diagram, the lines between management and control planes in cloud management platforms can become blurry, with some tasks offloaded to the local APs to execute.

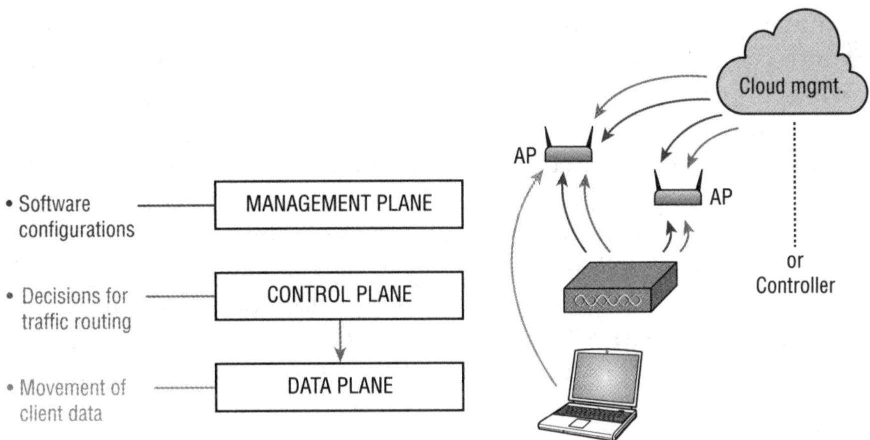

Figure 2.1: Management, control, and data planes

Data Plane

As already hinted previously, the data plane is simply in charge of moving the packets around, as prescribed by the control plane. Components of the data plane include the data paths for user traffic such as which AP the device is connected to, the AP-to-switch port connection, and the data path between the AP and the controller or gateway.

Separation of data, control, and management planes is not new, and in fact is a fundamental need not only in software-defined networking (SDN) technologies but also zero trust network architectures. Separating the control and data planes is how we're able to implement network virtualization and enhance security. Plus, separating the management plane allows flexibility of more distributed processing such as in cloud-managed architecture. Think of it kind of like out-of-band (OOB) management interfaces.

Table 2.1 offers an at-a-glance view of the three planes of operation and sample functions in wireless architecture.

Table 2.1: Example of planes in wireless

PLANE OF OPERATION	EXAMPLE OPERATIONAL TASKS
Management plane	Applying software updates and configuration files to wireless APs and controllers
Control plane	Decisions for segmentation, routing, and traffic filtering for client data
Data plane	Movement of client data along the path as prescribed by the control plane

Cloud-Managed Wi-Fi and Gateways

The cloud-managed revolution arguably started a little before 2010 as Meraki came to the scene, however it was almost a decade later before cloud-managed Wi-Fi architecture matured and became widely accepted as a viable option for enterprise environments.

The early days of cloud-based products offered a narrow subset of what's considered enterprise-grade Wi-Fi services, and an over-simplified user interface (UI) that was great for small environments but greatly hindered the ability of large and mature organizations to effectively design, deploy, and secure Wi-Fi at scale. At that time, there were few drivers for adopting a cloud-managed infrastructure aside from ease-of-use for moving to a cloud-based management system and elimination of in-house applications or hardware that needed to be maintained.

Today's Cloud-Managed Benefits for Enterprise

The latest generation of cloud-managed Wi-Fi edge networking products brings a new era Wi-Fi, and a new promise of heightened AI-driven operations (AIOps), API-managed infrastructure (a critical component of zero trust architectures), and machine learning (ML) for enhanced security, resiliency, and performance.

The driver for considering cloud-based products includes promises of easier deployments, more streamlined zero touch provisioning (ZTP), less complexity within the infrastructure (through eliminating or reducing on-premises management components), and the benefits of AI and ML available only through the computational resources in elastic and extensible cloud environments.

The other benefit of cloud-native products is their use of microservices, which allow the manufacturer to further separate services (beyond just control and data planes), enabling inter-related but independent service sets that can be maintained and restarted without impacting the other services. For example, with microservices, a cloud manufacturer could update the code and service set to tune and improve the radio resource management (RRM) features without impacting other operations or possibly without even requiring an AP reboot. Although on-premises infrastructures can also make use of microservices, few do, meaning traditional architectures would require a full platform operating system upgrade, which impacts all services and operations, not just the one(s) updated.

Because the control infrastructure is in the cloud, it also moves the onus of revisions, upgrades, and patching of the core management platform from the organization's IT team to the manufacturer. While this may sound insignificant (or even undesirable to some), the reality is that it can mean more uptime and resiliency, increased security through regular updates, and it frees up IT teams to focus on more strategic initiatives.

We'll cover this more later in Chapter 4, "Understanding Domain and Wi-Fi Design Impacts," but the processes of patching and upgrading on-premises controllers is lengthy and clunky at best. Namely, the IT teams have to prepare the environment by pre-downloading code versions, get approvals for maintenance windows (which is its own challenge), and then schedule upgrades after-hours, and deal with the cascade of inevitable fallout as there are glitches, bugs, or incompatibility with certain models of APs or endpoints in the environment. In larger organizations, this is all preceded by extensive lab testing to validate the code and operation in the environment.

Next is a glimpse at the architecture of cloud-managed solutions and the operations of control and data planes in this model.

MOVING TO THE CLOUD

Determining whether a cloud-managed solution is right for your organization is outside the scope of this book, but it's worth pointing out that today's enterprise solutions in this market bring benefits to the organization that can never be realized with legacy architecture. Cloud platforms (especially public, but also private in many cases) provide a level of agile compute and processing not attainable with traditional servers, services, and hardware.

Organizations across many industries, from global retail to universities, and even federal civilian agencies, are making the shift to cloud-managed environments and seeing unprecedented performance and return on investment. I've personally worked with scores of such organizations who are clients and spoken to the wireless and network architects from many others, who are all exceptionally impressed with these new ML-driven cloud offerings.

Whether you like the idea of cloud or not, cloud-managed and cloud-facilitated architectures are where most edge network vendors are headed. Of course, vendors like Cisco and Aruba will continue offering platforms compliant with the various federal and defense requirements, which may not include a public cloud component just yet.

Architectures with Cloud Management

In cloud-managed Wi-Fi deployments, the APs are deployed anywhere in the organization (that is, in any type of location—an office, remote site, or a user's home) and managed via applications hosted in the cloud. Fundamentally this means a separation of the management, control, and data planes. More confusingly, this also means distribution of control plane duties between the APs and the cloud platform.

Since, with cloud-managed solutions, the APs are physically (and to some degree, logically) separated from their management platform (which now resides in the cloud), there are blurry lines between whether a feature is managed within the cloud or locally at the AP. You'll find many features may be configured in the cloud, but then executed and managed (to some degree) by the AP. These may include functions such as VXLAN for segmentation, key distribution for secure roaming, and certain other filtering and radio management operations.

The separation of AP from local management also necessitates changes in the data plane architecture—something we cover more in depth later. In many environments—and especially in campus deployments—the IT teams are accustomed to the flexible data path options with legacy controllers, where traffic can be bridged (dropped locally at the switch edge), tunneled (tunneled through the switch/route infrastructure to the controller), or hybrid (a blend of bridged and tunneled).

However, in cloud-managed systems, there's not a controller (in the traditional sense) within the environment to terminate traffic tunnels. Instead, traffic is either simply bridged at the edge switch, or a gateway device is added to the local environment solely for the purpose of participating in the control plane and terminating those client data tunnels.

As you'll see, the significant difference between a traditional controller and a newer gateway model is that the controller participates in both the management and control planes, whereas the gateway device only serves the control plane. In the gateway model, the cloud-hosted application retains the management plane functions.

Figure 2.2 offers a side-by-side comparison of tunneled (left) vs. bridged (right) client traffic in a cloud management architecture. In the tunneled example, all client VLANs (12, 99 in the figure) are tunneled back to the tunnel gateway (for cloud deployments) or controller. In bridged mode, the client VLANs are configured at the edge switch and handed off there. In the tunneled example, only the AP management VLAN 3 has to be configured throughout the edge switch and route infrastructure. The client VLANs will terminate at the tunnel gateway appliance (just as they could to a controller).

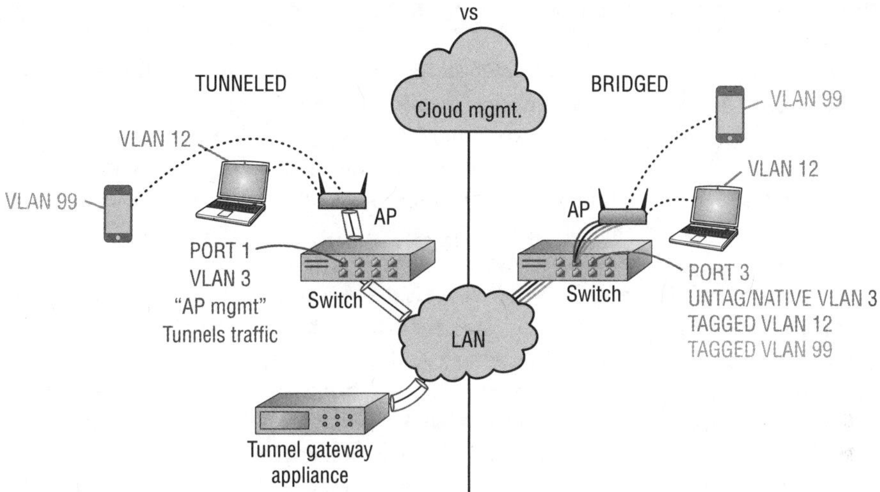

Figure 2.2: A sample architecture for bridged versus tunneled client data in cloud managed architectures

The Role of Gateway Appliances with Cloud-Managed APs

Think of gateway appliances as a pseudo-controller, an on-prem device that retains control plane function but not management function. In cloud-managed environments, the cloud is responsible for the management plane and orchestration, and the gateway is used for tunnel terminations for APs (hence the control plane).

Why would we want gateways? Well, in deployments with just APs and cloud management, there's no way to tunnel traffic to a central point. To address this limitation, many cloud-managed Wi-Fi vendors offer options to place virtual or physical appliances on-prem to act as tunnel terminators for the user traffic. This is also a key feature for organizations looking to migrate from legacy controllers to cloud management—something we'll discuss in the upcoming conversation on data paths.

These devices, utilized to terminate tunnels and participate in the data plane primarily, may be referred to as "edge appliances" (as with Juniper Mist Edge and Celona Edge); "gateways" as is Aruba's new deployment model starting in AOS 10, which uses gateways instead of controllers; or some other vendor-specific moniker. The important thing to know is that these options exist, and if your security architecture requires traffic to be tunneled and/or encrypted through the network infrastructure, then talk to your vendor about a "tunnel termination" option (versus a controller) if you're interested in moving to cloud-managed APs but want to tunnel some or all traffic to a central location.

INDUSTRY INSIGHT: VENDORS WITH CLOUD-MANAGED SOLUTIONS

For IT pros who may not be familiar with the various offerings from wireless vendors, cloud-managed solutions for Wi-Fi are offered by Cisco through the Meraki product line; Aruba Networks via its Central Cloud platform; Extreme (formerly Aerohive) with ExtremeCloud; and Juniper with Mist AI. Other manufacturers such as Fortinet have cloud-based offerings such as the FortiLAN, which encompasses (and replaced) FortiAP Cloud.

On the private cellular front, companies like Celona offer fully cloud-managed 5G LAN solutions. Amazon Web Services (AWS) announced its fully managed private 5G solution at the end of 2021, and the expectation is that many vendors will follow suit. In each case, these vendors also offer certain products and services that extend to or integrate with the existing on-premises infrastructure.

In addition to the cloud-managed solution, many vendors are moving toward cloud-facilitated models to take advantage of the flexible compute capabilities and machine learning. Cisco's DNA Center is one example of such a model, allowing organizations to maintain a traditional controller architecture while also taking advantage of the cloud's processing capabilities.

Table 2.2 summarizes the relationship of the management architecture to the three planes of operations and the related options for the client data path. In purely cloud managed environments, client data must be bridged at the edge switch. Cloud managed environments with a gateway appliance (to terminate tunnels) and controller environments both offer flexibility in the data path options and can manage combinations of bridged and/or tunneled client traffic.

Table 2.2: Cloud architecture and planes

ARCHITECTURE	MANAGEMENT PLANE	CONTROL PLANE	DATA PATH OPTIONS
Cloud-managed only	Cloud platform	Cloud platform	Bridged traffic only
Cloud-managed with gateway appliance	Cloud platform	Gateway appliance	Bridged, tunneled, or both
Controller-managed	Controller	Controller	Bridged, tunneled, or both

Controller Managed Wi-Fi

For years, controllers have served as the primary on-premises management devices (virtual or physical) with which APs would connect and be managed by. Controller-managed Wi-Fi blurs the lines with management, control, and data planes since most products support various data-forwarding models (detailed later) and blend management and control in a centralized model.

Over the years, the AP-to-controller relationship has varied with a heavy→light→heavy→hybrid cycle—where APs started off very autonomous and heavy (just like managing individual switches) then were at one point dumb or light APs incapable of any operations without explicit interaction from the controller. In these cases, all user traffic was typically tunneled back to the controller for processing, and options for a hybrid model with some traffic bridged locally to the switch was not even supported.

Later generations (including all controller products on the market today) offer hybrid models with granular options for per-group and per-SSID controls of data paths. And as you'll see in some solution offerings (covered in the following section), there's yet again a heavy offering where APs can be managed more autonomously.

Having said that, pretty much every enterprise wireless vendor is heading toward cloud-managed or cloud-facilitated architectures, almost exclusively. By including public or private cloud in the architecture, vendors can leverage powerful compute resources that facilitate machine learning. Even with the cloud movement, controllers remain a popular option for Wi-Fi management due to several factors.

Controllers have been attractive in enterprise network environments because of their deployment flexibility, including the option to terminate and tunnel certain client traffic from the APs all the way through the switching and routing infrastructure—called the distribution system (DS) in wireless—to the controller. This eliminates both the need to extend additional VLANs to the edge ports servicing APs, and to track and manage those configurations (which is still a largely manual process in many environments).

We'll dig more into the data path options soon. First, we need to cover two more common wireless management architectures.

In controller architectures, the topology looks the same as the samples in Figure 2.2, with the only difference being that the tunnel gateway appliance is the Wi-Fi controller itself.

Local Cluster Managed Wi-Fi

One of the lesser-known models of Wi-Fi management is something I call local cluster mode. In this mode of operation, the APs self-cluster and one or more APs in the cluster takes on the responsibilities of a controller or management platform, meaning it may perform management plane and/or control plane tasks.

With local clusters, the APs will usually require layer 2 adjacency, meaning with most products the APs will have to be in the same network or VLAN—specifically the same broadcast domain so they can find one another. There's usually a virtual IP with which the cluster can be accessed and managed, and in most cases it's possible to configure most of the features you need locally.

This mode is attractive to smaller organizations that can handle the more manual tasks involved in this type of deployment, and perhaps specifically for smaller organizations that are trying to avoid the ongoing operational expenses (OpEx) of subscription-based cloud models.

Because the deployment and management of these local clusters can be cumbersome and fall short in the areas of centralized management, centralized security policies and workflows, and visibility, some manufacturers offer options to connect local cluster APs to management software. Note that in this example management software is not a traditional controller, but an additional (or different) application.

INDUSTRY INSIGHT: VENDORS WITH LOCAL CLUSTER MODE PRODUCTS

Cisco offers a local cluster mode, as does Aruba with its Instant mode APs ("Instant" here is used by the manufacturer as a differentiator from campus AP modes that connect to controllers).

While several manufacturers offer similar features with local clusters, each vendor goes about it a bit differently and the operations, features, and functions of the various management, control, and data planes will vary. Talk to your vendor team to understand the options and ask for validated reference design guides to ensure you're following best practices for securing the clusters for both management and client data.

Figure 2.3 shows an example cluster management architecture. Note that most products require the APs to have layer 2 adjacency for the discovery protocol to work. APs across layer 3 boundaries would form a separate cluster. In the figure, all five APs are on the same IP subnet/VLAN and will form a single cluster. In this example, the cluster has elected AP2 as the coordinator, and the admin has configured the system to use the virtual IP of 192.168.3.250 to access the cluster.

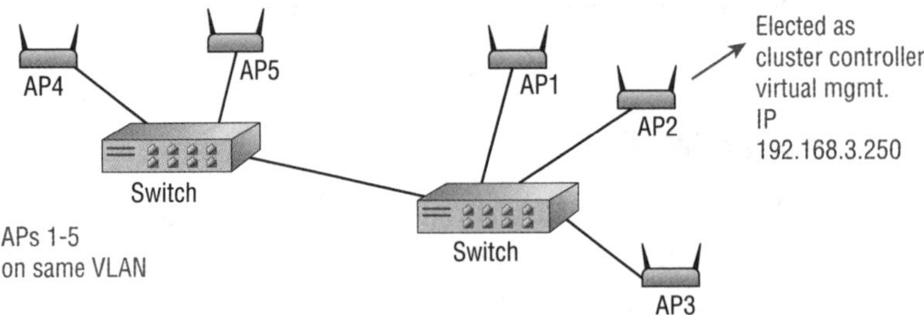

Figure 2.3: Sample local cluster management architecture

Remote APs

Earlier in our overview of wireless concepts, we touched on the use of remote APs for addressing "work from home" and "work from anywhere" user access models. From a management design perspective, remote APs introduce architectures that further obscure an already complex hierarchy of data paths and inter-relation of management, control, and data planes.

With these remote access options, additional infrastructure (such as a physical or virtual controller) is put in place to terminate tunnels; these added components are often configured differently (and separately) from the enterprise Wi-Fi system servicing on-premises users. They forward traffic differently; they manage radio resource coordination differently; and they interact with network and domain services (such as DNS and DHCP) completely differently. And of course, the data path in remote access now involves the Internet, meaning our infrastructure has another ingress point that must be properly monitored and secured.

Figure 2.4 shows an example of a remote AP architecture. The remote AP creates a tunnel over the Internet back to a corporate controller or tunnel gateway appliance. The corporate networks ("ACME-Secure" SSID in the figure) are then available to the home user. The office or datacenter will host a virtual or physical appliance to terminate the tunnel from the remote AP. In most architectures, this device will be placed in the DMZ of the firewall with policies to restrict only allowed connections.

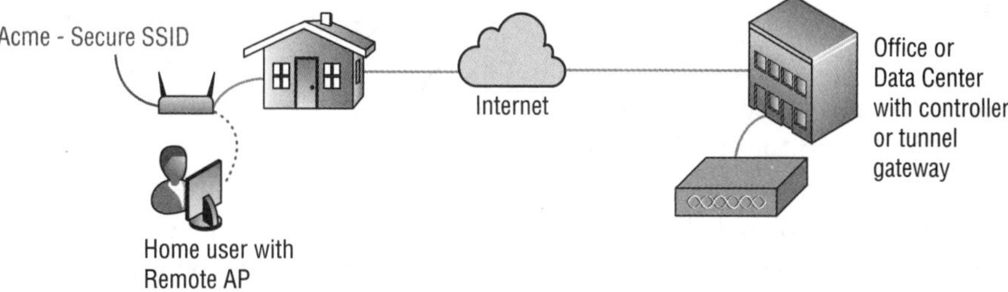

Figure 2.4: Sample remote AP architecture

Summary

With traditional Wi-Fi controllers comes a set of known elements as it relates to security—known management access through CLI, SNMP, NETCONF, and a GUI; known options for AP adoption and authentication; known data path control; and of course, known and well-defined options for configuring SSID security and user authentication.

As we saw, moving to cloud-based solutions influences drastic changes in the architecture and specifically the management, control, and data plane. Therefore, how we architect and secure these networks will (and must) fundamentally change. In cloud platforms, we have a new breadth of considerations for secure management including APIs (which we cover later) as well as new challenges in securing data paths.

Even if the complexity decreases, for some the unknowns may increase, and changes in management architecture will force organizations and IT teams to rethink their security architecture—from secure management all the way to client data privacy.

Understanding Data Paths

In some ways, I put the cart before the horse by covering the management models before working through the data paths section here. However, I think the understanding and visualizing those models is definitely helpful in understanding the data paths if you're not already familiar with them.

The previous section started with an overview of the management, control, and data planes in wireless architecture. Here, we specifically dive into the data plane and the forwarding paths of the client data traffic.

These next terms—tunneled, bridged, and hybrid—describe the relationship of the wireless clients' data as it leaves the AP and gets forwarded through the wired network (referred to as the distribution system, or DS, in wireless texts) and reaches the final destination, which may be a resource on the wireless network, on the wired network, or something outside the organization's network such as resources on the Internet.

It's worth noting that in most wireless systems, these forwarding path options are configured at the network (or SSID) level, so it's not an all-or-nothing global setting that impacts all traffic. As you'll see, there are some specific security considerations that will likely drive your decision to use tunneled versus bridged forwarding for certain classifications of networks.

Figure 2.5 offers a comparison of tunneled versus bridged traffic. This image is very similar to the view offered in Figure 2.2 comparing tunneled and bridged paths in cloud managed environments. Figure 2.5 further extends the view to demonstrate tunneled data paths (left) beyond the controller (or gateway appliance). This offers a visual representation of where client data is passed, making it easier to identify where segmentation rules should be applied. In the bridged mode (right) rules will have to be applied at the AP, switch, or router whereas the tunneled mode (left) allows policies to be applied more centrally at the controller.

Most architectures will use combinations of both tunnel and bridge modes, often referred to as a hybrid deployment. Figure 2.6 shows such a configuration, with a single AP serving a guest SSID that's tunneled to the controller, along with a secure SSID that bridges the traffic to the edge switch. In this example the secure SSID VLAN 12 must exist at the edge switch, and throughout the LAN infrastructure, whereas the tunneled VLAN 99 guest traffic need only exist at the controller. Further explaining the architecture, the edge port for this AP would be configured with untagged (native) VLAN 3 (the AP management VLAN) and be tagged (trunked) for VLAN 12.

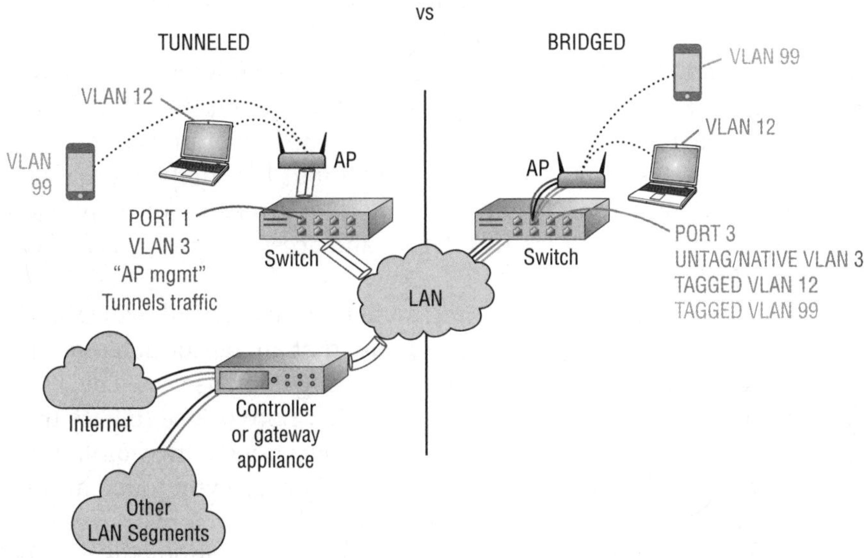

Figure 2.5: Comparison of tunneled (left) versus bridged (right) client traffic

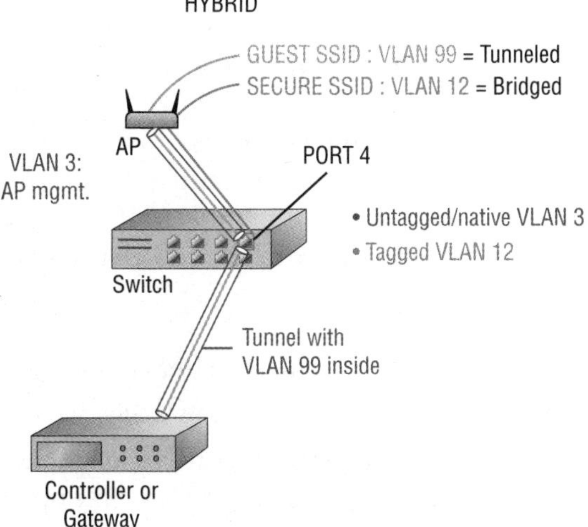

Figure 2.6: A hybrid data path model using both tunneled and bridged traffic from a single AP

Tunneled

In tunneled mode, an SSID will encapsulate (possibly with the option to also encrypt) the user's data and send it from the AP to the controller or tunnel termination device (such as an edge appliance or gateway we covered earlier). While there are obvious concerns of latency and hairpin routing in certain scenarios, there are also several unique security benefits for tunneling user traffic.

First, in a tunneled mode, it's not required to extend client VLANs through the network infrastructure to the edge switch ports servicing the APs. Aside from the obvious alleviation of operational tasks related to initial configuration and upkeep of those switch and router configurations, it also reduces the risk of security incidents through misconfigurations.

It's far too common that organizations end up with configuration artifacts as their edge networks evolve and change over time. Leaving VLANs on ports after an AP is moved, not removing VLANs that are no longer in use, and adding VLANs that are not required for that area are things I see daily in organizations of all sizes and industries. Another common occurrence is the confusion in configuring and maintaining traffic filtering and access control lists (ACLs) on the wired network.

Whether ACLs are applied in a routing switch or an internal firewall, it's common for them to be overlooked and not added at all, not updated as elements change, or configured but incorrectly, rendering them useless. The result is that one or more client networks can reach other network segments they shouldn't be allowed to. This causes obvious complications with maintaining the organization's security posture and will certainly cause you to fail audits, assessments, and pen tests. Depending on the severity, an incident or even failing security assessments and falling short of compliance requirements may be "resume-generating events" as my friend says. Our organizations rely on proper segmentation to mitigate malware and lateral movement by malicious actors.

It's not always the recommended model, but it's usually the one that's always safe. By tunneling traffic, we can alleviate some operational overhead and increase security.

In essence, tunneling traffic is our "easy button" when it comes to architecture and security planning, but there are several network architectures that don't fit well with that design and a few reasons we may want to drop traffic locally versus tunneling it back to a central termination point. The following lists explain why.

Pros of Tunneling Client Traffic:

- Enables strict control of data path through the wired network (no accidental routing or security failures due to misconfigured ACLs)
- Offers encapsulation with options to encrypt traffic over the wire (sometimes defined as a compliance requirement, e.g., as in PCI DSS over open networks)

- Reduces operational overhead by removing the need to extend client VLANs to the edge switch and maintain VLANs as Wi-Fi networks are added or changed
- Reduces security risk by not having multiple client VLANs on edge ports servicing APs
- Provides demonstrable and auditable client path security for guest and IoT networks that may have specific security control requirements
- Offers an easy path for upgrades and migrations from one product solution to another with similar architecture

Cons of Tunneling Client Traffic:

- Introduces hairpin routing and forwarding in some instances (e.g., if an authenticated employee is accessing a local printer on the same local VLAN and same office, tunneling would route the traffic out to the Wi-Fi controller, and then back into that switch)
- Possibly introduces latency across larger and distributed environments (for example, in deployments where the controller is deployed in regional datacenters and not co-located with the AP)

Bridged

Bridged mode is the exact opposite of the tunneled mode we just explored. In bridged mode, the AP hands off the client traffic at the edge switch port servicing the AP, and the traffic is then switched or routed over the wired infrastructure to its destination (which may be a resource on the local network or out on the Internet). Based on our tunnel conversation you can probably already guess the pros and cons, since they're inverse here.

Before we dive into the pros and cons, let's examine what happens with bridged traffic in a bit more detail because it may greatly influence your decision on whether to tunnel or bridge certain types of traffic, or perhaps traffic destined for certain classifications of data or resources.

Considerations of Bridging Client Traffic

When we bridge client traffic, the result is that as soon as the traffic egresses the local AP to the wired network, the remaining data path is exactly the same as if you were to connect any endpoint to that wired switch configured for an untagged (native) VLAN.

Specifically, bridging traffic means that—in the absence of some other control—data coming from the client device will be dropped locally on that switch port and will have access to any resources or destinations available through the switching and routing functions unless explicitly restricted on the wired infrastructure (methodologies of which we cover later in the section "Filtering and Segmentation of Traffic").

To the untrained network administrator, this may not seem like a big deal, but after working in hundreds (or thousands) of environments, I can tell you wired-side segmentation for wireless almost always fails in organizations without strict change management control. We'll talk about why and how later, but for now know that if you want client traffic segmented (due to operational reasons or compliance requirements) it's much harder to ensure that's being enforced when the traffic is bridged. In general, there's a lot more coordination, planning, documentation, and upkeep in security architecture as well as regular testing to ensure this enforcement remains. Later we'll talk about the complexity that often accompanies planning around proper segmentation when there's a blend of wired and wireless controls.

The other major consideration for bridging traffic is the operational overhead it entails. When client traffic is bridged, for any bridged network, you must extend each of those client VLANs through the entire wired infrastructure and to each switch port servicing an AP with that network.

If we imagine a small school, perhaps there are three SSIDs (one 802.1X-secured for students and staff, one portal for guests, and a one pre-shared key network for IoT and some Chromebooks), and each SSID is available on all 80 of the school's APs. The school IT team may decide to tunnel the guest traffic (something we'll look at next) but bridge all other traffic for the other two networks. In this case, if a switch port changes, or an AP is moved, removed, or added, the IT team will need to configure not only the AP management VLAN but also tag (or trunk, in Cisco) each edge AP port with the two client VLANs that are bridged. And then, of course, properly segment and secure those on the wired network.

What frequently (and by frequently, I mean 90 percent of the time) happens is that the correct VLANs don't get configured or maintained consistently. At some point change management falls through the cracks, the VLAN names aren't correct, an admin doesn't have a process document, or it just simply gets overlooked or fat-fingered and configured incorrectly. Regardless of why, it happens, and it inevitably causes issues that impact clients—they get disconnected; they can't roam; signal is bad; etc. In organizations with strong network management teams and processes, wired segmentation may be not only viable, but preferred; however, many lack the proper resources and change management processes to be successful in that endeavor.

Now that we've looked at some of the considerations and complications, let's recap the pros and cons you should consider when designing your wireless network.

Pros of Bridged Client Traffic:

- Removes hairpin routing by dropping traffic off at the edge switch port and allowing it to be locally switched or routed

- Reduces latency in many cases by eliminating the additional roundtrip path to and from a centralized controller
- Allows IT teams to take advantage of wired-side segmentation, which may be well-documented and understood by non-wireless teams (in organizations with strict change management)

Cons of Bridged Client Traffic:

- Moves traffic control and segmentation to the wired side, possibly out of view and control of the wireless architect and admins
- Enhances the chances of vulnerability through missing or misconfigured ACLs in the wired environment
- Removes the option for encryption of the client data back to a central point (without an overlay system such as those used in zero trust deployments)
- Adds operational overhead since IT admins will need to track and extend the proper VLANs through the infrastructure to each AP
- Adds vulnerability to attacks through the wired ports by having multiple client VLANs accessible through the AP port
- Lacks the native ability to deliver auditable data segmentation required by some compliance regulations

Hybrid and Other Data Path Models

Over the years, different vendor offerings have allowed for different configurations when it comes to client data path control. I can't recount them all here, but certainly in addition to straightforward bridging and tunneling options for SSIDs, vendors often add their own secret sauce and have hybrid offerings that fall somewhere in between, or more commonly, the provision for doing both tunneling and bridging in the same environment but applied to different SSIDs. And in some cases, there are vendor-specific implementations for dynamic decisions around the data path, which may depend on conditions of the endpoint, user, or location versus simply the SSID.

Data path models that don't fall strictly in tunneled or bridged categories include:

- Hybrid (a combination of both tunneled and bridged SSIDs)
- Vendor-specific (other proprietary and split-models that are vendor-specific)
- Remote APs and Wi-Fi VPN (while technically still tunneled, remote APs and Wi-Fi VPN client models follow a different data path than local tunneling and have different models for the distribution of control and data planes)

Many wireless deployments are a hybrid, or a blend, of the two client data architectures we've just covered, with the reminder that tunneling the client data requires some type of tunnel termination point opposite the AP—either a controller, a gateway, or some virtual or physical appliance to be that other end.

Hybrid deployments are not only common but are often the best solution for many environments. The ability to mix and match data path models to meet specific operational needs is a great benefit and gives us flexibility in architecture choices. On the flip side, hybrid deployments can create opportunity for confusion and misconfiguration since some data is tunneled but other data is bridged, meaning filtering and segmentation controls will be applied in different places.

We'll cover this topic more in the next section as well as in Chapter 5, "Planning and Design for Secure Wireless," where we get into the details of decision-making around tunneled, bridged, and hybrid deployments. For now, just know that with all enterprise-grade products, you'll have options and the ability to mix and match for the best security posture and user experience.

Filtering and Segmentation of Traffic

There could be an entire book just on network segmentation; it's a complex topic, with numerous options for achieving the mission, multiple places the policies can be configured, and multiple places those policies can be enforced.

Later, we'll talk through the decision-making process of when to use what, how, and where when it comes to segmentation. Not only do we factor this into the design of wireless networks, but we also consider it in hardening the infrastructure in "Securing Management Access" in Chapter 6, "Hardening the Wireless Infrastructure."

For now, our purpose is to understand filtering and segmentation as a concept, explore a few ways to enforce segmentation, and become familiar with the relationship of those options to our wireless architecture.

The Role of ACLs and VLANs in Segmentation

When we talk about segmentation, we're just describing any method (or combination of methods) to contain or separate certain devices (or data) from others on the network. Layer 2 segmentation uses VLANs which typically still require a layer 3 control. Layer 3 segmentation uses access control lists (ACLs) or other IP-based segmentation (such as Virtual Routing and Forwarding, or VRFs, described shortly). Filtering at layers above 3 happens at the port and protocol level up through the application layer and can be applied in several ways, including through ACLs and other policy enforcement options both wired and wireless.

ACLs in their traditional implementation are a set of rules or policies that dictate what traffic is allowed where, at layers 3 and higher. ACLs are most commonly associated with routers or routing switches, but as you'll see, we have several options for applying ACLs including within the wired or wireless infrastructure, and on firewalls. Note that Cisco and other vendors have proprietary access control list features that operate at layer 2 or the port level.

Filtering Traffic within Wireless and Wired Infrastructures

Before we delve into methods for filtering and segmenting traffic, the first big distinction to make is that, with wireless technology, there are times when we need to filter traffic within the wireless infrastructure (at the AP, controller, or gateway), and times when we need to filter traffic once it's on the wired network, after egressing the wireless infrastructure. Obviously, the wired infrastructure can't filter traffic over the air, and (in most cases) the wireless infrastructure can't filter traffic once it's left the wireless and is passing through the wired network. I say "in most cases" because there are some vendor-specific implementations that are bordering on this.

Let's first examine some ways we may enforce segmentation within the wireless infrastructure—that is, enforcing access rules within the APs and/or controller versus within the wired network.

Figure 2.7 compares segmentation on the wireless infrastructure versus the wired infrastructure. On the left, the wireless infrastructure is able to segment traffic between endpoint devices, such as not allowing the laptop to communicate directly to the smartphone, or the phone to the LAN. On the right, wired segmentation is applied and in this example traffic from the smartphone is also not allowed to the LAN, but that's enforced on the wired network, not at the AP or controller.

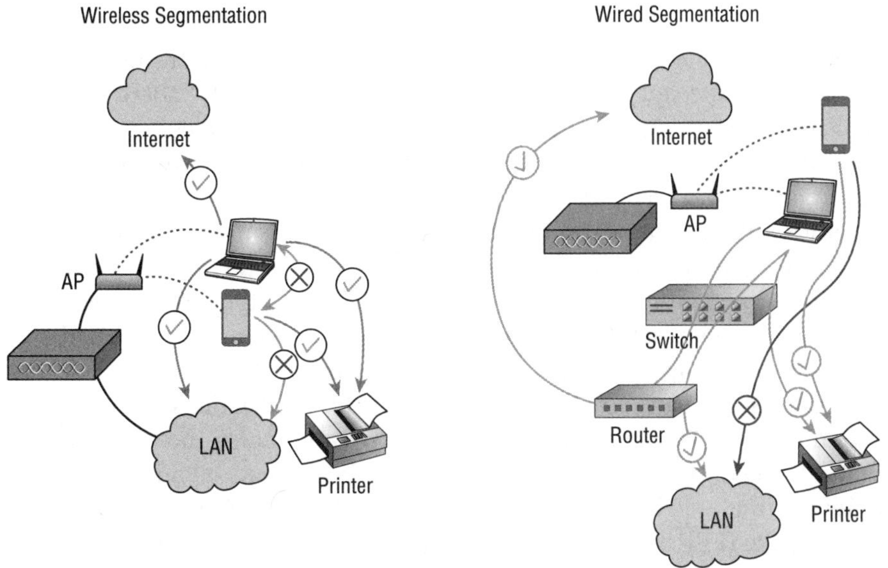

Figure 2.7: Sample overview of segmenting on the Wi-Fi infrastructure versus the wired. Some controls can be implemented on either, while others (such as client-to-client) can only be controlled one way

The trick to proper segmentation is to know where along the data path the traffic filtering needs to be enforced, with the understanding filtering may need to be applied in more than one place. In an SSID where the client data is bridged from the AP to the edge switch, there should be filtering not only within the wireless infrastructure, but also on the wired infrastructure since the traffic is dropped at the edge for forwarding and routing.

Conversely, in an SSID where the client data is tunneled from the AP to a controller, the filtering can be enforced fully within the wireless infrastructure, in most cases eliminating the need to manage ACLs on routers within the wired infrastructure.

Filtering with Inter-Station Blocking on Wireless

One of the most basic controls we have within the wireless infrastructure is the use of inter-station blocking. By necessity we cover this more in Chapter 6. Inter-station blocking simply prevents Wi-Fi clients on the same network or SSID from communicating with one another. It was a standard operation in Wi-Fi deployments for many years, but more recently became a troublesome setting as more and more protocols use peer discovery for proper function.

I'm using the phrase "inter-station blocking" here to avoid any confusion with other operations, but this feature may be referred to as peer-to-peer blocking in some vendors' products or documentation. Peer-to-peer communications are also often referred to in terms of ad-hoc networking, which is different in that ad-hoc networks bypass the enterprise infrastructure completely.

Specifically, enabling inter-station blocking impacts various direct connection protocols such as those required for direct printing, screen casting, and other applications that require client-to-client connectivity over the Wi-Fi infrastructure.

Note that inter-station blocking specifically means filtering traffic between clients that are connected to the same enterprise-managed SSID; there are other protocols that set up a true direct peer-to-peer (ad-hoc) wireless connections (such as Miracast), which bypass the managed infrastructure. We'll talk about the security risks and recommendations for these later.

The topic of these various multicast protocols warrants a sidebar. In general, Wi-Fi was designed to support unicast and broadcast and was not optimized for multicast. When multicast packets are sent over Wi-Fi, they have to be transmitted to match the capabilities of the lowest common denominator of client; the result was that many multicast protocols were unreliable and slow. For that reason, multicast often introduces a high error rate for applications like voice and video. Some products convert multicast to unicast and repeat the transmission to all targeted clients; this mode ensures the packet was received but negates the benefit of the entire multicast function. Challenges with multicast were addressed in the IEEE 802.11v amendment, which included Directed Multicast

Services (DMS), and networks with clients and APs that support this feature will have a better experience.

If multicast client-to-client communication is required for specific applications, set the Wi-Fi infrastructure to limit and block any that aren't required by business needs. Unfortunately, that's easier said than done in an era of consumerization where many users like their easy-to-use Apple products.

There are many use cases for enabling multicast though, such as supporting push-to-talk operations and various streaming needs.

Filtering with SSIDs/VLANs on Wireless

The most commonly used and straightforward approach to segmenting traffic within the wireless infrastructure is to use SSIDs that place endpoints on different client VLANs. This can happen through static configuration of the SSID—for example, "ACME-IoT" SSID may place endpoints on VLAN 99, which is Internet-only, whereas the "ACME-Secure" SSID may place endpoints on a production VLAN 20.

There are also options for a dynamic VLAN assignment. The most common implementation of this would be on the Enterprise/802.1X secured SSIDs where we simply instruct the authentication server to attach a standard RADIUS attribute along with the authentication approval. When the wireless infrastructure gets the "OK" from the RADIUS server, it would be accompanied with an instruction for the specific VLAN, which may be based on one or more user or device attributes. We'll cover this more in depth later.

Aside from standard RADIUS attributes on Enterprise/802.1X secured networks, there are also options to return dynamic VLANs for Personal/passphrase-based SSIDs. That operation is a non-standard vendor-specific feature that might be built into the wireless product or integrated with an external database and authentication product such as the various network access control (NAC) products.

Filtering with ACLs on Wireless

There are also options to filter traffic within the wireless infrastructure at a more granular level than VLANs. Although I'm referring to this section using "ACLs," this feature may come in many forms, including access control policies that look like router ACLs, or policies that mimic firewall rules. The point here is to demonstrate the ways of controlling traffic within the wireless infrastructure above layer 2, using layer 3 and higher methods such as specifying traffic based on IP address, subnet, port/protocol, or even application signatures. With these controls in place, traffic is being filtered by the wireless infrastructure, but within the wired components, not over the air, which is always layer 2.

Similar to how we specify rules in firewalls, the wireless vendors often have options to define labels or tags used like address objects, which allow you to specify policies using names instead of IP addresses. For example, in some platforms you can specify any traffic coming from a portal SSID as "guest" and apply rules to that guest traffic without ever referring to a subnet like 192.168.99.0/24. You could also create a label of "conference room printer" and specify the IP address (or hostname), such as 10.10.20.186 (or `conf01print.acme.com`) and then apply a policy to allow "guest" to "conference room printer." In many cases, you could even be more granular and only allow certain ports or protocols in that policy. Labels are a great way to simplify otherwise complex ACLs and reduce the potential for human error.

Wireless vendors often refer to their ACL-based filtering using their own terms, such as defining "roles" with Aruba Networks or "policies" using VXLAN in Juniper Mist. In addition to their self-defined names of features, vendors also offer some proprietary secret sauce to help make specifying and applying segmentation easier to manage.

Lastly, it's worth mentioning that we can also send dynamic or downloadable ACLs via RADIUS attributes, like the method described previously with dynamic VLANs. Downloadable ACLs have mixed results, are often messy to maintain, and are defined as a Vendor-Specific Attributes (VSA) versus a standard RADIUS attribute. Vendor-Specific Attributes send a non-standard instruction back to the wireless controller or AP that has specific meaning to that particular product. Because they're proprietary and very specific to the platform and version, I'm not covering them here in detail. There's a bit more later in Chapter 3, "Understanding Authentication and Authorization," but for more information on dynamic ACLs, contact your wireless vendor for resources.

Controlling Guest Portals with DNS on Wireless

In the preceding sections, we looked at ways to control traffic in the wireless infrastructure at layer 2 and above. The other common traffic control method in wireless is via Domain Name System (DNS) and it's used for controlling traffic during a captive portal experience. Whether the captive portal is forcing registration and authentication or simply asking the user to accept terms and conditions, the traffic from the client device is most often controlled by DNS during that process.

When the user connects to a portal-controlled SSID, the wireless AP or controller inserts itself in between the client and rest of the enterprise network. As such, it can serve the client Dynamic Host Configuration Protocol (DHCP) typically with itself as the IP gateway and DNS server, thereby controlling access of that client device to all other resources.

This covers all the common segmentation methods implemented within the wireless infrastructure (at the AP or controller/gateway). Next, we'll look at the options for segmenting or filtering traffic within the wired infrastructure.

Filtering with VLANs on Switches

Controlling traffic with VLANs is a blend of layer 2 and layer 3 enforcement. As you'll see in the tip, while VLANs are, by IEEE (Institute of Electrical and Electronics Engineers) definition, a layer 2 standard, they are most often attached to an IP-based interface, which means we must then implement layer 3 (ACL-based) controls.

UNDERSTANDING VLANS AND ACLS

Although VLANs and ACLs are considered 101-level for any network architect, I want to add a note to avoid any confusion that may be caused by various vendors' implementations of these two most basic functions.

By IEEE definition of standards, VLANs operate at layer 2 (OSI data link layer). However, in most cases, at some point the VLAN is associated with an IP-based subnet as an interface, at which point traffic can be routed in and out of that VLAN, and controls must be implemented at layer 3 (OSI network layer).

Most enterprise routing switches will by default perform inter-VLAN routing within themselves. For example, if I configure VLAN 20 with IP subnet 10.10.55.1/24 and then on the same routing switch configure VLAN 99 with IP subnet 192.168.99.1/30, traffic between VLANs 20 and 99 will be routed unless an ACL is applied. Similarly, if edge switches have gateways upstream that can route between VLAN interfaces, that traffic could be routed there unless filtering is in place. I mention this only to help readers avoid the all-too-common mistake of creating VLANs, assigning IP addresses and subnet interfaces to the VLANs, and believing traffic between those subnets is segmented. It's not.

Although not a hard-and-fast rule, routing switches that are commonly used as edge (plus often as distribution and core also) will most often perform the inter-VLAN routing as described, by default. Exceptions to this are pure routing devices (routers versus routing switches), which have different interface structures and routing operations. With the routing-only devices, it's unlikely there are any default routing behaviors. I mention this for clarification only; a great majority of organizations have infrastructures based on routing switches in the areas where segmentation needs to be applied.

Figure 2.8 demonstrates the subtle difference between pure IEEE standards-based VLANs, and implementations with IP addresses assigned to VLAN interfaces. With no IP addresses, traffic between VLANs is segmented at layer 2. If IP addresses are added to the VLANs, most platforms will automatically enable inter-VLAN routing, meaning filtering will need to be configured with ACLs at layer 3.

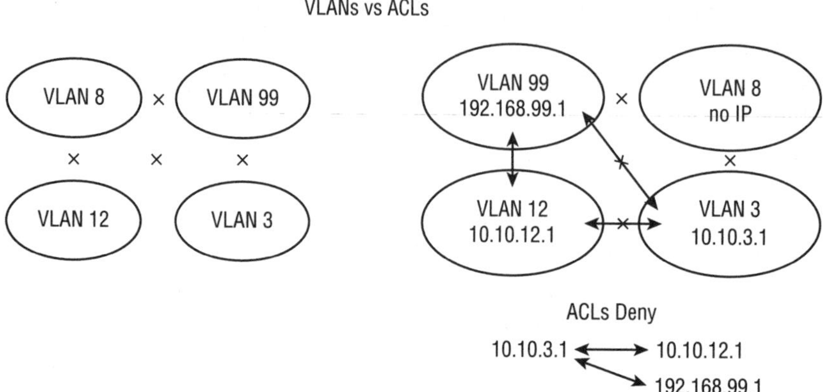

Figure 2.8: Comparison of the segmentation behavior of VLANs (left) versus VLANs with IP addresses (right)

Before we get into ACLs, let's cover those rare occasions when there can be pure VLAN-based segmentation. If we were to configure VLAN 20 throughout our infrastructure and not assign any IP addresses to the VLANs nor IP-based routing interfaces for VLAN 20 anywhere in the routing switches, then we will have successfully implemented layer 2 segmentation, and that VLAN's traffic will not be forwarded or routed to or from other VLANs. In practice, however, there will almost always be a need to route the traffic somewhere—if nowhere internally, something in that VLAN is sure to require Internet access.

In small environments, there are some fun tricks we can do by carefully architecting VLANs and gateways to avoid having to manage ACLs in the environment. Some samples of this will be covered in the Appendix C, "Sample Architectures." For context, one example is to create a VLAN and only specify an IP address for that VLAN at a firewall, which would serve as the gateway for all devices on that VLAN. By default, firewalls have an explicit deny policy, meaning traffic to and from that VLAN would have to be intentionally allowed by a firewall policy. This is one easy and simple way to segment at layer 2 on the LAN but provide Internet access and granular policies for any traffic needing to re-enter the enterprise LAN. An important item to note is that if anyone were to add a routable interface for that VLAN inside the network, it is possible traffic could be routed internally under certain circumstances.

Many of these challenges are overcome with network virtualization overlays, described later.

Filtering with ACLs on Routing Devices

As mentioned, most times we're not able to do pure layer 2 segmentation and we're going to rely on ACLs somewhere. In cases where traffic is egressing the wireless network and is then routed on the wired side, there will be a need for ACLs on wired routing devices.

Most commonly, this need arises when wireless client traffic is bridged at the edge switch. As described in "Understanding Data Paths," at that point the client traffic has left the wireless infrastructure and is being forwarded or routed just like the rest of the wired traffic on the network. In this architecture, the wireless client is frequently served a DHCP assignment from a scope specifying a gateway and DNS on the wired network, and as such that gateway routing device should have appropriate ACLs to control traffic to and from that client data subnet.

With some vendor controllers, it's also possible for the wireless client data to be tunneled back from the AP to the controller, and then egress the wireless infrastructure to the wired side while preserving the VLAN and subnet. While less common of an architecture, in these instances if the wireless controller is not applying the access control rules, it will again fall to the wired infrastructure to apply ACLs at the appropriate places.

APPLYING ACLs

I'm using the phrase "apply ACLs at the appropriate place(s)" intentionally. In all my years of helping clients through network architecture, a common occurrence is a poorly designed or poorly understood layer 2 and layer 3 infrastructure. Quite often, organizations have routing-capable switches throughout the network including at or near the edge, and mistakenly assign IP interfaces to VLANs where they're not needed (and not desired).

Any misconfiguration such as this, combined with inadvertent misconfiguration of a client gateway along the way will mean that traffic is being erroneously routed. While it sounds like a one-off situation, if I had to guess I'd say I've seen this at 75 percent of organizations I've worked with, spanning all sizes and industries. It's very common for a network admin to not understand the switching and routing infrastructure, add an IP interface to a VLAN where it doesn't belong, and for someone to specify that device as a client's gateway. Sometimes this occurs during the initial deployment and other times during expansions or through troubleshooting processes.

Figure 2.9 shows an endpoint with an IP address 10.10.12.126 and a 24-bit subnet mask. Its default gateway is 10.10.12.1, which is an IP address of the routing switch also depicted in the figure. This image tells us two things. First, for traffic not already filtered at the AP, any ACLs will need to be applied at the routing switch. Second, if this is a wireless client, we can infer that the traffic is bridged, not tunneled.

INDUSTRY INSIGHT: NOVEL SEGMENTATION PRODUCTS

Later in Chapter 8, "Emergent Trends and Non-Wi-Fi Wireless," we'll talk more about the zero trust concepts that relate to wireless. For now, know that some of the zero trust microsegmentation products on the market use granular routing control for zero trust

security enforcement. They vary in specifics but generally can inject themselves as the DHCP server, thereby gaining control of endpoints' gateways and DNS. Some issue an IP address with a /32 subnet mask, routing all traffic off-network (effectively all traffic) to itself as the gatekeeper and then allowing or restricting access to resources based on the configured policies.

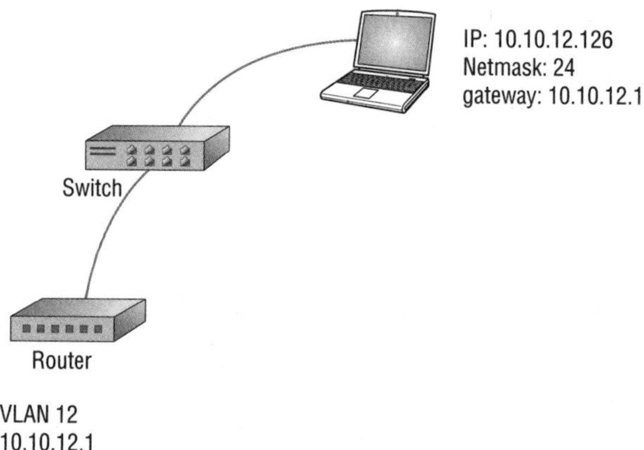

Figure 2.9: This endpoint's default gateway is on the wired network

Filtering with Policies on Firewalls

Earlier in this chapter in "Filtering and Segmentation of Traffic," I provided an example of using VLAN-based segmentation with a firewall as the gateway and routing device. In this architecture, the firewall is then where the traffic filtering happens and can be specified at a level of granularity not supported in all routing switches. In addition to granularity, firewalls can offer ease of management as most firewall policies can be name-based with address objects by IP or subnet, making it easier to specify a policy such as "allow guest to Internet," "deny guest to LAN interface," or "allow guest to conference room printer."

In smaller environments the firewall can also serve DHCP and DNS to a subset of endpoints, for example in a guest user scenario, eliminating the need to allow communication from the guest network to the organization's production servers. Typically, this architecture is reserved for a limited scope, not an entire organization, although in very small environments or remote branch sites, it's not terribly uncommon to see an IT team configure a firewall to service all endpoints (wired and wireless) in this way.

One other advantage with firewalls is the visibility from logging. It's possible to log allowed or denied traffic and offload that easily to a firewall management platform or other standard syslog or SIEM tool, which can be used for monitoring as well as forensics in incident response in the event of a possible breach.

Note that the firewall in these examples could be the enterprise's gateway firewalls or specially deployed LAN-based firewalls for internal segmentation. The architecture there will depend on the size of the organization, bandwidth of the traffic going through it, and processing capabilities of the firewall. In a larger organization, a firewall sized for Internet traffic only won't be able to sustain the overhead of all LAN-based traffic.

Using firewalls for traffic filtering is most common for subsets of traffic, users, or devices. For example, it frequently makes sense to direct all guest Internet-only traffic out directly through a firewall instead of routing it internally. It may also make sense to protect resources or subnets with sensitive data and applications behind a firewall and only allow specific traffic to and from them.

Filtering with Network Virtualization Overlay on Wired Infrastructure

The final method of traffic segmentation comes in the form of vendor-specific implementations of network virtualization overlay technologies such as Virtual Routing and Forwarding (VRF) and Ethernet VPN-Virtual Extensible LAN (EVPN-VXLAN).

I'll provide a brief intro to these technologies and then encourage you to explore in more detail with the manufacturers. VRF, VXLAN, and other virtualization technologies are supported by most enterprise network vendors including Juniper, Cisco, and HPE Aruba, among others, but may have vendor-specific implementations that don't allow integration across platforms.

VRF is most simply described as a layer 3 version of VLANs. It's a protocol used by routing devices that allows virtualized routing tables. Originally a carrier technology, it's been commonly used in LANs for years (usually as VRF lite) to segment traffic more deterministically without having to manage complex ACLs.

EVPN-VXLAN is a datacenter technology that is making its way toward the network edge. It extends layer 2 connectivity as a virtualized network overlay across an existing layer 3 physical network. Put more simply, it lets us extend VLANs across routed boundaries and keep the layer 2 properties.

Summary

In this section, we looked at the different data path models for client data egressing from the wireless infrastructure to the wired network, and we took a deep dive into the various options for segmenting and controlling data paths throughout the wireless and wired infrastructures.

Next, we'll take a closer look at the security profiles for Wi-Fi networks and then further explore the authentication and authorization mechanics.

Understanding Security Profiles for SSIDs

The concepts and architecture in this book cover many wireless technologies, but this section is specific to IEEE 802.11 WLANs and the various types of security profiles that can be specified for those wireless networks.

In Wi-Fi, a security profile is specified for each network or SSID. The security profile defines how an endpoint can associate and authenticate with that Wi-Fi network. Different security profiles offer different levels of security, and there are three classes of security profiles we'll explore:

- Enterprise/802.1X-secured (includes mutual authentication, key rotation, and encryption)
- Personal/passphrase-secured (no authentication, includes encryption)
- Open networks (no authentication, no encryption unless using Enhanced Open)

The loose rule of thumb is:

- Enterprise/802.1X secured networks are the standard for secure Wi-Fi for users and devices accessing production resources, connecting from devices that can handle certificates or user-based credentials for authentication.
- Personal/passphrase-secured networks are a bit of a stop-gap for devices that require access and encryption but can't perform an 802.1X authentication required for Enterprise-secured SSIDs.
- Lastly, open networks require no authentication or information to join the network and are most often used for captive portal experiences for guests, third parties, or BYOD registrations.

The security profile is important not only for security purposes, but also has a big impact on how the encryption keys are distributed and moved throughout the wireless infrastructure to support device roaming. The more authentication and encryption required, the more work the wireless system has to do on the backend; the end result being the potential for additional latency that will impact latency-sensitive applications. That topic and more will be covered later in Chapter 4.

Figure 2.10 shows an example of Aruba Central's security profiles for the Enterprise class. The WPA3-Enterprise networks support several cipher suites. CNSA stands for Commercial National Security Agency and is the 192-bit mode of WPA3-Enterprise approved for federal use. The CCM-128 is the default (and lowest supported) cipher suite for WPA3 secured networks.

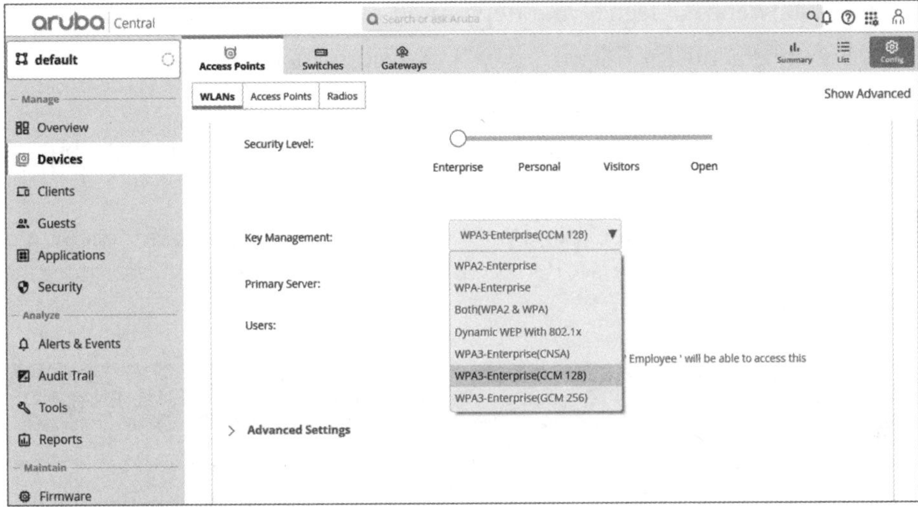

Figure 2.10: Aruba Central's security profile options for the Enterprise class

Before diving into the types of security profiles for SSIDs, an overview of Wi-Fi Protected Access (WPA) is provided—the suite of security functions for 802.11 WLAN networks that define the authentication, key exchange, key derivation, and crypto suites. Also covered is the newer Enhanced Open standard, which adds encryption for open authentication SSIDs.

UNDERSTANDING THE TERM *PERSONAL NETWORKS*

Before we proceed, I want to clarify a common misconception outside of wireless professionals. "Personal" networks in this context do not mean home use. "Personal" refers to a network that requires a passphrase to connect to, and that passphrase is used to derive an encryption key.

Personal class networks using passphrases are exceptionally common in enterprise organizations for a variety of reasons we'll explore later. Regardless of your feelings about the nomenclature or security concerns, passphrase-based SSIDs will be a part of enterprise Wi-Fi architecture for the foreseeable future.

WPA2 and WPA3 Overview

Wi-Fi Protected Access version 3 (WPA3) is the most recent security suite certified by the Wi-Fi Alliance. It replaces its predecessor, WPA2, and adds several significant security enhancements including:

- Required use of Protected Management Frames (PMF) to add integrity and support availability
- Protection from downgrade attacks

- Disallowance of legacy and deprecated protocols
- Enhancements for Enterprise/802.1X secured networks
- Enhancements for Personal/passphrase-based networks
- Enhancements for open networks (not technically part of WPA3 but its close cousin)

Upcoming sections will expand on the specific enhancements for each type of security profile (Enterprise, Personal, Open), as well as migration strategies and best practices for each.

INDUSTRY INSIGHT: WHAT'S IN A NAME? THE CONFUSION OF SSID SECURITY OPTIONS

One interesting point to note is that the SSID security options in many products including enterprise-grade and residential Wi-Fi routers aren't at all obvious. They can confuse even the most well-trained Wi-Fi professionals.

Today's most popular Wi-Fi vendors seem to just make up whatever names make sense at the time.

Residential products' use of non-standard terms and settings such as "auto" may be configuring a lower-capability (and more vulnerable) security suite without the user's knowledge.

Although out of scope for what's covered in this book, in the "work from home" era, residential connectivity and security are more important than ever. Be sure to use the product's help resources or configuration guides to ensure the configuration matches your desired security profile.

WPA3 is still new, and vendors are not consistent with their naming conventions for the WPA3 secured networks. Further complicating the matter, there's not even consistency between a vendor's naming convention of the WPA2 counterpart. For example, a vendor may show "WPA2-Personal," but the WPA3 equivalent, instead of being displayed as "WPA3-Personal," may show as "WPA3-SAE."

A certain amount of inconsistency is to be expected as the representations of "WPA3-Personal" and "WPA3-Enterprise" are reserved for products that have undergone Wi-Fi Alliance certification for those features. Products may be certified for WPA2-Personal, but haven't yet undergone WPA3-Personal certification, meaning the vendor can use "WPA2-Personal" but not "WPA3-Personal" yet. Table 2.3 covers the WPA2, WPA3, and Transition Mode features along with notes on other representations.

Table 2.3: Current SSID security options in products

WI-FI ALLIANCE CERTIFIED FEATURE	OPTIONAL ALTERNATE REPRESENTATION	DESCRIPTION AND NOTES
WPA2-Personal	WPA2-PSK	Vendors may have variations to designate multiple or personal PSKs.
WPA3-Personal	WPA3-SAE	SAE is specific to WPA3-Personal. At least one vendor erroneously may list this as WPA3-PSK, which is incorrect and misleading.
WPA3-Personal Transition Mode	WPA3-SAE + WPA2-PSK	Vendors with submenus may display this as Both (WPA2 & WPA3) after selecting Personal.
WPA2-Enterprise	WPA2-802.1X	At least one vendor displays this as "WPA2-EAP," which may be confusing.
WPA3-Enterprise	WPA3-802.1X	Several cipher suites are supported, and vendors may represent these by the acronym and key length; if no options are specified, assume CCM-128.
WPA3-Enterprise Transition Mode	WPA3-802.1X + WPA2-802.1X	Vendors with submenus may display this as Both (WPA2 & WPA3) after selecting Enterprise.
Enhanced Open	OWE	Some vendors may use OWE to designate Enhanced Open.
Enhanced Open Transition Mode	OWE Transition Mode	Enhanced Open Transition Mode may be named OWE Transition Mode or OWE + Open or something similar.

Security Benefits of Protected Management Frames

PMF (IEEE 802.11w) has been optional in WPA2 networks since 2018 and is now mandatory for all WPA3-enabled networks.

PMF ensures the integrity of the management data being sent between APs and wireless clients (also referred to as STAs for stations in the WLAN standards) through mutual authentication and encryption of the management traffic.

By ensuring the communications between the AP and client are authentic, authenticated, and (at times) encrypted, PMF enables security protections in several important ways:

- Protection from client spoofing
- Protection from AP spoofing
- Protection from some denial-of-service attacks
- Protection from replay attacks
- Protection from some man-in-the-middle attacks

Although there have been legitimate uses of spoofing management frames (such as using de-authentication messages to keep managed corporate devices from connecting to rogue APs), ultimately the industry has realized the benefits of securing the management traffic outweigh the few lingering use cases for keeping it unprotected. With PMF enabled, neither clients nor an AP will be able to send management packets as another entity, regardless of whether it's a legitimate use case or malicious attack.

From an availability perspective, use of PMF protects against a subset of denial-of-service attacks, including those which may incorporate spoofing. Obviously, PMF (or any protocol) can't protect against layer 1 denial of service, if a malicious user is jamming the RF, or if a piece of equipment is inadvertently interfering at the RF layer.

As with protection from spoofing, PMF mitigates against replay attacks where a malicious attacker resends packets often from a spoofed device. Lastly, by authenticating the endpoint and AP to one another, the system is further protected from certain types of man-in-the-middle attacks.

Transition Modes and Migration Strategies for Preserving Security

In Wi-Fi, migrating from a legacy security protocol to a newer one is always a bit messy because enterprise organizations are dealing with volumes of devices (hundreds, thousands, or often tens of thousands) that are of varying ownership (by departments, teams, business units, or personal vs. corporate-owned) and varying capabilities (device ages, features, generations, and software updates all impact options).

INDUSTRY INSIGHT: CHALLENGES WITH ENDPOINT CHANGES

Industry-specific environments can further complicate matters. In many commercial environments, subsets of endpoints are deployed, managed, and frequently locked for configuration by a third party, meaning the enterprise IT team may not be able to change the endpoint settings.

> **In retail, manufacturing, and warehouse environments, there are often handheld devices and scanners that require specific SSID settings for them to work properly, and any changes to the network can break their fragile little ecosystem.**
>
> **In healthcare environments, there may be biomedical devices with similarly precarious network settings, and there's certainly general opposition to modifying FDA-regulated devices, even if the change being requested doesn't require a re-certification process.**
>
> **The list of "Why we don't want to make any SSID changes" could go on for pages.**

This tapestry of endpoint diversity means the enterprise doesn't have the luxury of flipping a switch one day and upgrading the environment (or even a single network) to a new security suite; in this case WPA2 to WPA3.

In acknowledgment of this, there are Transition Modes that may ease some of the migrations from WPA2 to WPA3; however, there are many risks to be considered when using Transition Modes, and these features vary by security profile type. Because of that, we'll cover the pros, cons, and considerations of Transition Modes and the respective best practices recommendations for each.

CROSS-REFERENCE **Since WPA3 is planned for continuous improvement as the security landscape changes, please also visit the Wi-Fi Alliance site at `https://www.wi-fi.org/security` for the most up-to-date documentation on security considerations for WPA configurations.**

Enterprise Mode (802.1X)

WPA2-Enterprise and WPA3-Enterprise secured SSIDs use the IEEE 802.1X standard for authentication, key exchanges, and key rotations for encryption. You may also hear this written as "1X" or called "dot-1X" and it refers to a standard for port access control that uses the Extensible Authentication Protocol (EAP, or more specifically EAP over LAN or EAPoL). Note the standard is .1X (one-X) and not .11X (eleven-X). There will be public shaming and stoning if I see a blog post about "802.11X."

While the 802.11 suite of standards are for WLANs, this standard is not a wireless-specific standard and was originally created in 2001, and later updated to be used with other network types such as 802.11 WLANs. We'll cover 802.1X as an authentication method more in depth later.

Planning Enterprise (802.1X) Secured SSIDs

An 802.1X-secured SSID will require the endpoint to participate in 802.1X with EAP. The authentication mechanism is most often either a device certificate (from a PKI) or user credentials (such as a Windows logon username and password), although other methods including tokens are supported.

With an 802.1X-secured network, the endpoint has no meaningful connectivity until it's authenticated. The only packets that make it to or from that endpoint before it's authenticated successfully are EAP frames for the authentication process—which means the endpoint won't be visible from a monitoring standpoint; won't be able to request or receive DHCP; and therefore won't even have an IP address.

Planning for an Enterprise (802.1X) Wi-Fi network requires:

1. Wi-Fi infrastructure that supports Enterprise (802.1X) SSID security profiles
2. Endpoints that support 802.1X/EAP (natively in the OS or with add-on supplicant software)
3. A way to configure the endpoints for the specified connectivity (can be configured manually or pushed through Active Directory Group Policy or a mobile device management platform)
4. An authentication server that supports RADIUS (such as Microsoft NPS, Cisco ISE, Aruba ClearPass, FreeRADIUS, Pulse Secure Steel-Belted RADIUS)
5. An authentication directory repository such as Active Directory or other LDAP (which could be on-premises or in the cloud)
6. A certificate for the RADIUS server (even if the endpoint is authenticating with a username and password, the server must authenticate itself with a certificate)
7. Optionally, if the endpoints are configured to authenticate to the network with certificates, then there is also need for a Public Key Infrastructure (PKI) to distribute certificates to the endpoints

I know, I know—you're thinking "Dang, Jen, that's a long list!" Don't worry; it's not nearly as complicated as it may seem and in the upcoming sections, I'll bestow upon you all the 802.1X, RADIUS, and EAP knowledge you need to know, as well as tips and tricks for troubleshooting when it doesn't work (and yes, when you're getting started it won't work but you'll have the tools to resolve that quickly).

Figure 2.11 shows a high-level view of the 802.1X components mapped to the bulleted list including (1) the wireless infrastructure that support 802.1X, (2) an endpoint device that supports 802.1X, (4) an authentication server, (5) an authentication directory, and (6) a RADIUS server certificate.

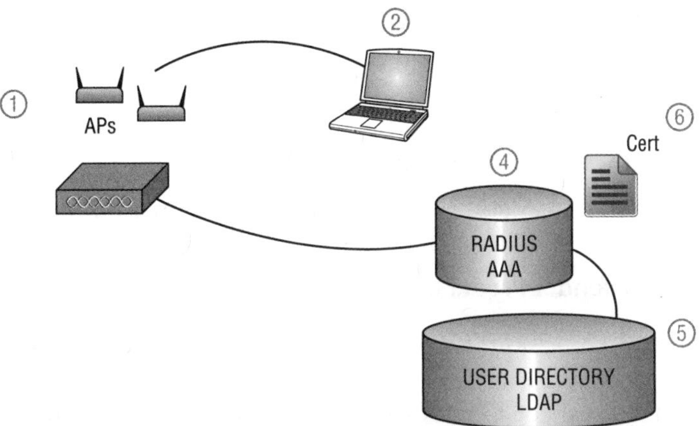

Figure 2.11: High-level view of required components for 802.1X in secure Wi-Fi

SIZING IT UP: 802.1X-SECURED NETWORKS

Small organizations using Microsoft infrastructure will usually opt for 802.1X secured SSIDs with user credentials, since it's very straightforward to configure the settings to pass through a user's Windows login information.

Large organizations and highly regulated environments tend to prefer device certificates (ideally layered with a user login or visibility of the logged-in user).

Schools and universities that service students with their own personal devices often standardize on 802.1X-secured networks with the Eduroam service, which can authenticate endpoints with certificates or user credentials.

Regardless of the size of organization or authentication method (usernames or certificates), using 802.1X-secured networks for anything other than guest users is an ideal scenario. You may still need a passphrase-based network for some devices, but your managed users on managed devices accessing production resources should be using 802.1X.

Untangling the Enterprise (802.1X) SSID Security Options

WPA3 defines sets of cipher suites as well as authentication and key management (AKM) suites. It's a lot to untangle, because the suites are referenced by numbers, and the list of cipher suite numbers overlaps the list of AKM suite numbers.

WPA3's Cipher Suites and AKMs

The cipher suites and AKMs can (in theory) be mixed and matched. In practice, some vendors don't expose the options to the administrators, and others expose some options, and make assumptions about the remaining configurations. I'm not going to lie—it's a hot mess.

The mix-and-match list specifies the following:

- An authentication and key management suite
- A pairwise cipher (for unicast traffic)
- A group data cipher (for certain multicast traffic)
- A group management cipher suite (for protected management frames)

802.1X Security Options in Vendor Products

The screenshots in Figure 2.12 include SSID security options from three different products, highlighting the inconsistency in naming conventions. See Table 2.3, shown previously, for a cheat sheet. Obviously, this is subject to change, and in fact I do hope it changes as we will be approaching vendors to request some standardization of the nomenclature.

The Juniper Mist web UI shows Enterprise class networks as "EAP (802.1X)". Its WPA3-Enterprise Transition Mode is displayed as "WPA-3/EAP (802.1X) (+WPA-2)." Note there are no additional options within WPA3-Enterprise (802.1X) with Mist; the default (minimum) cipher suite of CCM-128 is in use but not visible in the web UI. Cisco and Aruba Networks have more comprehensive options.

In Cisco's configuration options, once Enterprise mode is selected, the administrator can then specify whether the policy allows WPA3 only, WPA2 only, or WPA3 and WPA2 (aka WPA3-Enterprise Transition Mode). Once selected, all four WPA3-Enterprise cipher suites are available.

Similarly, Aruba Network's configuration in Central offers options once the security mode is set to Enterprise. In this view, only three cipher suites are made available and there is no WPA3-Enterprise Transition Mode shown here.

The allowed WPA3 Enterprise cipher suites are explained further in Table 2.4. Also shown are the different AKM suites for each mode with and without FT (Fast Transition roaming).

Table 2.4: WPA3-Enterprise cipher and AKM suites

Network Type	Selector	Hash Algorithm	Cipher Suite	PMF	Notes
WPA2-Enterprise	802.1x 00:0F:AC:1	SHA1	AES-CCMP-128	Optional	Default supported in WPA2
WPA3-Enterprise Transition Mode	802.1x 00:0F:AC:1	SHA1	AES-CCMP-128	Optional	WPA2/WPA3 transition mode; WPA3 requires PMF
	802.1x 00:0F:AC:5	SHA256	AES-CCMP-128	Optional	
WPA3-Enterprise Only Mode	802.1x 00:0F:AC:5	SHA256	AES-CCMP-128	Required	Default supported in WPA3
WPA3-Enterprise Only Mode FT	802.1x 00:0F:AC:3	SHA256	AES-CCMP-128	Required	Default supported in WPA3 with FT (11r)
WPA3-Enterprise Only Mode FT	802.1x 00:0F:AC:13	SHA384	AES-CCMP-128	Required	Optional mode for WPA3 with FT (11r)
WPA3-Enterprise-192 Mode	802.1x 00:0F:AC:12	SHA384	AES-GCMP-256	Required	Highest security in WPA3; Suite B, CNSA Compliant Suite
			AES-CCMP-256	Required	

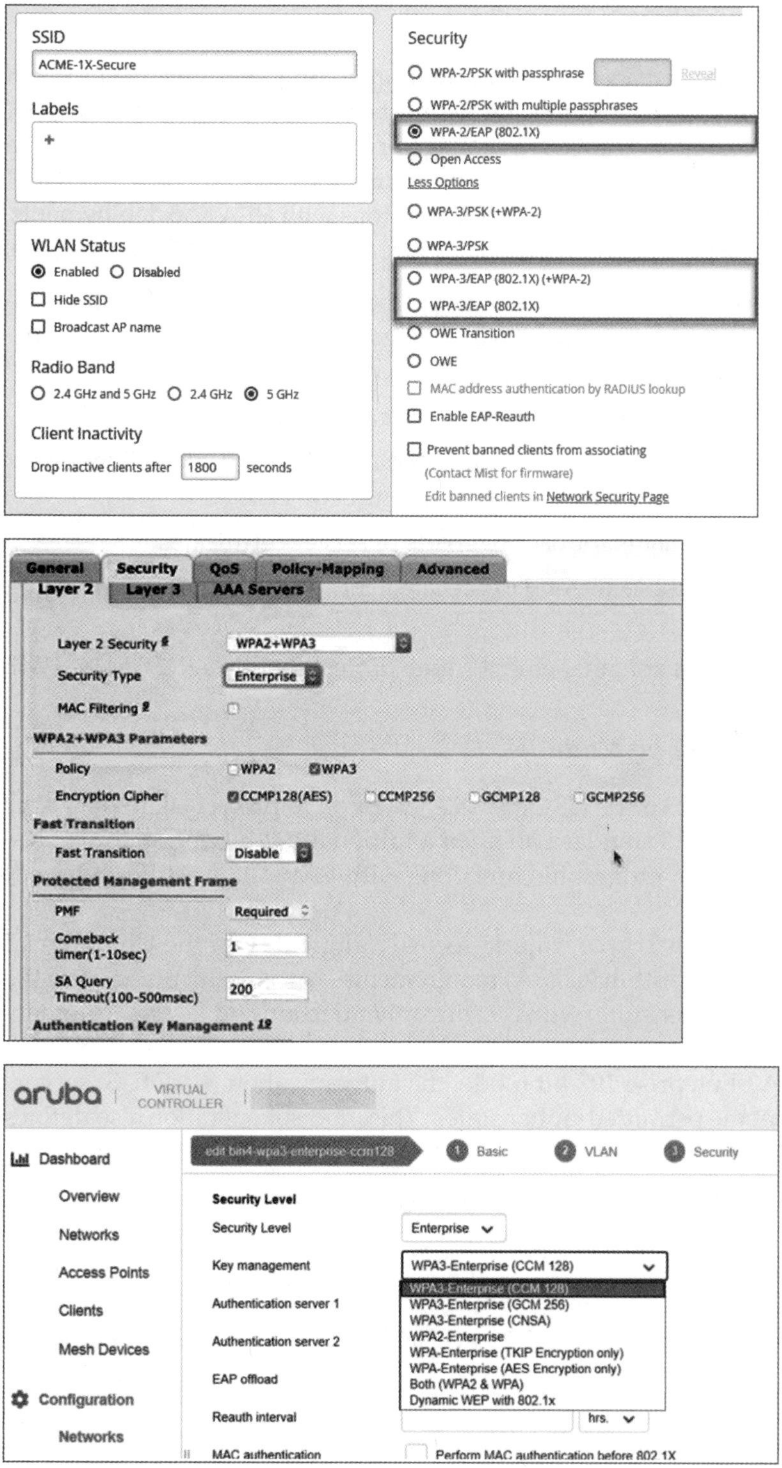

SSID
ACME-1X-Secure
Labels
+
WLAN Status
Enabled Disabled
Hide SSID
Broadcast AP name
Radio Band
2.4 GHz and 5 GHz 2.4 GHz 5 GHz
Client Inactivity
Drop inactive clients after 1800 seconds

Security
WPA-2/PSK with passphrase Reveal
WPA-2/PSK with multiple passphrases
WPA-2/EAP (802.1X)
Open Access
Less Options
WPA-3/PSK (+WPA-2)
WPA-3/PSK
WPA-3/EAP (802.1X) (+WPA-2)
WPA-3/EAP (802.1X)
OWE Transition
OWE
MAC address authentication by RADIUS lookup
Enable EAP-Reauth
Prevent banned clients from associating
(Contact Mist for firmware)
Edit banned clients in Network Security Page

General Security QoS Policy-Mapping Advanced
Layer 2 Layer 3 AAA Servers
Layer 2 Security WPA2+WPA3
Security Type Enterprise
MAC Filtering
WPA2+WPA3 Parameters
Policy WPA2 WPA3
Encryption Cipher CCMP128(AES) CCMP256 GCMP128 GCMP256
Fast Transition
Fast Transition Disable
Protected Management Frame
PMF Required
Comeback timer(1-10sec) 1
SA Query Timeout(100-500msec) 200
Authentication Key Management

aruba VIRTUAL CONTROLLER
Dashboard
Overview
Networks
Access Points
Clients
Mesh Devices
Configuration
Networks
edit bin4-wpa3-enterprise-ccm128 1 Basic 2 VLAN 3 Security
Security Level
Security Level Enterprise
Key management WPA3-Enterprise (CCM 128)
Authentication server 1
Authentication server 2
EAP offload
Reauth interval hrs.
MAC authentication Perform MAC authentication before 802.1X
WPA3-Enterprise (CCM 128)
WPA3-Enterprise (GCM 256)
WPA3-Enterprise (CNSA)
WPA2-Enterprise
WPA-Enterprise (TKIP Encryption only)
WPA-Enterprise (AES Encryption only)
Both (WPA2 & WPA)
Dynamic WEP with 802.1x

Figure 2.12: Sample screenshots of 802.1X security options from Juniper Mist, Cisco, and Aruba Networks. Note the inconsistency of naming conventions and security options across vendors

Enhancements with WPA3-Enterprise

Out of all the updates in WPA3 so far, the impact on the Enterprise mode of security is probably the least significant for most organizations. The three biggest security boosts come from Protected Management Frames, the introduction of a 192-bit encryption mode, and the shift away from deprecated protocols like TKIP.

In addition to the required use of PMF ubiquitous with all WPA3 deployments, the WPA3-Enterprise's main differentiators from legacy WPA2-Enterprise are:

- Addition of an optional 192-bit encryption mode for stronger authentication (for EAP-TLS only, primarily designed for security conscious customers including government, defense, and industrial)
- Support for additional EAP methods for authentication
- Required use of authentication (RADIUS) server certificate validation for certificate-based endpoint authentication (EAP-TLS)
- Stricter requirements for user override of server certificates
- Prohibition of less secure and deprecated encryption protocols (including TKIP)
- Addition of robust management frame protection through 802.11w PMF

WPA3-Enterprise 192-bit Mode

Both WPA2 and WPA3 offer 128-bit AES encryption to protect client data over the air. The new WPA3 standard adds an additional 192-bit option for consistency and stronger cryptographic functions with EAP-TLS (certificate-based) connections.

This mode includes cryptographic security aligned with the Commercial National Security Algorithm (CNSA) requirements—specifically prescribing the use of GCM-256 (cipher suite) with certain authentication and key management (AKM) suites.

To use WPA3-Enteprise 192-bit mode, the authentication (RADIUS) servers must use one of the permitted cipher suites. This mode of operation also defines allowed EAP methods:

- TLS ECDHE-ECDSA with AES-256-GCM SHA384
- TLS ECDHE-RSA with AES-256-GCM SHA384
- TLS DHE-RSA with AES-256-GCM SHA384
- 802.11 encryption using AES-256-GCMP and BIP with AES-256-GMAC

NOTE To geek out on the NSA's designated CNSA cryptographic suite, visit `https://apps.nsa.gov/iaarchive/programs/iad-initiatives/cnsa-suite.cfm`.

Deciphering the Acronyms of 192-bit Mode

Remember from Chapter 1, "Introduction to Concepts and Relationships," in "Cryptography Concepts" we talked about the use of asymmetric versus symmetric encryption and key exchanges. The alphabet soup above is simply describing the various cryptographic pairs and algorithms in use, along with the key length (which correlates to cryptographic strength).

Each TLS option describes the cipher suite with the asymmetric/symmetric pairs, the digital signature algorithm (DSA), and the hash-based message authentication code (HMAC). Other acronyms in the list of 192-bit operation modes include:

- **TLS:** Transport Layer Security—in this case, this is simply the EAP method (device certificate)
- **ECDHE:** Elliptic Curve Diffie-Hellman Ephemeral Key Exchange, a fast and low-overhead asymmetric key exchange based on ECC
- **DHE:** Diffie-Hellman Ephemeral Key Exchange, a key exchange with forward secrecy
- **RSA:** Rivest–Shamir–Adleman, an asymmetric algorithm based on public key cryptography, in this case used for digital signatures
- **ECDSA:** Elliptic Curve Digital Signature Algorithm, a low-overhead signature algorithm based on ECC
- **AES-256-GCM:** AES with Galois/Counter Mode, a method for authenticated symmetric encryption to provide confidentiality and authentication with message integrity check
- **SHA-384:** Secure Hash Algorithm, 384-bit used for message integrity

The 192-bit TLS cipher suites are detailed in Table 2.5 showing each accompanying algorithms for key exchange, digital signature, encryption, and message integrity.

Table 2.5: WPA3-Enterprise 192-bit TLS cipher suites

TLS CIPHER SUITE	KEY EXCHANGE	DIGITAL SIGNATURE	ENCRYPTION	MESSAGE INTEGRITY
ECDHE-ECDSA with AES-256-GCM SHA384	ECDHE- P-384	ECDSA P-384	AESGCM-256	SHA-384
ECDHE-RSA with AES-256-GCM SHA384	ECDHE-P-384	RSA >= 3072-bit	AESGCM-256	SHA-384
DHE-RSA with AES-256-GCM SHA384	DHE- >=3072	RSA >=3072-bit	AESGCM-256	SHA-384

INDUSTRY INSIGHT: ELLIPTIC CURVE CRYPTOGRAPHY

Over time, tools for cracking encryption have become more sophisticated, allowing faster and more efficient "hacking" methods to expose the keys. To combat this, our traditional approach has been to increase the key length, thereby increasing the time and effort required to break the encryption.

The problem with this approach is twofold: the computational power of attack tools is increasing while wirelessly connected endpoints are getting smaller and computationally more lightweight (less capable), especially in IoT applications.

As you learned earlier in Chapter 1, cryptographic functions can be very resource-intensive, and with algorithms using Finite Field Cryptography (FFC), longer keys take much more processing power. Just moving from 1,024-bit to 2,048-bit keys takes eight times the computational power. Soon, the power of the cracking tools will outstrip the ability of endpoints to handle the longer keys required for attack mitigation.

To solve this problem, the industry has moved toward elliptic curve cryptography (ECC). ECC key pairs are based on the algebraic structure of elliptic curves versus traditional key pairs based on large prime numbers that are mathematically related (and computationally intensive). For context, a 256-bit ECC key is as secure as a 3,072-bit RSA key. Surprisingly enough, ECC concepts have been around since 1985, but are only recently being adopted. Certain applications of ECC are also resistant to attacks from quantum computers.

By increasing the security with much shorter keys, ECC enables us to extend enterprise-grade security to less capable devices. The world of cryptography changes more often than you may think, so keep an eye on ECC and for trends in post-quantum cryptography.

As a metaphor, consider a bank safe that criminals may try to break into. Think of legacy approaches of longer key lengths as building a safe with thicker walls, whereas moving to ECC is akin to designing the safe with a thinner material that's harder to penetrate. The walls of the safe can only be so thick before it's too heavy or too large to manage.

WPA2 to WPA3-Enterprise Migration Recommendations

The WPA3 security suite offers transitional modes of operation to help enterprises move easily from WPA2 to WPA3. However, Transition Modes (as always) introduce significant vulnerabilities and can, in some cases, completely invalidate the newer WPA3 security and put the network at risk.

WPA3-Enterprise can be configured on an SSID in one of three ways:

- WPA3-Enterprise Only mode (WPA3 only)
- WPA3-Enterprise Transition Mode (WPA2 and WPA3)
- WPA3-Enterprise 192-bit mode (WPA3, 192-bit only)

WPA3-Enterprise Only Mode (and 192-bit)

This requires the endpoint to support and participate in Protected Management Frames, whereas the WPA3-Enterprise Transition Mode makes PMF optional—meaning the Transition Mode network will support both WPA2 and WPA3 Enterprise clients (but if a WPA3 client joins, it will be required to use PMF).

PMF is also required for organizations taking advantage of the new WPA3-Enterprise 192-bit mode, and that 192-bit mode does not support a Transition Mode of operation, meaning there's no option for backward compatibility to WPA2 on a WPA3-Enterprise 192-bit network.

WPA3-Enterprise Transition Mode

This is a mode that allows a single SSID to offer both WPA2- and WPA3-Enterprise services simultaneously. The WPA3-capable clients will be directed by the system to use the full WPA3 suite, whereas clients that don't support WPA3 can connect to the same network using the legacy WPA2 suite. In transition mode, only these two suites are supported, and deprecated security of WPA, WEP, and TKIP are not supported.

WPA3 includes protection mechanisms designed to ensure that WPA3-capable clients do not ever connect with a lesser secure WPA2 security suite. This enhancement prevents downgrade attacks whereby a malicious user forces or coaxes a more capable client toward a less-secured protocol (in this case, it's what prevents WPA3 clients from connecting to the SSID with WPA2, even if Transition Mode is enabled).

Should you choose to enable WPA3-Enterprise in Transition mode, the infrastructure and specifically any resources attached to or available from that SSID will not have the full benefit of Protected Management Frames.

WPA3-Enterprise Transition Mode pros:

- Allows the organization to migrate incrementally from WPA2 to WPA3 without adding a new SSID
- Eliminates the need to reconfigure endpoints to connect to a new (WPA3 only) SSID (only required if the EAP authentication methods are changed)

WPA3-Enterprise Transition Mode cons:

- Conflates endpoints that are using WPA2 vs. WPA3, making it harder to track progress through migration
- Creates less visibility and sense of urgency through a migration, often leading to a longer than necessary migration period and lack of buy-in to abandon legacy WPA2 security
- Lacks the benefit of PMF with the WPA2-connected endpoints, which can put the remainder of the network at risk
- Allows WPA2-connected endpoints using less secure algorithms such as SHA-1 (instead of SHA-256)

As a final thought on WPA3-Enteprise Transition Mode, each organization is unique. Whether an organization should or shouldn't use Transition Mode depends on its unique environment, which includes not only the wireless infrastructure but also the types of endpoints connected, the resources and availability of the IT teams to update endpoint devices, and the corporate culture as it relates to security.

As a general best practice recommendation, only use transition modes when absolutely necessary. Document process and progress, track the migration, and if needed put in additional monitoring to protect the system through the transition period.

Next, we present an alternative way to securely manage migrations from WPA2 to WPA3 without the use of WPA3 Transition Mode.

New Recommendations to Expand SSIDs

For years, the Wi-Fi industry has been preaching the praises of collapsing SSIDs to preserve airtime and streamline operations. In Wi-Fi, networks can be collapsed to as few required based on common security profiles. For example, all 802.1X-secured networks could be merged to one. The same is true for passphrase-secured and open networks with portals. The result of that has been a sweeping recommendation to consolidate SSIDs to as few as possible, with guidance ranging from maximums of three to six.

Well, folks. We're in a new era. While it's still important to be mindful of the beacon overhead from numerous SSIDs, that consideration needs to be balanced with security requirements.

The new recommendation is for security-conscious environments to not co-mingle networks of mixed security modes. Specifically, that means using separate networks (for example) for WPA2 vs. WPA3 Enterprise, and even for separate networks for pure 802.1X vs. 802.1X with MAC Authentication Bypass (MAB). Similarly (and especially) WPA2 and WPA3 Personal networks should be separated (and not use the same passphrases).

Personal Mode (Passphrase with PSK/SAE)

Personal mode secured SSIDs are such a misnomer. As someone who's worked in Wi-Fi for 20 years, it was only recently that I began to understand the general perception (even within technology and infosec circles) is that "personal" SSIDs are what's used at home, not something ever used in the enterprise.

Personal mode security profiles for SSIDs are the networks that allow users to connect with a passphrase, often referred to as a pre-shared key (PSK) network. Technically, PSK in WPA2 networks has been replaced with Simultaneous Authentication of Equals (SAE) in their WPA3 counterparts, so you'll see these networks referenced in this book as "personal" and "passphrase" based, unless the content is specific to WPA2, in which case you'll see "PSK" or WPA3, where you'll then see "SAE."

INDUSTRY INSIGHT: PERSONAL-SECURED NETWORKS IN ENTERPRISE

While there are organizations that claim to have no Personal/passphrase mode SSIDs in their environment, in 20 years and hundreds of clients, I've only personally worked in three; one of those was a large bank, and the others were both strictly controlled federal environments. It's worth mentioning that I don't work with small businesses. Most of my clients have had large and complex environments with various IoT or other connected devices that don't support 802.1X authentication.

I don't doubt there are other organizations that don't have any passphrase-based networks, but they are few and far between.

Planning Personal/Passphrase-Secured SSIDs

Personal/passphrase-secured SSIDs are the most straightforward of the three types of security profiles for Wi-Fi networks. Their ease of configuration is precisely why many IT admins will quickly provision a passphrase-secured SSID when in a jam.

When the SSID is set in Personal mode, one or more passphrases are configured, and endpoints can simply use a passphrase to connect to the network. The Wi-Fi drivers on the endpoints will simply pop up a standard prompt for the user to input the passphrase.

Unlike Enterprise (802.1X)-secured SSIDs, there is no requirement to configure anything external to the wireless system (such as authentication servers, certificates, external portals, or NAC and registration systems). And, unlike captive portals (covered next) there's no need to configure landing pages and rely on browser redirects.

To configure a Personal-secured network only requires:

- Wi-Fi infrastructure that supports Personal/passphrase SSID security profiles (that's all of them)
- Configuration of one or more passphrases in the Wi-Fi management platform
- Any certified Wi-Fi-capable endpoint

The standard implementation of this feature across products enables configuration of a single passphrase per SSID. Aside from this most basic standard operation of passphrase-secured networks, vendors do add additional features that vary by product and implementation. One such feature is multi- or personal-pre-shared keys (MPSK/PPSK). These vendor-specific implementations support configuration of multiple pre-shared keys and/or pre-shared keys tied to specific endpoints (usually by MAC address).

Support for MPSK and PPSK varies by product and some support dynamic MPSK generation via APIs and/or as part of a QR-code-based onboarding. Other products may support only static configuration of MPSKs or PPSKs within the wireless management platform, and at least one product currently implements MPSKs through integration with a separate NAC product.

Figures 2.13 and 2.14 show Personal network options from Aruba Central and a Fortinet FortiGate-managed AP (respectively). Products that are not yet Wi-Fi Alliance WPA3 Certified may use other labels such as "WPA3-SAE" shown in the Fortinet example.

Enhancements with WPA3-Personal

WPA2-Personal with its pre-shared key operation had more than a few shortcomings. Over time, this generation of PSK-secured networks became susceptible to an assortment of attacks including offline dictionary attacks (allowed for recovery of encryption keys at an attacker's leisure), decryption attacks (ability to decrypt peer endpoint traffic in real time and retroactively offline), and man-in-the-middle attacks (due to often unprotected protected management frames).

If you heard of the KRACK attack back in 2016 (`www.krackattacks.com`), you know the weaknesses of WPA2-Personal networks. In that attack, researchers Mathy Vanhoef and Frank Piessens created a key reinstallation attack that exploited WPA2's vulnerability to replay attacks.

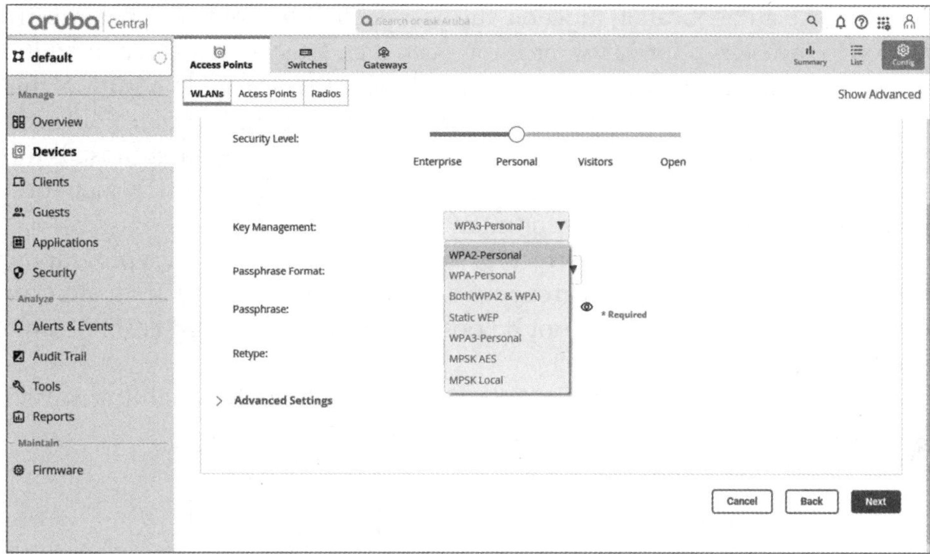

Figure 2.13: Screenshot from Aruba Central showing Personal network options

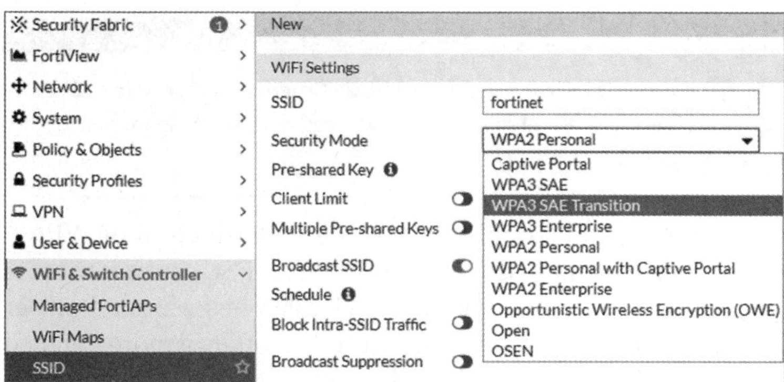

Figure 2.14: Screenshot from a Fortinet FortiGate-managed AP showing Personal network options

With WPA3-Personal, not only were the protected management frames introduced, but the pre-shared key (PSK) authentication function was replaced by Simultaneous Authentication of Equals (SAE), all of which greatly increases the security of Personal mode networks in several ways, including:

- Resistance to offline dictionary attacks
- Separation of cryptographic key derivation from user-entered passphrase
- Provisioning of unique cryptographic keys for each endpoint
- Perfect forward secrecy

In the SAE authentication function, the user-entered passphrase is not directly related to the encryption keys, meaning even if an attacker has discovered the passphrase, she/he would not be able to decrypt other traffic. In addition, this means the user-entered passphrases can be shorter and less complex while still providing the same cryptographic strength. Of course, if the passphrase is too simple or short, it would be easy to guess and allow an attacker on the network, which is obviously undesirable for many reasons.

Another benefit of SAE is perfect forward secrecy, which is a cryptographic feature that protects traffic from being decrypted retroactively. If an attacker were to derive keys, it would not be possible to have captured earlier traffic and then decrypt it after the fact.

As with the Enterprise mode, there are also options for enabling WPA3 Transition Mode on Personal secured networks.

WPA3-Personal modes supported are:

- WPA3-Personal Only Mode (WPA3 only)
- WPA3-Personal Transition Mode (WPA2 and WPA3)

SAE: HELP IT HELP YOU

SAE's security is leaps and bounds better than that of its PSK predecessor; but for all the cryptographic genius it entails, it's all for naught if you reuse WPA2 passphrases or choose short or easily guessed passphrases.

The SAE function of WPA3-Personal replaces the PSK function of WPA2-Personal. Figure 2.15 shows the main differentiators. With PSK, the passphrase is the master key used for encryption, whereas SAE derives a unique key per endpoint and rotates keys for forward secrecy. Figure 2.16 offers a more detailed view of the exchanges and key derivation used in WPA3-SAE.

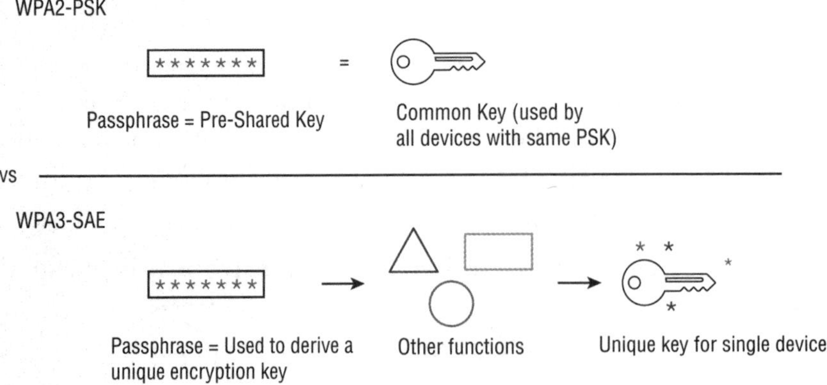

Figure 2.15: WPA2-Personal PSK versus WPA3-Personal SAE

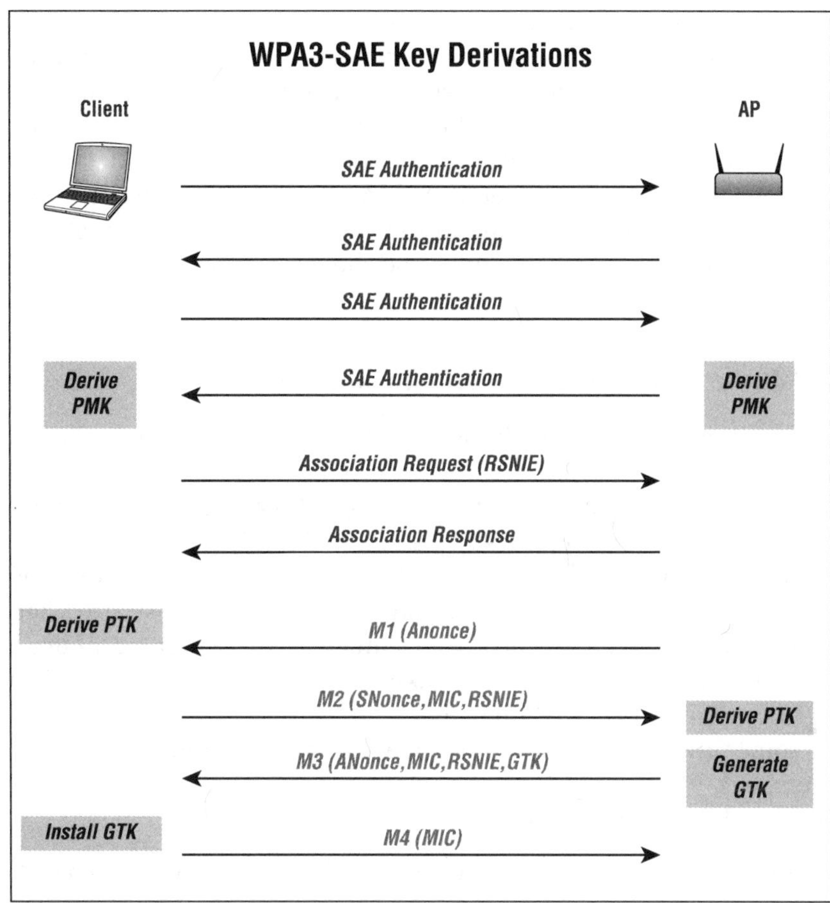

Figure 2.16: This image demonstrates the exchanges during SAE and the derivation of the pairwise and group keys, for unicast and broadcast, respectively

WPA3-Personal Only Mode This enforces PMF and up to AES-256 encryption. WPA2, WPA2, WEP, and TKIP are all prohibited on WPA3-Personal Only networks. It's obvious from the name that in this mode, endpoints must connect with the full WPA3 security suite.

WPA3-Personal Transition Mode Alternatively, this allows the SSID to offer both WPA2- and WPA3-Personal security services and allows clients to connect in either method using the same passphrase. If an endpoint connects as a WPA3-enabled device, it will have the benefit of protections from SAE (over PSK) and the benefits from PMF. For downgrade protection, starting in WPA3 revision 3, Personal networks include the transition disabled indicator function, allowing an AP to signal a client to disable transition to other (less secure) modes.

Even though WPA2-Personal is susceptible to offline dictionary attacks, WPA3-Personal endpoints still realize many benefits even on an SSID configured for transition mode. If an attacker is able to perform an attack to derive the cryptographic key and the passphrase, the malicious user can join the network (and possibly wreak havoc) but they would not be able to decrypt data from the WPA3-connected endpoints (even if they're using the same passphrase) and would not be able to decrypt any of the WPA3-connected clients' past packets, due to the unique key derivations and the forward secrecy offered by SAE (versus legacy PSK).

WPA2 to WPA3-Personal Migration Recommendations

From a security perspective, use of Transition Modes that allow comingling of differing levels of security capabilities are never recommended. However, moving to a WPA3-Personal Only Mode of operation would require each and every endpoint to support the new security suite, and that doesn't happen overnight.

Very few organizations have the luxury of flipping a switch and performing forklift migrations that impact endpoints, and with passphrase-secured networks, the endpoint must be touched or in some way re-configured to use a new network and/or a new passphrase, making it much more disruptive than upgrading from WPA2 to WPA3-Enterprise.

Here are some pros, cons, and recommendations for migrating from WPA2 to WPA3 Personal.

WPA3-Personal Transition Mode pros:

- Allows the organization to migrate from WPA2 to WPA3 without reconfiguring endpoints (if same passphrases are reused)
- Removes the need to provision a second, possibly temporary SSID
- Affords some protection for WPA3-connected endpoints even if WPA2-connected endpoints are compromised

WPA3-Personal Transition Mode cons:

- Conflates endpoints that are using WPA2 vs. WPA3, making it harder to track progress through migration
- Creates less visibility and sense of urgency through a migration, often leading to a longer than necessary migration period and lack of buy-in to abandon legacy WPA2 security
- Lacks the benefit of PMF with the WPA2-connected endpoints, which can put the remainder of the network at risk

- Lacks proper downgrade attack protection for WPA3-Personal endpoints
- Exposes network to risk with reused passphrases since an attacker can perform on offline attack and connect to the network, enabling lateral movement and related attacks

WPA3-Personal Transition Mode recommendations:

- Only use Transition Mode if absolutely required, and only for as long as required
- Where possible, configure different passphrases for WPA2 and WPA3 endpoints (which may be challenging to do and track)
- Couple unique passphrases with dynamic VLAN assignments and proper segmentation so that WPA2 connected endpoints can't access WPA3 connected endpoints
- Where applicable, further restrict access to WPA2 endpoints to only the required resources on the network

Ultimately, it's preferable, much more secure, and easier to manage a migration by creating a new WPA3-Personal Only SSID, and systematically moving capable endpoints to that network, while planning mitigations for endpoints that are not WPA3-capable.

Recommendation for migrating to WPA3-Personal:

- Create a new SSID configured for WPA3-Personal Only
- Do not reuse any passphrases from the WPA2-Personal networks
- Configure the WPA3-Personal Only network with a different VLAN and appropriately segment traffic to and from the WPA2 network to it
- Configure monitoring and alerting for the WPA3-Personal Only network to identify active online brute force attacks
- Identify WPA3-capable endpoints and configure them for the new network, continue moving endpoints as capabilities arise (e.g., through software updates on endpoint NICs)
- Create a plan for endpoints that are not WPA3-capable, which should include replacement, further segmentation, and additional security monitoring

WPA3 is always changing and evolving. For the most up-to-date recommendations please visit the Wi-Fi Alliance security page at `https://www.wi-fi.org/discover-wi-fi/security`.

Open Authentication Networks

Open Authentication and newer Enhanced Open networks are designed to allow an endpoint to connect without formal authentication or even identification of the endpoint or user. There's also no passphrase or anything for the user to input—they just click to join and they're on.

Because of their simplicity, open networks don't require any configuration (aside from naming the network and selecting "open" or "enhanced open" from a drop-down menu for the security), but most organizations use open networks to offer a captive portal experience for the end user, which of course requires additional configuration addressed later in authentication and onboarding sections.

Legacy Open Authentication Networks

For many years, open networks have given security professionals heartburn because there has been no mechanism for encrypting data over the air between the endpoint and the AP. In the early days, this was significant since Internet traffic and even most internal applications didn't use encrypted protocols—meaning, if a user was connected to an open network, most or all their data was easily sniffed through passive eavesdropping and could be read in plain text without any special tools or password cracking.

As the industry evolved, encryption became omnipresent for both internal applications and Internet traffic. In fact, the Google Transparency Report, which tracks HTTPS encryption on the web across all its services, shows the percentage of encrypted web traffic worldwide has risen from 50 percent in 2014 to a whopping 95 percent as of the end of 2021. This is demonstrated in Figure 2.17 and you can view real-time data at `https://transparencyreport.google.com/https`.

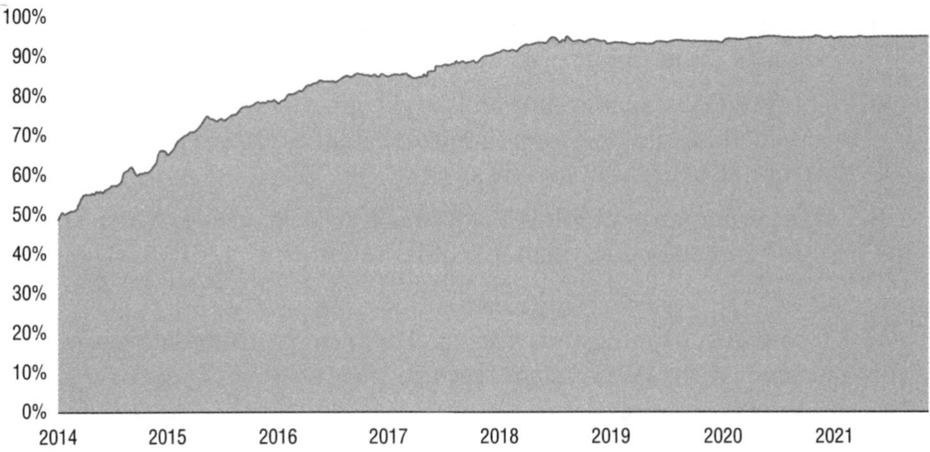

Figure 2.17: Adoption of encrypted Internet traffic from Google
`https://transparencyreport.google.com/https/overview`

However, even with those statistics and the maturation of applied encryption, sending Wi-Fi packets unencrypted over the air simply isn't a best practice. Users have a terrible habit of reusing passwords, and if a user happens to have just one website or application they're using that's not properly secured, an attacker could capture that over the air and have access to credentials and possibly other personal information.

IEEE 802.11 ASSOCIATION STATES

As alluded to in an earlier sidebar, although open networks are not authenticated in the traditional sense, the process for endpoints connecting to an AP in all 802.11 networks includes what's called an "Open System Authentication" followed then by the association to the AP. This is simply an artifact of language from old legacy standards for 802.11 that predate newer Wi-Fi security operations.

I mention this only because if you're studying Wi-Fi technologies, you will have been taught the three 802.11 connection states as described in the following list. These describe the relationship of the endpoint to the AP through the association process. For Enterprise and Personal secured networks there are then additional sequences for network-based authentication and key exchanges for encryption.

- Not authenticated or associated
- Authenticated but not yet associated
- Authenticated and associated

The Open System Authentication is a legacy 802.11 term that occurs as part of the process of a client associating to an AP. Figure 2.18 shows the exchanges and the three states during association. Afterwards, the client can complete the 4-way handshake and any network authentication.

Wi-Fi Enhanced Open Networks

To solve the problem of unencrypted traffic, the Wi-Fi Alliance created a program for Enhanced Open. Announced in 2018, most enterprise Wi-Fi vendors began supporting the feature in mid to late 2021. Enhanced Open adds encryption to open authentication networks, offering protection from passive eavesdropping.

Based on Opportunistic Wireless Encryption (OWE, RFC 8110) the feature uses an anonymous Diffie-Hellman exchange to securely negotiate encryption keys without requiring user or device authentication. Like the WPA3 suite of security, Enhanced Open also requires PMF to add integrity and confidentiality to Enhanced Open networks.

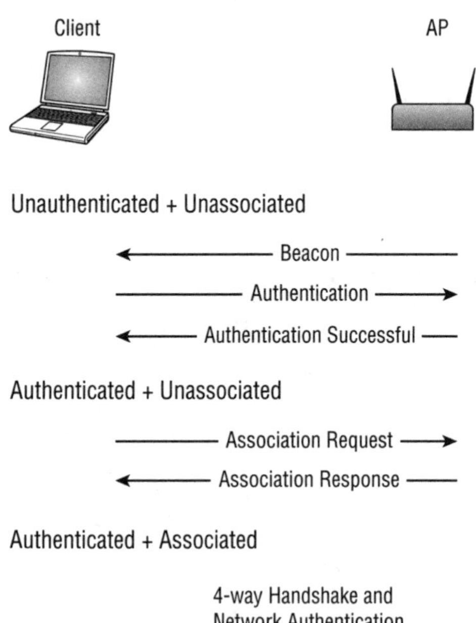

Figure 2.18: The 802.11 Open System Authentication

It also offers a Transition Mode of operation, allowing organizations to seamlessly migrate from legacy open authentication networks to more secure Enhanced Open. Unlike the WPA3-Personal and WPA3-Enterprise Transition Modes, the Enhanced Open Transition Mode doesn't present the same security risks and is appropriate for use. In fact, one major driver for this was to address quirky behavior of some clients who didn't understand an open network defining security (via the Robust Security Network Element information element).

The Transition Mode for Enhanced Open operates much differently than WPA3-Enterprise and WPA3-Personal Transition Modes; OWE Transition Mode will actually broadcast two unique SSIDs—one for legacy open, and one for OWE. Trying to expedite adoption of Enhanced Open, if OWE Transition Mode is supported, when an Open Network is configured on an AP it will automatically enter Transition Mode, and an OWE SSID will be configured. Note that this dual-SSID twin network mechanism creates some quirky behavior with various types of endpoints and should be tested thoroughly before deploying.

A new OWE Transition Mode element was created that is included in beacon and probe response for both the Open and OWE BSS. The OWE transition element is an optimization that will help endpoints that support OWE more quickly associate to the OWE BSS instead of the Open network.

The Transition Mode element when advertised on the Open network includes information on the OWE network:

- OWE BSSID
- OWE SSID

- Band
- Channel

Subsequently the Transition Mode element when advertised on the OWE network includes information on the Open network:

- Open BSSID
- Open SSID
- Band
- Channel

Enhanced Open Transition Mode networks will broadcast two SSIDs, one hidden, one visible. Depending on the vendor, this may be a manual configuration of two SSID profiles, or the vendor may generate the second SSID for you. Figure 2.19 demonstrates this dual-SSID behavior.

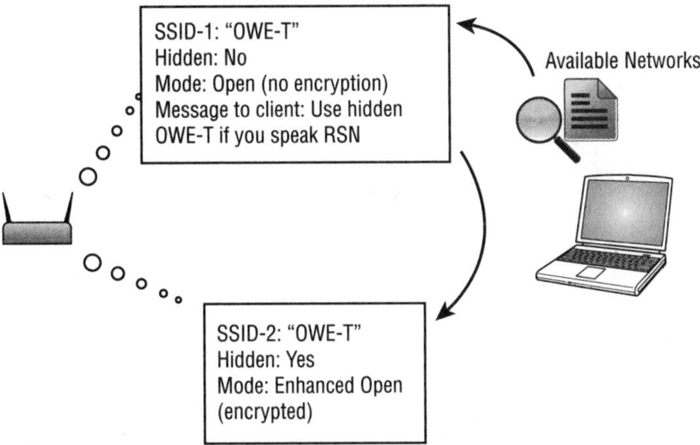

Figure 2.19: Enhanced Open Transition Mode uses two SSIDs

Otherwise, the transition from legacy open to Enhanced Open is transparent to the user. There's no different or additional action on the users' part; they simply join an Enhanced Open network (from an Enhanced Open–capable endpoint of course) and proceed as usual. When they join the network, they have the added security benefits of encryption over the air and protected management frames.

It's important to note that the OWE process uses an anonymous Diffie-Hellman exchange to generate the secret for the 4-way handshake. Meaning, while a certain amount of integrity is added with PMF and confidentiality added via encryption—the enhanced features in no way protect the infrastructure from man-in-the-middle attacks. While OWE offers encryption (privacy), it does not offer authentication in either direction (to or from the AP and endpoint).

There's no mechanism or control to prevent a device from initially connecting to an Enhanced Open network from a rogue AP (malicious or otherwise). Having said that, for corporate-managed devices, there are controls that can be applied to prevent enterprise assets from attaching to unsecured or open networks; these usually come in the form of policies applied by mobile device management (MDM) tools or directory services.

Figure 2.20 shows the two SSIDs created for an Enhanced Open Transition Mode network, as viewed by a NetAlly EtherScope nXG tool. The hidden SSID is broadcast without the SSID name. Endpoints that connect to the legacy Open network (OWE-T) will be redirected to the hidden network by the infrastructure.

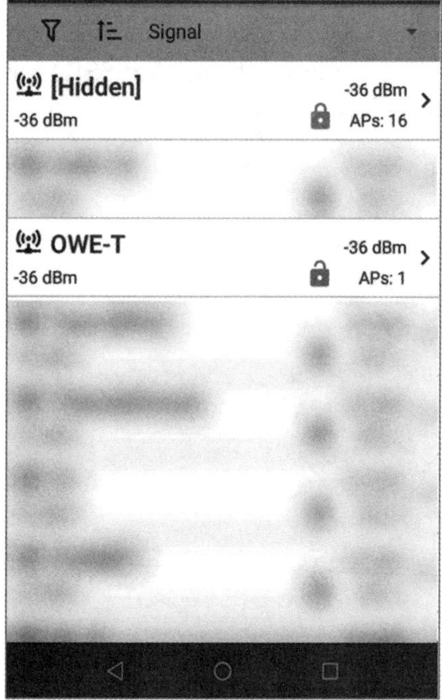

Figure 2.20: Snapshot from NetAlly EtherScope nXG showing the two networks for OWE-Transition

Summary

If you're new to wireless, this chapter gave you a solid foundational knowledge of the various Wi-Fi management architectures, an overview of the evolution of cloud-managed solutions, and an introductory look at Wi-Fi's three classes of security profiles—Enterprise/802.1X, Personal/passphrase, and Open/Enhanced Open.

The concepts of wireless management architectures play a role in the data paths, and therefore the segmentation options available. Later, in Chapter 4, you'll also see how this plays a role in planning the domain services such as DHCP.

This chapter also took a hard look at the latest WPA3 security suite and how it impacts each of the classes of network—Enterprise and Personal, as well as the extension of encryption to Open Networks by way of the Enhanced Open feature. Throughout the WPA3 tour, we covered differences between WPA2 and WPA2 for each profile type. In that, you've seen that the WPA3-Personal with SAE brings far superior security than its WPA2-Personal PSK predecessor. And that WPA3-Enterprise ups the 802.1X game with stronger encryption, depreciation of legacy protocols, and the 192-bit mode for support of CNSA suites for security-conscious organizations.

This chapter also offered updated recommendations related to migrating from WPA2 to WPA3, specifically that all security-conscious organizations should begin migrating both Personal and Enterprise class networks to WPA3 for the added security and protected management frames which bring integrity to the system. In addition, profile-level transition modes should be avoided when possible to reduce security risk, and instead recommendations are to add SSIDs and separate networks and endpoints participating in different classes or levels of security—as in, don't comingle WPA2 and WPA3 on the same SSID.

Lastly, you've gotten a taste of the complexity in understanding cipher suites and AKM in WPA3 and have now seen the current state of chaos and confusion as vendors offer differing subsets of the suites with different nomenclature. I do hope that will change soon, and I won't mind at all if some vendor screenshots are out of date by the time you read this.

The next chapter is dedicated to authentication and authorization, with a deep dive into concepts of 802.1X, RADIUS, and untangling EAP methods.

CHAPTER

3

Understanding Authentication and Authorization

Refill your coffee or Mountain Dew® because this chapter may require extra concentration. Here, we cover all the gory details of the authentication schemas used in Wi-Fi (and in other wired and wireless technologies). The majority of this chapter's content is most applicable to Wi-Fi, but many of the concepts translate to other wireless technologies.

Along with the authentication process of validating devices and/or users comes authorization, which defines what the user or device has access to once connected. We've already covered the mechanisms of segmentation, detailing exactly how the authorization policies may be enforced on the network.

Out of the three main types of security profiles for Wi-Fi, the most challenging and misunderstood concepts are related to the Enterprise (802.1X) secured networks because of their complexity. As you saw in Chapter 2, "Understanding Technical Elements," for 802.1X to work configurations are required on the Wi-Fi infrastructure, the endpoints, and one or more authentication servers.

Because of that, the first three topics apply specifically to the configuration of 802.1X-secured networks—IEEE 802.1X, RADIUS, and EAP methods.

After that, certificates offer a good segue to the other types of security profiles since they apply not only to 802.1X networks, but also open networks servicing captive portals.

INDUSTRY INSIGHT: THE INTRICACIES OF 802.1X

This portion of the book goes into a bit more detail than others. There's an unfortunate pattern of a lack of information and misinformation around the topics of 802.1X, EAP, and RADIUS floating around the Internet. At best, some are lacking clarity or are incomplete in the explanation, but in many cases the content is simply incorrect.

I have, myself, re-read many IETF RFCs in support of writing this, and am fortunate to have further vetting from amazing technical editors, one of whom (Stephen Orr) serves on the Wi-Fi Alliance Security Working Group and is intimately familiar with both the WPA security suites, as well as IEEE and IETF work. Though it involved a painful process of research, I hope this body of work can serve the community as a definitive guide on these topics.

Of course, none of us is perfect, so if there are errors or omissions here, we certainly hope you'll report them to us for tracking errata.

The IEEE 802.1X Standard

An introduction to 802.1X and a slightly deeper dive were provided in Chapter 1, "Introduction to Concepts and Relationships." With that foundation, we can continue the journey of 802.1X's deeper technical dive.

NOTE I've personally worked with, taught, and spoken on 802.1X for more than 15 years, and have read every single page of every version of the standard since its inception (and yes, that's a lot of pages, and no I had no life). I was fortunate to have access to (and mentorship from) some members of that working group for many years, which afforded me access to additional information and their blessing to go speak on 802.1X and 802.1X-REV publicly at conferences nationally and internationally.

It's a technology I feel passionate about, but it's also a complex one, even to me. My history with 802.1X plays a role in how I've chosen to present this topic, the order, the level of depth, and the vendor-neutral concepts. This chapter isn't a walk-through of configuring authentication on specific products; it explains the underlying technology. Where appropriate, specific products or features may be called out to help guide you in creating a vendor-specific architecture.

While I could write a book dedicated solely to the topic of 802.1X, for this book's purposes I'm going to give you the most pertinent information to enable you to make decisions about security architecture, understand how to properly configure secure 802.1X in wireless, and how to effectively troubleshoot issues with 802.1X and RADIUS authentication.

In Wi-Fi practice, 802.1X is the standard we use for secure wireless authentication of users and/or devices, as well as the related key exchanges. 802.1X is prescribed when we configure SSIDs of the "enterprise" flavor of security—specifically WPA2-Enterprise and WPA3-Enterprise.

Terminology in 802.1X

For readability I'll be using common language so you don't have stop and recall or research the meanings of unfamiliar words. Depending on the type of document you're reading, endpoints may be referred to as "STAs" or stations, "supplicant," "client," "device," or something else. And technical STA can refer to either an endpoint or an AP in Wi-Fi standards. It can all get very confusing if Wi-Fi isn't your daily job. Heck, sometimes it's even confusing to me.

For that reason, throughout this book I've been intentional about using consistent, accurate language that removes ambiguity (as can happen with "STA" and "device").

However, 802.1X and EAP have their own vernacular and if you're reading other documents and specifications, it's helpful to understand these terms. You may also see these words and phrases referenced in graphics here (see Figure 3.1 on the next page). At times in this chapter, you'll see these represented in parentheses along with the common term, such as "Wi-Fi infrastructure (802.1X Authenticator)" when it makes sense or helps clarify another part of the explanation.

Supplicant The 802.1X Supplicant is the part of an endpoint that's participating in the EAP authentication. Back in the day, Supplicants used to be a separate agent or piece of software we had to install to provide that functionality. Now, Supplicant functions are built into most common operating systems. You may see an endpoint labeled as the "supplicant" in diagrams, and while not completely accurate, it's close enough.

Authenticator The 802.1X Authenticator is the infrastructure device the endpoint is connecting to for network connectivity. In wired applications that would be the edge switch, and in Wi-Fi it will be either the controller or AP. The Authenticator is the liaison between the endpoint (or Supplicant) communicating with EAP and the backend authentication server that's communicating with the RADIUS protocol.

Authentication Server or AAA Server The Authentication Server is what makes the decision about whether an authentication request is approved or denied. This is most commonly a RADIUS server, but other protocols such as Diameter are supported in the EAP specifications.

The Authentication Server could be purpose-built, or part of an integrated platform such as a network access control (NAC) product. AAA stands for authentication, authorization, and accounting. Authentication is the process of validating identity; authorization involves assigning access rights; accounting is simply logging. When referenced in the 802.1X specification, "Authentication Servers" will appear with capitalization. In the remainder of the text when generally referencing authentication servers, lowercase will be used.

EAP and EAP over LAN (EAPoL) Extensible Authentication Protocol (EAP) defines an authentication framework in IETF RFC 3748. EAP over LAN (EAPoL) is simply an extension of EAP encapsulated for transmission on a LAN. Within the EAP framework, there are numerous EAP methods that prescribe specific types of authentications—such as EAP-TLS and EAP-MSCHAPv2.

The overall EAP framework stays consistent while various EAP methods are added as security requirements and use cases evolve. In the text to follow, EAP and EAPoL will be used to describe the exchanges during the 802.1X processes.

RADIUS Remote Authentication Dial-In User Service (RADIUS) is a protocol for authentication, authorization, and accounting (AAA). In daily use, "RADIUS" may refer to the RADIUS protocol, or to a server that is offering RADIUS protocol services. The RADIUS protocol defines processes for AAA. In this context, accounting is really just logging of the activity, and there are different levels of logging such as basic start/stop or more frequent interim logging.

Port Entities To clear up any confusion due to the use of "port," in 802.1X there are two logical Port Entities: a "Controlled Port," which serves as the gatekeeper for all packets to and from the endpoint based on whether it's been authorized or not; and an "Uncontrolled Port," which is the pathway for the Authenticator (such as an AP) to pass authentication (EAPoL) packets to and from the endpoint (or Supplicant). These are not physical ports but something more abstract, and a Controlled Port in 802.1X does not translate to "a port which is under 802.1X enforcement."

Figure 3.1 offers a high-level view of the 802.1X components with formal 802.1X/EAP terms as labels. Included is a wireless endpoint (with a Supplicant), wireless infrastructure components including APs and a controller (the 802.1X Authenticators), and the RADIUS server (the AAA or 802.1X authentication server). The communications shown include EAPoL between the endpoint and the APs, and RADIUS protocol between the APs and Authentication Server. Also depicted is a standard LDAP-based user directory.

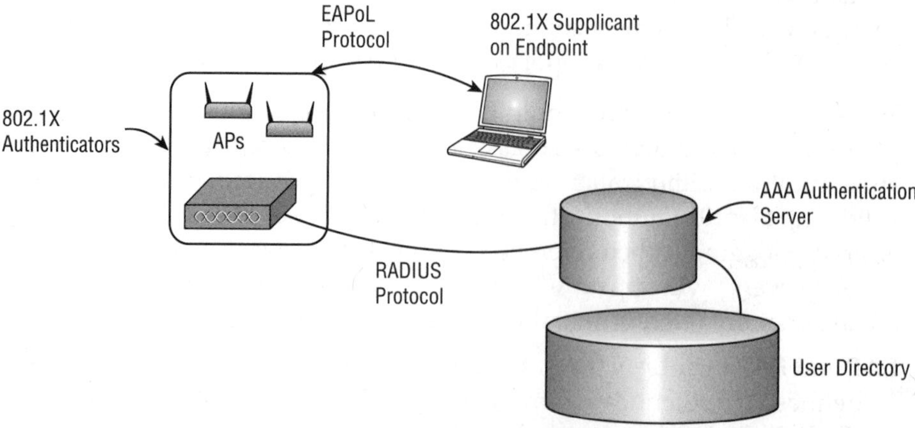

Figure 3.1: High-level view of 802.1X components

THE HISTORY OF 802.1X

While 802.1X is most frequently used in wireless networks, it's not a wireless-specific protocol. The original standard came about in 2001, with an update in 2004 to address other network mediums (including Wi-Fi).

Then, in 2010 the 802.1X-REV (major revision) was finally released. In that, several new features were added to support the .1X standard. The potentially most disruptive new feature was layer 2 encryption through the MACSec feature (802.1AE), something Cisco leverages in its TrustSec technology.

Along with MACSec were additions for Secure Device Identifiers (DevID, 802.1AR, a standard for cryptographically unique device identities based on X.509 certificates) and network advertisements, along with a new standard for key agreements for MACSec.

MACSec and its cast of supporting protocols were part of the puzzle for delivering a standards-based Wi-Fi-like experience to switch ports. Using 802.1X-REV protocols, a device could connect to a wired port and be presented with several networks of various security levels, just like Wi-Fi SSIDs, using the network advertisements feature, then secured with MACSec and DevID.

Around 2012, I was sure MACSec would take off. Layer 2 encryption gives network architects a powerful new tool for securing networks, eliminating the need for costly firewalls to encrypt wired traffic, in addition to the advantages for securing edge ports. The sad truth is that now, more than ten years later, MACSec is still just eking its way into enterprise edge switches. Because of the cryptographic advances, adding the MACSec suite of protocols is a hardware upgrade. In an era where most organizations see switches as a commodity product with a 15-year life, there hasn't been enough momentum to get this new technology off the ground. However, MACSec has already found its foothold in WAN use cases, and I remain eternally optimistic about the future of 802.1X for edge networking connectivity.

High-Level 802.1X Process in Wi-Fi Authentication

At a coarse level, in Wi-Fi the 802.1X protocol (along with EAP) manages the authentication of the endpoint or user to the network along with the initial key derivations, distributions, and ongoing key rotations for secure encryption.

The endpoint requests to join the network, the infrastructure device initiates the authentication process with the endpoint, and then repackages the authentication request and forwards it to the specified authentication server. After that, the data path is reversed with either a "yes-authenticated" or a "no-not authenticated" reply, which come in the form of access-accept or access-deny, respectively. Depending on the authentication methods used, there may be other intermediary exchanges. A successful authentication reply may also be accompanied by additional authorization information such as a dynamic VLAN assignment.

During authentication, the Wi-Fi infrastructure will use the EAPoL protocol to communicate with the endpoint, and will repackage those as RADIUS protocol packets to send to the RADIUS/AAA server, as shown in Figure 3.2.

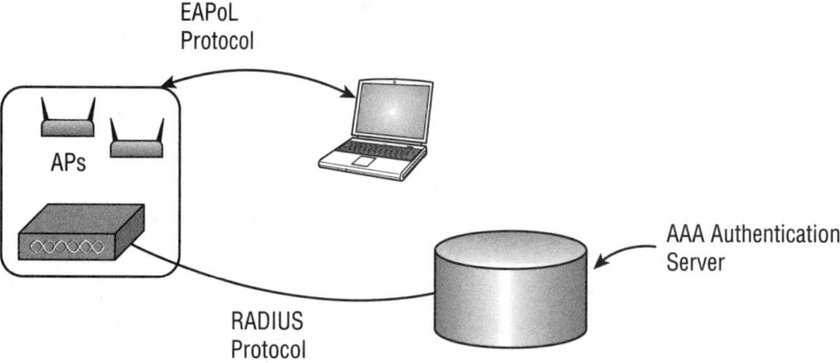

Figure 3.2: The use of EAPoL and RADIUS protocol during authentication Wi-Fi

The infrastructure devices (such as wireless controllers or APs in this case) aren't participating in the authentication process except for the repacking and forwarding of the request and response, and later for key management. That means the negotiation of all authentication parameters (including EAP methods) involves only the endpoint and the authentication server; the infrastructure device is just an intermediary.

With certain implementations of WPA3-Enterprise (specifically using EAP-TLS), the Wi-Fi infrastructure will play a role in validating the server certificate. This is a novel function that gives the infrastructure (the 802.1X Authenticator) a bit more of a role during authentication.

802.1X as the Iron Gate

When using 802.1X, whether on the wired port or wireless, the "port" is considered closed until the endpoint is authenticated. During that time, the only packets allowed to and from the endpoint are EAP packets for authentication.

Consider 802.1X the thickest metal medieval gate imaginable; nothing gets through. There's no pinging, no DHCP requests—nothing—until that port is "opened" with a successful authentication. This is described in the function of the two (non-physical) 802.1X port entities: the *Uncontrolled Port* passes the EAP traffic so the endpoint can authenticate, and the *Controlled Port* function serves as the gatekeeper preventing any other packets to or from the endpoint until it's authenticated.

If you didn't pick up on this from that description, 802.1X operates at layer 2. Endpoints using DHCP won't even have an IP address during the 802.1X/EAP processes. This is nuanced but it also means that Wi-Fi products and security and network monitoring tools don't have visibility into the endpoint through that process. The lack of visibility makes troubleshooting 802.1X more challenging, but there are several tips and tricks covered in Chapter 7, "Monitoring and Maintenance of Wireless Networks." Figure 3.3 offers a simplified representation of 802.1X logical port entities with the Uncontrolled Port used for EAPoL authentication packets, and the Controlled Port for standard network traffic that becomes accessible after successful authentication.

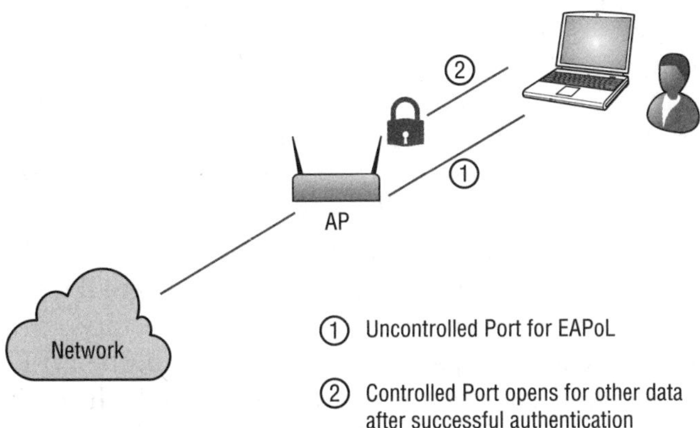

Figure 3.3: The two logical 802.1X port entities

RADIUS Servers, RADIUS Attributes, and VSAs

RADIUS is both a protocol (defined by IETF RFC 2865 `https://datatracker.ietf.org/doc/html/rfc2865`) as well as the common name for the authentication server providing the RADIUS services.

There's a lot more to RADIUS servers and protocols, but for the purposes of Wi-Fi (and any LAN-based) network authentication, there are just a few relevant parts to be familiar with. In Appendix A, "Notes on Configuring 802.1X with Microsoft NPS," you'll find a how-to guide with tips for setting up RADIUS on Microsoft NPS. Here in this chapter, we focus on the elements of RADIUS that are pertinent to this section on authentication and authorization.

TIP **When discussing RADIUS servers and services, you may see it represented as an authentication server, RADIUS server, or AAA server or services. These are all different ways to describe the same thing. A RADIUS server is an AAA server, and in the 802.1X and EAP specifications (covered earlier) it's referred to as an authentication server.**

RADIUS Servers

By definition, RADIUS servers provide AAA services—authentication (identity validation), authorization (access rights), and accounting (logging).

The RADIUS server is simply a purpose-built server or service that handles the authentication requests coming from the infrastructure devices (switches, APs, controllers, etc.). RADIUS servers can authenticate users for the purpose of accessing network resources (such as in an 802.1X-secured SSID) and also can authenticate users for management of network devices (similar to TACACS+ but without the same control granularity). The RADIUS server will often be servicing authentication requests from other parts of the infrastructure such as SSL-VPN and device management.

The authentication (RADIUS) server is most often connected to a directory service (such as Microsoft Active Directory) where the user groups and user accounts with credentials are stored. It is possible to create local user accounts in RADIUS servers, but domain user accounts will always be in a directory. And local user accounts on RADIUS servers are most often in service of non-domain accounts such as certain BYOD and guest registration use cases where it's not desirable to have the users or devices in the enterprise domain directory.

Figure 3.4 demonstrates the relationships between an endpoint, the Wi-Fi infrastructure, RADIUS server, and directory server. In this example the endpoint is connecting to the 802.1X-secured network named "ACME-Secure1X". The Wi-Fi infrastructure is forwarding the authentication request to the RADIUS server, which is in turn performing a lookup in the user directory.

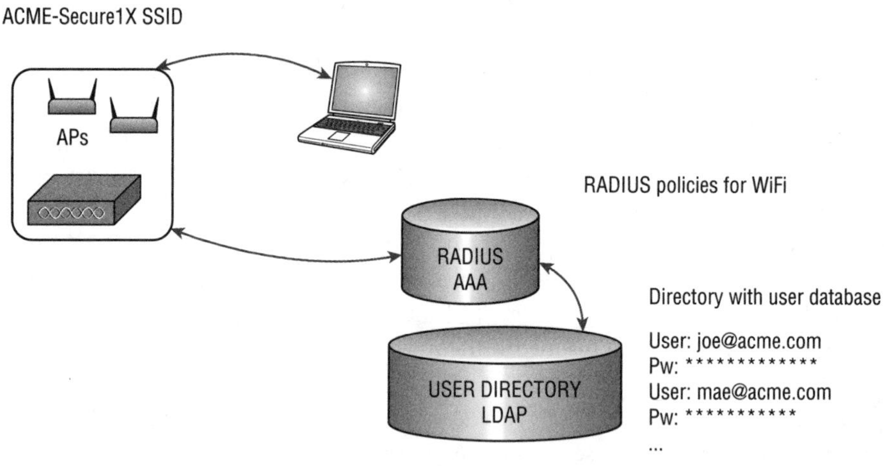

Figure 3.4: Relationship between endpoints, Wi-Fi infrastructure, and the RADIUS server

RADIUS Servers and NAC Products

Products that offer RADIUS services include Microsoft Server Network Policy Server (NPS), Cisco Identity Services Engine (ISE), Aruba ClearPass Policy Manager, Fortinet FortiAuthenticator, FreeRADIUS, Pulse Secure Steel-Belted RADIUS, and the list goes on.

You may recognize several of the RADIUS-enabled products as falling into the category of NAC solutions. Out of this list, I'd loosely categorize them as going from basic features to full features in order of Microsoft NPS, FreeRADIUS, and then the more advanced NAC products such as Cisco ISE and Aruba ClearPass Policy Manager, among others.

NOTE The line between NAC products and RADIUS servers is blurred because many professionals think of "NAC" as indicating posturing and profiling, whereas in practice many organizations use NAC products only for the authentication features and not the additional functions. Don't get hung up on this—it's exceptionally common for organizations to have tools like Cisco ISE or Aruba ClearPass and use it for RADIUS and/or TACACS+ services only. In those cases, ISE or ClearPass is simply the RADIUS server or authentication server in the architecture.

While 802.1X only requires the most standard RADIUS protocol support for operation, NAC products add additional feature sets including more granular policy control for authentication and authorization. One great example is that natively, Microsoft NPS doesn't offer a straightforward option for building out logic AND policies for evaluation—meaning you can't create a single policy that says to authenticate, the user *and* the device must both be authorized and members of the AD groups. It's going to process it as an OR statement. The operation is extremely limiting, resulting in an inability to restrict (for example) employees from using personal devices with their domain credentials (with EAP-PEAP with MSCHAPv2).

Many of the other products allow for exceptionally complex and granular policies with nested logic functions (combining AND and OR statements in a hierarchy) as well as provide additional policy inputs such as profiling (identifying a device type) and posturing (checking an endpoint for security posture).

NAC products offer a suite of features beyond the basic yes/no operations of RADIUS and EAP. Common use cases and features include:

- Ability to layer and nest logic statements such as AND and OR for granular policies
- Options to incorporate endpoint security posture into authorization decisions
- Support for dynamic assignment to classes of networks including registration, quarantine, and remediation
- Enhanced ability to enforce policy decisions based on several inputs such as the user authorization, device posture, location, and connection type
- Added support for third-party integrations for management including mobile device management (MDM) tools
- Added support for third-party integrations for security automation such as with security information and event management (SIEM) tools

TIP Some vendors add their own secret sparkles to Wi-Fi products to augment certain NAC-like functions. For example, most versions of Aruba products support both user and machine authentication without an external NAC product or advanced authentication server. Additional controller feature licenses may be required (such as the Policy Enforcement Firewall or PEF in Aruba's case).

Whether your environment uses Aruba or another product, check with your vendor account team and/or the product documentation and see if there's an option for "user and machine authentication"—it's a security feature that should be part of almost every architecture.

Throughout this chapter and others, you'll see mention of using NAC products or *advanced RADIUS servers* to meet very common security needs such as authenticating both a user and device (something not supported natively in all RADIUS servers). What I mean by "advanced RADIUS servers" is simply a product that can offer more granular control with nested logic statements, versus a simple user or device lookup along with an accept or reject message.

MORE ON NAC

For more articles, whitepapers, and recorded webinars on topics of NAC, visit my blog at `www.SecurityUncorked.com`.

Relationship of RADIUS, EAP, and Infrastructure Devices

Although everything related to EAP in the 802.1X authentication process is negotiated between the endpoint and the authentication (RADIUS) server, the RADIUS protocol packets are passed between the infrastructure device (such as an AP or controller) and the RADIUS server. Between the infrastructure device and the endpoint, the protocol in use is EAPoL (EAP over LAN). It's the infrastructure device (referred to as the 802.1X Authenticator) that takes the EAP packets for authentication and wraps or repackages them as RADIUS to communicate to the authentication server.

At its most basic operation in 802.1X, RADIUS is the protocol that facilitates the EAP authentication framework and returns either an "access accept-authenticated" or "access reject-not authenticated" response for the endpoint trying to join the network.

In Wi-Fi, the authentication server (most often a RADIUS server) performs several critical functions for 802.1X, including:

- Defining the EAP methods available for the network
- Authenticating the endpoint

- Creating the pairwise master key (PMK) and delivering it to the 802.1X Authenticator (the Wi-Fi APs or controller)
- Setting the idle timer
- Determining the session timer for re-authentication

TIP For authentication, timer settings are critical for the proper operation of the system and may require adjustments from the default depending on the architecture.

The infrastructure uses the value of the `Session-Timeout` attribute to determine the duration of the session, and it uses the value of the `Termination-Action` attribute to determine the device action when the session's timer expires.

If the `Termination-Action` attribute is present and its value is `RADIUS-Request`, the infrastructure re-authenticates the endpoint. If the `Termination-Action` attribute is not present, or its value is `Default`, the infrastructure terminates the session.

RADIUS Attributes

RADIUS attributes are the discrete elements involved in the RADIUS authentication, and there are about 255 standard RADIUS attributes defined in the IETF RFCs (`www.iana.org/assignments/radius-types/radius-types.xhtml`, plus extended attributes added later) that are understood and should be honored by any infrastructure equipment, regardless of brand, model, or version. There's an emphasis on "should be" because mileage will vary with certain products.

Common RADIUS Attributes

There are attributes for the user, the credential challenge components (such as EAP methods), the requesting infrastructure (such as controller IP address and SSID), session information (e.g., timeouts and logging), as well as attributes returned for authorization instructions (which is what's relevant and covered here). RADIUS attributes may come in the form of standard RADIUS attributes (the original 255) or extended RADIUS attributes.

For example, attribute value 61 of `NAS Port Type` with a value of 19 means it's 802.11 WLAN medium. There are hundreds of attributes; Table 3.1 shows a sample of a few common attributes used in Wi-Fi authentication and authorization, loosely grouped by theme. Pointing out a few that might not be as intuitive, in RADIUS "NAS" is the Network Access Server, which is the infrastructure device acting as the 802.1X Authenticator.

NAS-IP-Address The IP address of the infrastructure device (AP or controller) sending the authentication request. This can usually be configured in the Wi-Fi devices.

NAS-Port More useful in wired 802.1X deployments and may or may not appear in Wi-Fi logs. It may contain a 16-bit association ID.

NAS-Port-Type Defines the medium type; in this case type 19 is "Wireless - IEEE 802.11."

Called-Station-ID The MAC address of the AP the endpoint is authenticating through, appended with the SSID. It may appear with dashes or colons such as "5C-5B-35-00-11-22:ACME-SECURE1X". Note this may be different than the NAS-IP-Address from earlier.

Calling-Station-ID The MAC address of the endpoint attempting to authenticate. Depending on the vendor product it may appear delineated or not such as "7C-76-35-00-11-22" or "7C7635001122".

Table 3.1: Examples of standard RADIUS attributes in Wi-Fi

ATTRIBUTE	DESCRIPTION	DATA TYPE	IETF RFC
1	User-Name	text	RFC2865
2	User-Password	string	RFC2865
4	NAS-IP-Address	ipv4addr	RFC2865
5	NAS-Port	integer	RFC2865
61	NAS-Port-Type	enum	RFC2865
30	Called-Station-ID	text	RFC2865
31	Calling-Station-ID	text	RFC2865
7	Framed-Protocol	enum	RFC2865
44	Acct-Session-Id	text	RFC2866
49	Acct-Terminate-Cause	enum	RFC2866
95	NAS-IPv6-Address	ipv6addr	RFC3162

`https://www.iana.org/assignments/radius-types/radius-types.xhtml`
with Data from IANA, 2021.

Assuming a successful authentication, the RADIUS server may return specific RADIUS attributes related to authorization of the device—defining what it's allowed to access. The RADIUS protocol is used for many purposes ranging from legacy dial-in services to carrier operations, so not all are used for Wi-Fi network authentication.

RADIUS Attributes for Dynamic VLANs

By far, the most common use of a returned custom RADIUS attribute in Wi-Fi is a dynamic VLAN assignment. With dynamic VLAN assignments, instead of a static VLAN attached to an SSID, the infrastructure can be told to accept a dynamic VLAN from the RADIUS server. The RADIUS policy may map or prescribe a VLAN for a device or user based on any number of properties and is commonly attached to a user role as defined by group membership.

As described earlier in the section, "RADIUS Servers and NAC Products," implementations with NAC products may have extended use cases for dynamic VLAN assignment including pushing an endpoint to a network designated for remediation or quarantine.

Standard RADIUS attributes used in defining a dynamic VLAN are shown in Table 3.2. Data types of "enum" indicate the value is selected from a fixed list. In the right column you'll see examples for the three bottom rows.

Table 3.2: Standard RADIUS attributes and values for dynamic VLAN assignment

ATTRIBUTE	DESCRIPTION	DATA TYPE	EXAMPLE DETAILS
81	`Tunnel-Private-Group-ID`	text	Text value of VLAN ID
64	`Tunnel-Type`	enum	Value 13 = `VLAN`
6	`Service-Type`	enum	Value 2 = `Framed`
65	`Tunnel-Medium-Type`	enum	Value 6 = `802 (wired and Wi-Fi)`

Figure 3.5 shows an example of the standard (non-vendor-specific) RADIUS attributes used to assign a dynamic VLAN. In this example, a successfully authenticated device would be placed in VLAN 99 using standard RADIUS attributes. You'll also see options in the same screen to define Vendor-Specific Attributes, covered next. The `Tunnel-Pvt-Group-ID` attribute is returning the value of "99" accompanied by the `Tunnel-Type` attribute of "`VLAN`". Regardless of the brand of Wi-Fi infrastructure, this simple set of attributes for dynamic VLAN assignment *should* be understood and implemented.

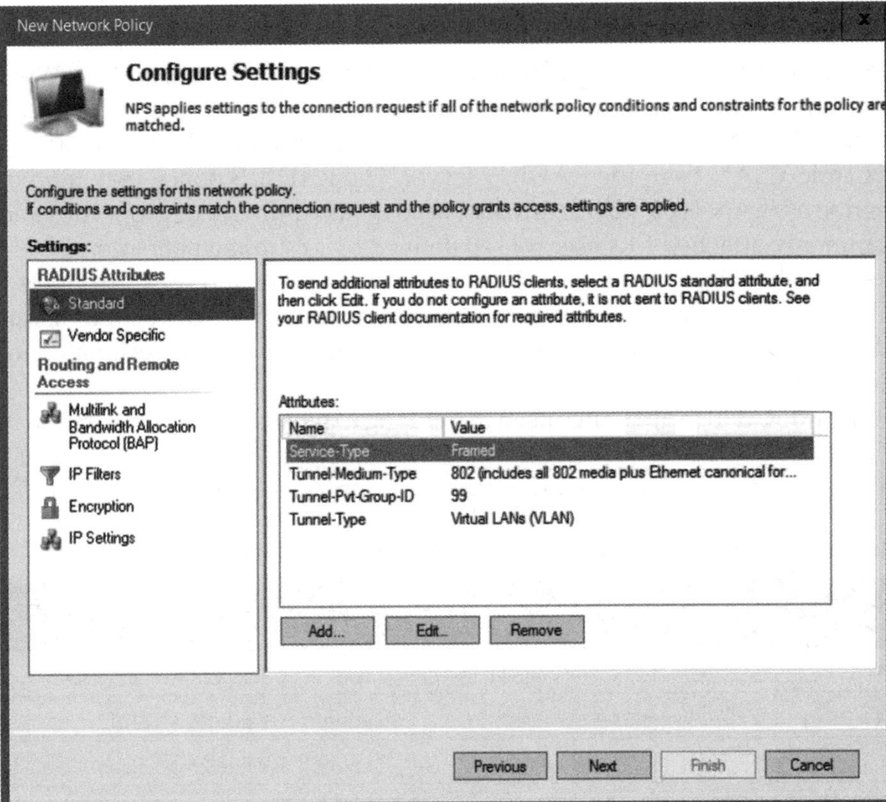

Figure 3.5: Sample attributes for dynamic VLAN assignment

INDUSTRY INSIGHT: VENDORS AND STANDARD RADIUS ATTRIBUTES

Twice in this topic of "RADIUS Attributes" it has been noted that vendor products *should* understand and honor standard RADIUS attributes. In my experience the majority of enterprise-grade solutions such as those from Cisco, Aruba Networks, and Juniper have consistently honored attributes such as these for dynamic VLAN assignments.

I've personally had mixed success with other platforms such as Ubiquiti and certain code versions of other vendors' products. The point here is not to call out specific vendors out of ill will but to point out there are specific cases where you may run into some challenges and need to request vendor support or select a different product.

Mostly, don't beat yourself up if your RADIUS authentications aren't working properly, or at all. There are troubleshooting tips in Chapter 7, but the point of this note is also that vendor support varies, and you may be attempting something that's not supported.

RADIUS Vendor-Specific Attributes

The previous example uses *standard RADIUS attributes*—the ones supported by default with any RADIUS server and understood by the infrastructure devices participating, regardless of the vendor or code version.

Additional attributes returned can also include custom *Vendor-Specific Attributes (VSAs)* that send product-specific instructions—for example, telling a Cisco WLC to apply a downloadable access control list (ACL) to that user, or instructing an Aruba controller to return an Aruba *role* attribute.

Some authentication servers may come with a suite of VSAs pre-installed, while others only offer the standard RADIUS attributes out of the box. In either case, if VSAs are needed, they can simply be imported to the RADIUS server as a dictionary file for additional functionality.

The VSAs used in Wi-Fi are often related to authorization and access control and may come in the form of a downloadable ACL, or mapping to a vendor-specified role or security profile within the wireless infrastructure

Table 3.3 offers a few examples of VSAs from Cisco and Aruba. Most vendors will have a list of VSAs that can be enabled in most RADIUS servers.

Table 3.3: Examples of Vendor-Specific Attributes

VENDOR	ATTRIBUTE	EXAMPLE VALUE	DESCRIPTION
Cisco	Cisco-AVP	CiscoSecure-Defined-ACL; value=guest_acl	Sends a downloadable ACL to the WLC from RADIUS
Aruba Networks	Aruba-AVP	Aruba-User-Role	Sends instructions from RADIUS to Aruba Wi-Fi to enforce access control based on a named role
Aruba Networks	Aruba-AVP	Aruba-Location-ID	Sent from Aruba to RADIUS to identify AP name

NOTES ON USING VSAs

Consider this tip Jen's personal opinion on using VSAs vs. standard RADIUS attributes. I'm not a fan of VSAs in most Wi-Fi use cases for a few reasons:

- **They can be cumbersome to code correctly—imagine trying to create a granular ACL that's being pasted into a tiny input box.**
- **They're not just product-specific but may be version-specific, meaning when the infrastructure device has a code upgrade, the VSA will likely have to be updated, or the operation may fail.**

- **Many infrastructure devices become quickly overloaded with downloadable or dynamic ACLs. I've seen even the beefiest new routers crippled by downloadable ACLs with various NAC products performing the same operation.**
- **Standard RADIUS attributes offer most of the features required in enterprise use cases.**

If the mission can be accomplished with standard RADIUS attributes, it's advisable to use those instead of VSAs to save yourself headaches later. There are always exceptions and I'm not telling you not to use VSAs, but consider the pros and cons and be sure to include monitoring and testing to ensure proper operation (especially since VSAs are typically used for segmentation) and add workflows for testing in a lab or staging area to be sure the VSAs persist through a code upgrade of the APs or controllers.

If your architecture calls for VSAs, it's highly recommended to add more regular (ideally, persistent) testing to ensure the authorization and access controls are working as designed. Chapter 7 offers suggestions for tools and alerting for scenarios such as this. As an example, if your VSAs are enforcing dynamic ACLs, have a monitoring solution in place to alert on instances where the policy is not properly enforced.

RADIUS Policies

RADIUS policies define the conditions for evaluating when to use the policy, the authentication methods allowed, and any authorization parameters to be returned along with the successful authentication.

Each brand of RADIUS server will have its own labels for the components of a policy, but they all have the same basic construct. In this example, these are the three elements of a Microsoft NPS network policy. I've selected NPS for this book's examples because it's readily available, it's common in home labs, and it offers a simple structure that's easy to build on and correlate other products to.

Conditions The incoming request has to match the specified conditions for the policy to be applied. Conditions may include any/all processing for all incoming Wi-Fi requests or matching to requests from specific controllers (using the attribute for NAS IPv4 address), or incoming requests from specific locations, or membership in a specific user or computer group in the domain directory. Most RADIUS servers evaluate against a list of active/enabled policies in a top-down fashion, using the first policy that matches the criteria, similar to legacy firewall rule evaluation.

Constraints The policy constraints define parameters such as the allowed authentication types (EAP methods and inner authentication methods), session timeouts, and day and time restrictions. In RADIUS-enabled NAC products, these may also include profiling and/or posturing criteria that take into consideration what an endpoint is and its current security posture.

Settings In the settings area the RADIUS attributes to be returned with an access-accept (successful authentication) are applied. As in the previous example, this may include a dynamic VLAN assignment or other parameters using standard or vendor-specific RADIUS attributes.

Figure 3.6 shows the three main policy components of conditions, constraints, and settings. In Figure 3.6 the conditions are set to only process incoming requests from the IP address 10.10.33.252 (a Juniper Mist Edge appliance in this example), and only for users in the directory group "SecureWiFiUsers." The constraints shown specify allowed EAP methods of PEAP with MSChapv2 or EAP-TTLS. Finally, the settings (in the third image) are what tells the infrastructure to put the endpoint in VLAN 99.

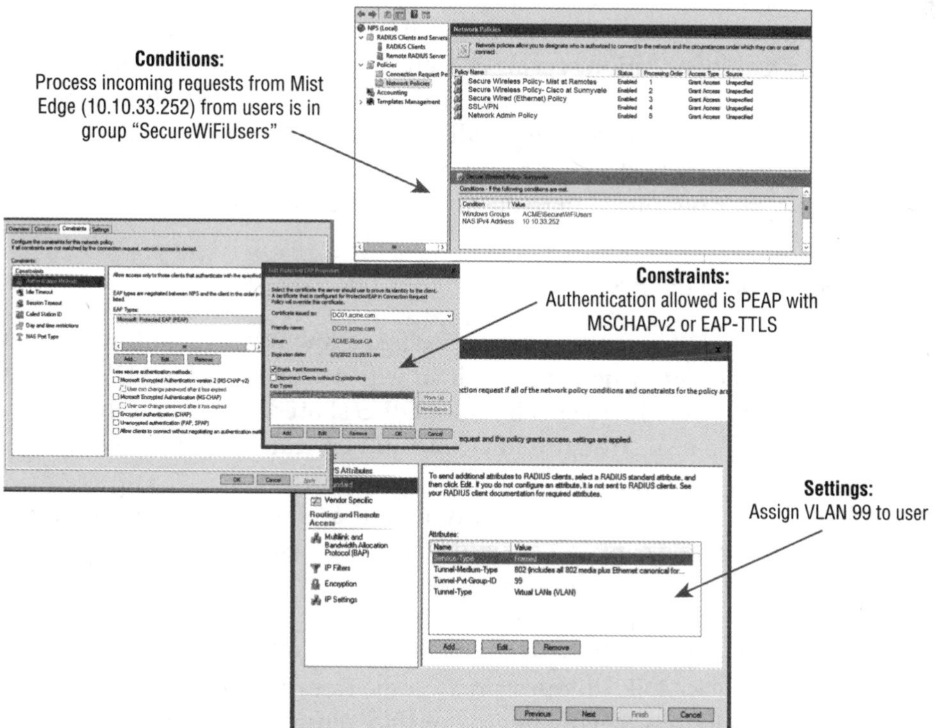
Conditions:
Process incoming requests from Mist Edge (10.10.33.252) from users is in group "SecureWiFiUsers"
Constraints:
Authentication allowed is PEAP with MSCHAPv2 or EAP-TTLS
Settings:
Assign VLAN 99 to user

Figure 3.6: Example of RADIUS policy conditions, constraints, and settings

INDUSTRY INSIGHT: RADIUS, DIAMETER, AND RADSEC

The RADIUS protocol has been around since the 1990s and, understandably, has several deficiencies especially with security features. Since then, a new AAA protocol aptly named Diameter has emerged. Diameter was designed to solve the security gaps in RADIUS by upgrading AAA to a connection-oriented protocol (specifically moving from UPD to TCP) and adding encryption.

While I love the idea of a more reliable and secure AAA protocol, the reality is Diameter is not widely adopted and almost absent in enterprise environments. Although it's not covered in this book, I recommend you keep it on your radar. Diameter operates similarly to RADIUS with several added features so the RADIUS quick start you're getting here will translate to the core AAA concepts of Diameter.

In the meantime, we do have RADSEC, which is RADIUS using TCP over TLS or DTLS. Meaning, we get a connection-oriented protocol and encryption, just as Diameter promised. More on RADSEC later.

RADIUS Servers, Clients and Shared Secrets

Confusingly named, the *RADIUS clients* (802.1X Authenticators) are the infrastructure components that will be sending authentication requests to the RADIUS server (802.1X authentication server) on behalf of the endpoints. In most environments this will be the wireless controller(s), AP(s), or wireless gateway devices, and of course switches, firewalls, or other network devices for wired and remote access requests. When you configure RADIUS clients, it's just an allowlist of what's permitted to make an authentication request.

Specifying RADIUS Clients

In all products, RADIUS clients can be specified a few ways, including by host IP addresses, FQDN, or by networks defined with a subnet. Each RADIUS client entry (by IP or network) will specify a RADIUS shared secret; simply a passphrase used by the infrastructure device to authenticate itself to the RADIUS server as an allowed requestor.

Authentication servers support many RADIUS clients, so there's no need to configure one single catch-all or wildcard entry. Instead, plan the RADIUS clients to be as specific as reasonable based on the environment and degree of network management segmentation.

What's required here will depend heavily on how the Wi-Fi infrastructure is configured—the RADIUS requests may come from a centralized source like a wireless controller or gateway device, or they may come from each AP, or a designated AP acting as a virtual mini-controller. On the Wi-Fi infrastructure, these settings are specific to each product's implementation and are frequently configurable. In addition, in many Wi-Fi products, you have the option to specify a fixed (real or virtual) IP address as the RADIUS client source IP. This maps to the RADIUS attribute of `NAS IP`, *Network Access Server IP address*, in RADIUS terms.

For example, in a cloud-managed architecture where individual APs are sending RADIUS requests, if you have an AP management VLAN you could add that network as an approved RADIUS client network. For a local cluster

AP architecture with no controller and no cloud, you may prefer to specify the cluster's shared or virtual IP as the RADIUS client source, in which case you'd enter that IP as the RADIUS client. If the RADIUS server receives a request from an unauthorized RADIUS client, it will log that event and be available to you in troubleshooting (assuming logging is enabled).

Table 3.4 uses three fictional scenarios to depict how RADIUS clients may be specified. In the first two lines, specific IP addresses are defined—this is common for specific known, fixed, and limited infrastructure devices. The IP address 192.168.33.250 in the first line defines the configured static virtual IP used in a local AP cluster. In this imaginary architecture, the organization has APs that aren't managed by a controller and the product has been configured to send RADIUS requests using the 192- address.

The second line of Table 3.4 also specifies a single RADIUS client, in this case a controller. Finally, the third line provides an example of a RADIUS client entry using a network, in this case a /25 netmask that would support up to 126 devices. This is appropriate if there's a network or Wi-Fi management network, and the APs are sending RADIUS requests directly (common in many cloud-managed products). Instead of adding a single entry per-AP (which would be tedious), the entire network is authorized. This is also preferred since the APs will likely have DHCP-served addresses which may change.

Table 3.4: Sample list for planning RADIUS clients

ENTRY	SUBNET	SHARED SECRET	DESCRIPTION
192.168.33.250	/32	**********	Virtual IP for local AP cluster at Maintenance Building
`rwifi01.acme.com`	/32	**********	Controller for Remote APs
172.16.36.1	/25	**************	Wi-Fi Management Network at North Campus

Refer back to Table 3.1 for a refresher on the use of NAS-IP-Address, which is the attribute that will be matched against the approved RADIUS clients list you create. Tying that information with the aforementioned examples for creating RADIUS clients, you'll start to see the benefit of assigning the NAS IP address in infrastructure devices, which often have numerous interfaces and IP addresses. When you set the NAS IP address, you're telling the AP or controller to use that specific IP address when sending RADIUS requests to the server. It's a great way to maintain visibility for troubleshooting and to enable the proper RADIUS client entries.

RADIUS Shared Secrets

For any RADIUS clients you configure, be sure to record the shared secret(s) somewhere safe (like in a password vault) as some products (like Microsoft NPS) will not let you view the shared secret after it's created. Oddly enough, while the server may not let you see the shared secret after it's created many of the Wi-Fi products will allow you to view it in the CLI or GUI.

You'll need to enter the shared secret on the infrastructure device (the 802.1X Authenticators) as well as the RADIUS server, or the RADIUS server will discard all authentication requests from those RADIUS clients. Figure 3.7 demonstrates the options to add RADIUS clients in Microsoft NPS by IP address, IP network, or hostname. Most authentication servers will have the same options.

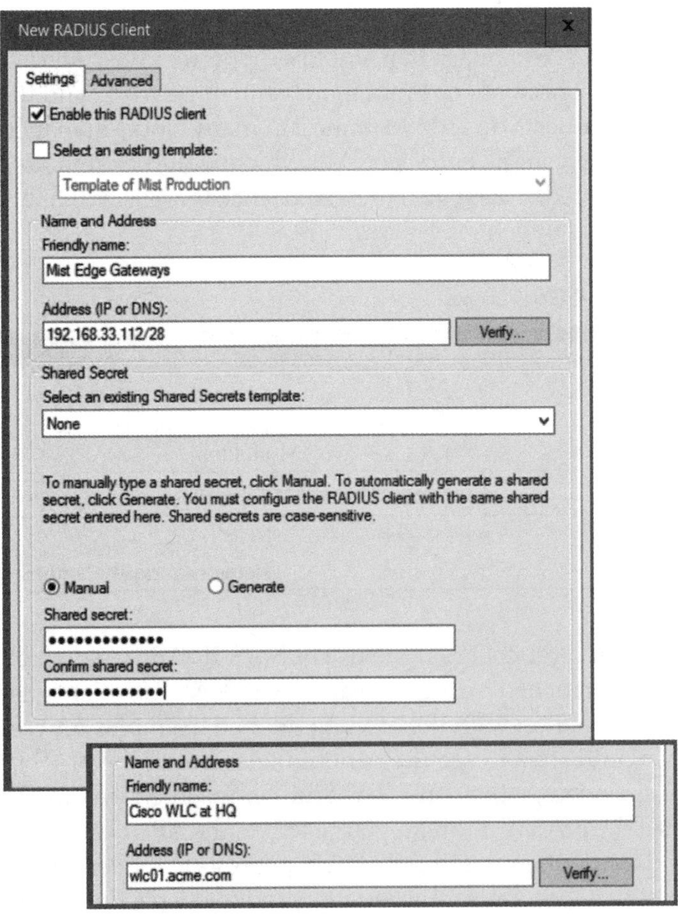

Figure 3.7: RADIUS clients can be added by IP address, IP network, or hostname

RADIUS SECURITY

The RADIUS services were designed to be deployed on trusted networks and therefore do not encrypt data, including the shared secret, meaning it will be sent over the network in plain text. Because of this, you should never specify a shared secret that's a real password for anything else.

Because of the lack of encryption and confidentiality, newer implementations of RADIUS over TLS (RADSEC) are becoming more popular. RADSEC has been around for over a decade but hasn't been as widely adopted within enterprises, possibly because it hasn't been natively supported in many popular products such as Microsoft's IAS and NPS. RADSEC is most commonly used for authentication across untrusted networks, and/or for network managing authentication when roaming between networks of different types or ownerships (such as cellular to Wi-Fi). One of the largest deployments I know of that uses RADSEC is the Eduroam service popular with universities in the United States.

The beefier (and usually more expensive) RADIUS-capable products like Cisco ISE, Aruba ClearPass Policy Manager, and Fortinet FortiAuthenticator (among others) do support RADSEC.

Other Requirements

In addition to policies and RADIUS clients, RADIUS servers require some additional inputs and integrations to work with the infrastructure. Specifically, connection to an authentication directory, logging (accounting), and certificates.

Certificates are covered in depth shortly in "Certificates for Authentication and Captive Portals," and more detailed configuration requirements for each of these is provided in Appendix A.

User Directories

Aside from NAC products that extend features beyond basic RADIUS functions, the user and device repository is typically external to the RADIUS server. The most common connection from RADIUS would be to a Microsoft Active Directory or similar LDAP. Regardless of the RADIUS server, a connection to one or more authentication sources will be required. For NAC products, there may be an internal database for user or device registration for purposes of guest and BYOD management.

Server Certificate

A simple concept repeated throughout this book is that, regardless of how the endpoint is authenticating (username/password or certificate), the server will always authenticate itself with a certificate. This is the foundation of 802.1X's mutual authentication precept.

Logging/Accounting

The third "A" in "AAA" is accounting, which just means logging. In some RADIUS products like Microsoft NPS, logging must be configured (either to a local and admittedly very ugly text file or to a SQL database).

Additional Notes on RADIUS Accounting

Properly enabling RADIUS accounting (logging) is a critical part of security for both audit purposes as well as troubleshooting. For that reason, additional notes and tips are included here.

Logging can (and should) not only be enabled but also configured to send data to the security tools for monitoring including SIEM tools—a topic covered more in Chapter 7.

Many NAC products with RADIUS services will have their own logging enabled by default. However, if you're using Microsoft NPS, you'll need to manually configure logging.

With NPS, you'll need to enable logging and choose whether to log to a local text file or to a SQL database. You'll also configure the level of logging (start/stop and authorization changes versus intermittent logging) and what the server should do if the logging destination is not available (see Figure 3.8).

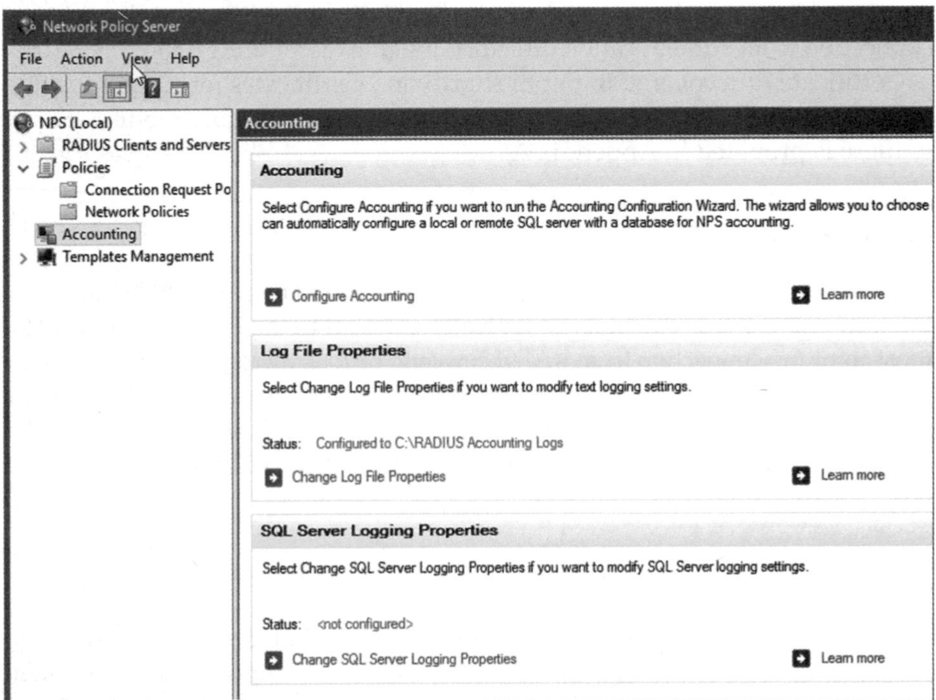

Figure 3.8: Microsoft NPS options for logging targets

For most deployments, it's recommended to log accounting and authentication requests, and omit periodic status logging. There's also an option to specify what to do if logging is not available—allow authentication or discard them. This setting should be applied based on the security requirements of the organization. For security-conscious and regulated industries, discard on logging failure should be selected (see Figure 3.9).

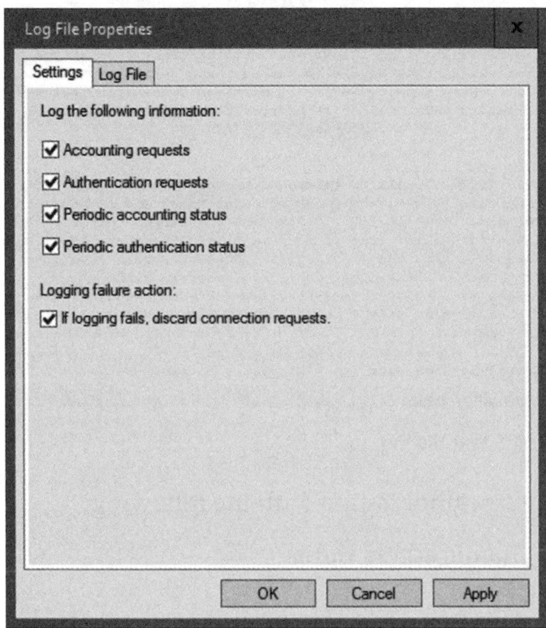

Figure 3.9: Microsoft NPS options for level of logging detail and logging failure action

Figure 3.10 is a marked-up Microsoft NPS raw text log. It's not meant to be legible here, it's purely to demonstrate that this type of raw logging can be challenging to parse through. It's one of the reasons I personally favor NAC products for authentication when possible. They almost always have better interfaces for viewing logs, which is especially helpful in troubleshooting.

Change of Authorization and Disconnect Messages

In the scope of standard RADIUS operation, the server will (ultimately) either accept or reject the endpoint's authentication request. It may send additional parameters (such as a dynamic VLAN) back along with a successful authentication, but once a user or device is authenticated with RADIUS, it's good to go until it reconnects or the infrastructure forces a re-authentication. What isn't supported in that model is a way to handle scenarios where the authenticating infrastructure wishes to remove or change an endpoint's authorization status.

```
<Event><Timestamp data_type="4">02/14/2020 09:35:46.642</Timestamp><Computer-Name data_type="1"
data_type="1">IAS</Event-Source><User-Name data_type="1">TESTfakeuser</User-Name><NAS-IP-Addres
data_type="0">0</NAS-Port><NAS-Identifier data_type="1">192.1.0.252</NAS-Identifier><NAS-Port-Ty
data_type="1">6649dda4d2c6</Calling-Station-Id><Called-Station-Id data_type="1">c8b5adc56394</C
data_type="0">1100</Framed-MTU><Vendor-Specific data_type="2">000039E705084341442D3158</Vendor-
data_type="2">000039E7060F4341442D436172792D41503033</Vendor-Specific><Vendor-Specific
data_type="2">000039E70A104341442D436172792D4941505650</Vendor-Specific><Client-IP-Address data
data_type="0">0</Client-Vendor><Client-Friendly-Name data_type="1">CaryAruba-IAP-VC</Client-Fr
Connections</Proxy-Policy-Name><Provider-Type data_type="0">1</Provider-Type><SAM-Account-Name
data_type="1">CADLAB\TESTfakeuser</Fully-Qualifed-User-Name><Class data_type="1">311 1 192.1.0
data_type="0">5</Authentication-Type><Packet-Type data_type="0">1</Packet-Type><Reason-Code da

<Event><Timestamp data_type="4">02/14/2020 09:35:46.642</Timestamp><Computer-Name data_type="1">
data_type="1">311 1 192.1.0.15 01/16/2020 22:54:36 11749</Class><Authentication-Type data_type="
data_type="1">CADLAB\TESTfakeuser</Fully-Qualifed-User-Name><SAM-Account-Name data_type="1">CADL
data_type="0">1</Provider-Type><Proxy-Policy-Name data_type="1">All Wireless Connections</Proxy
data_type="1">CaryAruba-IAP-VC</Client-Friendly-Name><Client-IP-Address data_type="3">192.1.0.
data_type="0">3</Packet-Type><Reason-Code data_type="0">8</Reason-Code></Event>


<Event><Timestamp data_type="4">02/14/2020 09:37:17.641</Timestamp><Computer-Name data_type="1"
data_type="1">IAS</Event-Source><User-Name data_type="1">jj-testuser</User-Name><NAS-IP-Address
data_type="0">0</NAS-Port><NAS-Identifier data_type="1">192.1.0.252</NAS-Identifier><NAS-Port-T
data_type="1">6649dda4d2c6</Calling-Station-Id><Called-Station-Id data_type="1">c8b5adc56394</Ca
data_type="0">1100</Framed-MTU><Vendor-Specific data_type="2">000039E705084341442D3158</Vendor
data_type="2">000039E7060F4341442D436172792D41503033</Vendor-Specific><Vendor-Specific
data_type="2">000039E70A104341442D436172792D4941505650</Vendor-Specific><Client-IP-Address data
data_type="0">0</Client-Vendor><Client-Friendly-Name data_type="1">CaryAruba-IAP-VC</Client-Fr
Connections</Proxy-Policy-Name><Provider-Type data_type="0">1</Provider-Type><SAM-Account-Name
data_type="1">CADLAB\jj-testuser</Fully-Qualifed-User-Name><Class data_type="1">311 1 192.1.0.1
data_type="0">5</Authentication-Type><NP-Policy-Name data_type="1">Secure Wireless Policy - Int
data_type="0">0</Reason-Code></Event>
```


Figure 3.10: Markup of default Microsoft NPS text file log

Wi-Fi operations involve change of authorization with the following:

- NAC products that assign dynamic access rights
- Most captive portal operations
- Onboarding and device registration for BYOD
- Manual or automated security-related access rights changes

Reasons for a change of authorization may include:

- Change in a user or endpoint VLAN assignment, ACL, or other access rights after initial connection
- Change in posture of the endpoint, when used with a NAC product checking for security posture
- Change in identity of the endpoint, such as profiling operations in NAC that may (for example) provide different authorization for a printer versus a security camera
- Change in MAC address or any other feature used for identification, typically in a NAC scenario

- Change in authorization of a user or endpoint manually or automated through security integrations as part of user or endpoint life-cycle management or security monitoring
- Change in authorization through initial endpoint registration or BYOD onboarding

Change of Authorization (CoA, defined in IETF RFCs 5176 and 8559) is an extension of RADIUS that allows the RADIUS server to ask the network infrastructure to either disconnect or change the authorization of an endpoint.

The CoA function is used extensively in NAC deployments, where an endpoint may undergo several changes of authorization as it is identified and/or registered, on the production network, or possibly quarantined. It may also undergo changes dependent on whether a user is logged on, and what access rights that user has. Further, security posture and profiling may be used for NAC products to make decisions about changes in an endpoint's level of authorization.

In addition, the CoA functions are often used in various onboarding workflows such as those for device registration (often via a NAC product) or a BYOD workflow where a user interacts with (for example) a portal during onboarding of certificates and is then redirected to the secured network afterwards.

Although extended by various vendor-specific proprietary features, the basic operation of CoA is that the RADIUS server sends either a disconnect or CoA request to the infrastructure device (e.g., Wi-Fi controller or AP) and that device communicates with the endpoint and returns a message of either ACK (acknowledged) or NACK (negative acknowledgment) to the authentication server (see Table 3.5 and Figure 3.11).

Table 3.5: CoA and DM RADIUS code descriptions

COA RADIUS CODE	COA PACKET	DESCRIPTION
40	`Disconnect-Request`	Request sent from RADIUS server to infrastructure device to disconnect an endpoint, used for VLAN changes
41	`Disconnect-ACK`	Acknowledgment of successful disconnect sent from infrastructure device to RADIUS server
42	`Disconnect-NAK`	Negative (unsuccessful) acknowledgment of disconnect sent from infrastructure device to RADIUS server

Continues

Table 3.5 (*continued*)

COA RADIUS CODE	COA PACKET	DESCRIPTION
43	`CoA-Request`	Request sent from RADIUS server to infrastructure device to issue a change of authorization to an endpoint
44	`CoA-ACK`	Acknowledgment of successful change of authorization sent from infrastructure device to RADIUS server
45	`CoA-NAK`	Negative (unsuccessful) acknowledgment of change of authorization sent from infrastructure device to RADIUS server

TIP **In packet analysis tools such as Wireshark, you can use the filter `radius.code` == XX, where XX = 40 (Disconnect-Request) or other RADIUS and CoA codes. Wireshark and other tools are covered in Chapter 7.**

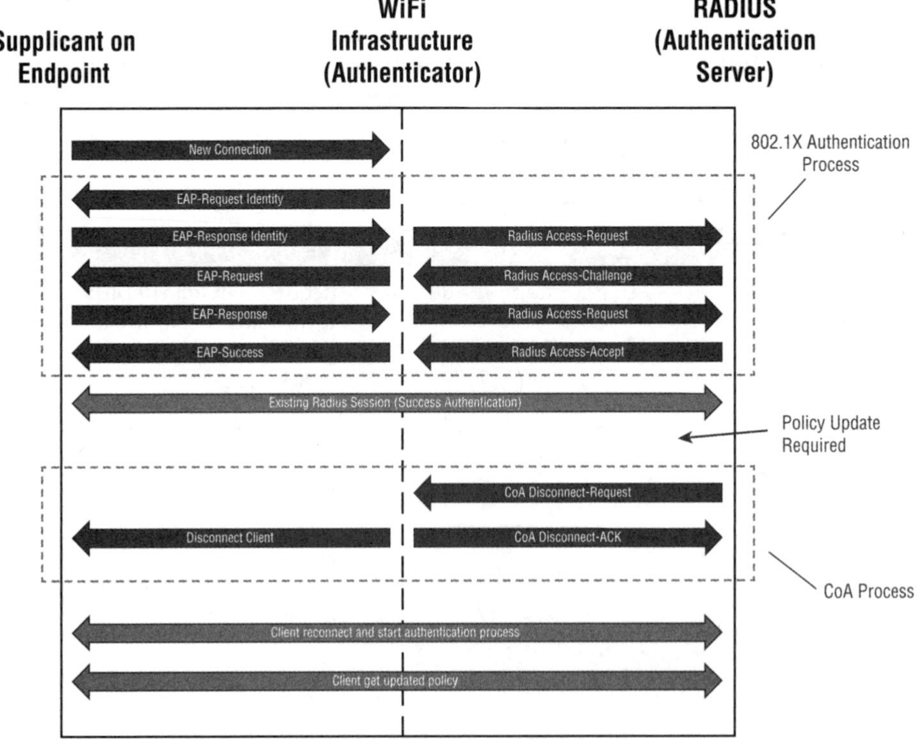

Figure 3.11: View of CoA operations within 802.1X authentication

The difference between a full disconnect and a CoA for an endpoint becomes evident in situations were the endpoint VLAN is changing, and a full disconnect/reconnect may be required to force the endpoint to make a new DHCP request. Otherwise, the endpoint may be left in a VLAN with an IP address and gateway that effectively isolates it from everything.

Under other RFCs, there are further extensions for communications related to (for example) why a disconnect or CoA was not successful. These exchanges use the RADIUS attributes to identify which session it is that's being modified. If the disconnect or CoA request does not match a session, the request is unsuccessful. If it matches more than one session, then all sessions are impacted by the disconnect or CoA operation.

We could fully dissect the CoA protocol here, but what's worth knowing for architecture purposes is that not all products support all CoA functions equally. Some products extend the basic CoA suite with Vendor-Specific Attributes while other products offer no support, or very limited support for CoA operations.

Some products such as this Cisco example (Figure 3.12, top) have the option to enable CoA within an already-defined RADIUS server, while others such as Juniper Mist (Figure 3.12, bottom) have CoA/DM server parameters configured separately.

INDUSTRY INSIGHT: VENDOR CoA SUPPORT

Although I avoid making vendor-specific comments around feature support (since it's always changing), I feel comfortable with the general statement that traditional enterprise-grade products such as the Cisco WLC-based products, Aruba Networks, and Juniper Mist all support at least the expected CoA operations defined in the RFCs.

Some further extend the operations with Vendor-Specific Attributes. The products that grew from small-medium business use cases, such as Meraki and similar classes of products, have had limitations with CoA on wired and/or wireless networks. Having said that, it's fair to say vendors are almost always adding to features and support for this and other standards, so just ask about CoA support before selecting a product. And, if the CoA isn't working as expected, check with your vendor support before spending too much time banging your head against a wall.

EAP Methods for Authentication

As described earlier in "Terminology in 802.1X," the Extensible Authentication Protocol (EAP) is a framework within which we can fit several methods of authentication such as username-password credentials or device certificates.

Figure 3.12: Comparison of CoA configurations

www.cisco.com/c/en/us/support/docs/security/identity-services-engine/115732-central-web-auth-00.html

Because RADIUS and EAP protocols are used for multiple purposes and classes of networking (including carriers and cellular), there are myriad EAP methods defined in various IETF RFCs, and that list will continue to grow and change with use cases. For our purposes, this book focuses on the most common EAP methods used in network and wireless authentication.

The presentation of EAP methods here may be a bit different than what you're used to seeing. To provide the most robust explanation and foundational knowledge that will extend beyond today's most common EAP methods, this content is presented in two sections: "Outer EAP Tunnels" and "Inner Authentication Methods."

EAP-PEAP and EAP-TLS are the two primary standards-based EAP methods in today's network and Wi-Fi authentications. Most often it's considered that "PEAP is a Microsoft thing that uses username and password" and that "TLS uses device certificates." While that's not entirely wrong, it's also not entirely accurate.

There are many combinations of outer and inner mechanisms that are most often combined for the best security (see Figure 3.13). As you'll see, many of the inner authentication methods (EAP-TLS aside) weren't designed for confidentiality, meaning they may pass some or all of the username and credential exchanges in cleartext. Those authentications then are best protected by being passed in a secured tunnel (similar to VPNs). Although not all widely used, just about every combination of inner and outer is supported by the IETF RFCs—for example it's completely possible to send EAP-TLS within an EAP-PEAP tunnel.

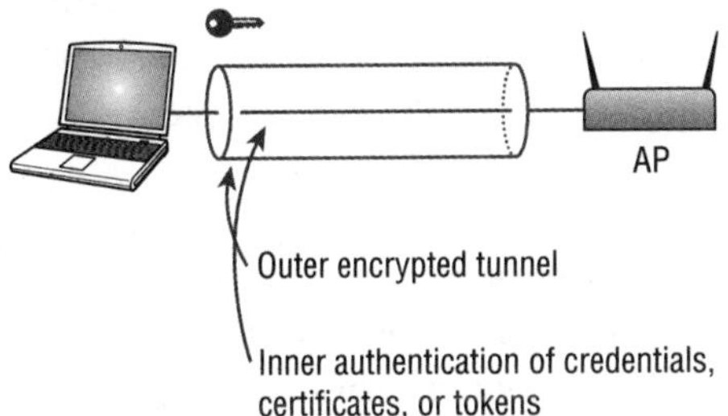

Figure 3.13: Conceptual representation of the EAP framework using an outer tunnel and inner authentication

Outer EAP Tunnels

Because the authentication exchanges themselves are not always encrypted by default, the EAP inner authentications are (ideally) wrapped inside of TLS-secured tunnels in order to offer confidentiality and integrity of the credential exchanges between the endpoint and the authentication server. This TLS tunnel is created using the authentication server's certificate (another reason it's required).

The purpose of secured tunnels (in all implementations) is to ensure the integrity and confidentiality of the user credentials including the username or identity as they're passed from the endpoint. Prior implementations of EAP that don't offer this protection have been deprecated and shouldn't be used.

The following sections describe the most common EAP tunnel methods used in enterprise Wi-Fi networks.

EAP-PEAP

PEAP (Protected EAP) was originally a joint project by members from RSA, Microsoft, and Cisco, and is considered the most widely supported tunneled (technically encapsulated) type for 802.1X authentication. It's embedded in all major operating systems and can be used with several inner authentication methods including MSCHAPv2 and EAP-TLS as well as non-Microsoft EAP methods.

Microsoft PEAP offers a fast reconnect option which, when enabled on both endpoint and authentication server, facilitates faster roaming on an 802.1X-secured network by caching credentials.

EAP-TTLS

EAP-TTLS (Tunneled Transport Layer Security) is an extension of EAP-TLS that adds an outer encrypted tunnel but unlike EAP-TLS does not require device certificates for endpoints. A secured TLS tunnel is established and then an inner authentication method can be used, such as EAP-TLS or even EAP-MSCHAPv2.

If the endpoint is authenticating with a device certificate (via EAP-TLS as its inner method), mutual authentication can happen in one phase without additional phases for exchanging username-based credentials and challenges through the secured tunnel. EAP-TTLS supports fast reconnect as well.

TIP **The difference between EAP-TTLS and EAP-PEAP is exceptionally subtle and has to do with the nuance of PEAP being defined as an encapsulation of the full EAP exchanges, whereas EAP-TTLS is an encryption of the distinct RADIUS attributes (specifically EAP-TTLS defines use of the extended RADIUS attributes, which you may see referenced as AVP, or Attribute-Value Pairs).**

From a security perspective (until someone creates a novel attack) consider them equivalent and if you're looking for a standards-based tunneled EAP method, use whichever is best supported in your environment.

EAP-FAST

EAP-FAST (Flexible Authentication via Secure Tunnel) was a Cisco proprietary implementation similar to the more common EAP-PEAP but with a different mechanism to facilitate roaming. EAP-FAST uses a cookie-style token in the form of a Protected Access Credential (PAC). The EAP-FAST tunnel is now a standard, defined in IETF RFC 4851, `https://datatracker.ietf.org/doc/html/rfc5421`.

The PAC token is considered proof of a successful device authentication, avoiding the latency of re-authentication while roaming on an 802.1X-secured network. EAP-FAST also supports credential chaining, to validate both user and device. Like TTLS and PEAP, EAP-FAST can be used with several inner authentication methods such as EAP-MSCHAPv2, EAP-GTC, and EAP-TLS.

EAP-TEAP

EAP-TEAP (Tunneled EAP) is another standards-based EAP tunnel similar to Cisco's EAP-FAST with support for EAP credential chaining to facilitate device and user authentications (per IETF RFC 7170 `https://datatracker.ietf.org/doc/html/rfc7170`). In addition, EAP-TEAP incorporates several security enhancements and mitigations against man-in-the-middle attacks absent from virtually every other common EAP tunnel method.

It's been ratified since 2014 but still has virtually no traction in the industry. I mention it here only because it would offer a native, standards-based opportunity to deliver the level of chained authentications only available in costly NAC products or advanced AAA servers currently. It's been slow going on adoption but would be a great addition to the arsenal for security architecture in wireless. It's worth noting that enterprise AAA and NAC products such as Cisco ISE and Aruba ClearPass Policy Manager currently support EAP-TEAP and credential chaining; however, this class of products can use chained logic statements in policies to obtain the same effect without requiring EAP-TEAP.

A comparison of the EAP outer tunnel methods can be found in Table 3.6. These are very loosely organized in order of least to most secure, with the understanding that all encrypted tunnel methods listed, if properly implemented, are appropriate for enterprise use. EAP-PEAP is perfectly acceptable if that's what the environment supports. EAP-FAST and EAP-TEAP offer the added benefit of native credential chaining (to authenticate both user and device) without requiring additional tools or products. The security column references more info covered next in "Securing Tunneled EAP."

Table 3.6: Comparison of EAP outer tunnels

EAP TUNNEL	INDUSTRY ADOPTION	CHAINING SUPPORTED	SECURITY
EAP-PEAP	Widely adopted, supported on almost every standard operating system	No	Encrypted TLS encapsulation
EAP-TTLS	Widely adopted, supported on most platforms	No	Encrypted TLS tunnel
EAP-FAST	Was Cisco proprietary, ratified standard but lower adoption	Yes	Encrypted + Cryptographic binding with MSK (some protection)
EAP-TEAP	Ratified standard but low adoption	Yes	Encrypted + Cryptographic binding with EMSK (best protection)

SIZING IT UP: CUSTOMIZING EAP SUPPORT

With Microsoft and most other popular operating systems, if you're using any method other than EAP-PEAP or EAP-TTLS, you will need to add support for the other methods to the endpoint through an installer or third-party supplicant, pushed to each endpoint.

In addition, even if you've added support for additional EAP methods to the endpoint, you may not be able to push those settings through Group Policy or MDM tools. For some non-standard EAP methods in a Windows environment, the EAP methods will not be available as an option in domain-joined endpoints if the EAP method is not supported in the domain's Group Policy. There are workarounds for Microsoft Group Policy through exporting, modifying XML content, and re-importing. Use of Microsoft PowerShell DSC (Desired State Configuration) is another option.

Lastly, not all RADIUS servers natively support all EAP methods. Verify support throughout the 802.1X-connected elements before jumping into an EAP method in your planning and design.

Securing Tunneled EAP

Although simplified here, there are many layers to tunneled EAPs, and varying levels of resiliency to attacks. For example, many tunneled EAP processes don't offer cryptographic binding between the outer tunnel and inner authentication method, making it possible for man-in-the-middle attacks where a malicious user injects them in the middle as an endpoint, an infrastructure device, or an authentication server. Long-standing work within the IETF has been in progress to resolve many of the remaining security gaps in the EAP authentication methods, especially for Wi-Fi networks where data is transmitted over the air and therefore more susceptible to attack by eavesdropping.

EAP-TEAP already has several mitigations for man-in-the-middle attacks such as the more stringent EMSK-based cryptographic binding to ensure the inner and outer authentication parties are the same, and one isn't an attacker. EAP-FAST, although offering cryptographic binding, does so with MSK-based cryptographic binding, which has been proven to be ineffective against man-in-the-middle attacks where the attacker is spoofing the infrastructure device acting as the Authenticator (RADIUS NAS device).

Additional vulnerabilities are introduced when an inner method is also allowed as an outer method. This is one of the reasons to pay special attention to the recommendations for allowed EAP methods when configuring the server. As an example, many IT professionals don't understand the subtle difference between configurations that define MSCHAPv2 as an inner authentication method (which is fine) and an authentication method without a tunnel (not secure). The latter opens up the environment to a host of attacks.

The guidance here is to not let perfection be the enemy of progress. A move from non-802.1X-secured networks to any of the secure EAP methods presented in this book will be a huge increase in security. The minutia of the vulnerabilities present in various EAP tunnel methods and the relationships of the inner and outer methods pale in comparison to the vulnerabilities introduced by misconfigurations or poor design. Meaning, unless you're in the type of organization or industry that's likely to be targeted by malicious actors, start with any EAP method here that's attainable for the environment.

Following the best practices for strong mutual authentication and the other recommendations in this book will put you in something like the 97th percentile of security.

INDUSTRY INSIGHT: EAP IN HIGH SECURITY ENVIRONMENTS

If you *are* architecting a solution for an organization that is likely to be targeted, such as certain global financial institutions and government agencies, then additional research and testing is warranted. The research of vulnerabilities in EAP and TLS protocols evolves with time and for that reason, professionals architecting for highly sensitive implementations should evaluate risk based on then-current data. Meaning, I'm not going to provide specific recommendations here that may be out of date within a few weeks or months.

The IETF RFCs and informational memos provide a wealth of information about the operation of these authentication protocols, as well as detailed analysis of their shortcomings and vulnerabilities. That, coupled with the most current data from security researchers in this space will inform decisions on architectures and protocols. Any solutions implemented should be penetration tested (ideally by a third party who specializes in such testing, not using off-the-shelf tools) and known vulnerabilities should be documented and mitigated in other ways or monitored and alerted on if no mitigation is available.

Alternatively in many cases, other technology is combined for a higher level of security, such as using VPNs for Wi-Fi connectivity.

Inner Authentication Methods

Inner authentication methods can be wrapped inside one of the outer tunnel methods just described, or in some cases used on their own. The tunnel protects the credentials (especially usernames and passwords) from being exposed during the authentication process.

As a reminder, these inner methods describe how the endpoint is authenticating to the server, which may be credential- or certificate-based; however, the authentication server is always authenticating to the endpoint with a server certificate.

These common inner methods are supported by each of the outer tunnels described previously per the RFC specifications, but that doesn't mean all vendors support every combination. Manufacturers of authentication servers and endpoint operating systems are regularly updating software versions to add new EAP support. As always, do your research, and don't rely on an eight-year-old blog or knowledge base article for up-to-date information.

To set the stage, Table 3.7 offers an at-a-glance comparison of the EAP inner authentication methods covered. Note that most current inner EAP methods are strongly recommended to be used with an outer tunnel, and those with exposed (unencrypted) credentials should only be used within a secured tunnel.

Table 3.7: Comparison of EAP inner authentication methods for Wi-Fi

INNER AUTHENTICATION	VALIDATION USING	ENCRYPTED WITHOUT TUNNEL	RECOMMENDED OUTER TUNNEL
EAP-TLS	Certificate	Yes	EAP-TTLS
EAP-MSCHAPv2	Username-password credentials	No, username sent in cleartext	EAP-PEAP or EAP-TTLS
EAP-GTC	Token	No	EAP-PEAP or EAP-TTLS
EAP-POTP	Token	Yes	Not required

Out of the standard inner authentication methods, there is really only one that is viable on its own, and even then, it's recommended to be tunneled. That protocol is EAP-TLS.

EAP-TLS

EAP-TLS (Transport Layer Security) is based on mutual certificate authentication and includes encryption, making an outer tunnel optional (but still recommended). EAP-TLS can be tunneled in any of the outer methods mentioned here. The outer tunnel is recommended with EAP-TLS because the protocol supports weak ciphers such as TLS 1.0, and it's up to the endpoint to select the version. Unfortunately, there's not a way to force the version from the server side, meaning you'll want to add the mitigation of either a secured outer tunnel and/or configuring the endpoints to use secure TLS versions. See the sidebar on "Forcing TLS Versions in Microsoft Clients" for more.

EAP-TLS does require device certificates and therefore a PKI infrastructure. The PKI infrastructure could be as simple as enabling Microsoft AD Certificate Services within the domain and setting up auto-enrollment for both servers and domain endpoints, or the PKI infrastructure could be a third-party solution that may require additional configuration and integration to the infrastructure.

FORCING TLS VERSIONS IN MICROSOFT CLIENTS

Since the TLS version is selected by the client, you may want to explore options to allow only the latest versions. If you're working with a Microsoft environment, you can configure the endpoints to use specific TLS versions by defining the `DWORD` value that's named `TlsVersion`. The value to define for TLS version 1.2 is "0xC00". TLS version 1.3 should be rolling out by the time you're reading this. Check Microsoft's documentation for updated info.

EAP-MSCHAPv2

EAP-MSCHAPv2 (Microsoft Challenge Handshake Authentication Protocol version 2) authenticates with username-password credentials and is one of the EAP methods that can be specified on its own or as an inner method with tunneled protection described earlier—but should *always* be used with a tunneled protection. Recall from Table 3.7 that EAP-MSCHAPv2 does not encrypt the username portion of the credentials during authentication.

MSCHAPv2 in Microsoft implementations allows for the user to update their password when required by an expiration policy.

One point of confusion is in understanding where MSCHAPv2 fits in the RADIUS server and endpoint configurations. You'll see this again in Appendix A, but the sample screenshots in Figure 3.14 show the correct and incorrect configuration for an EAP-PEAP deployment with an MSCHAPv2 inner authentication.

Notice in the correct configuration (top left) EAP Types window, only EAP-PEAP is present, and in the list below, MSCHAPv2 is not selected; instead, the inner method of MSCHAPv2 is defined in the pop-up dialog box on the top-right screenshot. This relates to the security comments earlier, where we specifically want to avoid allowing any method as both an inner and outer.

Finally, the bottom screenshot shows an improper configuration, where MSCHAPv2 is selected in the EAP Types window along with PEAP, and/or is selected in the check box options below that window. Neither of these are correct for the desired secure configuration.

EAP-GTC

EAP-GTC (Generic Token Card) is an inner method that supports generic authentications to an assortment of identity stores including tokens and one-time password (OTP) platforms as well as LDAP. Although flexible, GTC is not commonly supported or used. If needed though, GTC can support databases with passwords that are not only cleartext or Microsoft NT hashes but also salted and unsalted MD5 and SHA1 hashes.

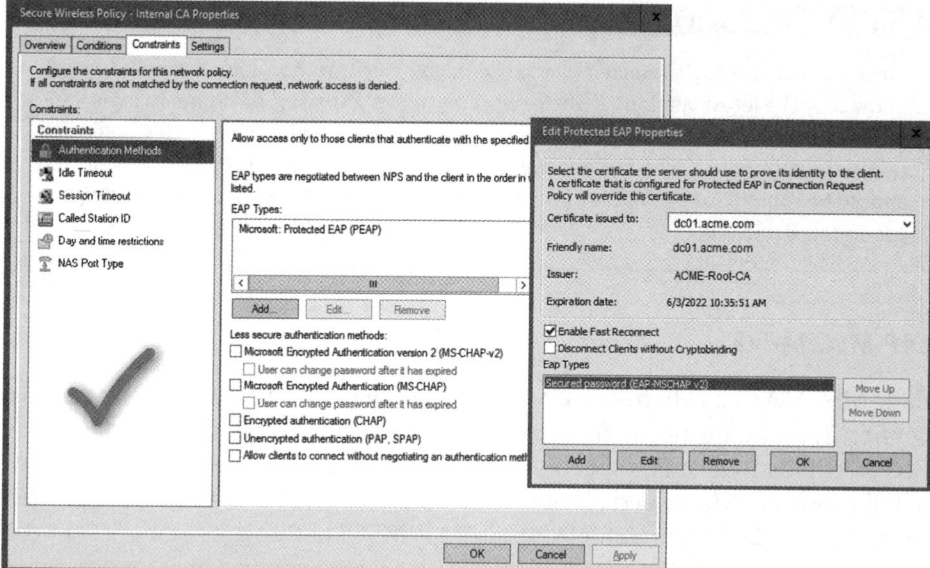

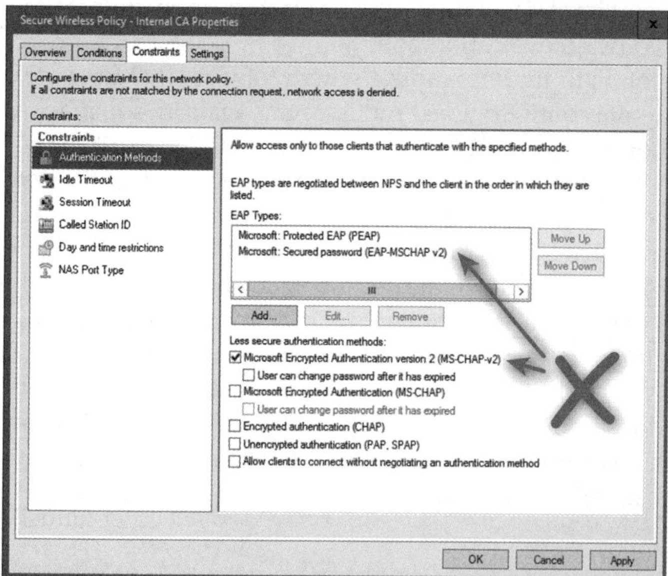

Figure 3.14: A correct PEAP with MSCHAPv2 configuration on the top, and an incorrect (not secure) configuration that would allow MSCHAPv2 authentication without being encrypted, on the bottom

EAP-POTP

There are also specialized EAP methods for similar functions, such as EAP-Protected One-Time Password (POTP) developed by RSA with features favorable to hardware-based USB-connected tokens. Newer specialized protocols like EAP-POTP may bring additional security benefits (over EAP-GTC) such as channel binding and mutual authentication for additional protection against attacks, as well as the elimination of human intervention to enter a passcode (see Figure 3.15).

Figure 3.15: RSA SecureID USB token
`https://rsa.com`

Legacy and Unsecured EAP Methods

There are volumes of EAP methods defined in various IETF RFCs, some for carrier use, some for cellular use, and others that were used for LAN and Wi-Fi authentication but are now deprecated or should only be used inside a tunneled method.

These include (but aren't limited to):

- EAP-MD5 (fully deprecated)
- EAP-LEAP (fully deprecated)
- EAP-PAP (unencrypted, do not use without outer tunnel, required for LDAP)
- EAP-CHAP (encrypted but vulnerable, do not use without outer tunnel)
- EAP-MSCHAP (encrypted but vulnerable, do not use without outer tunnel)

In general, avoid these deprecated and unsecured EAP methods; they all have known vulnerabilities and with prevalent support of the other methods offered here, there shouldn't be a need to support any of these in your infrastructure.

I'll make the blanket statement to never use the deprecated protocols (MD5 and LEAP) under any condition.

For the others, if there is ever a case where there is a business requirement to use an older authentication method be sure it's protected in a tunnel, and in highly sensitive environments, I recommend treating those devices as though they were unsecured, and segmenting and monitoring as such. Most often the

EAP-PAP and CHAP classes of authentication may be used for other parts of the infrastructure (such as SSL-VPN or device admin authentication)—and therefore may be present in RADIUS policies, but would not be used for users authenticating to the wireless network.

Even though 802.1X-secured networks can accommodate as many EAP methods as you like, don't combine secured and unsecured protocols on the same SSID as it opens the environment up to a battery of insider attacks including those that exploit group keys.

Recommended EAP Methods for Secure Wi-Fi

Even though the EAP methods were pared down to the core set of the most widely used, supported, and recommended, there's still a lot there to digest and organize. Here's a quick rundown of recommendations based on the prior sections, starting with easiest/fastest to implement and progressing to a bit more involved integrations:

- At time of writing, this is a longshot, but if EAP-TEAP is supported throughout the environment, that is the most secure and flexible option regardless of whether you want to authenticate with credentials, certificates, or both.
- If you prefer to use username-password combinations and are in a Microsoft environment, EAP-PEAP outer with MSCHAPv2 inner and fast reconnect enabled is your easiest and most secure option.
- If you're in a position to use device certificates, EAP-TTLS outer with EAP-TLS inner is ideal; with EAP-TLS solo (no outer tunnel) as a second choice. If using EAP-TLS alone, see the earlier notes for forcing secure TLS versions on the client side.

  Device certificates are considered the most secure implementation by many, but there are still vulnerabilities if a malicious user is in possession of a device with a certificate, and therefore I prefer dual or chained authentications when possible. Obviously, there are also vulnerabilities associated with only authenticating a user and not also the device, which is again why chained authentications or a RADIUS server that can add logic statements is strongly preferred.
- If you need to verify or authenticate both the user and device and you don't have a PKI infrastructure or prefer not to issue device certificates, use a NAC product or advanced AAA server that supports chained or nested logic statements in the policy, such as "verify the computer is a member of the domain, AND the user in the group 'AllowedWiFiUsers'". Remember that this AND function is not supported in Microsoft NPS currently.

- If you need to authenticate both the user and device and you have a PKI infrastructure, you should again choose EAP-TEAP if supported in your environment; or alternatively EAP-FAST, and use a combination of EAP-TLS for device certificate-based authentication plus EAP-PEAP/MSCHAPv2 for user credential-based authentication.
- If you have special use cases that involve tokens, use EAP-POTP or EAP-GTC either alone or in a chained or logic configuration using EAP-TEAP, EAP-FAST, or a NAC product or advanced AAA server as described in the earlier bullet. One other EAP method not covered earlier that can also be considered is EAP-FAST-GTC, defined in IETF RFC 5421.

Recommended combinations of EAP inner and outer methods for enterprise Wi-Fi are summarized in Table 3.8 along with use cases and additional notes regarding support and security.

Table 3.8: Recommended EAP outer and inner combinations for enterprise Wi-Fi

OUTER TUNNEL	INNER EAP	USE CASES	NOTES
EAP-TEAP	Any	Any, but currently not widely supported	Includes strong cryptographic binding of inner/outer plus supports credential chaining
EAP-PEAP	MSCHAPv2	Easy basic credential-based for environments with Windows AD	Supported on all major platforms including laptops, tablets, smartphones, as well as printers and other IoT devices; does not authenticate the device.
EAP-TTLS	EAP-TLS	Certificate-based where PKI is in place	Very widespread support; does not authenticate the user.
none	EAP-TLS	Certificate-based where PKI is in place	Encrypted without the outer tunnel, but TTLS is still recommended.
EAP-TEAP or EAP-FAST	EAP-TLS and/or EAP-MSCHAPv2	For credential chaining without a third-party NAC product	EAP-TEAP is a standard and most secure via cryptographic bindings, but not widely supported.
Any	Any	Where NAC product is present, chained authentication and logic statements are possible without TEAP or FAST	NAC products have varying features but all support features that allow validating both user and device, among other things.

Continues

Table 3.8 (*continued*)

OUTER TUNNEL	INNER EAP	USE CASES	NOTES
EAP-PEAP or EAP-TTLS	EAP-GTC	For no-frills token implementation	Broad support but no inherent security, use POTP if possible.
none	EAP-POTP	More secure token implementation	More robust features, no manual user intervention, and includes encryption without an outer tunnel.

MAC-Based Authentications

Harking back to the authentication concepts of Chapter 1, remember that use of a MAC address is a form of identification, not user or device authentication. In the world of infosec, it's widely accepted that identification is the process of asserting an identity, while authentication extends that by validating (proving) identity.

With that caveat and clarification, within the industry and technology frameworks these methodologies are referred to as "MAC authentication," and that language is honored here.

There are a few common ways MAC authentication appears within wireless architecture:

- *MAC Authentication Bypass (MAB) with RADIUS* allows a fall-through mechanism for non-802.1X endpoints to authenticate on an 802.1X-secured network using RADIUS.
- *MAC Authentication without RADIUS* is typical in network access control (NAC) deployments with products that support non-RADIUS-based authentication and enforcement.
- *MAC filtering, denylisting, and allowlisting* speaks to the ability to filter (allow, disallow) specific endpoint MAC addresses from the network infrastructure.

MAC Authentication Bypass with RADIUS

You've likely heard the use of 802.1X with MAC Authentication Bypass (MAB) for wired and/or Wi-Fi deployments. This method allows endpoints that aren't participating in 802.1X to connect to an Enterprise/802.1X-secured network and authenticate. In a perfect world, all endpoints would be able to authenticate with 802.1X. In an almost-perfect world, only devices that absolutely couldn't authenticate with 802.1X would use a method such as MAB. And in the real world, MAB is used as the "easy button" to avoid complex configurations on non-traditional endpoints that can support 802.1X such as printers and VoIP phones.

With any vendor, when 802.1X is configured on a port, there may be follow-on commands that specify allowing MAB and the parameters accompanying it such as re-authentication timers, authentication timeouts, and allowed states (whether only a single MAC address is allowed or multiple, and how to handle multiple). A port in this case can mean either a physical edge port on a switch or a logical SSID port on a Wi-Fi system.

Over the years I've found MAB to be exceptionally prevalent in Cisco environments, where it's just assumed to be part of the function of 802.1X networks, especially when Cisco ISE or other RADIUS-based NAC products are in use (including Aruba ClearPass). Conversely, in my experience working with security-conscious organizations, I've supported but rarely designed MAB as part of an architecture. There are more secure and less troublesome options, which will be addressed next.

INDUSTRY INSIGHT: VENDORS THAT FAVOR MAB

You may be wondering why Cisco (and a few select other vendors) favor MAB. 802.1X/EAP deployments rely on RADIUS authentication servers, and many vendors (including Cisco and Aruba Networks) offer robust authentication services in their NAC products (a la ISE and ClearPass Policy Manager, respectively).

While exceptionally robust, these two products specifically aren't designed for non-RADIUS-based authentication and enforcement. Technically, Aruba CPPM has a feature called OnConnect designed for this purpose, but customers are steered away from using it, and I've yet to see it work at scale. Cisco ISE and its predecessors also don't offer non-RADIUS-based authentication.

Meaning, these (and a few other products) rely on RADIUS authentication even for non-802.1X devices, and therefore need MAB to address the numerous exceptions of non-802.1X endpoints in enterprise environments. To clarify, products that rely on RADIUS-based authentication and enforcement don't have a mechanism to interact with the network infrastructure other than via RADIUS.

Alternatively, there are products that support and excel at non-RADIUS-based enforcement, and these include ForeScout and Fortinet FortiNAC. I mention these two specifically because in all the years of working with dozens of NAC products, these are the only two I've seen effectively and securely handle MAC authentication without RADIUS. Products that offer non-RADIUS-based enforcement support options to interact with the infrastructure through SNMP, CLI, and/or APIs, in addition to RADIUS. This is less important in Wi-Fi networks, but a critical differentiator for port access control on wired ports.

Overview of Typical MAB Operations

At a high level, MAB works by having the infrastructure device (such as an AP or controller) create an authentication request on behalf of the endpoint (using its MAC address) and sends that to the RADIUS server for authentication.

For it to work, the network device will need to be configured to allow MAB operations, something applied at the port or SSID context.

The details of MAB's operations get a bit fuzzy after that because MAB, unlike 802.1X, is not a standard. Each vendor has its own implementation of MAB, uses different RADIUS attributes for MAB, and offers different auxiliary settings around MAB. There are a few common themes and security considerations across vendors' MAB implementations.

When enabled, the endpoint's MAC address is used for authentication—the username and password both equal to the MAC address. And of course, for that to work means there's a user account in the directory for RADIUS to authenticate against.

As shown in Figure 3.16, here's the order of operations:

1. Endpoint connects to 802.1X-secured network and is prompted to start EAPoL with the network.
2. Endpoint does not respond to EAPoL request, and usually after three attempts with no reply the network device will switch to MAB, if configured.
3. Network device uses the MAC address of the endpoint as a username and password and crafts an authentication request to the RADIUS server.
4. The RADIUS server looks up the credentials against the directory specified in the policy (Active Directory or a NAC product, for example).
5. If the MAC-based user account matches, a successful authentication is returned, and the endpoint is allowed on.
6. In the event of a failure, a dead end or Internet-only VLAN may be specified.

Vendor Variations of MAC Operations

Since MAB isn't defined in any standards, vendors implement it in whatever method they see fit, within the general parameters described previously in "Overview of Typical MAB Operations."

Some common variations in MAB among vendors may include:

- Use of different RADIUS attribute fields (such as calling-station-ID, password, login)
- Use of different authentication protocols (PAP, CHAP, EAP-MD5, or proprietary)

- Addition of NAC features for decision-making (such as device profiling or security posture analysis)
- Communication wrapper between the network device and authentication server (can include proprietary implementations in homogeneous environments)
- Configuration options for authentication timeouts, re-authentication settings, and mode (e.g., single or multiple MAC addresses)
- Varying repositories of the endpoint database, possibly within domain infrastructure, external database, or database within a NAC products

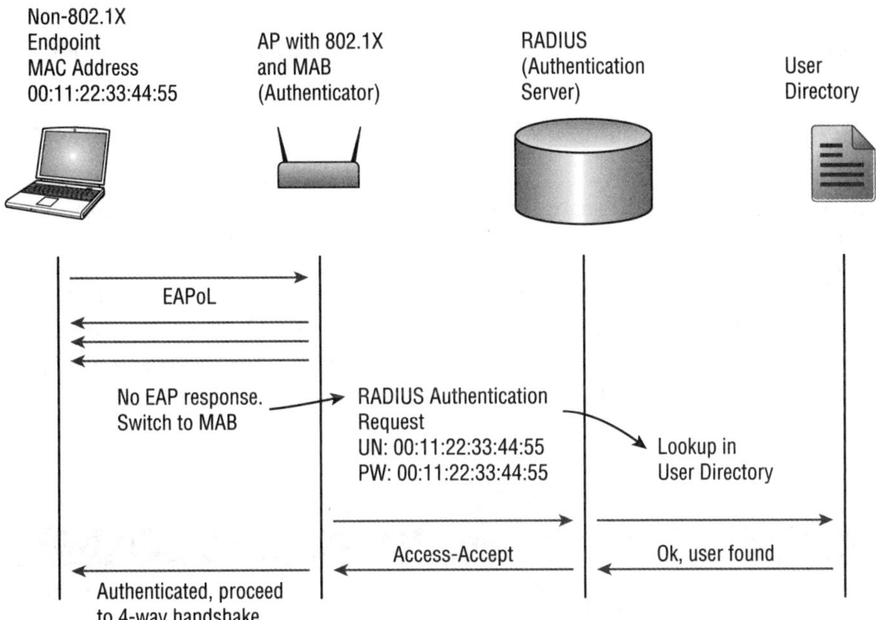

Figure 3.16: MAB authentication process

Security Considerations for MAB

Speaking personally, I have strong feelings about MAC-based authentication, and a particularly strong aversion to most MAC Authentication Bypass (MAB) deployments. As throughout this book, security considerations are provided, and it is up to you as the architect to make the decisions best for your environment.

It's not that MAB is a bad option, but it's often not implemented in a way commensurate with security expectations of an enterprise network. Certainly, I don't consider MAB to be a no-brainer addition to any security architecture, but if you do plan to incorporate MAB, the following are some considerations to factor in the architecture.

- MAB uses unprotected authentication protocols (e.g., CHAP and MD5) that are not secured unless used with RADSEC or a vendor-specific implementation
- MAB requires a user password policy that allows the same username and password
- MAB is not authentication by infosec and compliance definitions
- Attackers can easily force a MAB override by withholding EAPoL responses and connect using a spoofed MAC address
- MAB may not handle dynamic VLANs well
- MAB does not protect against MAC spoofing
- Vendor security features for MAB may require additional products or licensing
- Many endpoints support 802.1X including printers and other network-connected devices
- MAB does not benefit from the same encryption and key management as 802.1X

Table 3.9 highlights some of the differences in 802.1X and MAB that impact security. This illustrates why it's advisable to separate 802.1X-secured assets from endpoints connecting with MAB. Remember, if they're on the same wireless network, they're sharing a broadcast domain over the air, even if they're on separate VLANs.

Table 3.9: Security differences in 802.1X vs. MAB

FEATURE	802.1X	MAB
Authentication	Strong mutual authentication	Endpoint identification by MAC address only
Encryption	Strong encryption and key rotation	Equivalent to passphrase-secured network
Integrity	Credentials are secured during authentication if properly implemented	Credential is MAC address, passed in cleartext in Wi-Fi packets over the air
Standard	Defined by IEEE and IETF standards for 802.1X and EAP	No standard, vendor-dependent implementations and variability
Security	High if properly implemented	Low

AN IMPORTANT NOTE ON SECURITY AND ENCRYPTION FOR MAB

First, let's talk about MAB's encryption. It's a short conversation: there is none. Because MAB is a subconfiguration on an 802.1X-secured port, it's common to assume that the MAB connections are encrypted the same as the 802.1X connections, but this isn't the case.

MAB endpoint communications are secured over the air with the equivalency of a Personal or passphrase-secured network.

Further, because of how 802.11 works, the MAC addresses of endpoints are transmitted in cleartext over the air and easily discovered by simple eavesdropping. This means any user or application within proximity of the endpoint can easily capture the MAC address, spoof the endpoint's MAC address, and gain access to the network.

There are some mitigations for this in NAC products and authentication servers that can incorporate profiling rules into access policies, but in most environments, those layered controls are not standard.

Recommendations when Using MAB

MAB should really be a last ditch effort to apply some level of security controls to a Wi-Fi network. Given its prevalence in many environments, I don't expect network architects to be given carte blanche to remove MAB in one fell swoop. If MAB is a requisite part of your architecture, these recommendations will offer the most security possible and protect your 802.1X-secured assets.

Add Additional Segmentation from Production Networks for MAB Authorized Devices MAB connected devices should not share an SSID nor VLANs with 802.1X-authenticated endpoints. MAB offers no form of authentication, profiling, or posture assessment, and endpoints connecting with a MAC address identity should not be comingled on production networks with fully 802.1X-secured endpoints. Just because the endpoint was allowed to connect through an 802.1X network does not mean it meets the security requirements. While RADIUS can return dynamic attributes such as dynamic VLANs and ACLs, remember that MAB devices (if they don't have a supplicant) may not properly release and renew the IP address.

Ensure MAB Uses its Own RADIUS Policy Since MAB uses less secure authentication protocols, preserve the integrity of your 802.1X-authenticated endpoints by configuring a different policy for MAB devices, and only allow the lesser secure protocols on that policy. Ensure that policy only allows the MAB devices, and not regular user or computer accounts that could and should be authenticating with 802.1X.

Create a Separate Directory Group for MAB Endpoints Placing MAB devices in a dedicated directory group provides a container you can reference in the RADIUS policy and a way to be granular with other changes, like the password policy changes required. Remember, MAB endpoints will need to be user (not computer) accounts, and in most implementations the username and password are both the endpoint's MAC address.

Be Granular with Any Password Policy Changes in Directory Services Enterprise domain policies will demand strong passwords for users, and most will specifically disallow passwords that match the username (in whole or part). Meaning, if you're adding MAB user accounts to the domain, you'll need to create a very granular policy for that group that has decreased password security requirements.

Test the Vendor's Implementation for Security Vulnerabilities Each vendor's flavor of MAB will vary and vendors may introduce behavior changes in firmware updates. MAB should be tested and retested through the network's life cycle to ensure the operation meets the organization's security expectations. In addition to in-house testing, third-party pen testing is also highly recommended, and results of these tests can help inform the additional monitoring and alerting to configure.

Monitor MAB All networks should be monitored, but especially those with known vulnerabilities such as MAB. Monitor for anomalies with SIEM or user and entity behavior analytics (UEBA) products.

Use a NAC Product for Profiling and to Prevent MAC Spoofing For a higher level of security and to prevent MAC spoofing, use a NAC solution to layer dynamic endpoint profiling rules. Profiling rules examine endpoint profiles by comparing open ports DHCP fingerprints, HTTP headers, SNMP, or other parameters to the expected value. If a Honeywell IP camera suddenly looks like a Linux server, there's a problem, and obviously this is the type of event that should trigger an immediate security alert.

Upgrade Everything Legacy RADIUS servers required passwords for MAB devices to be stored using reversible encryption, and some network devices use deprecated EAP-MD5. If you must use MAB, update and upgrade the infrastructure components as much as possible for better security.

Check Endpoint Support for 802.1X Although many resources on the Internet cite MAB as a use case for common endpoints like printers and VoIP phones, the truth is these devices have supported 802.1X for many years. Before you get MAB-happy, check the endpoint inventory and migrate endpoints that support 802.1X off MAB. Some may require software or firmware upgrades, but chances are a high percentage of these devices don't need MAB.

Avoid MAB for Wired Port Control Wired network security is outside the scope of this book, but this is a key point. 802.1X and MAB are both configured at the port-level context. In Wi-Fi, that's an entire SSID, but on a switch those are discrete edge port commands. Managing 802.1X on a wired network is profoundly more complicated than on a wireless system. In fact, most NAC implementations fail when organizations attempt to implement 802.1X on wired ports—a situation avoided with either proper expectations and planning or by using non-RADIUS-based enforcement solutions (covered next).

MAC Authentication Without RADIUS

While MAB relies on 802.1X and RADIUS, there are MAC authentication solutions that do not require RADIUS at all. Products that support non-RADIUS-based authentication and enforcement use the infrastructure devices and a policy engine (most often a NAC server) to make decisions about authentication and authorization.

When properly configured, NAC-based solutions bring several security benefits over MAB deployments:

- They don't compromise enterprise 802.1X deployments
- They don't require directory user accounts with MAC addresses; the endpoint directory is stored on the NAC server
- They don't require creating less secure password policies in the directory policies
- They don't require less secure authentication protocols to be enabled
- They offer granular enforcement not binary 802.1X open/closed behavior
- They support VLAN changes seamlessly even for MAB endpoints
- They offer advanced endpoint profiling and mechanisms that greatly reduce the chance of MAC spoofing
- Most products offer onboarding and device registration
- They can easily be used for wired enforcement since no port-level configurations are required

MAC Filtering and Denylisting

Filtering MAC addresses through denylists is a common feature of enterprise Wi-Fi products. Another "not authentication" protocol, it's worth mentioning here since there are security applications for MAC denylisting. Most often these are applied dynamically through integration with a security product such a SIEM, IPS, or other detection and response engine.

With MAC denylists, the wireless infrastructure enforces the filtering within the wireless system.

A use case of MAC filtering is to add controls to limit the endpoints that can connect to (for example) a passphrase-secured network. This used to be common for networks that supported devices like handheld scanners and other non-traditional endpoints that didn't support authentication.

Static MAC filtering has been slowly fading from implementations, in preference of better controls such as NAC, or other MAC-based authorization like MAB.

We're at the dawn of the era of MAC randomization, which will inevitably hamper any MAC-based controls including MAC filtering (along with MAB and certain NAC authorizations).

Certificates for Authentication and Captive Portals

When, where, how, and which types of certificates to use with network authentication seems to be a twisty and often misunderstood topic. If we rewind to the prior chapter, you'll remember there are about five use cases for certificates and authentication related to wireless infrastructure: authenticating end users, authenticating endpoint devices, authenticating administrative users, authenticating servers, and authenticating the wireless infrastructure components to one another.

This portion focuses on certificates related to authenticating the end users, endpoint devices, and servers, which applies to Wi-Fi security profiles using Enterprise (WPA2/WPA3) and browser-based captive portals most often used with Open System Authentication networks (Open/Enhanced Open).

RADIUS Server Certificates for 802.1X

The 802.1X standard defines mutual authentication—that means both the authentication server and the endpoint need to authenticate to each other. There's a common misconception that if the endpoints aren't using certificates, then the server also doesn't require a certificate. That's incorrect.

The RADIUS server performing AAA functions will always, absolutely, 100 percent of the time require a server certificate to authenticate itself to the endpoints. Meaning, even if the endpoints are authenticating through EAP-PEAP (outer) with MSCHAPv2 (inner) via user credentials (username and password), the server still (and always) authenticates itself to the endpoint using a certificate (see Table 3.10). I realize this statement is repetitive, but I assure you I'm only referencing it here in the book once for every 2,000 times it's been incorrectly stated elsewhere (see Figure 3.17).

Table 3.10: Server certificate requirements for various EAP methods

INNER EAP AUTHENTICATION	ENDPOINT AUTHENTICATION	SERVER AUTHENTICATION
EAP-TLS	Certificate	Certificate
EAP-MSCHAPv2	Username-Password Credentials	Certificate
EAP-GTC or EAP-POTP	Token	Certificate
any	any	Certificate

As you'll see in "Best Practices for Using Certificates for 802.1X" coming up, the server certificate used here should be a dedicated certificate (as in not a wildcard cert) and in most cases, should be an internal domain-issued certificate (as in not a publicly signed certificate).

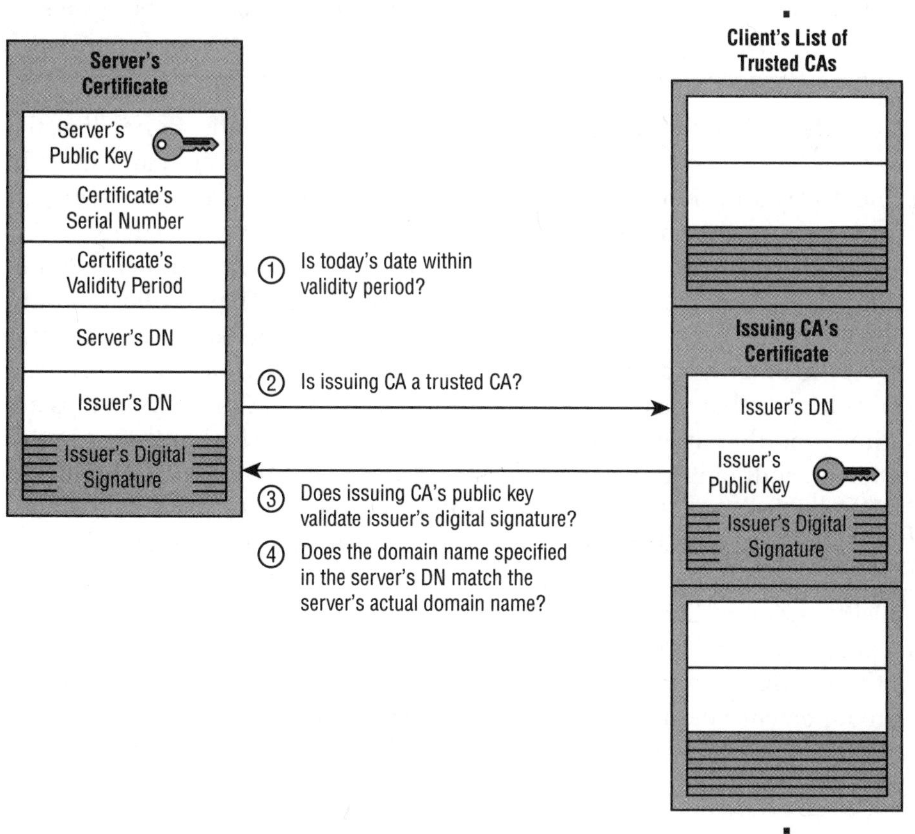

Figure 3.17: Process of an endpoint validating a server's certificate

INDUSTRY INSIGHT: STEPS IN A CERTIFICATE REQUEST

Certificates based on X.509 use the asymmetric (public/private) keys (covered in Chapter 2) and can come in many formats. The process of creating a certificate will include several components and files. Follow instructions from your authentication server (RADIUS or captive portal) for generating a certificate signing request (CSR) and installing certificate(s).

Through that process you'll get a certificate along with a private key file. The private key file you download is the only copy, anywhere. Store it in a safe place just as you would a password—a credential vault or secured storage area is ideal. Chapter 6 "Hardening the Wireless Infrastructure" includes suggestions for credential vaulting.

1. Generate a CSR from (or for) the server needing a certificate. Certificates are issued with a fully qualified domain name (FQDN); IP addresses are no longer supported.
2. Use CSR to request a certificate from a certificate authority (CA).
3. The CA will issue a certificate file along with a public and private key.
4. Install the certificate to the server using the certificate file and private key.
5. Store the private key in a safe place. If you lose it and need it later, you'll have to repeat this entire process.

The only trick with server certificates for 802.1X authentication is that the server certificate needs to be trusted by the endpoints that are authenticating. In a standard Windows environment where the endpoints are corporate-managed (and therefore members of the domain), this is a trivial task because the RADIUS server is commonly Microsoft NPS, which would have been auto-enrolled in the Microsoft certificate authority and issued a certificate. That certificate would automatically be known and trusted by any endpoints joined to the domain.

If, on the other hand, you're using an internally issued domain certificate with endpoints that are not already joined to the domain, then extra steps are required, and that internal certificate will need to be pushed to the endpoint through an endpoint configuration system such as an MDM tool, or manually installed by the user. This scenario is most common in bring your own device (BYOD) use cases, or any scenario where endpoints aren't already domain members—extending authentication to other domains during mergers and acquisitions, supporting non-domain OS platforms (such as Mac, Chromebooks, Android, and Apple IOS phones).

If the environment is not all Windows, the take-away here is simply there are two requirements: the authentication server must have a valid server certificate, and that certificate must be known and trusted by the endpoints authenticating on that network.

SIZING IT UP: USING CSRs AND SAN CERTS

In smaller to mid-size organizations with a Windows environment, it's possible to get up and running with an internal domain certificate authority, server certificates, and even device certificates in a matter of minutes to hours because of the auto-enrollment features in Windows Server platforms.

In larger environments and in environments using a third-party RADIUS server or NAC product, there will be additional steps whether supporting 802.1X or captive portal needs.

Larger and distributed deployments may require support for multiple RADIUS servers or captive portal servers. Regardless of the product in use and its processing or load balancing algorithm, one common desire is to use a single certificate to cover the network across all possible authentication servers. This is especially helpful when we deploy captive portals, which often use expensive publicly signed certificates.

Wildcard certificates are a no-no. Don't ever use them. If you want a certificate that covers multiple servers, there's an option to issue a certificate for a group of servers, using what's called Subject Alternative Name (SAN) certificates. To get a SAN certificate requires that the certificate signing request (CSR) specify the desired alternate names. Some products natively support requesting SAN certs in the CSR process, while others don't. If your product doesn't support that, there are many readily available free Linux tools that will serve that purpose.

As an example, ACME Enterprises may want to request and issue a single SAN cert for a cluster of guest registration portals on servers acme-guest.acme.com, acme-wifi01.acme.com, and acme-wifi02.acme.com. This is much easier and less expensive than purchasing certificates for the three servers separately. For internal domain use as with 802.1X, there's usually not the same cost benefit but it's great to have as an option.

Endpoint Device Certificates for 802.1X

As you've now learned, while the server must always have a certificate, the endpoint devices could authenticate to the network with either a certificate or alternate means such as a token or username-password credentials.

If EAP-TLS or similar certificate-based authentication is used, the endpoint will, of course, need a certificate. Just as with the server certificates, the only two items absolutely required are that the endpoint have a valid certificate, and that the authenticating server have access to the repository to validate that certificate. There are other options and best practices for certificate security such as revocation parameters, which are covered later.

Use of device certificates requires a Public Key Infrastructure (PKI). As already covered, in a basic Microsoft environment, there are many ways to automate enrollment for both servers and endpoints, making the process of issuing and installing certificates pretty easy.

However, there are often situations where the organization is of a size or industry where a straightforward homogenous Microsoft deployment isn't possible or desirable. The organization may support non-Windows operating systems, there may be BYOD or contractor support requirements, and there may be endpoints that are not centrally managed and/or non-standard OSs such as IoT, industrial IoT (IIoT) and operational technology (OT), or Internet of Healthcare Things (IoHT) including a broad assortment of biomedical devices.

INDUSTRY INSIGHT: HOSPITALS TAKE THE IoT CAKE

Healthcare environments are the quintessential IoT test beds. Hospitals were IoT-ing before it was even mainstream technology, and they serve as a great example for complex endpoint ecosystems—with biomedical devices that may be certified with a specific configuration by the Food and Drug Administration (FDA, in the United States), at times attached to legacy operating systems, and often supported (and locked) by third parties contracted to manage or monitor the system.

In these cases, it's not only possible but extremely likely some of the biomedical devices may be issued certificates by a purpose-built CA. These often autonomous systems manage the entire life cycle of the device certificate including issuance and revocation.

Here is some food for thought, but worth mentioning here as it pertains to authentication security. While it's true mutual certificate-based authentication (such as with EAP-TLS) is the most cryptographically secure, remember there are still vulnerabilities that arise from authenticating an endpoint without validating the user attached to it. The vulnerabilities surrounding password-based authentication methods like MSCHAPv2 have to do with specific attacks that target the detachment of the outer tunnel and the inner authentication, making it possible for some specific man-in-the-middle attacks. As the EAP methods are updated for cryptographic binding of inner with outer, these vulnerabilities will fade (and I'm sure new ones will arise). To make the blanket statement that "certificates are always better" is not necessarily true, despite the cryptographic integrity.

Ultimately my suggestion will be to authenticate both the endpoint and the user, but when that's not possible, the organization has to make a risk-based decision on whether it's more important to validate the user or the device. There are pros and cons with each.

Best Practices for Using Certificates for 802.1X

There's a lot to digest with issuing and managing certificates, but there are some easy best practices applicable to most, if not all, 802.1X deployments.

Never Use Wildcard Certificates

Wildcard certificates are exactly what they sound like—an * (asterisk) used in part of the FQDN, which indicates any subdomain of the domain would be valid with that certificate. So, a wildcard cert of *.`acme.com` would be valid for "`guest.acme.com`" and "`vpn.acme.com`" as well as "`wifi01-acme.com`," "`portal.acme.com`," and basically "`anythingyoucanthinkof.acme.com`."

Aside from being a terrible security practice, wildcard certs are usually not honored (accepted) for most RADIUS servers. The server will allow you to install the certificate and specify it in the RADIUS policy, but the authentication will fail without any specific indication of why in the logs. Among others, Microsoft NPS is a server that (as of writing and to date) does not allow the use of wildcard server certificates. It's quite frustrating and not secure, so simply avoid wildcard certificates, always. Instead, if the goal is to have a single certificate for several components, use a SAN certs. See "Sizing It Up: Using CSRs and SAN Certs" covered earlier.

Never Use Self-Signed Certificates

Self-signed certificates occur when the server (in this context usually third-party software) issues itself a certificate. In doing so, the certificate is not issued from any trusted root CA—not internal or public. Many operating systems have begun systematically removing the option for a user to bypass security warnings from self-signed certs. Options for self-signed certs usually appear in third-party software and infrastructure devices.

If a wildcard cert is bad, using a self-signed certificate is even worse. Even in a lab, it would be better to spin up internal root CA services or use public signed certificates. Self-signed certs likely won't work and even if they do, you're opening yourself (or your organization) to additional vulnerabilities.

One distinction I'll make for clarity—in considering your internal domain infrastructure and specifically a Windows Server environment where you've enabled Certificate Services on the same platform as your RADIUS and/or AD services, issuing the RADIUS NPS a certificate from the domain root CA is not considered a self-signed certificate. That's a legitimate certificate with a root CA the endpoint can trust and validate.

GET FREE CERTIFICATES FROM LET'S ENCRYPT

If you don't have the option to create your own internal certificates, check out Let's Encrypt—a free, automated, and open certificate authority, run for the public's benefit. It's a service provided by the Internet Security Research Group (ISRG), and it's sponsored by tech industry heavyweights like Cisco, Mozilla, EFF, Chrome, Meta, AWS, Akamai, and dozens of others.

You can find out more and get started at `https://letsencrypt.org/`.

Always Validate Server Certificates

Since the inception of 802.1X, we've had the luxury of selecting the option to "not validate server certificate" on most endpoints. This selection would ignore the server certificate altogether and not validate whether the server certificate was valid or trusted. Originally a great option for network admins working through troubleshooting and initial 802.1X deployments, it quickly became the source of vulnerability to man-in-the-middle attacks.

Since around 2019, endpoint platforms have slowly been removing the option for users to ignore the server certificate. That trend will only continue throughout 2022–2025 as we see wider adoption of WPA3 security and additional requirements for validating server certificates on 802.1X-secured networks.

Even in testing, it's a bad idea to bypass the server certificate. The only valid time I'd suggest using this is for some extreme troubleshooting or emergency. Because the RADIUS server will force you to select a certificate during the policy creation, there should be a valid certificate on the server. The gap that may exist is that the endpoint might not know or trust that certificate, which happens frequently with personal devices, BYOD, or connections from devices not in the domain (such as some tablets or smartphones).

Even with the stricter WPA3-Enterprise deployments, there are mechanisms supported to allow the user to manually accept and install the server certificate if it's not already trusted. That should be more than enough for testing and troubleshooting.

Under no circumstances should a production environment allow endpoints to bypass the server certificate validation—not for normal operations, not as a BYOD workaround, and not even for testing.

PEN TESTERS' TAKE ON CERTIFICATE VALIDATION

When I ask the penetration testers I work with about the top five vulnerabilities they see in Wi-Fi, high on the list is organizations that deployed 802.1X networks and bypassed the validation of the server certificate.

In every case, organizations thought they had the best 802.1X security, but this one overlooked setting introduced a vulnerability that allowed testers to easily launch a man-in-the-middle attack and bypass the security.

Most Often, Use Domain-Issued Certificates for RADIUS Servers

There are two primary drivers for using a domain-issued certificate for RADIUS servers. One is that publicly signed certificates most often have a cost associated with them, and the maximum validity period for those certificates is ever-decreasing. That translates into high costs for annual renewals. The only circumvention of cost I've seen is with certain public sector deployments such as schools and governments that have often negotiated bulk purchasing for little to no cost. For the rest of us, we're shelling out the moolah for publicly signed certs.

The second driver is security—there are security benefits to using domain (internal CA) issued certificates to the authentication servers. Endpoints can be instructed to trust only those certificates, or any certificates issued by the domain CA for that network connection. In the cases where third-party or publicly signed certificates are used by the server, it's possible (likely) the endpoint will already have a trust chain established for that public CA. Meaning, out of the box, endpoints are pre-configured to trust the most common root CAs used today, and would therefore inherently trust a certificate issued by a public CA.

By using a publicly signed cert for internal 802.1X/EAP authentication, you've opened the network to another man-in-the-middle attack scenario where an attacker may simply launch an evil twin of the corporate network and use a certificate from a root CA already trusted. If the endpoints were locked to only trust specific certificates (from the publicly signed CA) that mitigates this attack but can become unruly to manage as the certificates are renewed or reissued.

For these reasons, it's usually easier, more secure, and much more cost effective to issue RADIUS server certificates from your internal CA and ensure the endpoints connecting to that network only trust certificates from the appropriate CA.

Like everything in technology, there's an "it depends" attached to this recommendation, and it's not meant to be a "thou shalt commandment." There are specific use cases when using an internally issued certificate is not possible or not preferred, such as with schools and universities that use the Eduroam service, which issues certificates from a third party, and whose certificates are honored and valid at all participating entities.

In addition, there may be standard enterprise deployment scenarios that have preference for using a publicly signed certificate such as to allow an endpoint to connect to Wi-Fi and then be joined to the domain—something helpful during unplanned work from anywhere (WFA) models. In Windows AD, the typical process to join a computer to the domain or to connect the computer to an 802.1X-secured network the first time involves IT staff or the user provisioning the device first on a wired network. This allows the credential caching used in subsequent secure Wi-Fi connections. Obviously, in the current climate dictated by pandemic operations, that may not be possible.

Having said that, in most deployments (and definitely in Microsoft's ecosystem) there are safer workarounds for one time bootstrap connections that involve an IT admin team member and the user working together for a temporary wireless profile configuration that's then overwritten by the group policy update once the user is on the corporate domain network .

CONFIGURING TRUSTED CERTIFICATES IN GROUP POLICY

I'm not saying this has happened to me, but. . . it's definitely happened to me. One thing that can come back to bite you later is setting the directory or MDM group policy to look for a specific authentication server certificate.

If, later, you add another server or update/renew the original server certificate, the endpoint will fail authentication. You may, instead, choose to set the policy to look for a specific root certificate or issuing root CA. If the certificate settings are controlled and locked by a Group Policy or MDM, the user will not be able to override or change those settings. If this happens, the user(s) will be up a creek until you modify the policy and get them connected in a manner in which they can receive the updated policy. Although there are bootstraps for new devices, if an endpoint already has a group policy that restricts access to the settings, it may not be possible to follow the bootstrap processes.

On the flip side, if no servers or root CA trust is defined in the policy, the endpoint will allow connections with all root CAs in the endpoint's trusted root CA store.

Figure 3.18 shows Microsoft AD Group Policy settings for server certificate validation in PEAP. You can specify a specific server or one or more trusted root CAs.

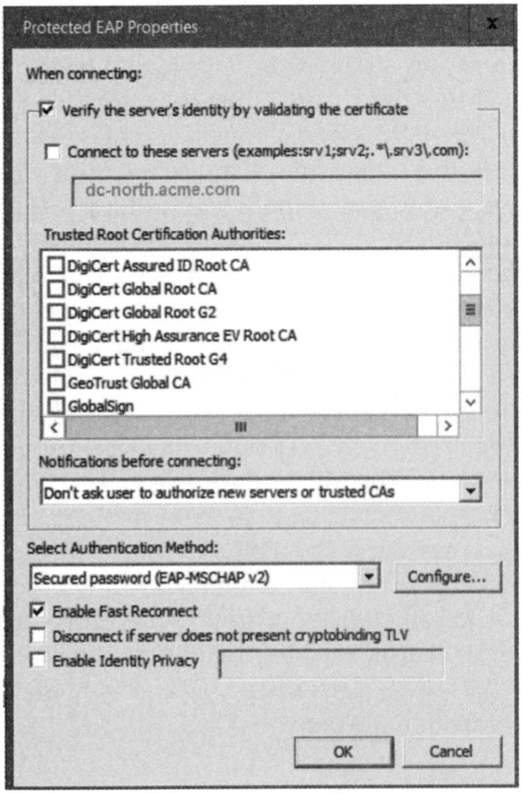

Figure 3.18: Microsoft Active Directory Group Policy options for validating server certificate(s) or trusted root CAs

Use Revocation Lists, Especially for Endpoint Certificates

With PEAP and MSCHAPv2 or other token and password-based authentication, the credentials are current and valid, or not. The RADIUS server is connected to the various directories to provide an immediate accept or reject. However, with device certificates on endpoints, such as with EAP-TLS, the certificate is valid for the duration it was issued and is good unless otherwise revoked. For that reason, it's critical to enforce validation and use a method to check for certificate validation/revocation so the authentication servers know not to allow that connection.

The likelihood of having an issue and revocation of an internally issued server certificate is generally a lower risk, but managing certificates securely requires managing the entire life cycle including revocation.

The endpoint with a device certificate may need to have the certificate removed and/or revoked for a number of reasons: the endpoint may be deprovisioned; the employee may be separating from the company; the user's role and access rights may be changing; or there may have been an exposure by the issuing CA that invalidates the security of the endpoint certificate.

If using endpoint device certificates, follow your product's guidance for certificate revocation. Typically, the RADIUS server will use certificate revocation lists (CRLs), which can get to be large and bulky, or delta CRLs, which use a "diff" process to only transfer the differences in the CRLs at certain intervals. The PKI architecture may instead use online certificate status protocol (OCSP), which offloads the CRL checking. Note that expired certificates are not considered revoked.

INDUSTRY INSIGHT: THE CONTRACTION OF CERTIFICATE VALIDITY PERIODS

As if dealing with certificates wasn't complicated enough, there have been recent changes in certificate validity that can further confound the process of planning and tracking certificates. First, the industry forced us to use FQDNs and stopped supporting IP address–based subject names. Then industry-wide changes to SSL/TLS constricted the validity period down to a max of two years for publicly issued certificates.

More recently, though, some manufacturers (currently including Google and Apple) unilaterally decided their platforms will only trust SSL/TLS certificates that are valid for 398 days or less (roughly 1 year plus a 10 percent grace period). That new policy has not been adopted industry-wide but certainly impacts a large population of users and devices. That restriction applies to certificates created on or after September 1, 2020.

Currently this 398-day policy is a browser-based restriction only, which won't impact domain certificates such as those issued for 802.1X-based authentication, however it will impact captive portals and similar restrictions may creep into other areas of certificate usage. The industry-wide two year max is imposed on any publicly issued certificate, regardless of application.

That concludes everything you ever (or never) wanted to know about certificates used in 802.1X authentication, including the certificates for both the RADIUS server (always required) and device certificates (required if endpoints are authenticating with certificates). Next up we'll look at certificates used for guest portals, why they're different, and some best practices.

Captive Portal Server Certificates

I (and probably the rest of the Wi-Fi architects out there) have a love-hate relationship with captive portals. Let's face it—they're finicky, never configured the same two ways on any infrastructure products, and never handled in a consistent way by the various endpoint OS versions. Today an Apple iPhone may do "this" (whatever "this" is) and tomorrow it may do "that," which is always decidedly different than "this." The same goes for any operating system, although I'll say traditional full operating systems such as Windows tend to be a bit more consistent. Captive portals' moody temperament is one of the (many) reasons onboarding processes for BYOD can be so frustrating.

I imagine you're thinking to yourself, "How hard can it possibly be to get an endpoint to pop up a browser page?" Well, as it turns out, the answer is "very hard!" There's a delicate dance of DHCP, DNS, redirects, access control lists, and certificate validation—all of which is handled a bit differently from platform to platform and Wi-Fi product to Wi-Fi product. It can become a lot more confusing in the presence of a NAC-type product that is handling both 802.1X requests as well as hosting the captive portal pages.

But alas—you'll probably have to configure a captive portal at some point and if you want it to work properly you'll need certificates (well, at least one).

Take everything I've said about 802.1X certificates and tuck them away in a little box for later. Almost none of that logic applies here, and the basic recommendations are different.

While 802.1X-secured networks are most often authenticating managed users on managed devices—captive portals are most often used for guest registration, presenting guests with an acceptable use policy, BYOD device registration (for unmanaged endpoints), and other personal device registration for things like gaming systems (very common not only in universities residential halls but also in healthcare and other facilities with housing or long-term care).

Captive portals may be hosted directly on the Wi-Fi infrastructure (on controllers or APs) or may use an external captive portal (such as redirecting to NAC products like Cisco ISE, Aruba ClearPass Policy Manager, or Fortinet FortiNAC, to name a few).

Best Practices for Using Certificates for Captive Portals

Although captive portals are most often associated with (served up by) Open System Authentication networks, it's perfectly possible to force a captive portal with a Personal/passphrase-secured or Enterprise/802.1X-secured network. Regardless of the specific use case, the guidance for captive portal certificates is consistent.

In Most Cases, Use a Public Root CA Signed Server Certificate

For any device connecting to the captive portal, if it's not a corporate managed device (either a member of the domain or managed by a corporate MDM), then you'll want to use a publicly signed certificate. This is the exact opposite recommendation for most 802.1X-secured networks.

Those guest or unmanaged devices won't have your internal domain root CA certificate downloaded and trusted. They will, however, have all the common public root CAs already installed and trusted. As a sidebar, this can be resolved by enrolling an endpoint (smartphone, tablet, or laptop) in an MDM tool and pushing the server certificate. If you're doing that and have a way to directly manage the endpoint, then this section doesn't apply. There will be more on MDM later.

Understand the Impact of MAC Randomization on Captive Portals

I'd be remiss if I didn't also call out the impact of MAC address randomization on captive portal experiences. I'm not going to debate the validity of using MAC addresses for network identification or usefulness of MAC randomization for privacy here, but know that most captive portal backends rely on the MAC address and MAC caching to identify the endpoint and make decisions about the workflow (e.g., whether to present the portal page or understand the user already did that and move to the next step).

MAC caching simply attaches a successful portal completion to the MAC address of the endpoint, and during that or subsequent portal experiences, the infrastructure identifies the endpoint as having already completed the process. If the MAC address changes at any point, though, the infrastructure will think it's a new endpoint joining, and restart the process.

The portal with MAC caching experience can be broken depending on the endpoint's settings. The MAC randomization is configurable, and the user (or IT team for managed endpoints) may select to use it, not use it, or select how often the MAC address changes. Going any deeper than that won't be fruitful since those options and default settings change regularly on all the platforms. MAC randomization may appear as "Hardware Address" or "Private Address" in settings for that specific Wi-Fi network.

With a traditional captive portal system if my Windows laptop has been set to use a new MAC address daily, and I've registered through a system as a guest (such as in a hotel) or a BYOD portal with an expected connection duration of more than a day, as a user, I may be frustrated that I keep getting forced to a portal when I think I'm registering for several days (as in a hotel) or a year (as with many BYOD policies). Figure 3.19, Figure 3.20, and Figure 3.21 show current MAC randomization options on Windows, Apple iPhone, and Android, respectively.

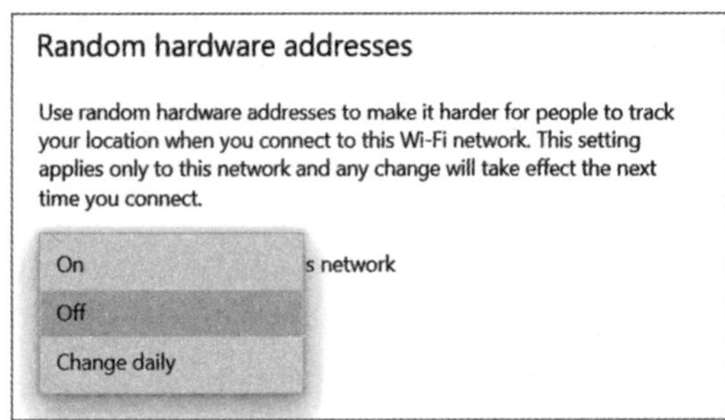

Figure 3.19: A Windows 10 device, showing MAC randomization options

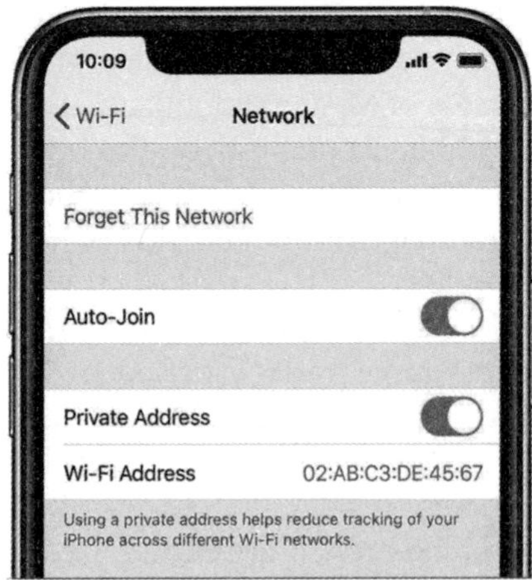

Figure 3.20: An Apple iPhone private address (MAC randomization) settings
https://support.apple.com/en-us/HT211227

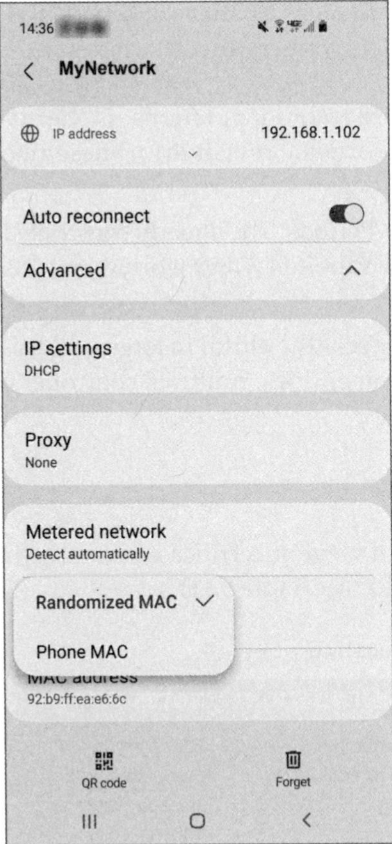

Figure 3.21: An Android device MAC randomization settings

Captive Portal Certificate Best Practices Recap

The best practices for certificates on captive portals can be summarized as follows:

- Use a public root CA signed server certificate.
- Know that for most use cases, the endpoint will need to trust the server certificate without pre-installing it through other means.
- Know that there may be a need for multiple server certificates even with one authentication server or cluster. For example, when using an external portal (such as with a NAC product) as well as certificates on the Wi-Fi infrastructure that's redirecting to the external portal.
- Remember that MAC address randomization may impact user experience and backend operations.

- Remember that captive portal operation involves a battery of behind-the-scenes network operations such as DHCP, DNS, redirects, access rules, and critical timers/timeout settings.
  Plus, if you skipped the 802.1X certificate recommendations in "Certificates for Authentication and Captive Portals," revisit it for these two relevant notes:
  - The Contraction of Certificate Validity Periods. The already constricted validity is two years max, and some endpoints won't work with certs valid for more than a year.
  - Using CSRs and SAN certs. This is especially helpful in larger organizations with a cluster of portal hosting entities, multiple controllers, APs, or other authentication servers.

Summary

Just as a quick run-down, here's the high-level view of certificate recommendations for 802.1X and portal-secured networks (see Table 3.11).

Table 3.11: Summary of server certificate recommendations by use case

USE CASE	SSID SECURITY	RECOMMENDATION	NOTES
Internal domain users on managed domain device	Enterprise/802.1X	Use an internal CA-issued certificate	The domain computers will already trust the internal root CA
Internal domain users on managed non-domain device	Enterprise/802.1X	Use an internal CA-issued certificate and push the certificate via an MDM tool	Personal devices, BYOD, or corporate devices that are not a member of the domain (e.g., smartphones) will need to have the internal certificate installed
External federated users (e.g., Eduroam)	Enterprise/802.1X	Use a public CA-issued certificate	Transient and federated users won't have inherent trust of the domain root CA
Guest portal	Open or Enhanced Open	Use a public CA-issued certificate	For a smooth experience, the portal should use a certificate the guest user's device will already trust

USE CASE	SSID SECURITY	RECOMMENDATION	NOTES
Unmanaged BYOD registration portal	Open or Enhanced Open	Use a public CA-issued certificate	For the best onboarding, the portal should use a certificate the device will already trust

Captive Portal Security

Captive portals are used for many purposes, such as:

- User or guest registration
- Acknowledgment of acceptable use policies
- Endpoint registration for BYOD
- Presentation of payment gateways for paid access

With each use case comes different security and architecture needs. While it's possible to present a captive portal to a user connecting to Personal/passphrase or Enterprise/802.1X networks, they're most often used in combination with Open System Authentication SSIDs, which is the focus here.

Captive Portals for User or Guest Registration

The most common use case, captive portals are often the mechanism for allowing guest users access to (ideally) an Internet-only network. The term "guest" here covers a variety of use cases though, as there are undoubtedly fifty shades of guest, from pure guest visitors to contractors to college students and more.

Guest registration workflows are typically configured in one of a few ways, depending on the security requirements of the organization.

Guest Self-Registration Without Verification

Captive portals of this type allow guests to register themselves without any verification that the information is real. This is sufficient for many organizations but lacks details for audit and attribution in the event of an incident.

Guest Self-Registration with Verification

These captive portals allow guests to register themselves but will send credentials (password, link, or logon code) via email or SMS text to verify at least one identifiable piece of information.

The benefit here for security-conscious organizations is that they have at least an email address or phone number that is real and traceable to the user. Although it may not be easy to trace, requiring an email or phone number is typically enough to prevent people from signing up as Mickey Mouse or their favorite Marvel superhero. And while they could create a burner Gmail account, that's a lot of trouble to go through for guest access.

Guest Sponsored Registration

Sponsored workflows on captive portals allow a guest to request access, which is then approved by one or more allowed sponsors through an automated workflow (most often email-based).

This is the most secure of the registration options; the guest user is making a request in real time, and the sponsor is approving it in real time when the access is needed. Most of today's enterprise Wi-Fi solutions offer sponsored workflows either natively in the product or via add-on software such as NAC solutions.

Guest Pre-Approved Registration

There are also options to pre-provision with captive portals that allow sponsors within the organization to pre-approve a guest visitor. Credentials are sent to the visitor ahead of time, usually by email or SMS text.

Similar to the sponsored registration workflow, the pre-provisioning of access is usually specified for a user by email address, and for a specific date and duration (e.g., for Tuesday, April 5 from 9am to 5pm). Compared to on-demand sponsorship, this model has two drawbacks. The account may be provisioned for a longer duration than required, or the visit may be cancelled or rescheduled, opening an opportunity for abuse if the credentials are captured. Secondly, and not really security related—it's common for there to be time zone mismatches or other quirks in the provisioning system that may require reprovisioning at the time of visit. Obviously this negates any added benefits of pre-approving accounts.

Guest Bulk Registration

Captive portals can also allow the organization to create bulk guest user accounts for events where issuing one-to-one credentials may not be desirable. Most often guest bulk registration uses the pre-approved registration applied in bulk. In many cases, there are unique credentials for each user that can be automated through API or email integrations.

Personally, I see this less often and feel bulk registrations are usually a corner case. There are times this model makes sense but there are often simpler and easier options such as allowing guests to self-register during an event or offering an event-specific SSID.

Captive Portals for Acceptable Use Policies

For a bit lighter touch, captive portals can also be used simply to present and ask a user to acknowledge the organization's acceptable use terms. Acceptable use policies commonly address how the user is expected to behave on the network and absolves the organization from any liability.

When a captive portal is used for this purpose, it can be combined with any of the registration methods just discussed, or presented for the user to click an accept button, or simply present the page and tell the user by continuing they are agreeing to the terms.

Activities and topics to consider when creating an acceptable use policy include:

- Illegal file sharing, including sharing or access of data that violates copyright laws
- Illegal content, including content that violates the Child Protection Act
- Copyright Infringement and the Digital Millennium Copyright Act
- Digital reconnaissance and attacks such as IP port scanning, DoS, and distribution of malware
- Sending of unsolicited email (spam)
- Any activity that may result in legal action or denylisting of the organization's IP space

In addition, the acceptable use policy should also address expectations of access or availability of services:

- The organization does not guarantee services (user may not be able to make or maintain a Wi-Fi connection and the organization does not guarantee availability or quality of the service)
- Disclaimer that data transmitted may not be encrypted or otherwise secured
- The organization is not responsible for any damages to the user's endpoint or any data that may result from use of the network

INDUSTRY INSIGHT: TO PORTAL, OR NOT TO PORTAL. . .

Whether or not a captive portal is required is always a hotly contested topic among IT and infosec professionals. Many Wi-Fi pros consider it an unnecessary nuisance, while others deem it mandatory to protect the organization from legal liability. In most organizations, I recommend at least a presentation of acceptable use policies (AUPs) to the user, but there are cases where that's not advised.

Healthcare is a great example of one such situation where it makes sense to bypass any portal experience including AUPs. The Hospital Consumer Assessment of Health Providers and Systems (HCAHPS) surveys for patient satisfaction impact not only a hospital's reputation (think Yelp reviews for hospitals), but scores are also tied to millions of dollars in reimbursements. After wait times, quality Wi-Fi was the second most significant factor for millennials' evaluation of a clinic (source `https://www.hfmmagazine.com/articles/2893-millennial-patients-want-less-waiting-more-wi-fi`). For that reason, healthcare will almost exclusively use open networks sans the often frustrating captive portal experience for patients and visitors.

For other organizations, it's a mixed bag. Recently I asked a few CISOs, lawyers, and a friend in the FBI's cybercrimes division for thoughts on whether captive portals and acceptable use terms are required.

Our U.S.-based CISOs mostly felt it was window dressing, with the exception of guidance from lawyers who unanimously felt the organization benefits from at least some legal protection, not only for outbound traffic but also against complaints from users that may have been infected with malware while using the guest network.

The FBI cybercrimes teams note they need "something" legal to open a case and prosecute, such as acceptable use clauses or allowed access banners on network resources.

CISOs in Europe have stricter requirements; organizations providing public Internet access fall under Internet service provider (ISP) laws and therefore must be more diligent about guest registration and acknowledgment of acceptable use terms.

Captive Portals for BYOD

Captive portals are also used for endpoint registration (as opposed to user registration) and onboarding for BYOD.

In these scenarios, the portal pages and workflows may look similar to the guest registration options covered earlier with the notable exception that endpoint registration and BYOD are most often tied to a managed user. In corporate environments, that may mean Ed Employee is registering his personal cellphone under the company BYOD policy, and in a university scenario Sally Student may register her PlayStation and printer by MAC address with the college's portal system.

In each of these cases, Ed and Sally will provide their user credentials for authentication and then attach their endpoints to their corporate account.

This feature is most often found in the various NAC products, but some Wi-Fi products may have limited capabilities natively integrated, most often through cloud services such as Microsoft Azure or Google.

Captive Portals for Payment Gateways

A piggy-back of the guest registration suite, captive portals may also be used for payment-based gated Internet access.

Certain hotspots such as those in hospitality and airplanes tend to use captive portals for this purpose. By virtue of collecting payment, collection of certain personal information is necessitated, which is why these captive portal types are an extension of the guest registration model.

Although not common in most enterprise environments, many enterprise Wi-Fi products offer native support for payment processing for access.

Security on Open vs. Enhanced Open Networks

As you've already learned, the industry has at its disposal now two flavors of open networks: legacy Open System Authentication (Open) networks circa WPA2 era, and the newer Enhanced Open profile, which offers encryption.

- Legacy Open System Authentication (Open) offers no encryption over the air.
- New Enhanced Open offers unauthenticated encryption over the air, but no authentication.

For more, revisit in Chapter 2.

Access Control for Captive Portal Processes

Captive portals, because of their intended limited access for users, do require additional security planning. Configuring access control policies for captive portals varies by product and implementation.

At a minimum, captive portals intended for Internet-only use should have strict policies and segmentation to prevent guest users from accessing the production resources on the network.

The access should be tested at regular intervals to ensure proper operation of the controls and proper segmentation.

In Chapter 4, "Understanding Domain and Wi-Fi Design Impacts," you'll find additional information on captive portals' relationships to DNS and DHCP, which includes recommendations for planning secure captive portal networks.

LDAP Authentication for Wi-Fi

In almost every case, enterprise Wi-Fi will authenticate against a user directory store via RADIUS. However, there are some small business products that can authenticate directly against an LDAP directory such as Microsoft Active Directory, without using an external RADIUS server. Here are two of those models:

Wi-Fi with Built-in RADIUS Services for 802.1X Products that allow the configuration of 802.1X-secured networks directly to an LDAP are simply hosting a basic RADIUS server within their Wi-Fi product, since RADIUS and EAP are how 802.1X operates. In this case, the Wi-Fi infrastructure is simply acting as the RADIUS server.

Wi-Fi with Captive Portal to Non-802.1X LDAP Authentication Products that offer different (non-802.1X) connections to LDAP often present the user with a captive portal experience hosted on the product that then has an LDAP (versus RADIUS) connection for authentication. Since this is not 802.1X, the authentication can be LDAP instead of EAP/RADIUS.

Enterprise-grade products don't support this feature, so while it's worth mentioning, that's as deep as we'll go.

The 4-Way Handshake in Wi-Fi

As we wrap up Chapter 3 with this topic, I imagine every Wi-Fi professional is asking "Why the devil didn't you cover the 4-way handshake first? It's so important in Wi-Fi!" Yes, that's true, and I agree, but not everyone reading this is a Wi-Fi pro, and I don't want readers to get bogged down or confused when attention should be focused on 802.1X and RADIUS. And, hey—the 4-way handshake comes after the 802.1X authentication so this helps construct a perfect mental timeline of the events.

Having said that, I've made a point in this book not to delve too deeply into packet-level operations of Wi-Fi, but it's important to understand this very basic operation of the 4-way handshake that occurs between the endpoint and the AP. This exchange is important in security context for troubleshooting and also because of its role in how the various encryption keys are generated and distributed.

The 4-Way Handshake Operation

The 4-way handshake occurs in both passphrase-secured and Enterprise/802.1X-secured networks.

During the 4-way handshake the endpoint and the AP each start with a known value that's not transmitted over the air, and they then exchange four messages with input values used to derive the key sets used for data encryption (see Figure 3.22).

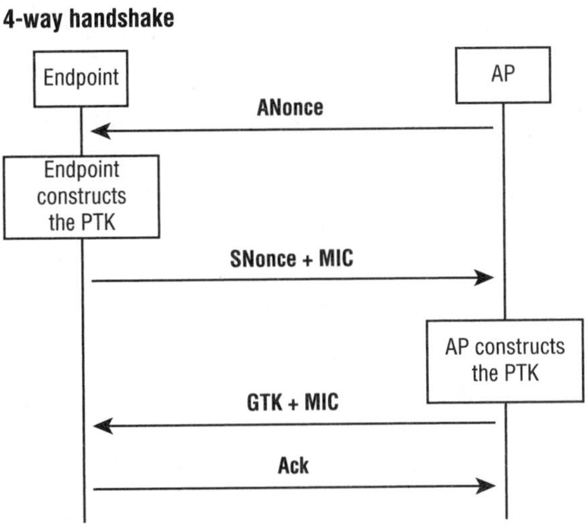

4-way handshake

Endpoint → AP sequence:

- AP → Endpoint: ANonce
- Endpoint constructs the PTK
- Endpoint → AP: SNonce + MIC
- AP constructs the PTK
- AP → Endpoint: GTK + MIC
- Endpoint → AP: Ack

Figure 3.22: A view of the 4-way handshake between an endpoint (defined as a station or STA in 802.11) and the AP

The 4-way handshake also leverages the EAPoL (EAP over LAN) protocol for the exchanges—but remember this applies to both passphrase-secured and 802.1X-secured networks. Usually seeing "EAP" or "EAPoL" is a trigger that we're talking about 802.1X authentication specifically.

Here's the 4-way handshake process, step by step:

1. The endpoint will have completed association to the AP, and 802.1X authentication (if applicable). Both the AP and endpoint have a known value for a pairwise master key (PMK) from that process.
2. In message 1, the AP sends the endpoint a value called ANonce (Authenticator Nonce, simply a pseudorandom generated value that's used once in cryptographic functions). The endpoint then has the inputs it needs to calculate the pairwise transient key (PTK) that will be used in unicast encryption with the AP.
3. In message 2, the endpoint reciprocates by sending an SNonce (the endpoint's or supplicant's pseudorandom generated value) plus a message integrity check (MIC) value for the AP to validate integrity. The AP will then have all the values needed to derive the initial keys including the

pairwise transient key (PTK) used for unicast encryption, as well as the group transient key (GTK) used for broadcast encryption. At this stage both the endpoint and AP have derived the PTK.

4. In message 3, the AP will provide the endpoint with the group transient key (GTK) along with a message integrity check (MIC) parameter.
5. Lastly, in message 4, the endpoint simply responds to the AP with an acknowledgment.

This is a bit of a simplification since there are sets of keys derived from these initial exchanges. The pairwise transient key (PTK) is comprised of five different keys, used for different purposes, whereas the group temporal key (GTK) has three unique keys for various broadcast encryption uses.

Without going in to too much detail, here's a quick key cheat-sheet:

- *Pairwise keys* are used for unicast traffic between an AP and single endpoint
- *Group keys* are used for broadcast and multicast traffic between an AP and all endpoints in the same BSSID
- *Master keys* are the top of the hierarchy, immutable, and used for a chain of key derivations
- *Transient keys* are used to derive other keys
- *Temporal keys* are used to encrypt the data

The 4-Way Handshake with WPA2-Personal and WPA3-Personal

With the WPA2-Personal passphrase-based SSIDs the handshake happens right after the process of Open System Authentication and association—which is the ubiquitous method all endpoints use to connect to an AP. In WPA3-Personal SSIDs, the SAE process includes SAE-Authentication commit and confirm exchanges before proceeding to the association request and response. It's during these exchanges the pairwise master key (PMK) is derived.

This Open System Authentication was introduced earlier in Chapter 2, and there you'll also find Figure 2.18 showing these exchanges.

From that same section, you may also recall that there are differences in how WPA2-Personal with pre-shared key (PSK) and WPA3-Personal with newer Simultaneous Authentication of Equals (SAE) derive the keys. In WPA3-Personal with SAE the encryption key derivation is not a function of the length of the passphrase, whereas with WPA2-Personal with PSK, the data encryption keys are directly derived from the passphrase. This is one of the reasons WPA3-Personal is much more secure than its legacy counterpart.

The 4-Way Handshake with WPA2-Enterprise and WPA3-Enterprise

With 802.1X-secured networks (WPA2-Enterprise or WPA3-Enterprise) the handshake happens after the endpoint completes a successful EAP authentication (which itself occurs after the Open System Authentication).

Because the 4-way handshake is such a basic and pervasive operation in Wi-Fi, understanding its operation is useful knowledge for advanced troubleshooting outside of the network-based authentications. Figure 3.23 shows an abstracted overview of the 4-way handshake after a successful 802.1X authentication. The authentication occurs on the 802.1X Uncontrolled Port, and once the endpoint has authenticated, the 4-way handshake can proceed on the now opened Controlled Port. Revisit "Terminology in 802.1X" at the beginning of this chapter for more on the 802.1X port entities.

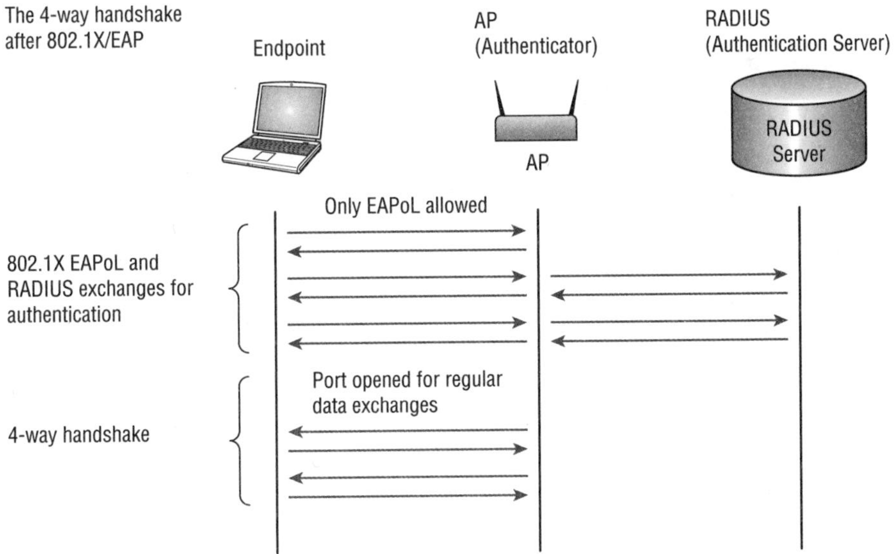

Figure 3.23: The 4-way handshake after 802.1X authentication

Summary

This chapter took a much deeper dive into the world of 802.1X authentications, but with the scope focused on wireless and most specifically 802.11 WLAN networks. Aside from 802.1X, authentication and authorization for captive portals was also covered, along with the ever pervasive 802.11 4-way handshake. This information should set you on a solid path for evaluating and implementing authentication in Wi-Fi networks. You'll see many of these themes repeated in the sample architecture guidance found in Appendix C.

A professional friend recently said they'd love to have a book on RADIUS and EAP for Wi-Fi, to which I replied, "It's really not complicated enough to write a book on." In hindsight, that was probably a poor evaluation of the situation on my part. In this chapter, I've done my best to distill down the essential concepts of 802.1X, RADIUS, and EAP for network authentication, but there's a lot more knowledge out there.

More detailed walk throughs with specific examples using Microsoft NPS are provided in Appendix A for those of you interested in testing out some 802.1X policies. There's also detailed troubleshooting guidance in Chapter 7, which will further solidify how and when to use the authentication and authorization schema you've seen here.

CHAPTER

4

Understanding Domain and Wi-Fi Design Impacts

Any system is best when designed with the entire ecosystem in mind, and security architecture for wireless is no exception. Proper design balances business objectives with security for a holistic approach.

As such, security can't be planned in a vacuum. The wireless infrastructure is one piece within the greater enterprise environments of networked systems and security programs. This chapter highlights just a few of the many related domain services and non-security Wi-Fi design elements that impact or play a crucial role in security architecture and covers the following:

- Understanding Network Services for Wi-Fi
- Understanding Wi-Fi Design Impacts on Security

Understanding Network Services for Wi-Fi

IP-based wireless technologies rely on a multitude of fundamental networking and domain services including Domain Name System (DNS), Dynamic Host Configuration Protocol (DHCP), and time services to name a few. Yeah, I know—you think all of that's boring. Well, it may be, but it's also now remarkably complex and something that needs to be addressed.

It's probably a safe generalization to say those of us that began working with Wi-Fi early on (for me it was mid-to-late-'90s) all had a relatively accomplished level of networking background as a foundation. Learning Wi-Fi (at that time) only required a bit of additional knowledge around RF as a different layer 1 medium, and the added layer 2 protocols to facilitate passing IP data over the air.

In more recent years, Wi-Fi technology (and certainly Wi-Fi products) have become markedly more complex, driving a new generation of technologists with a strong focus on Wi-Fi who excel at wireless but lack expertise with many of these (often more foundational) networking operations.

From my perspective, the increase in Wi-Fi complexity makes it even more compelling to understand the underlying networking services. With the numerous types of Wi-Fi management architectures (such as cloud, controller, remote, and gateway) plus the vendor-specific sprinkles on top, having a mastery of the system holistically offers us the best chance for success.

Unless there's a product bug or an over-the-air (RF) issue, chances are great that issues with Wi-Fi can be traced back to one of these boring ol' services. In addition, proper security architecture for Wi-Fi depends heavily on managing and documenting the tedium of DNS, DHCP, and similar services.

This book assumes basic networking knowledge of switching, routing, and domain functions; as such, the following topics don't include introductory information. The topics here focus on:

- Time Sync Services
- DNS Services
- DHCP Services
- Certificates

Time Sync Services

The first thing I do with any new network device or application is set the proper time and time sync services. Aside from soothing my borderline obsessive-compulsive tendencies, it serves three critical functions for security purposes:

- Accurate time helps in troubleshooting.
- Accurate time is needed for event correlation for security monitoring and during response.
- Accurate time is requisite for proper operations of anything based on certificates.

These days, many Wi-Fi functions rely on certificates—especially for cloud-managed architectures, and the burst of connected devices including Wi-Fi and other IoT devices means securing monitoring is more critical now than ever.

Time Sync Services and Servers

It doesn't matter what time sync service you use; whatever the device supports—use it. Here are the two common time sync services for networking:

Network Time Protocol (NTP) A time synchronization protocol that keeps devices synchronized within milliseconds of one another.

Simple Network Time Protocol (SNTP) A simplified time protocol that is intended for servers and clients that do not require the degree of accuracy that NTP provides.

As for the time synchronization servers, they can be internal within the organization, and/or external on the Internet. It's advised to set infrastructure devices to use internal domain-hosted time sync within the organization, which in turn syncs with an external time server from the ISP or a manually configured trusted source such as the pools from `ntp.org`. Individual devices within the environment should only use external time servers directly in a jam.

Using an internal time server ensures all the devices are in tune, so to speak, and not searching for time updates from different Internet services. Internal servers reduce the likelihood of a free public time service becoming unavailable or moving. Lastly, certain types of devices demand time sync servers be input as IP addresses or hostnames, and IP addresses are not always an option with public servers.

Microsoft Server infrastructure supports both NTP and SNTP, which can be easily enabled and configured on any server. There are also free SNTP applications you can download and install from online, but always be leery of free online tools that aren't from a known, reputable source.

TIP **If devices aren't regularly synchronized with time services, *clock drift* is an eventual and undesirable outcome. Clock drift happens when the time on one device isn't incrementing at exactly the same rate as the reference clock, and/or the other devices in the environment. It's expected, but if clock drift isn't mitigated and corrected with time sync services, the devices or services will diverge and eventually cause issues.**

Time Sync Uses in Wi-Fi

Aside from the obvious benefit of having accurate timestamps for troubleshooting, time synchronization services play a critical role in many Wi-Fi deployments. Security event correlation engines rely on synchronized and accurate timestamps for event monitoring and management. For security operations center (SOC) analysts, logs and alerts without the correct time are borderline useless. Whether the SOC monitoring workflows rely on human review, or automated engines, accurate time is king.

In addition, certificates of all kinds (for SSL, 802.1X authentication, etc.) rely heavily on time services. Without correct time, certificates and certificate services are rendered useless.

Use your imagination a little for this one. Pretend there's an endpoint device that doesn't have date and time configured out of the box, and it's connecting to a network with 802.1X. Of course, the endpoint is authenticating the server based on the server certificate, which will (for example) have been issued January 1, 2022, for a period of validity of 1 year, expiring December 31, 2022. The endpoint (still unconfigured) thinks it's 2018. When it sees the server certificate date, it will reject the certificate for not being valid, because the endpoint thinks it's 2018.

That was a silly example, but many services and especially cloud-based microservices rely on much more narrow scopes of time, and validity of certificates and other credentials such as Application Programming Interface (API) tokens may have life spans of days or even hours, versus a year or more. Meaning, if the time is off by even a little bit, some functions may fail. As depicted in Figure 4.1, time differences of even a few milliseconds can throw off certain operations in today's microservice-based architectures.

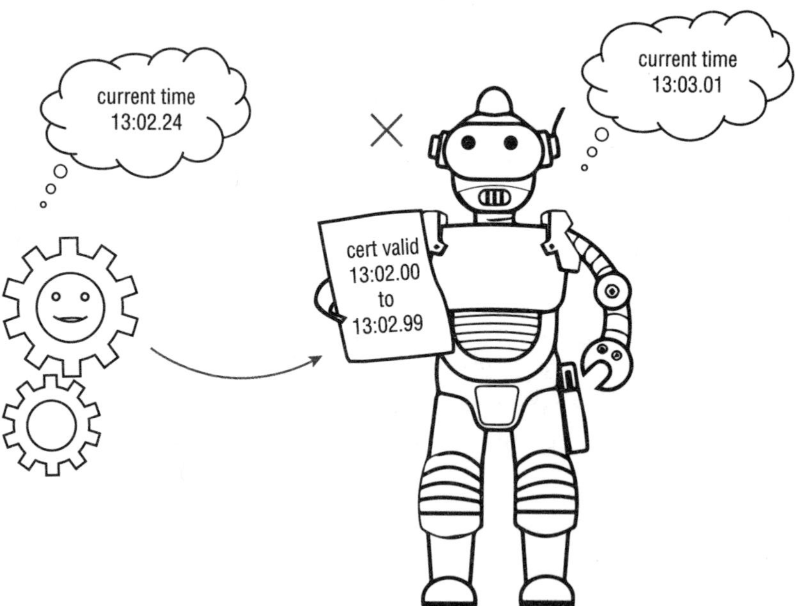

Figure 4.1: Machines and microservices use certificates, signatures, and tokens with short validity periods to authenticate to one another.

As use of APIs and certificates expands in network operations, time services will become increasingly critical. In addition to traditional certificate services, most vendors are moving to fully certificate-based authentication for infrastructure components, not only in the cloud, but also for controller architectures.

In Wi-Fi, time sync issues and clock drift can impact everything from user authentication through AD to Kerberos tokens, captive portal experiences, and proper function of 802.11 target wake time functions.

In other wireless technologies such as private cellular, clock drift is exceptionally impactful to even the most basic data transmissions because of its use of scheduled transmission coordination (versus Wi-Fi's contention-based transmission).

INDUSTRY INSIGHT: NETWORK TIME SECURITY DRAFT RFC

Time sync services, like so many of our fundamental networking services, were not designed with security in mind. All of these services were designed to be operated within trusted networks at a time when zero trust and insider threats weren't yet of concern. The IETF has a new RFC 8915 for Network Time Security for NTP, found at `https://datatracker.ietf.org/doc/rfc8915/`. The project adds TLS encryption and authentication to NTP.

You can also read a bit more behind-the-scenes info at this blog by Cloudflare, which recently announced a free Internet time service based on the new Network Time Security (NTS) protocol: `https://blog.cloudflare.com/secure-time/`.

DNS Services

DNS is one of those things that just "works" on a network—well, in most cases. Serving as a phonebook of sorts, DNS resolves hostnames to IP addresses, which may be static or dynamic. Public DNS services maintain records of publicly available domain names reachable from the Internet, whereas private DNS services are internal to an organization's domains.

There used to be an easy delineation between public and private DNS services, but like many other technologies, cloud and software-as-a-service (SaaS) models have really blurred those lines. SaaS offerings such as Microsoft 365 and network vendors' various hosted platforms (Cisco Umbrella, Meraki, ExtremeCloud IQ, Juniper Mist, Aruba Central, and Aruba Activate) create an environment that necessitates a delicate orchestration of both internal and external DNS.

DNS services for Wi-Fi architecture can be grouped into two main categories—DNS for Wi-Fi clients and DNS for AP provisioning. Regardless of the implementation, there are some security considerations around DNS covered here also.

DNS for Wi-Fi Clients and Captive Portals

DNS is, of course, used by all endpoints to resolve domain names, whether public or private. The entire Internet runs on DNS and hiccups in those services cause devastating and often widespread outages.

When the endpoint requests, and is given, an IP address via DHCP, it's accompanied by DNS server entries, among other things. Most often, when clients inside the network are offered up DNS entries, they'll be using internal (private) DNS servers. However, there are cases where public DNS servers may be assigned for use with Internet-only clients such as guest users.

Because DNS is one of the links in the kill chain for malware, and also because DNS is itself not secure and susceptible to manipulation, there are security considerations attached to the DNS services we use for various classifications of networks and users—a topic covered in "DNS Security" coming up.

The other important application of DNS with Wi-Fi clients is the use of DNS hijacking and redirects as a means to enforce the captive portal experience. In this case, the hijacking isn't malicious—the Wi-Fi infrastructure will insert itself in the process and, when an endpoint makes an Internet request, regardless of what website it asked for, the Wi-Fi infrastructure will redirect the endpoint to the specified captive portal.

This mechanism is accompanied by HTTPS redirects and sometimes TCP interception by the wireless infrastructure (if the endpoint is allowed to make a successful initial DNS query and initiate a connection to the originally requested website). Different products and implementations vary slightly in exactly what they do, when, and how, so consult your vendor documentation for configuration and troubleshooting guidance.

TIP **HTTPS and DNS redirects and TCP interceptions are basically man-in-the-middle attacks; they're just being used for non-nefarious purposes here.**

DNS redirects for captive portal have been prominent, especially in traditional retail hotspots and hospitality environments. With the advent of DNS security products and services, some vendors are pushing toward a shift to HTTPS versus DNS redirect.

Reminders and requirements for DNS with Wi-Fi clients:

- Endpoints that need to resolve internal resources should use an internal DNS server.
- Endpoints that only need to resolve Internet-based resources can use a public DNS server; however, this is not advised for managed/corporate endpoints due to lack of security and control over public DNS.
- Captive portals use a combination of HTTP redirect and DNS redirect.
- The captive portal server certificate must match the portal's FQDN(s).
- The client device must know and trust the portal server certificate, or the user will be prompted with a warning.
- Most portal implementations will require a DNS entry for the portal server using that same FQDN.

- For captive portals to work, the client device must be able to resolve DNS.
- It's common for clients to handle captive portal redirects inconsistently.
- In addition to normal inconsistencies, endpoints using DNS security services may experience issues with DNS redirect-based portals.

DNS for AP Provisioning

Wi-Fi management architectures that are controller or cloud managed will (or can) rely on DNS during the AP provision process, specifically for the APs to phone home and find the appropriate management platform.

In internal deployments with controllers, DNS is usually one of several ways APs can be quickly configured to find their controller(s). For most vendors, it's as simple as adding their specified DNS entry to the domain. For example, in ArubaOS the APs come preconfigured to come online and resolve the hostname "aruba-master" so an internal DNS entry of `aruba-master.acme.com` is what would be needed. Other vendors work similarly.

For cloud-managed APs, they, too, will use DNS to phone home and contact their cloud management. The difference is that the cloud-managed AP will need to route and tunnel through the Internet to connect to the service, and therefore the DNS entry is something the vendor has configured within the public DNS system. As an example, Juniper Mist APs will resolve the DNS entry `ep-terminator.mistsys.net` to locate the cloud platform. As mentioned, that DNS entry is in public DNS; you won't need to configure internal DNS for that entry, but the internal DNS will simply resolve queries for external domains through its next hop.

DNS Security

Because everything relies on DNS to resolve (and then connect to) resources both inside an organization and on the Internet, it remains a critical—yet very vulnerable—component of network services. You'd think something so vital to networks would be well protected, but unfortunately, DNS (like so many other basic network services) wasn't originally created with security in mind because it was designed for use within trusted networks.

Without an upfit, on its own, DNS is not authenticated, encrypted, or even monitored. And, aside from the usual glitches and hiccups that may occur, DNS is the principal network amenity malicious actors use and abuse—all while hiding in plain sight.

Ransomware, other malware, and phishing attacks rely on DNS at various stages of an attack—reconnaissance, infiltration, and exfiltration:

- DNS cache poisoning allows attackers to redirect users to websites, often to deliver malicious payloads or collect personal information.
- DNS hijacking allows attackers to redirect DNS queries to a different DNS server.
- DNS tunneling allows attackers to exfiltrate data by tunneling SSH, TCP, or HTTP disguised in a legitimate DNS query; the attack can be reversed and also used to deliver malware, both methods likely undetected by any security tools including firewall inspection.
- DNS beaconing allows malware to establish a connection to command-and-control servers using DNS, bypassing most security controls and inspection.

The list of DNS vulnerabilities and attacks goes on, and these aren't theoretical attacks; they're all in the wild.

In fact, it wasn't too long ago that the U.S. Department of Homeland Security (DHS) issued a warning and emergency directive with guidance for securing DNS after a wave of hijacking incidents by foreign cyber espionage actors (`www.zdnet.com/article/dhs-issues-security-alert-about-recent-dns-hijacking-attacks/`).

All of that is really to say that organizations will (or should) be dialing in DNS security in the coming years. With those changes will be new tools, new services, and new processes for configuring and using DNS services, both at the infrastructure level and the client level. How DNS servers are configured, deployed, and accessed will evolve, as will the means by which clients request DNS.

TIP **The suite of DNS security options is evolving quickly, but expect to keep an eye on DNSSEC, DNS over TLS, DNS over HTTPS, added DNS security within infrastructure solutions, and the growth of DNS inspection services (such as Cisco Umbrella and modern zero trust and SASE solutions). As you can see, DNS plays a crucial role in Wi-Fi architecture and security.**

DHCP Services

If you thought DNS was boring, surely you're thinking, "what could possibly be complicated with DHCP?" Yet, DHCP is always easier said than architected, especially in today's world of hybrid and cloud data paths.

Although this book assumes basic networking knowledge and has omitted introductory concepts of such, misunderstanding of DHCP operations seems to be prevalent in many communities. To bridge any gaps and address misconceptions, this section covers more 101-level content than others.

Selecting from where and how devices receive DHCP responses is not only a basic design element for networking, but also a critical component of the security architecture. Most (if not all) of the segmentation technologies covered earlier rely on devices having the correct IP address, default gateway, and DNS. Missteps with DHCP can result in not only unintended asymmetric routing, but also gaps in access controls, especially when ACLs are applied in a way not commensurate with the device's default gateway settings.

DHCP for Wi-Fi Clients

At a glance, there are drastic differences in how and from where DHCP can be served, and the security controls around that decision. Options will vary depending on the management architecture (controller, cloud, cluster) and data path (tunneled or bridged), but in the end, a Wi-Fi client's IP address comes one of a few ways:

- DHCP from a domain DHCP server, proxied through the wired infrastructure
- DHCP from a domain DHCP server, proxied through the Wi-Fi infrastructure
- DHCP from wired infrastructure (scopes on router, routing switch, or firewall)
- DHCP from Wi-Fi infrastructure (scope on controller or AP)

Notice the subtle difference of "from" and "proxied through" as in, proxied by, in the first two bullets.

Figure 4.2 shows a DHCP request path using proxies. On the left, an endpoint on VLAN 12 is connected to a secured network, the client data is bridged to the edge switch, and therefore the wired infrastructure is sending the DHCP request on behalf of the client. On the right, the same process is happening via the Wi-Fi infrastructure.

Figure 4.3 shows the DHCP request paths when the infrastructure devices are serving DHPC (and therefore not proxying requests to an external server).

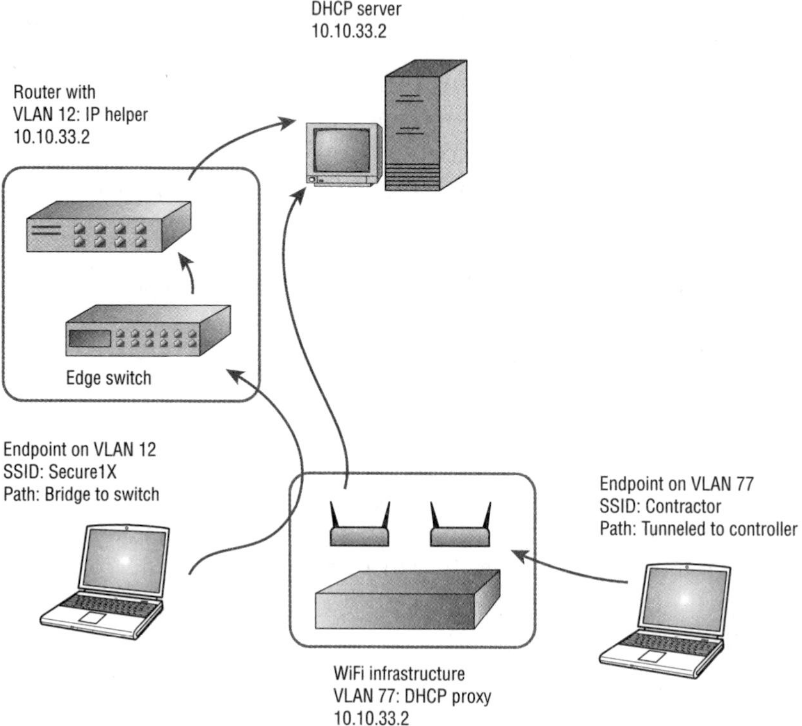

Figure 4.2: DHCP request paths using proxy from wired and wireless infrastructures

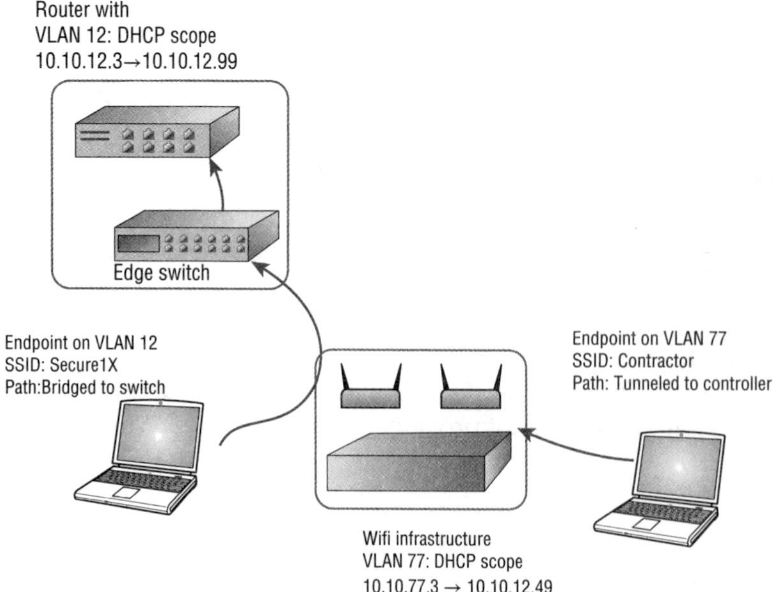

Figure 4.3: Alternate configurations demonstrating ability of certain network devices to serve DHCP to endpoints

Regardless of what exactly is serving the IP addresses, DHCP requests happen in one of two ways—through DHCP proxy (IP helper) or through a broadcast:

DHCP through Broadcast In default operation, when an endpoint joins a network, it will broadcast a request for an IP address, and the configured DHCP server (hopefully only one) will respond. For the broadcast operation to work, the endpoint and the server both have to be in the same broadcast domain (same VLAN and IP network).

For DHCP servers offering up scopes for multiple networks, this architecture would require the server have network interfaces on each network. While I have seen that in smaller environments, it's not ideal. To solve that limitation, we have a second option via DHCP proxy.

DHCP through Proxy With the proxy (or IP helper) configuration, the infrastructure device (routing switch or Wi-Fi controller or AP) is configured manually to send requests for a specific network to an external DHCP server—most often a domain server or IPAM product. This eliminates the need for layer 2 adjacency and the broadcast function.

DHCP proxies (helper addresses) are configured on the network devices serving as the endpoints' IP gateways. In the wired environment, that's the routing switch with an IP in the client's network that is also the IP specified as the client's IP gateway. For a Wi-Fi client, the gateway could be the controller (if in tunnel mode) or a wired device (if in bridged mode). In scores (probably hundreds) of client environments, I've seen IP helper addresses sprinkled all over the place. You can save yourself time and energy with proper planning and documentation.

Figure 4.4 shows a Windows `ipconfig` output and a default gateway of 10.0.0.1. For DHCP proxy from the wired network, an IP helper address pointing to the external DHCP server would need to be added to the routing switch of 10.0.0.1.

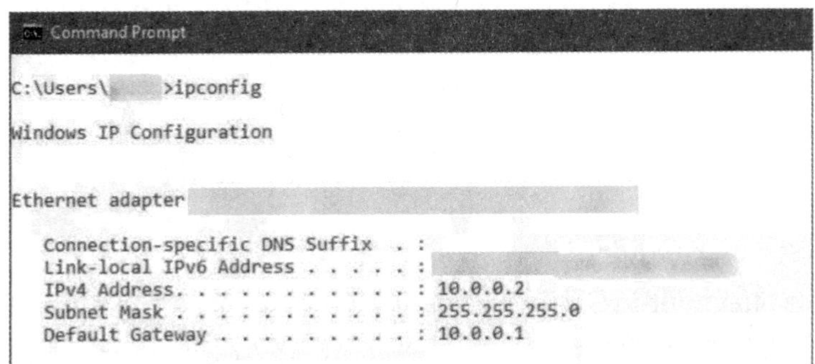

Figure 4.4: Windows ipconfig output showing the endpoint's default gateway of 10.0.0.1

Most enterprise architecture will rely heavily on the DHCP proxy model, and that's where most of the confusion stems. If the controller or AP isn't providing DHCP services (directly) to all clients—and in most environments, they're certainly not—the architecture calls for DHCP proxy, which is at times configured on the wired infrastructure and other times on the Wi-Fi.

To add more confusion, DHCP is specified per-network (per-SSID or per-VLAN), and different networks and data paths necessitate different DHCP configurations. For example, a guest network on VLAN 99 that's tunneled will need to have DHCP served or proxied by the Wi-Fi controller, whereas a bridged client VLAN 12 would require a proxy or helper configuration on a routing switch.

Planning DHCP for Wi-Fi Clients

Here are considerations for planning DHCP depending on the management architecture, network type, and data path.

DHCP considerations based on management architecture

- Cloud-managed APs (unless connected to a tunnel termination gateway) frequently require DHCP proxy to be configured on the wired infrastructure.
- Controllers and gateways can be configured to send DHCP proxy to an external server.
- Controllers can most often serve DHCP directly, without restriction.
- Some products can serve limited DHCP scopes, such as those for guest, and some will use non-existent NAT'd IP addresses for security (to avoid accidental routing to or from the guest network).

DHCP considerations based on network use and users

- Networks for internal employees and production resources most often have DHCP served from a designated server or IP address management (IPAM) system in the domain (such as Microsoft Server running DHCP services, Microsoft IPAM services, Infoblox, BlueCat, and others).
- Networks for guest and Internet-only use may have DHCP served from a domain DHCP server, but also may rely on DHCP from the Wi-Fi or firewall to further facilitate segmentation and avoid having to allow (and monitor) guest traffic to internal servers.
- Some products offer built-in DHCP services with NAT, designed to support guest and Internet-only networks, which is a great option to enhance security but will interfere with some external security tools such as NAC products.

DHCP considerations based on data paths

- Networks with a tunneled client data path most often terminate at a controller, which can be configured to serve DHCP or proxy DHCP, meaning no helper is required on the wired infrastructure.
- Networks with a bridged client data path most often rely on the wired infrastructure for DHCP proxy.
- Networks servicing users on remote APs operate the same as controller or gateway model with a tunneled client data path, meaning the Wi-Fi infrastructure manages DHCP or DHCP proxy.
- Remember the DHCP proxy (helper) should be configured wherever the client's IP default gateway is.

SIZING IT UP: WHAT SHOULD SERVE DHCP

It's not uncommon for small organizations to use part of the network infrastructure to serve DHCP—routing switches, firewalls, etc., instead of a server. I'm not fundamentally opposed to this especially for remote branch offices and very small businesses, but if the organization has a domain server onsite, then it's best to leverage that for DHCP services and centralize them as much as possible.

Most mid-market-sized organizations have full domain services and are running DHCP services, for example, in Windows Server. Others may offload certain DHCP functions to a specialized product like IPAM services in WhatsAppGold or SolarWinds.

Larger organizations will have much more complex IPAM systems for handling the volumes of IP networks. This enterprise class of IPAM includes products such as BlueCat and Infoblox that manage other services in addition to DHCP.

DHCP for AP Provisioning

Aside from servicing clients, DHCP is (or can be) used for AP provisioning. In Chapter 6, "Hardening the Wireless Infrastructure," the topic of AP provisioning and security is covered in much more depth.

In enterprise deployments—really in any size deployment—the APs should be getting dynamically assigned IP addresses from DHCP. Although I certainly have seen organizations deploy APs with static IP addresses, that's rare (even in the smallest environments) and there's never been a reasonable explanation for it; it was just an artifact of a legacy architecture.

Along with getting an IP address via DHCP, controller-managed APs can also use DHCP vendor-specific information (VSI) option codes (a little bit like vendor-specific RADIUS attributes) to identify themselves as APs and request special instruction to find their controller. Usually, DHCP options only support sending a response with one controller IP, while DNS supports multiple.

Most vendors use two standard VSIs or option codes in the process: Option 60 and Option 43. IPv6 implementations use Option 52. The vendor class request (option 60) is the field APs use to tell the DHCP server what vendor they are. The DHCP server is configured to match that request with a response (option 43 for IPv4 networks), and in that response is the IP address of the controller:

- DHCP Vendor Class Request (Option 60)
- DHCP Response (Option 43)
- DHCP Response (Option 52) (for IPv6)

DHCP options can be configured on most DHCP servers, certainly any full-featured server such as Microsoft Server running DHCP services or any IPAM solution. There may be limitations with "light" DHCP services running from infrastructure devices (routing switches, routers, and certain firewalls).

SIZING IT UP: USING AN AP MANAGEMENT VLAN

It's an industry-wide best practice to use an AP management VLAN for provisioning APs. If not an AP management VLAN, some VLAN for shared network management separate from client data on the wired network is acceptable.

Even in small and midsize organizations, a management VLAN is recommended (as you'll see in Chapter 6). However, if your organization is so small—with a handful up to a few dozen endpoints—and you don't have IT staff to manage it, then you may opt to scratch the management VLAN.

In those cases, it's recommended at a minimum to use DHCP reservations in the AP scopes so the APs will have consistent IP addresses for management and accessibility.

Certificates

Most of the certificate excitement in Wi-Fi centers around user and device authentication, which we've already covered. Along with that use case and the use case of certificates for authenticating the infrastructure components (mentioned earlier), there are also use cases to secure administrative logins to the infrastructure with certificates.

There are many uses for certificates in Wi-Fi, including:

- Certificates for endpoint and server authentication
- Certificates for infrastructure authentication (including cloud)
- Certificates for management authentication

The first bullet was covered in Chapter 3, "Understanding Authentication and Authorization," and the latter two bullets are covered in Chapter 6. Regardless of the function, certificate use within our infrastructures will continue growing, likely exponentially, in the coming years. Proper management and maintenance of certificates will be a new part of our role as network, Wi-Fi, and security architects.

INDUSTRY INSIGHT: STATS ON THE IMPACT OF CERTIFICATES IN IT

In one recent study, 60 percent of organizations experienced certificate-related outages that impacted critical business services within the last year. The report, which included feedback from over 550 CIOs from the U.S., France, Germany, and Australia, yielded other alarming statistics (Source: `https://www.venafi.com/resource/WhitePaper_CIOStudy_TLS_Risk_Mitigation`):

- 85 percent believe the increasing complexity and interdependence of IT systems will make outages even more painful in the future.
- Nearly 80 percent estimate certificate use in their organizations will grow by 25 percent or more in the next five years, with over half anticipating minimum growth rates of more than 50 percent.
- While 50 percent of CIOs are concerned that certificate outages will have an impact on customer experience, 45 percent are more concerned about the time and resources they consume.

Understanding Wi-Fi Design Impacts on Security

To round out this chapter, I'll just mention there are many design elements in Wi-Fi not specifically related to security, but which have an impact on security from the perspective of confidentiality, availability, and/or integrity.

The most prominent of such designs is the relationship of roaming protocols and SSID security profiles, which is covered here in some depth, followed by a shorter foray into design impacts on availability and resiliency, and lastly more nuanced SSID settings such as rate limiting.

The topics here include:

- Roaming Protocols' Impact on Security
- Fast Roaming Technologies
- Recommendations for Fast Roaming in Secure Wi-Fi
- System Availability and Resiliency
- RF Design Elements
- Other Networking, Discovery, and Routing Elements

TIP The WPA version 1 suite of security is deprecated. You will see WPA2 and WPA3 referenced here. In portions of the text that use just "WPA," it indicates the statements refer to all versions of that class of network. For example, "WPA-Personal" will mean WPA2-Personal and WPA3-Personal, and "WPA-Enterprise" indicates both WPA2-Enterprise and WPA3-Enterprise. When there are differences between WPA2 and WPA3, the specific version is included.

Roaming Protocols' Impact on Security

The 802.1X protocol, EAP authentication, and the 4-way handshake used in Wi-Fi for association were covered in Chapter 3. The result of those operations in concert introduces some special considerations during roaming—when an endpoint is moving, physically or logically, between APs.

Everything in those authentication and 4-way handshake sequences was specific to the particular AP (specifically the BSSID) and the endpoint associated with it. When the endpoint roams to another AP, the process has to be started from scratch unless a fast roaming mechanism is used to speed the process. The topics detailing roaming protocols' impact on security are:

- Roaming Impact on Latency-Sensitive Applications
- Roaming and Key Exchanges on WPA-Personal Networks
- Roaming and Key Exchanges on WPA-Enterprise Networks

SSIDs AND BSSIDs

We were doing so well not getting in the weeds, but since "BSSID" has just been referenced, here's the down-and-dirty on what you should know about the relationship of an SSID to the Wi-Fi infrastructure.

SSID and BSSID are the two to focus on, but here's the suite of service sets in Wi-Fi:

- *SSID* is the service set identifier, or simply the name of the Wi-Fi network.
- *BSSID* is the basic service set identifier, and it's the MAC address of the SSID from a specific AP's radio. Meaning there's a unique BSSID per-SSID, per-AP, per-radio. A dual-band AP with three SSIDs will have six BSSIDs.
- *BSS*, or basic service set, is the collection of SSIDs from a single AP.
- *ESS*, or extended service set, is the collection of the SSIDs across multiple APs in the system.

A picture speaks a thousand words, and the following image demonstrates the difference between the SSID and BSSID. The first two rows of the client list show AP-01 serving two different networks—one a WPA3-Enterprise and the other a WPA2-Personal. You'll notice the SSIDs are "ACME-1X" and "ACME-IoT," and the BSSID is unique for each network. In the third line, AP-03 has a device on the same "ACME-IoT" SSID as the second row, but the AP-03 has its own BSSID for that network.

Device Type	AP Name	BSSID	SSID	Protocol	Security
Intel Corporate	AP-01	5c:5b: 71:81	ACME-1X		WPA3-EAP-SHA256/CCMP
Resideo	AP-01	5c:5b: 71:a2	ACME-IoT		WPA2-PSK/CCMP
AIRTAME ApS	AP-03	5c:5b: 63:53	ACME-IoT		WPA2-PSK/CCMP

Functionally, a slow (hard) roam such as that would require the endpoint to disconnect from the existing AP before reassociating and connecting to the next one—creating a brief moment in time the endpoint has no network connection.

Imagine Tarzan for a moment, swinging from vine to vine among the treetops of the African jungle. If that were me swinging around, I'd feel a little safer having a grip on a second vine before I let the first one go—and that's what roaming protocols help us do.

Roaming Impact on Latency-Sensitive Applications

Why does this matter? Well, it might not matter for all endpoints, but latency-sensitive applications such as voice and video will certainly be impacted. With many services digitized, and volumes of users accessing streaming videos and meeting applications, roaming and latency are on the front burner. Poor roaming means poor user experience.

NOTE **Voice and video applications aren't just for leisure. With remote work and remote education models, many functions have moved to real-time remote collaboration, making these services business-critical. Even when on campus or in the same building, many organizations are relying on Internet-hosted voice and video.**

Historically only WPA-Personal and WP3-Enterprise have used any form of encryption, and attention is focused on these two. With the new Enhanced Open security, we'll also have a need for facilitation as they're now providing encryption.

Without some help, endpoints will naturally perform a full (slow) roam between APs, which repeats the full sequence of authentication (on Enterprise/802.1X SSIDs) and key exchanges (on both Enterprise and Personal SSIDs).

The exchanges for WPA-Personal of course don't require the lengthy 802.1X EAP authentication sequence, making passphrase-secured networks more desirable in supporting latency-sensitive applications. Because of that, it's been the recommendation of most application owners and Wi-Fi manufacturers to use passphrase-secured networks for VoWi-Fi (voice over Wi-Fi) and similar needs.

On a well-designed network, any endpoint that can authenticate with 802.1X should be doing just that, and the upcoming sections will demonstrate why and how to accomplish both security and performance with Fast BSS Transition.

NOTE **Fast Transition is an optional feature and not required for WPA3 certification and may not be supported by all endpoints.**

Roaming and Key Exchanges on WPA-Personal Networks

For WPA-Personal/passphrase-based SSIDs, the roaming process is simple and fast, because the encryption keys are derived from the passphrase, which is the same across all APs.

While the key exchanges are streamlined, the standard 4-way handshake still takes place during the roam.

One difference to note, the WPA2-Personal networks using PSK use the passphrase to directly derive the encryption key, whereas WPA3-Personal using SAE derives the keys indirectly from the passphrase, meaning there are additional steps in the derivation process, and additional keys to be shared among the infrastructure.

Because of their simplicity and speed (compared to the lengthier authentication process of 802.1X), personal-class networks have been the go-to for networks serving latency-sensitive applications. The exact times will vary, but in general a hard 802.1X roam will take about three to four times longer than a basic passphrase-based roam.

Figure 4.5 shows a hard roam between APs on a WPA2-Personal PSK network. Remember from Chapter 2, "Understanding Technical Elements," that the newer WPA3-Personal with SAE has a slightly different operation that includes four additional exchanges—two SAE commits and two SAE confirm messages.

Compare Figure 4.5 then to Figure 4.6, which shows an equivalent hard roam with an 802.1X-secured network. In both cases, the endpoint has to repeat everything from the association request to the authentication and 4-way handshake, but in the 802.1X example, the volume of exchanges and the additional roundtrip communication to the RADIUS server adds much more time to the process.

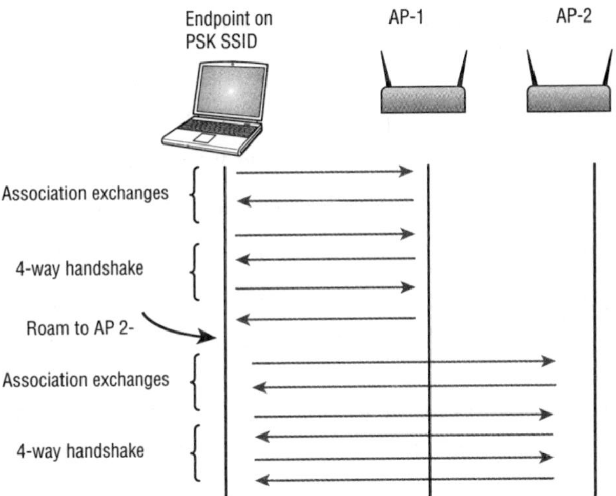

Figure 4.5: Exchanges required for a hard roam on a WPA2-Personal with PSK network

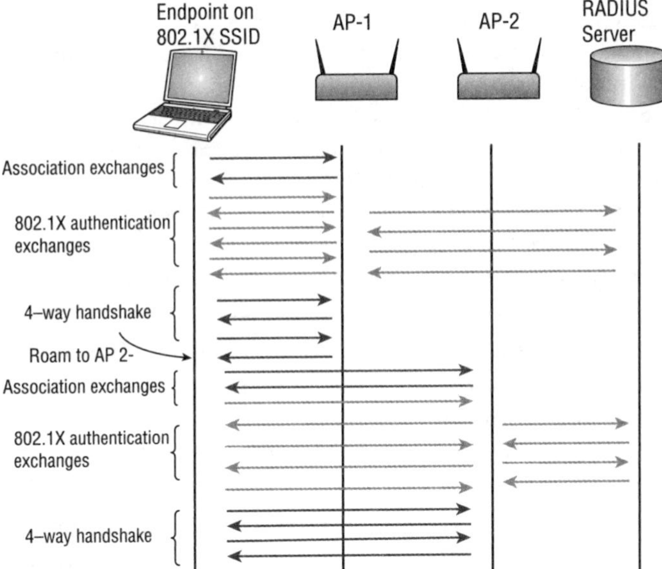

Figure 4.6: Exchanges required for a hard roam on an 802.1X-secured network

Roaming and Key Exchanges on WPA-Enterprise Networks

Over time, a few protocols have been devised to reduce latency in roaming on Enterprise/802.1X networks. Not all are created equal, and there are specific recommendations after the description of each.

The facilitation protocols leverage one or more of these functions to speed the key exchanges during roaming:

- Pre-authenticate the endpoint (in whole or part) to nearby APs
- Exchange keys among the APs over the air or via the wired network
- Partially or wholly eliminate the need for 802.1X re-authentication
- Reduce overhead by embedding the 4-way handshake in other frames

As you'll remember from earlier coverage of the 4-way handshake, with 802.1X the endpoint first completes the EAP authentication and then performs the 802.11 association process.

Fast roaming protocols have evolved over time to be more efficient, eliminating the need to repeat many of the processes of association and authentication after the original AP connection.

INDUSTRY INSIGHT: AN IN-DEPTH LOOK AT KEY EXCHANGES

A detailed, packet-level analysis of key exchanges is beyond the scope of this book. The goal here is to describe enough of the processes to give you an idea of appropriate use cases, pros, cons, and security considerations.

The world of Wi-Fi key derivations and key exchanges is a topic all unto itself; there's a mélange of keys used for varying purposes, all of which are derived and exchanged differently among controllers, APs, and endpoints in the network.

If you're interested in learning more, I suggest considering Certified Wireless Network Professionals (CWNP) Certified Wireless Analysis Professional (CWAP) training or learning materials from `https://www.cwnp.com/certifications/cwap`.

Any protocol selected must be supported by the endpoint, the RADIUS server, and the Wi-Fi infrastructure. In all cases of 802.1X-secured networks, a pairwise master key (PMK) is generated from the RADIUS server and distributed to the original controller or AP; the endpoint involved in the authentication also derives the PMK in parallel.

After the initial authentication, when an endpoint roams to a new AP, it must either perform a slow roam—repeating all steps of the original 802.1X authentication and 802.11 association—or use a protocol to eliminate some of those steps.

During the 802.1X authentication, the RADIUS server and endpoint will each derive the PMK. The RADIUS server will then send the PMK to the original authenticator (AP or controller). In fast roaming protocols, the PMK and/or other keys are then shared with other APs to speed roaming This behavior is shown in Figure 4.7.

As you'll see, the IEEE standard for Fast BSS Transition is the latest and greatest standard for facilitating roaming, but it comes with a few caveats. Any reference to roaming here supposes roaming within the same SSID, and among

APs connected to the same controller (or virtual controller function within a local cluster).

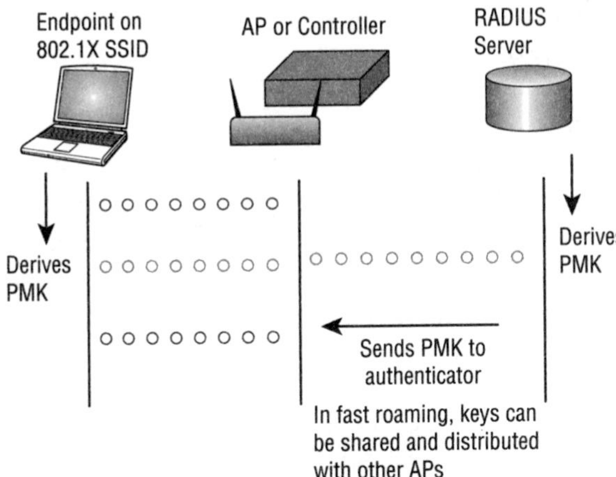

Figure 4.7: PMK derivation and key distribution for roaming facilitation

Fast Roaming Technologies

Fast roaming technologies are designed to ease the challenges of securely roaming between APs. There are four common technologies still in use, although only one (Fast BSS Transition) is an IEEE standard.

The four fast roaming technologies covered here are:

- Fast Reconnect
- PMK Caching (Roam-back)
- Opportunistic Key Caching
- Fast BSS Transition

After covering the four fast roaming technologies, you'll also find these topics:

- Summary of Fast Roaming Protocols
- Support for Fast Transition and Other Roaming
- Changes in Roaming Facilitation with WPA3 and Enhanced Open Networks
- Recommendations for Fast Roaming in Secure Wi-Fi

Fast Reconnect

The Fast Reconnect mechanism was originally specified in the Protected EAP (PEAP) IETF draft of 2003. Paraphrasing that language in the context of Wi-Fi roaming, it was loosely defined as the ability to reduce the number of round-trips when roaming between APs. More recently, additional EAP methods (including EAP-SIM and EAP-AKA for cellular technologies) have been updated to make use of Fast Reconnect.

Fast Reconnect allows a roaming endpoint to skip a portion of the 802.1X authentication when moving among APs. While it speeds the transition some, the endpoint still must complete a portion of the 802.1X process along with the full 4-way handshake when roaming.

Figure 4.8 shows an endpoint with Fast Reconnect. When roaming from AP-1 to any other APs in the system, portions of the 802.1X exchanges are eliminated to streamline the roam. The 4-way handshake still occurs for each roam. Although described in IETF documents, this protocol is not well defined. You'll notice Fast Reconnect options in most Microsoft infrastructure, where it may be most commonly found if Fast BSS Transition is not being used.

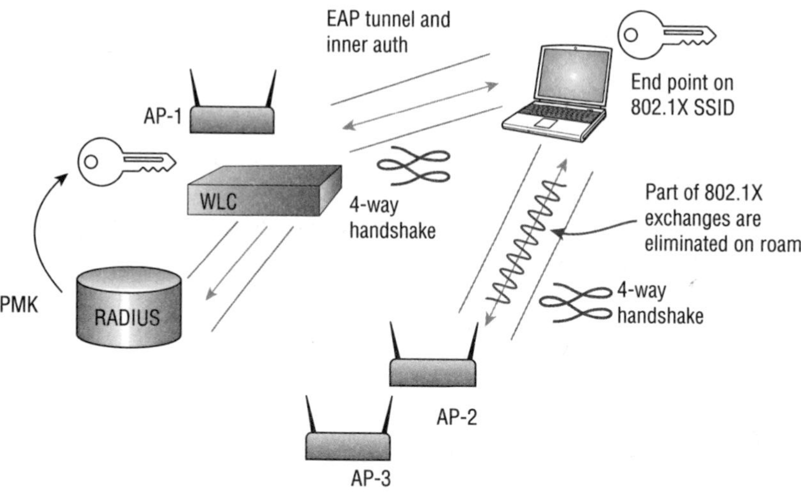

Figure 4.8: Conceptual view of Fast Reconnect

NOTE **Roaming protocols that rely on the RADIUS server to distribute keys will only work across APs that are using the same RADIUS server.**

PMK Caching (Roam-back)

Pairwise master key (PMK) caching allows an endpoint to roam back to an AP it was previously authenticated with and skip the 802.1X EAP authentication. Just as the name indicates, the original AP keeps (caches) the PMK used, and the returning endpoint can bypass the EAP authentication, use the previously generated PMK, and proceed straight to the 4-way handshake. While the PMK is reused, new session keys will be generated during the 4-way handshake.

It's quick and easy but limited to APs the endpoint has already been associated with, and the PMK is only cached for a period of time. In some environments where the same endpoints are moving among the same APs, it's helpful; but of course, with new endpoints or newly-roamed-to APs, the benefit is nullified.

In Figure 4.9, PMK caching is being used. The endpoint is attached to AP-1, roams to AP-2, then back to AP-1. During the roam back to AP-1 (assuming it's within the cache period), the endpoint skips the EAP authentication and then performs the 4-way handshake.

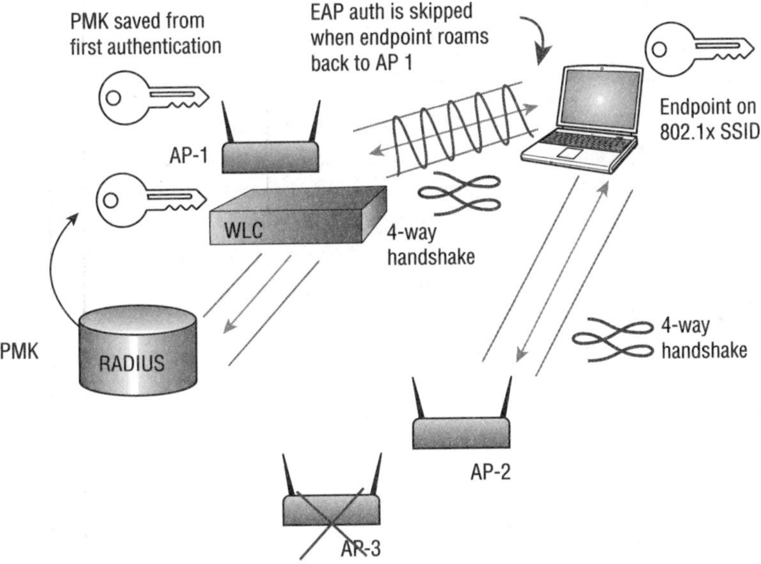

Figure 4.9: PMK caching, or roam back, offers benefits only in cases where an endpoint is roaming back to an AP it was previously authenticated with.

You may also see references to PMKSA (PMK Security Association) caching. The security association data includes the PMK along with other elements. There's also PMKID caching and other flavors of PMK caching that send or forward different elements from within the PMKSA, which operate outside of traditional PMK caching or OKC operations (covered next).

NEW PMKID ATTACKS ON WPA2-PERSONAL NETWORKS

WPA2-Personal is susceptible to many attacks, including a relatively new PMKID attack. Whereas legacy attacks on WPA2-Personal required an attacker to eavesdrop, wait for a new client association, and capture the full 4-way handshake, the new attack bypasses that and uses a tool to pull the PMKID directly from an AP.

This particular attack pulls RSN IE information from target APs that have cached PMK data, which applies not only to this fast roaming technology but also FT and OKC as well.

The PMKID includes the hash of the passphrase, which can typically easily be cracked with various tools (including hashcat) after the PMKID is retrieved.

In one example, a researcher from Cyberark walked down a street with a USD $50 kit, harvested 5,000 passphrases, and cracked 70 percent of them. For more information, look up hashcat and hcxdumptool.

Mitigations against this attack include:

- Use of any WPA-Enterprise network vs. WPA-Personal
- Use of WPA3-Personal instead of WPA2-Personal
- Use of WPA2-Personal without fast roaming enabled (see note to follow)
- Use of WPA2-Personal with fast roaming and a long, strong passphrase

Keep in mind that WPA2-Personal networks are still susceptible to the prior generations of passphrase brute force attacks even if fast roaming is not enabled. The original attacks require a bit more work but not a significant amount of time since they can simply force an endpoint to disconnect and reconnect with a de-auth attack.

Avoiding passphrases that are short along with any standard indexed data such as company names, phone numbers, and song lyrics, or other text will also provide greater protection against dictionary attacks.

Opportunistic Key Caching

Opportunistic key caching (OKC) works much like PMK caching with the added benefit of the PMK being distributed to more APs. Specifically, OKC preemptively sends the PMK to all APs in the client's immediate roaming table.

Just as with PMK caching, OKC eliminates the EAP authentication, allowing the endpoint to roam securely with just the 4-way handshake (see Figure 4.10).

OKC was never an official standard, although widely supported in the industry. Even with its improvements over PMK caching, OKC has gaps in key distributions and doesn't address the added overhead of QoS.

In Figure 4.10 the endpoint is roaming with OKC. What's different here from PMK caching is the endpoint could fast roam from AP-1 to either AP-2 or AP-3 in this diagram. The Wi-Fi infrastructure will distribute PMK key to APs the endpoint is likely to roam to.

NOTE It was already mentioned OKC is not standard, so vendor implementations and support will vary. This also means there's no prescribed way to distribute the keys to other APs. Extrapolating that statement means it's possible the keys could be sent unencrypted or in a manner that is not secure. This is one of many reasons Fast Transition should be the go-to technology for fast roaming.

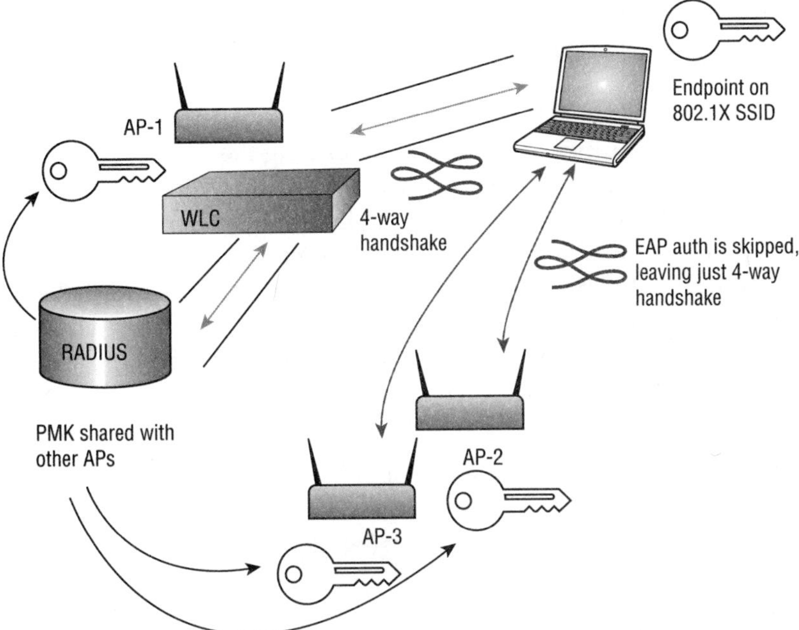

Figure 4.10: OKC eliminates the 802.1X re-authentication when roaming to nearby APs

Fast BSS Transition

Formerly defined in IEEE 802.11r, the Fast BSS Transition, or FT, has been rolled into the IEEE 802.11 standard as of 2008. FT further enhances roaming beyond PMK caching, OKC, and all prior methods. Fast BSS Transition defines a mobility domain and new key hierarchy—two derivations of the PMK, plus the pairwise transient key (PTK), versus only the PMK.

Fast Transition further streamlines roaming by combining the 802.11 Open System Authentication and reassociation with the 4-way handshake information embedded in a Fast Transition Information Element. This results in a roaming process for 802.1X-secured networks that's even more streamlined than that of a WPA2-Personal/PSK secured network.

NOTE **Fast BSS Transition should be referred to as FT and not 802.11r. It was originally defined in the 802.11r specification but has since received many updates and been absorbed into the IEEE 802.11 standard. I've included 802.11r here as a point of reference to those familiar with its prior incarnation.**

Not reflected in Figure 4.11, Fast Transition supports the preemptive exchange of keys both over the air, as well as over the wired network (referred to as the distribution system, or DS in Wi-Fi standards). When passed over the wire, the first part of the 802.11 Open System Authentication is also completed along with the key distributions—leaving only the reassociation to be completed by the client when it roams. It's a pretty slick function, but not supported by all Wi-Fi vendors.

As shown in Figure 4.11, FT further streamlines the roaming process by eliminating the 802.1X re-authentication and embedding the 4-way handshake in the FT exchanges. This simplified view doesn't quite do FT justice in showcasing its enhancements over prior fast roaming technology. With FT, there's a three-tier key hierarchy, and the first-level pairwise master key R0 (PMK-R0) is only known to the RADIUS server and the Wi-Fi controller. The APs use a second-level PMK-R1. PMK-R0 is used to derive PMK-R1, which in turn is used to derive the pairwise transient key (PTK) used to encrypt the data.

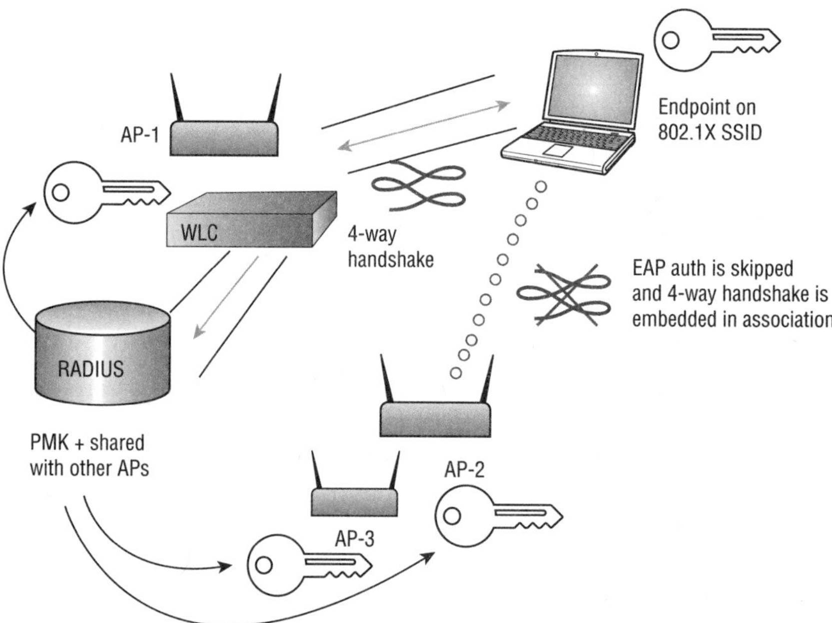

Figure 4.11: An endpoint roaming on an 802.1X network with FT is faster than even a WPA2-Personal roam

Summary of Fast Roaming Protocols

Table 4.1 provides an at-a-glance view of the different generations of roaming protocols and how each manipulates the 802.1X authentication process and/or 4-way handshake.

If supported and implemented properly, Fast BSS Transition on an Enterprise/802.1X-secured network offers a roaming experience even faster than with a basic Personal/passphrase secured network.

Table 4.1: Summary of key exchanges with roaming facilitation

FACILITATION PROTOCOL	SECURITY PROFILE	802.1X AUTHENTICATION AT ROAM	4-WAY HANDSHAKE AT ROAM
None	WPA2-Personal with PSK	N/A	Yes
None	WPA2/ WPA3-Enterprise	Yes, full	Yes
Fast Reconnect	WPA2/ WPA3-Enterprise	Partial; eliminates inner re-authentication	Yes
PMK Caching	WPA2/ WPA3-Enterprise	No; eliminates re-authentication when roaming back to original AP	Yes
OKC	WPA2/ WPA3-Enterprise	No; eliminates re-authentication when roaming to nearby APs	Yes
Fast BSS Transition (FT)	WPA2/ WPA3-Enterprise	No; eliminates re-authentication when roaming	Yes, but embedded in another exchange
Fast BSS Transition (FT)	WPA3-Personal with SAE	N/A	Yes, but embedded in another exchange

You may have noticed one scenario missing from the table. The new Enhanced Open networks use encryption and therefore keys. At the time of writing, the Enhanced Open FT doesn't exist yet, so it's not included in Table 4.1, but keep an eye out for that in the future.

Support for Fast Transition and Other Roaming

While FT may seem like the Holy Grail of fast, secure roaming, like its predecessors, there's still spotty support for the standard in the industry. Although always in flux, here are some of the long-standing notable exceptions for FT support:

- Apple MacOS does not support FT; instead, it relies on PMKID caching (Source `https://support.apple.com/en-us/HT206207`).
- At the time of writing, Windows 10 only supports FT on WPA2-Enterprise secured networks (although that's expected to change with WPA3-Personal SAE and Enhanced Open). (Source `https://docs.microsoft.com/en-us/windows-hardware/drivers/network/fast-roaming-with-802-11k--802-11v--and-802-11r`).

- Fast Transition may not be supported between APs of different models or versions.
- In cloud-managed deployments (e.g., no controller), key exchanges happen with proprietary protocols, which may be split between cloud management and AP roles.
- Cisco and Apple have developed a proprietary Adaptive 802.11r/FT protocol.
- Legacy endpoints may need to be updated to work within FT-enabled networks, even if they will not participate in FT.
- Some Wi-Fi manufacturers may implement controls to avoid denial-of-service (DoS) attacks, which often limit the number of APs an endpoint can create Fast Transition exchanges with.
- Some Wi-Fi vendors may not support their proprietary functions with FT.

PACKET ANALYSIS OF FAST TRANSITION

There's really none better than Peter "Magic Packets" Mackenzie to explain packet level analysis of Wi-Fi. For more on fast roaming protocols and a comparison to Fast BSS Transition, check out his post at `http://mackenziewifi.com/index.php/2019/01/02/over-the-air-fast-bss-transition/` and other blogs and videos by Peter (see the following illustration).

Packet	Source	Destination	BSSID	Channel	Protocol
1	Lab Phone	Motorola AP2	Motorola AP2	64	802.11 Auth
2	Motorola AP2	Lab Phone	Motorola AP2	64	802.11 Auth
3	Lab Phone	Motorola AP2	Motorola AP2	64	802.11 Assoc Req
4	Motorola AP2	Lab Phone	Motorola AP2	64	802.11 Assoc Rsp
5	Motorola AP2	Lab Phone	Motorola AP2	64	EAP Request
6	Motorola AP2	Lab Phone	Motorola AP2	64	EAP Request
7	Lab Phone	Motorola AP2	Motorola AP2	64	EAP Response
	Motorola AP2	Lab Phone	Motorola AP2	64	EAP Request

`http://mackenzieWi-Fi.com/index.php/2019/01/02/over-the-air-fast-bss-transition/`

Changes in Roaming Facilitation with WPA3 and Enhanced Open Networks

With traditional Open System Authentication networks, there was no encryption and therefore no need to exchange or distribute encryption keys. With Enhanced Open's addition of opportunistic wireless encryption (OWE) to provide unauthenticated key exchanges, expect to see Fast Transition support for these networks in the form of FT-OWE. The IETF draft RFC can be found at `https://datatracker.ietf.org/doc/draft-henry-ft-owe/`.

A similar situation will occur as we move from WPA2-Personal with PSKs to WPA3-Personal with SAE. While WPA2-Personal was encrypted, the PSK function for key derivation from the passphrase was more straightforward than the newer (and more secure) SAE functions. While the current specification for FT has been supported in WPA2-Personal networks, it's not available in all endpoints and Wi-Fi products. With WPA3-Personal and SAE, expect to see and implement Fast Transition support by way of FT-SAE.

Recommendations for Fast Roaming in Secure Wi-Fi

As mentioned earlier, due to the complexity of the key exchanges with 802.1X-secured networks, the "easy" button recommendation from most application owners and Wi-Fi manufacturers has been to simply use passphrase-secured networks instead of 802.1X-secured networks in support of latency-sensitive applications.

The WPA2-Personal with PSK security has plenty of security gaps and vulnerabilities such as those of offline dictionary attacks. While WPA3-Personal with SAE mitigates some of those risks, the best practice for secure wireless architecture still entails using 802.1X-secured networks for any production networks.

Sorting through the menagerie of key exchange details can be dizzying but can be summarized in a few basic recommendations.

For Production Networks (Corporate Data/Assets with Managed Devices) Use 802.1X with FT where supported. Key distribution over the wired network (DS) offers the slightest benefit over those using over-the-air; either is recommended. Use OKC if FT is not supported with the understanding that OKC is not a formal standard and therefore there exists no interoperability testing by the Wi-Fi Alliance.

For Networks Dedicated to Latency-Sensitive Applications VoWiFi, nurse calling systems, and networks supporting autonomous guided vehicles (AGVs), automated forklifts, and the like should use 802.1X with FT if at all possible. If 802.1X with another protocol such as OKC doesn't meet the business need, the third best option is to use WPA3-Personal with SAE and proper segmentation. As a last resort for low-security needs, WPA2-Personal with PSK can be used, but should be segmented and monitored closely. WPA2-Personal with PSK should not be used for any critical systems, sensitive data, or by devices that have access to sensitive data.

General-Use Networks Whether WPA3-Personal, WPA3-Enterprise, or Enhanced Open, general-use networks can take advantage of FT but likely will offer just as healthy an end-user experience with OKC for devices that support it.

In Fixed Environments with IoT and Legacy OT For connected devices running legacy protocols in warehouses, yards, manufacturing floors, and other fixed areas, it's best to consult the application manufacturer's documentation as handheld scanners, IoT devices, and legacy OT devices may not have support for newer secure roaming protocols. If FT is supported, always use that.

Upgrade what you can and use PMK caching or OKC as a second and third choice. If some devices support 802.1X, you'll benefit from migrating those from passphrase-based networks to 802.1X and segmenting them appropriately (assuming proper support for FT).

Check Wi-Fi Vendor Support for Your Specific Configuration Not all data path modes and product configurations will support FT.

Roaming requirements extend well beyond our security architecture, and in fact roaming is a lengthy topic when considering wireless architecture and designs as a whole. When designing roaming, there are myriad considerations around QoS, buffering, and neighbor transitions, to name a few.

Aside from the protocols and use cases covered here, there's an ecosystem of standards and operations related to Fast BSS Transition such as 802.11k and 802.11v for Neighbor Reports and BSS Transition Management Frames, respectively.

The security considerations here should be balanced with the organization's business needs and remainder of the design architecture for the wireless system.

FAST BSS TRANSITION IS THE TECHNOLOGY TO USE

Fast BSS Transition is *the* technology the industry should be using for fast roaming. It's an optional certification with the Wi-Fi Alliance, and therefore many vendors don't opt-in to testing, and others don't support it at all.

I'm giving a quick shout-out to Fortinet, Juniper Mist, and Ruckus, which each have several Wi-Fi products certified for FT with WPA3. From the endpoint side, Zebra Technologies gets kudos for certifying dozens of its enterprise handheld devices. You can view the most current list at `www.wi-fi.org/product-finder-results?sort_by=default&sort_order=desc&certifications=663,617`.

System Availability and Resiliency

Availability is one of the three legs of the IAC (integrity, availability, and confidentiality) triangle of cyber security. Although often overlooked, system availability and resiliency play a crucial role in secure wireless architecture.

Uptime, High Availability, and Scheduled Downtime

Every wireless architecture should include planning for the appropriate uptime based on the business needs of the networks, noting that different networks often have different requirements.

For example, it's not uncommon in healthcare for organizations to manage a Wi-Fi infrastructure for critical devices and a separate infrastructure for guest services. The separation allows IT administrators to perform updates on the individual platforms as needed, and to iterate the non-critical systems on a schedule that may not be suitable for the critical systems (such as biomedical devices and patient care devices).

Scheduled Maintenance and Testing

Availability planning should encompass high availability of the controllers (physical, virtual, or virtual cluster) with regularly scheduled testing to prove proper operation of failover scenarios.

TIP **Yes, that means you should schedule time to force failures to test the high availability—as painful as it is, it's always orders of magnitude less painful when it's scheduled versus unplanned. Just quit whining, drag yourself out of bed at 5a.m. every once in a while, and you can probably avoid a nasty 2a.m. outage call.**

In addition, managing controller-based systems should include regularly scheduled maintenance windows to apply critical security patches. Unfortunately, Wi-Fi vendors have made even minor upgrades such a painful process that Wi-Fi administrators often wait until their hand is forced with a bug fix or support case before scheduling upgrades or patches.

REFERENCE **Additional recommendations for patching and maintenance are covered in Chapter 6.**

In this case, every vendor's baby is ugly; if it's a controller, the upgrade and maintenance process is just buggy and frustrating. At times, a patch that fixes one issue breaks three others. Other times, the images may be pre-downloaded to APs, but not applied properly at reboot.

On at least one occasion, a vendor had an undocumented incompatibility of a controller version with a specific version of AP BootROM code. Attempts to upgrade the APs with that BootROM resulted in manual intervention (having to physically retrieve the AP, console to it, and initiate a recovery). Fellow Wi-Fi professionals could fill a book just with controller upgrade war stories. This is one reason organizations and IT pros alike have become quite fond of the more advanced cloud-managed solutions.

The point is—most IT professionals I know are dealing with a litany of projects on top of the daily fires, and adding a lengthy, painful task to the already-abundant list is not a pleasant thought. Plus, organizations are more likely to have stricter control over outage requests than they do control over mandating security updates, meaning it's often hard to request (and be approved for) a planned outage, and there may not be requirements for patching. More accurately, there are likely compliance requirements for patching and updates, but those rarely make their way down to the IT administrators.

RECOMMENDATIONS FOR MAINTENANCE IN HIGH AVAILABILITY

I'd love to provide more specific guidance on best practices for managing system maintenance and patching in high-availability environments. The challenge is that the best practices and options vary not only by vendor but also by implementation.

As a brief example, Cisco and Aruba both have multiple deployment models including controller, controller-less, and cloud-based products. Just within Aruba's ecosystem, products may be managed by a controller, a local cluster, AirWave, or Central. Should I even mention there's Central Cloud as well as Central On-Prem?

Zooming in further, controller deployments with Aruba vary with AOS 6 and 8, and with version 8 vary depending on whether there's a Mobility Conductor (a controller coordinator) or not. The algorithms Aruba uses for load balancing and hitless upgrades in its AOS 8 controllers is drastically different than other vendors' solutions, and their own prior versions of code. Cisco similarly has its own options and special features for upgrades. Meaning, there's just no longer a one-size-fits-all recommendation.

There are best practices from vendors for each of their architectures, and at any given time, there will be known bugs that may render some options unavailable or undesirable.

Your goal as architect is to work with what you have to meet the uptime objectives and ensure it's working as designed with scheduled testing.

AP Port Uplink Redundancy

Other non-security wireless design elements that should be considered include AP port uplink configurations and redundancy. From an availability standpoint, many organizations write requirements for dual Ethernet connections from each AP to the edge ports.

My personal opinion on this is that it rarely makes sense. As always, there are exceptions, but few organizations have the requisite redundancy throughout the system for dual AP ports to be effective as a redundancy tool:

- The cable paths are often co-located (no redundancy in pathing).
- They're often terminated to the same edge switch (likely no redundancy depending on power supply or backplane configuration).
- When they are terminated to different switches or power supplies, those devices are almost always in the same networking closet and not serviced by different power circuits (even if they're available).

The list goes on. If there are two cable connections, the IT administrators have to configure the switch ports and/or the AP ports appropriately to prevent network loops. Depending on the vendor and product, dual AP ports often introduce a lot of unnecessary expense and operational overhead for little or no benefit.

The use of dual uplinks for the sake of increased bandwidth (link aggregation) also hasn't been a reasonable use case. Historically up through Wi-Fi 6 (802.11ax), the chances of over-subscribing a 1 Gbps edge port were slim to none. After RF management overhead is stripped away, the conditions under which the data throughput egressing to the switch exceeds the switch port capability are rare.

With Wi-Fi6E's (Wi-Fi 6, 802.11ax in 6 GHz) increase in spectrum (and therefore increase in channels and bandwidth potential), APs will absolutely experience conditions that exceed a 1 Gbps uplink. For that reason, it's recommended to consider multi-gigabit Ethernet technology (IEEE 802.3bz) for edge ports moving forward.

If your organization is requiring redundant AP uplinks, be sure there's a reasonable business case, and that the connection paths and power configurations are such that the extra work and expense has afforded the expected benefits.

RF Design Elements

There are countless RF-related design elements that impact availability. We're just hitting the highlights here.

AP Placement, Channel, and Power Settings

Together, placement of the AP along with its channel and power settings will dictate the resiliency of the RF, or coverage. Just as with high-availability designs using dual network connections from an AP to the switch, RF coverage redundancy is a dodgy topic.

While most organizations believe their Wi-Fi solution is installed and configured to handle gaps in coverage automatically and dynamically, in practice, few deployments are designed to meet this goal. Wi-Fi vendors all offer dynamic radio resource management (RRM) protocols—each with their own secret sauce to make it better. Adaptive radio management handles the channel and power settings of the APs, eliminating the need for a Wi-Fi administrator to perform that very mundane task of configuring radios to avoid channel overlap (thereby introducing interference).

It's a feature vendors will tout as magical and wonderous. Yet, still in 2022, those same vendors will tell you to disable RRM protocols in many scenarios when you call support for help.

Even assuming a perfect world and unblemished operation of RRM, the likelihood that the Wi-Fi RF design was performed to withstand any major outages of APs is very low. Not impossible, but low. If your organization used a reputable Wi-Fi design company or architect and clearly defined your coverage redundancy needs, then you're the exception, and you likely have a redundant RF design.

For the rest of you, don't assume this degree of resiliency. Proper high-availability designs for RF coverage demand an intricate level of planning of AP placement and radio power to meet that need, while balancing the coverage and endpoint requirements. It's not easy, and it's not common. Most organizations also notoriously under-scope Wi-Fi projects, meaning they eke by with the minimum number of APs for then-current coverage requirements. When there is a refresh that includes an AP placement redesign, organizations tend to use those AP placements for several generations of Wi-Fi, with the end result being improper coverage and no wiggle room for providing redundancy.

RF COVERAGE AND ROAMING

Earlier, you learned about the impact of roaming on latency and user experience. Aside from the security considerations, there are fundamental layer 1 requirements for smooth roaming. In a professional Wi-Fi design, you'll see not only a "1st AP view" that indicates the modeled signal strength by a client in any given area; you'll also see a "2nd AP view" showing the signal strength from the second closest AP.

The "1st AP view" shows basic coverage, while the "2nd AP view" tells us more about anticipated roaming support. If areas of a floorplan don't have a high enough quality signal from the neighboring APs, clients in those spaces won't be able to roam seamlessly.

The result may be a dropped connection, which negates any of that awesome fast roaming you configured (such as Fast BSS Transition). When endpoints experience a full disconnect, they'll have to rinse and repeat the full series of Open System Authentication, network authentication (e.g., 802.1X/EAP), and then the 4-way handshake.

Endpoints have differing radio capabilities and expectations of signal quality and roaming, but there are widely accepted industry best practices that apply to laptops, tablets, and smartphones.

If your architecture and business requirements specify RF coverage overlap, just as with earlier guidance, ensure you understand the business objectives, the technical requirements, outage scenarios, and then use a qualified Wi-Fi integrator or professional to perform a survey of the environment. A survey will evaluate the current capabilities, and a professional design (greenfield or brownfield) will detail recommendations to add resiliency if needed.

Wi-Fi 6E

Wi-Fi 6E is 6th generation Wi-Fi (802.11ax) "extended" to the 6 GHz spectrum (as opposed to the 2.4 GHz and 5 GHz spectrums we're accustomed to). Wi-Fi 6E promises a new world of Wi-Fi, with more bandwidth and greater security.

How will Wi-Fi 6E deliver on these promises? Let's look at the bandwidth increase. In Figure 4.12, you'll see representations of channels in the spectrums for 2.4 GHz, 5 GHz, and 6 GHz from the top down.

Without an entire lecture on RF theory, the simple foundation here is that for many years, we've been using 20 MHz wide channels as the default go-to channel width. Twenty is the building block of channel "chunks" and we can aggregate or bond them together (similar to switch port aggregation) for more bandwidth over the air. Although not exactly linear because of overhead, a 40 MHz wide channel is roughly twice the bandwidth as a single 20 MHz wide channel, and so forth.

More bandwidth means we can transmit a fixed amount of data over a shorter time period, which is always a key objective in Wi-Fi, where the aim is to keep airspace open and reduce the battery drain of constant radio activity by endpoints. The different spectrums have a direct relationship to our ability to bond channels for more bandwidth.

The 2.4 GHz spectrum is old news, limited, and crowded. With it, we only get three non-overlapping 20 MHz channels (in most countries, four in a couple geographies but mostly the world sticks to channels 1, 6, and 11). A three-channel plan is the bare minimum to squeak by with 20 MHz—we certainly can't bond those channels for more bandwidth without overlapping channels, so we'll move on.

The 5 GHz spectrum eventually delivered 25 non-overlapping 20 MHz wide channels, but with caveats. Referring to Figure 4.12, the lightest grey (3 channels, labeled "TDWR") and surrounding 13 channels labeled "DFS Channels" in the middle are part of the dynamic frequency spectrum (DFS) channels, with the inner three being dedicated to weather radar. To the far right are 5 more U-NII-3 designated channels. Not all Wi-Fi clients will support DFS, weather radar, and/or U-NII-3, meaning our count of 25 channels could be reduced to 4 just like that! Four is better than three, but barely, and if we're looking at 40 MHz wide channels, best case 5 GHz offers 12, and worst case is 2.

Now, along comes 6 GHz spectrum, bringing a whopping 59 20 MHz wide channels, 29 40 MHz wide channels, 14 80 MHz wide channels, and even 7 mega 160 MHz wide channels. Of course, not all endpoints will support channel widths that high, but even staying at 40 and 80 MHz widths we have some room to work now.

What that means for the future is Wi-Fi networks capable of much greater bandwidth.

Figure 4.13 shows a closer look at the latest 6 GHz channels.

REFERENCE Resources including links to these graphics in full color are available at `www.securityuncorked.com/books`.

With great bandwidth comes great responsibility, and we have to secure all this data, which is why Wi-Fi 6E requires WPA3 security. If you missed it earlier, revisit the sections on security profiles in Chapter 2.

Rate Limiting Wi-Fi

Over the years, many IT professionals, CIOs, and CTOs have asked about throttling or limiting guest user traffic over Wi-Fi. The thought being the organization doesn't want the guest users (or employees on the guest network) overtaxing the infrastructure and consuming all the Internet bandwidth.

To meet this objective, many organizations unknowingly make the dreadful mistake of enabling rate limiting over the air.

Never, ever specify rate limiting over the air. There may be a Wi-Fi pro out there that's seen a legitimate use case, but I have never. The only outcome of this is that you'll slow down the entire Wi-Fi network (and other networks in the same airspace). If you care for more details, here's why.

Decisions around when to send traffic over Wi-Fi (specifically 802.11 WLANs) is contention-based; over the air, endpoints don't know if there's been a collision, so they use Carrier Sense Multiple Access / Collision Avoidance (CSMA-CA) to avoid interference from the get-go. Endpoints all wait impatiently for their opportunity to transmit and have no central coordination. They may all try to speak up (transmit) at once, and one will win, and the others will use a backoff timer and continue waiting impatiently for the next round.

As an interesting aside, cellular wireless uses a guaranteed scheduling mechanism, whereby the infrastructure (the eNode, a cellular AP) acts as air traffic control and tells each endpoint when to transmit.

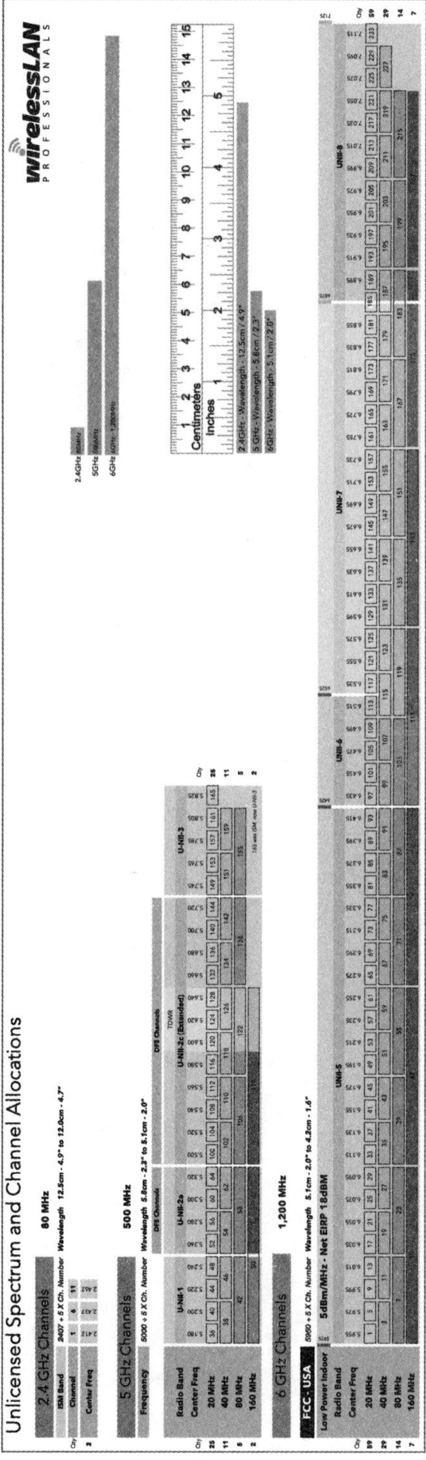

Figure 4.12: A high-level comparison of spectrum availability and channels in 2.4 GHz, 5 GHz, and new 6 GHz

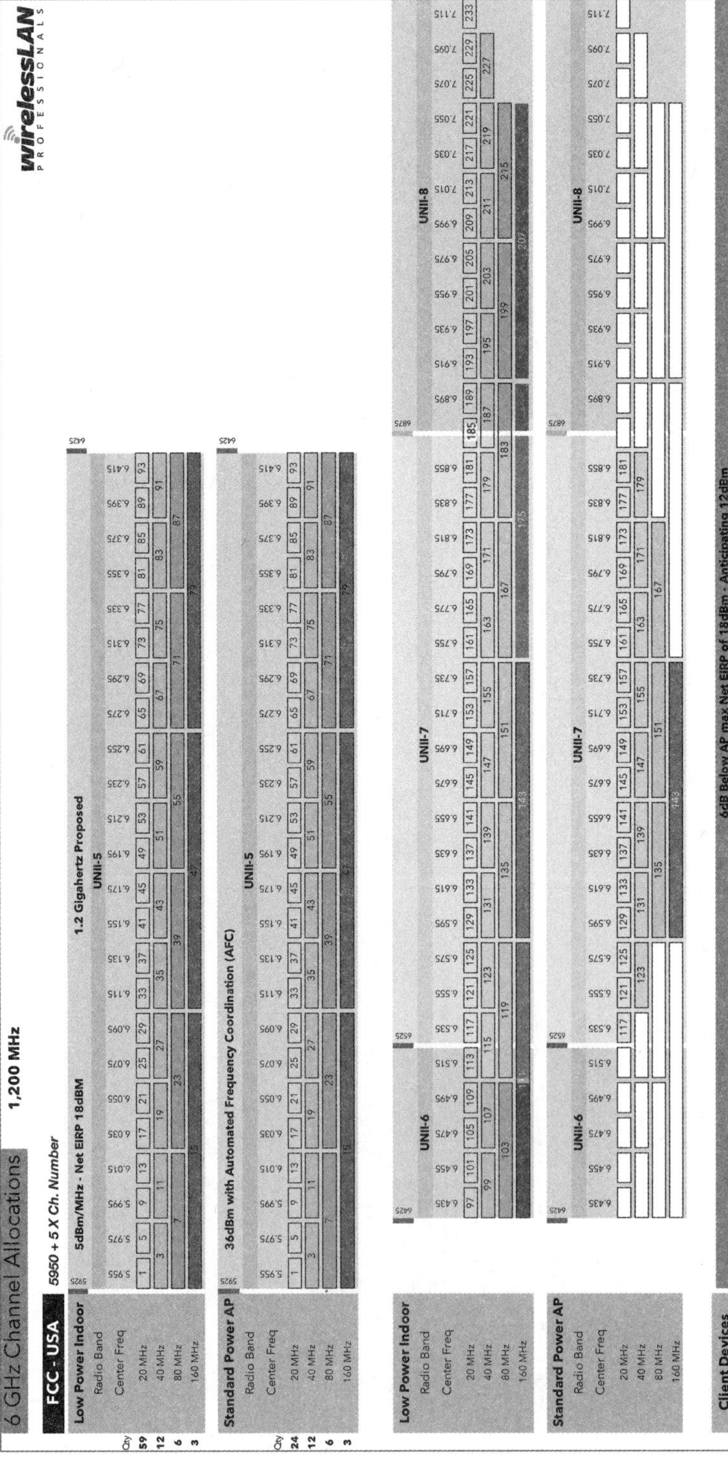

Figure 4.13: 6 GHz channel allocations in detail

Without the benefit of scheduling, Wi-Fi's contention-based mechanism can get a bit bonkers, with no standard way to prioritize faster, more capable endpoints. The result with Wi-Fi is that slower devices are afforded the same opportunity to talk as more capable devices. That speaks more to issues with support for legacy devices. That issue, though, is compounded if the capable, fast clients are also transmitting slowly. When you rate limit over the air, you're slowing down everyone. Because of the contention mechanism, any use of the same RF airspace (same physical location and Wi-Fi channels) is impacted, not a single SSID or single group of clients. Translated, that means your guest users can (and will) impact your production networks. If everything is moving slowly on Guest-SSID on AP01-radio 2 (5 GHz, channel 36) then everything is moving slowly on Secure-SSID being served by AP01-radio 2 (5 GHz, channel 36).

Instead, if your intention is to limit bandwidth of certain classes of users, networks, or specific VLANs, rate limit on the wired infrastructure—when traffic is egressing a controller (or AP), at a router, or firewall through bandwidth or application traffic shaping policies.

Here's the really nasty part. If planned poorly, rate limiting on the wired network will also slow down the wireless network(s). When Wi-Fi clients transmit and the packets can't get through to the intended destination (let's say a YouTube service on the Internet), the result is a lot of retransmissions, and consumption of air space.

If you're interested in a visualization of this phenomenon, see Figure 4.14 and its source link, which shows the impact of wired rate limits on streaming media. The graph is an image from the blog post "Wi-Fi and the Netflix Effect aka Don't Rate Limit your Wi-Fi," which includes a detailed analysis of the impact of wired rate limiting.

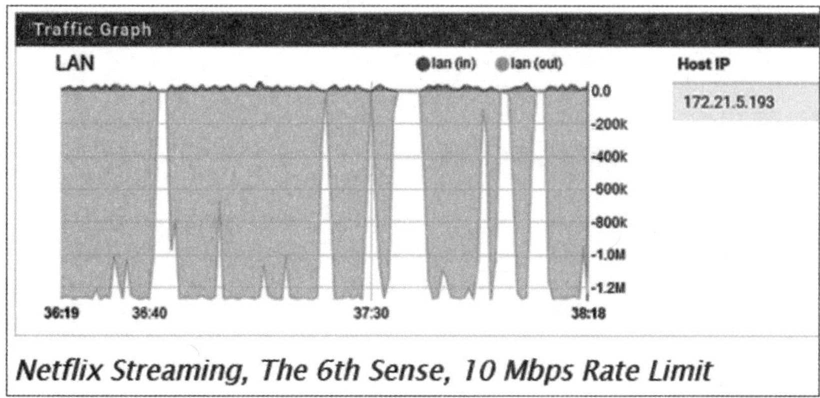

Figure 4.14: In Wi-Fi, the white spaces are the only times the RF medium is free for other clients to transmit.

`https://jimswirelessworld.wordpress.com/2019/12/19/wi-fi-and-the-netflix-effect-aka-dont-rate-limit-your-wi-fi`

My personal preference is to use application-layer traffic shaping, which discriminates more accurately critical traffic versus leisure. Shaping traffic instead of rate limiting is always smarter and gives you more bang for your buck, plus all around better user experience on wired and Wi-Fi networks.

There are also options for rate limiting traffic within the Wi-Fi infrastructure that don't apply the policy over the air, and those are perfectly fine to use (if designed correctly).

Having said that, most IT leaders in organizations find that the guest traffic is not impacting the production networks and users in any significant way, and the additional work of throttling or shaping traffic isn't warranted.

Whatever your decision, remember that availability is a critical component of secure wireless infrastructure and avoid shooting yourself in the foot with this seemingly innocuous feature.

Figure 4.15 shows rate limiting options on Juniper Mist. All enterprise Wi-Fi products have similar options. Be sure you understand how the rate limiting is being applied and the potential impact to the environment.

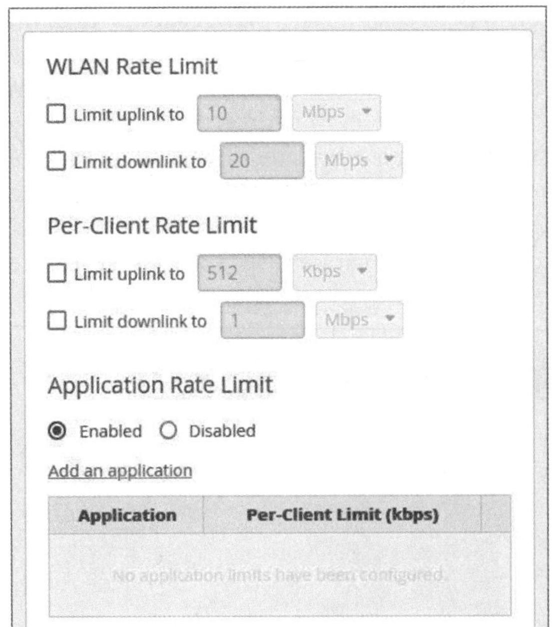

Figure 4.15: Sample rate limiting options on Juniper Mist

Key take-aways on rate limiting Wi-Fi clients:

- Rate limiting over the air is always bad.
- Rate limiting on the wired infrastructure is better, if done properly.

- Improper rate limiting on the wired infrastructure can cause backups that will impact Wi-Fi.
- Impacts to Wi-Fi users impacts all users on the same channel in a physical space, meaning your non-guest users are also affected.
- Shaping traffic based on prioritization instead of rate limiting is always better.
- Upon analysis, most organizations find guest traffic is not negatively impacting production traffic.

Other Networking, Discovery, and Routing Elements

Lastly, there are other networking architecture decisions for Wi-Fi that will impact security. Not an exhaustive list, these are a few topics that may require some additional investigation and planning for your security architecture.

Discovery Protocols

Discovery protocols such as Link Layer Discovery Protocol (LLDP), Cisco Discovery Protocol (CDP), and the LLDP Media Endpoint Discovery (MED) extension are all used for infrastructure devices to advertise attributes, capabilities, and (in the case of LLDP-MED) offer extended details. Each of these work at layer 2.

LLDP is the vendor-neutral discovery protocol that offers details such as the device's identity, capabilities, port information, and management address. LLDP is based on the IEEE standard 802.1AB.

CDP is Cisco's proprietary discovery protocol that operates similarly to LLDP with some additional features. Although proprietary to Cisco, CDP functions are often available in other vendors' products through licensing agreements. It's worth noting that some extensions to CDP are reserved for Cisco use only, such as an extension for phone handset location details used with Cisco VoIP solutions. These reserved CDP functions are not licensed and not available outside of Cisco products.

LLDP and CDP allow network devices such as switches, routers, and APs to discover adjacency and additional details of system capabilities. These protocols are also used by a host of networking management and monitoring tools for building out topologies and populating device inventories.

LLDP-MED is an extension of the vendor-neutral LLDP; it enables endpoint devices such as VoIP phones, certain compute and storage devices, and APs to communicate location and capabilities, and to automate discovery of policies for VLAN assignments, Power over Ethernet (PoE) negotiation, and layer 2 and layer 3 quality of service (QoS). Although an extension of the LLDP standard, LLDP-MED was originally developed by the Telecommunications Industry Association (TIA) and brought into the world as a standard by that organization and American National Standards Institute (ANSI).

The following figures show sample output of CDP (Figure 4.16), LLDP-MED from an AP (Figure 4.17), and LLDP-MED from a VoIP phone (Figure 4.18). Notice the additional information about the class of device and PoE in Figure 4.17, and the MED details include the tagged VLAN of 88, a layer 2 priority of 7, a layer 3 DSCP policy of 46, and PoE requirements in Figure 4.18.

```
CDP neighbors information for port B1

 Port : B1
 Device ID : 5c 5b 35 50 06 59
 Address Type : IP
 Address      :
 Platform     : Mist Systems 802.11ax Access Point.
 Capability   : Switch
 Device Port  : ETH0
 Version      : Mist Systems 802.11ax Access Point.
```

Figure 4.16: Sample output of CDP on the same AP and switch

```
LLDP Remote Device Information Detail

 Local Port   : B1
 ChassisType  : mac-address
 ChassisId    : 5c 5b 35 50 06 59
 PortType     : mac-address
 PortId       : 5c 5b 35 50 06 59
 SysName      :        -AP43JJMFD
 System Descr : Mist Systems 802.11ax Access Point.
 PortDescr    : ETH0
 Pvid         : 1

 System Capabilities Supported  : bridge, wlan-access-point
 System Capabilities Enabled    : bridge, wlan-access-point

 Remote Management Address
    Type    : ipv4
    Address :

 MED Information Detail
   EndpointClass          :NetworkConnectivity

 Poe Plus Information Detail

   Poe Device Type          : Type2 PD
   Power Source             : Only PSE
   Power Priority           : High
   Requested Power Value    : 25 Watts
   Actual Power Value       : 25 Watts
```

Figure 4.17: Sample output of LLDP-MED of an AP connected to a switch

```
LLDP Remote Device Information Detail

 Local Port    : A3
 ChassisType   : network-address
 ChassisId     : 192.168.1.115
 PortType      : mac-address
 PortId        : 08 00 0f 5f a8 ce
 SysName       : regDN 101,MITEL 5330 IP
 System Descr  : regDN 101,MITEL 5330 IP,GigE,Cordless,h/w rev 1,ASIC rev ...
 PortDescr     : LAN port
 Pvid          :

 System Capabilities Supported  : bridge, telephone
 System Capabilities Enabled    : bridge, telephone

 Remote Management Address
    Type    : ipv4
    Address : 192.168.1.115

 MED Information Detail
   EndpointClass          :Class3
   Media Policy Vlan id   :88
   Media Policy Priority  :7
   Media Policy Dscp      :46
   Media Policy Tagged    :True
   Poe Device Type        :PD
   Power Requested        :8.8 W
   Power Source           :Unknown
   Power Priority         :High
```

Figure 4.18: Sample output of LLDP-MED on a VoIP device

INDUSTRY INSIGHT: SECURITY AND DISCOVERY PROTOCOLS

Of course, discovery protocols can also make it easier for an attacker to perform reconnaissance on a network, and there may be a time and place these should be disabled for security purposes.

Having said that, for most organizations, the benefit of these protocols on a properly architected and secured network will greatly outweigh the security benefit of disabling them.

Yes, they introduce some risks. But in my experience and the experience of penetration testers I work with, there are almost always more glaring security holes and an easier path of attack and entry.

Properly architected and secured means proper segmentation controls, restricted access to the management network(s), and hardening of the infrastructure devices. With these in place, most organizations should feel comfortable enabling and using discovery protocols to their fullest potential.

As always, though, this is implementation-dependent. There are organizations that, due to the nature of their business, should opt to harden the infrastructure beyond what I consider to be normal and reasonable means. Given the resources available to organizations, this is most often limited to certain federal agencies and commercial institutions that may be targeted and need additional protection.

More information on securing discovery protocols is covered in Chapter 6.

Loop Protection

Loop protection is an often overlooked feature to aid in meeting availability requirements. I'm also going to use it as a small soapbox moment to remind everyone that spanning tree protocol is not, and never was, designed to be used for loop protection. Spanning tree is designed for calculating the best path in the presence of multiple (planned) redundant links.

There's almost no faster way to take down a network than to introduce a loop to an environment with no loop protection. I think many of us know that from painful firsthand experiences.

As such, planning loop protection is a critical part of ensuring availability and uptime—especially for architectures that include bridged client traffic from APs.

Best practices and specific how-tos vary by vendor implementation, so I'll just leave you with this: the best practice and recommendation is to use it—whatever that means for your environment.

CAUTION **Every network admin probably has a story (or twelve) about how spanning tree crashed a network. In fact, Stephen Orr suggested we simply title this section "How to Crash a Network in Two Easy Steps."**

IPv6

I've struggled with whether to cover IPv6 here, and to what depth. Architecting and securing IPv6 infrastructures is a topic not easily piggy-backed onto this book's content. The variations of feature support for, and implementations of IPv6 alone make it challenging to cover all scenarios.

A majority of this book's technical content is IPv4-centric and covering IPv6 would necessitate further abstraction of concepts—an option I felt would drastically reduce the usefulness of the information here. Lastly, there are simply not many organizations implementing IPv6 internally. A topic we've spoken about and planned for over a decade, IPv6 is in use as the primary network protocol in a very few environments I've worked in. A few more have IPv6 running in parallel with IPv4 for the foreseeable future, and of course a larger number are exploring and testing IPv6 in subsets of the network.

Having said all that, some of you may be architecting wireless on an IPv6 network—possibly for Wi-Fi services or an IoT deployment, and that should be factored into the security architecture.

For organizations that are not yet readily using and monitoring IPv6, the suggestion is to disable that protocol on the network, which means disabling it on endpoints as well as infrastructure devices.

Dynamic Routing Protocols

Dynamic routing protocols are often used within wireless infrastructures (especially controller-based) to advertise client networks to upstream devices, and to route between branch office locations over the WAN.

As part of the planning and documentation process, your security architecture should address any routing protocols including dynamic routing protocols such as Open Shortest Path First (OSPF, an IEEE standard and the most common protocol supported in Wi-Fi products), Enhanced Interior Gateway Routing Protocol (EIGRP, a Cisco proprietary protocol), and Border Gateway Protocol (BGP, popular for branch office deployments).

Whatever routing protocol is used (if any) for these purposes, the protocol should be properly secured according to the vendor's recommendations and should include segmentation, documentation, and regular testing, with an emphasis on "regular testing."

Layer 3 Roaming Mobility Domains

Earlier, we covered portions of Wi-Fi roaming protocols, specifically for the purpose of understanding their impact on encryption key exchanges. That roaming content addressed layer 2 roaming, but many Wi-Fi vendors also offer layer 3 roaming—especially common in controller-based deployments.

Layer 3 roaming allows an endpoint to roam among layer 3 subnets—for example, allowing Lucy Student to move seamlessly from the ACME-University network at the library (VLAN 86, subnet 10.10.86.1/24) to the same ACME-University SSID in the math building (VLAN 44, subnet 10.10.44.1/24).

While that scenario is supported via the industry-standard layer 2 roaming, Lucy would simply have disconnected and reconnected, receiving a new IP address somewhere along her walk from the library. If, on the other hand, your architecture demands Lucy make the journey and retain her original 10.10.86.x address, you'll need to configure layer 3 roaming.

This type of roaming is vendor-dependent, and capabilities vary from product to product. If layer 3 roaming is part of your architecture, certain aspects of other design elements may be impacted, meaning your security architecture may also need to be adjusted.

Summary

Networked architectures are always better when approached holistically. The purpose of this chapter is to highlight just a few of the many interdependencies of Wi-Fi design and Wi-Fi security architecture. From channels and radio power to roaming protocols and rate limiting—almost every design element has a place in a security conversation, and vice versa.

Take-aways include the importance of the many network and domain services that are critical to not only the operation of Wi-Fi but also to its security. Time sync services ensure our hyper-connected world of microservices and cloud services are working as planned. You've also seen the important role of DNS, DHCP, and certificates and the impact DHCP has on wireless security as a key factor in enforcing segmentation.

Within the realm of wireless, just about every design element impacts security either directly or indirectly. This chapter covered the vital role of fast roaming in secure Wi-Fi and explained why Fast BSS Transition (FT) is the protocol we should be using on all enterprise networks. Our hope for the industry is better alignment toward FT and an increased interest by both Wi-Fi and endpoint vendors to support essential features and demonstrate interoperability through testing and certification, even for optional functions.

Maintaining system availability with resilient designs both over-the-air and the wire were addressed here as well, along with the likely unpopular recommendation for scheduled failover testing and more frequent maintenance windows for security patching.

I've also called out several cautionary tales related to loop protection, spanning tree, and rate limiting of Wi-Fi networks. In these, there's certainly opportunity to learn from the mistakes of others.

As a prelude to the hardening topics of Chapter 6, we've also laid some groundwork around expectations related to IPv6, discovery, and routing protocols.

Just as RF parameters shouldn't be designed in a vacuum, nor should security. This book and this chapter contain a breadth of flexible guidance for making decisions and contains very few commandments of "always" or "never" using certain architectures.

Next in Chapter 5, "Planning and Design for Secure Wireless," you'll not only find technical information for planning your architecture but also a dedicated section summarizing the concepts in Part I for non-technical leadership.

Part

II

Putting It All Together

In This Part

CHAPTER 5

Planning and Design for Secure Wireless

Chapter 1, "Introduction to Concepts and Relationships," offered an introduction to the foundational concepts and relationships of the security and wireless networking realms. Chapters 2 through 4 included deep technical dives in the disciplines of segmentation, wireless security profiles and standards, authentication, authorization, and the domain services impacting wireless security.

This chapter begins our journey into Part II, "Putting It All Together," in which knowledge from the prior chapters comes together in actionable guidance.

Instead of following a model of traditional design methodology, this chapter walks through my own methodology—including the planning processes, documentation templates, and decision trees I've used to help hundreds of clients architect secure wireless.

What I've noticed over the years is that most networking professionals within an organization tend to wing it when it comes to planning, often bypassing any formal scoping and documentation and skipping to configuring products. However, I think it's fair to say these teams have enjoyed the process and benefited from the more structured planning that accompanies having a third party involved in the design architecture.

It's natural to feel so familiar with an environment that you dismiss any formal planning, but even a slightly structured model with some basic documentation is extremely helpful in tasks of planning, communicating, monitoring, and troubleshooting. You don't have to hire a consultant to get the advantage of planning; this step-by-step guidance offers a great model for you to do it yourself!

This chapter's content is organized in the following topics:

- Planning and Design Methodology
- Planning and Design Inputs
- Planning and Design Outputs
- Correlating Inputs to Outputs
- Planning Processes and Templates
- Notes for Technical and Executive Leadership

Planning and Design Methodology

You've likely heard of design methodologies such as *4D (Discover, Design, Develop, and Deploy)*. The Wi-Fi world has its own set of design steps addressing the many phases of RF design and validation. While all valid, these traditional models don't focus on design, nor do they address the complexity of architecture that crosses disciplines and domains.

My design methodology incorporates five interconnected phases, unabashedly borrowed from the constructs of the Design for Six Sigma (DFSS) framework. For any Six Sigma professionals out there, I hope you'll extend a bit of latitude and allow me to exercise some artistic license.

These five phases are not always linear in nature, but they do link to two concrete processes of inputs and outputs of a design architecture and can be grouped into three stages: *discover*, *architect*, and *iterate*.

The five phases for designing a secure wireless architecture are (see Figure 5.1):

Discover Stage

- Phase 1: *Define* (scoping)
- Phase 2: *Characterize* (requirements mapping)

Architect Stage

- Phase 3: *Design* (functional mapping)

Iterate Stage

- Phase 4: *Optimize* (design adjustment)
- Phase 5: *Validate* (validate design against requirements)

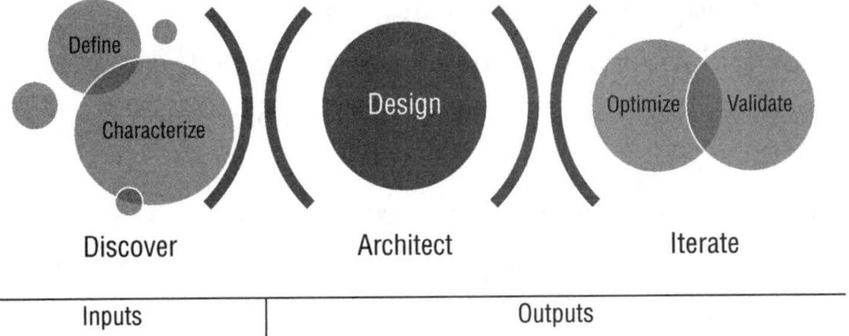

Figure 5.1: The five phases of the planning and design methodology

Discover Stage

The *discover* stage includes the tasks that serve as inputs into the architecture design. This entails scoping and requirements mapping with the first two phases:

- Phase 1: *Define* (scoping)
- Phase 2: *Characterize* (requirements mapping)

Once these two phases are complete, you'll move onto the *architect* stage, which encompasses the third phase, *design*.

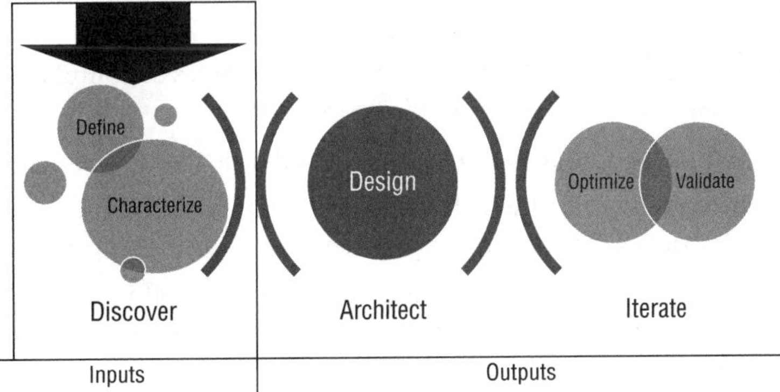

Phase 1: Define

The *define* phase includes identifying project requirements, elements of scoped environment, and scope limits.

During this time, the architect should perform activities such as:

- Identification of the teams and roles involved in the project
- Discovery of the environment (wired and wireless network infrastructure components, capabilities, and topology)

- Scope of user and endpoint population and capabilities
- Identification of applications to be supported over the wireless network
- Scope of geography/coverage areas (e.g., campus, branch offices, home users)
- Identification of security and compliance requirements
- Discovery of additional supporting policies or guidance for security
- Documentation of discovered items

This exercise of the *define* stage of discovery is enhanced by the *characterize* phase, which aligns requirements to the scoped elements.

Phase 2: Characterize

The *characterize* phase addresses the discrete elements for requirements mapping. In this phase the architect captures both qualitative and quantitative security characteristics mapped to the individual classes of networked elements such as endpoints, applications, and users. Those characteristics are then used for functional mapping in the *design* phase.

The architect correlates items from the *define* phase such as:

- Identify elements (endpoints, users, infrastructure, or assets) that need specific security controls to meet business objectives or compliance requirements (e.g., network segments in scope of PCI)
- Group and categorize elements with similar needs or characteristics
- Identify and document which scoped elements have requirements dictated by policy or regulation, such as authentication or encryption
- Document requirements for cases requiring elevated controls such as additional monitoring or inspection, security posturing, multi-factor authentication

The *define* and *characterize* phases together comprise the *discovery* tasks and are the inputs to the architecture tasks of *design*, *optimize*, and *validate*.

Architect Stage

The *architect* stage (architect being an action here) involves only the *design* phase, where the inputs from the discover stage are used for functional mapping.

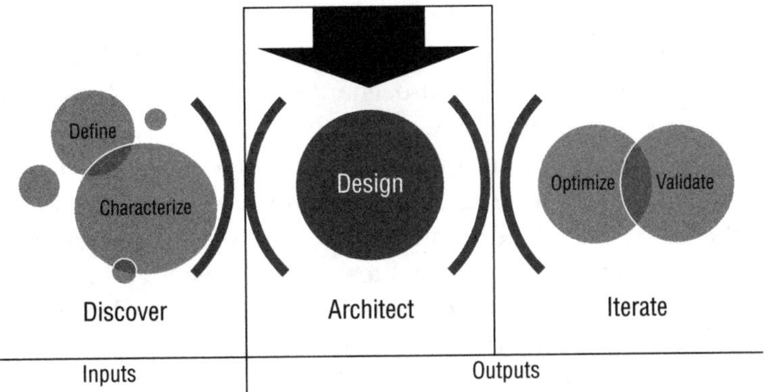

Phase 3: Design

The *design* phase encompasses the heavy lifting of taking the discovery inputs and performing functional mapping for requisite security controls and monitoring. As part of this work, the architect should also document conditions, variables, and known or anticipated design gaps.

During the *design* phase, an architect will:

- Begin mapping defined requirements to planned designs for scoped elements (wired and wireless infrastructure, endpoints)
- Document conditions and variables that may impact the expected outcomes and security posture (such as unknowns of planned but unscoped projects based on wireless connectivity such as digital transformation or IoT programs, or unknown variables of endpoint support for WPA3, or an upcoming merger or acquisition)
- Evaluate current infrastructure and tools to determine if they can meet the objectives
- Identify vendors, products, and configuration options to meet the security and connectivity objectives
- Define metrics and outputs for monitoring and testing against mapped elements
- Produce documentation for as-built designs of the infrastructure devices

Iterate Stage

Maintaining security requires continuous improvement, and the *iterate* stage helps meet this need with the final two phases:

- Phase 4: *Optimize* (design adjustment)
- Phase 5: *Validate* (validate design against requirements)

The *iterate* stage is focused on design iteration and ensuring the architecture is updated to meet changes including those related to new vulnerabilities, changes in the network infrastructure, changes in endpoints and applications, and changes in use cases, among other things.

The *design*, *optimize*, and *validate* phases are iterative, with *optimize* and *validate* phases often being interconnected and non-linear.

During these tasks, it's reasonable to expect a proof of concept (PoC). PoCs may be as basic as having the internal team create test SSIDs and validate the operation against the design architecture or as complex as a lengthy structured plan with a vendor that includes installation of hardware and/or software.

NOTE **In this model *optimize* and *validate* phases refer to optimizing and validating the design architecture, not the implementation. This is a subtle difference from other network design and deployment methodologies.**

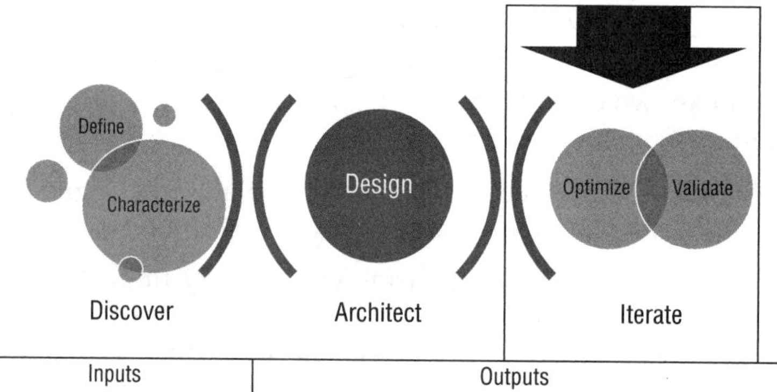

Phase 4: Optimize

During the *optimize* phase, the design is refined to enhance robustness of performance and security.

With industry standards evolving at an unprecedented rate, wireless endpoint capabilities always in flux, and security threats changing daily, wireless networks are no longer set-and-forget. For purposes of security, the architecture tasks are iteratively optimized and recurringly validated.

As part of recurring *optimize* phases, architects should be:

- Researching changes in security protocol standards and implementing enhancements in the architecture
- Evaluating new vendor product features for additional security benefits
- Consuming output from validation to further refine the architecture

- Communicating to stakeholders any major changes in guidance for security best practices
- Updating internal standards and process documents to reflect changes as needed

Phase 5: Validate

As part of the *validate* phase, the architect will verify capabilities and expected outcomes of the design against the originally scoped requirements from the *discover* stage tasks (*define* and *characterize*). The architect should also plan to communicate with other teams regularly and request feedback from stakeholders to ensure the scope hasn't changed and expectations are met and documented satisfactorily.

After an initial deployment and as part of ongoing improvement, the *validate* phase will include testing and validation of the system including security assessments and penetration testing, possibly along with compliance audit outputs.

In the iterative *validate* phases, the architect will:

- Evaluate the planned design against the requirements defined in *design* and *characterize* phases
- Document gaps to be addressed in an iterative *optimize* phase
- Communicate findings to participating teams
- Present findings to stakeholders and request feedback
- Incorporate data from identified metrics in *design* phase

The five phases facilitate the collection and organization of the data for planning in the form of inputs and outputs. Inputs being data consumed and factored in planning, and outputs being the actionable requirements for the infrastructure design.

Planning and Design Inputs (Define and Characterize)

The *define* and *characterize* tasks of the *discover* stage provide the planning inputs. During this stage you'll gather information around the business and technical scopes (*define*) and correlate scoped items to the security requirements (*characterize*).

In the following pages there are comprehensive notes on questions and considerations of project inputs. Given there is expected overlap, the content will naturally taper with successive topics as overlapping bullets are addressed in the earlier sections and not duplicated.

Scope of Work/Project

The first task is to understand the overall project scope. With wireless projects, the scope may be as simple as adding a single SSID, or it may entail a full system design as part of a greenfield deployment or organization-wide upgrade to a different product platform.

Even in the scenario of only adding an SSID, there can be a host of complexity depending on the use case, ownership, and applications to be supported.

The following is a sampling of questions designed to tease out some of the more intricate elements of a scope. Of course, not every project will warrant asking each question but consider this a fairly comprehensive starting point. I've been told I'm nothing, if not thorough. With each question is a selection of scenarios from various client projects and explanations on the relevancy.

- *Who are the stakeholders for the project?*

  This could be a person, a team, or department. For example, it may be the CIO of a retail chain, or clinical engineering team in a hospital, or facilities manager in a manufacturing organization. Understanding what you're doing, for whom, and what they care about is an instrumental part of planning.

- *What are the definitions of success or operational parameters?*

  The stakeholders likely have a high-level objective to meet, such as they need a wireless infrastructure to support autonomous guided vehicles (AGVs), or a network is needed to support a new nurse paging platform. Frequently the scope may be to fix or resolve an existing issue—such as the push-to-talk application isn't working, and they'd like it resolved. Or the guest network captive portal is being flaky and needs to be redesigned.

- *Is the scope organization-wide or a subset of networks/devices/geographies?*

  The scope may specifically include (or exclude) the PCI environment, or the scope may be to add a single SSID to support a new biomedical device set at a single location, or possibly the scope is to upgrade an existing PSK network to 802.1X.

- *What is the timeline for this project?*

  Non-technical stakeholders don't always understand the impact of the scope they've outlined and asking early provides the opportunity to set expectations early. Physically replacing thousands of APs is a much different time requirement than adding an SSID.

- *Is there budget allocated for this project?*

  Just as with timelines, stakeholders may not have a concept of whether the ask requires budget, and how much. Even if you don't believe you'll need to purchase anything at the outset, it's possible you'll need additional hardware or licenses and it's great to understand if that's an option. Also, whether budget is allocated or not can be a good indicator of how serious the organization or stakeholders are about solving a problem.

- *Do the stakeholders have specific concerns or requirements?*

  Often security-related projects are driven by concern following an incident at a peer organization, or a general increased threat level for an industry.

  As an example, when a local government agency has experienced a breach, the peer agencies will usually tighten down security controls. Similarly, healthcare (among other industries) often experiences cycles of being targeted by malicious actors, and as such those organizations will have heightened sensitivity to security needs during those periods.

  Fear and uncertainty often lead organizations toward reactions that aren't necessarily thoughtful responses. It may manifest as a project or request that isn't commensurate with the threat. Put more directly—you have an opportunity to help ensure whatever you're tasked to do will help solve the concern in a meaningful way and not just add busy work with no tangible security benefit.

- *Are there any known constraints?*

  The project may have some fixed parameters you don't have control over. For example, if you were tasked with creating a new network for connected sensors for facilities monitoring, a constraint may be that the endpoints and monitoring system has already been purchased and your design must work with those devices.

  Other times constraints may be operational, such as you're tasked with creating a secure network for home users with remote APs, but the organization is now operating 100 percent remote, and you can't touch the endpoints to provision them.

- *Who owns the project?*

  Aside from the stakeholders who have a vested interest in the project, there may be a different person or group who has final technical authority over the project. Or there may be specific guidance on who should have input into the process or project.

  I tend to like a simplified RACI (Responsible, Accountable, Consulted, Informed) approach, which will help capture who owns the project, who has input, and who needs to be updated about the progress or outcome. A simple RACI matrix may indicate who's involved in the project, and a more detailed matrix can detail discrete project phases or tasks.

  Figure 5.2 is a RACI chart for a sample wireless project with just four project tasks (left column) and a model with both internal and external teams. For each project task, ideally one person or team is identified as accountable, and others may be responsible, consulted, or informed, indicated with the letters A, R, C, and I, respectively. Completing some or all of the information here can help clarify who is responsible, who has input, and who needs to be informed throughout a project.

	Internal Teams					External Teams	
Project Task	CIO	Compliance Office	Clinical Engineering	Network Architect	NOC and Help Desk	Structured Cabling	Vendor Engineer
Scoping and Design	A	C	C	R	I	I	C
Installation and Configuration	A	I	I	R	I	R	I
Testing and Validation	A	C	C	R	I	I	C
Ongoing Support	A	I	I	C	R	I	I

Figure 5.2: Example of a simplified RACI matrix for a wireless project

The project tasks can be as granular as needed and in more complex projects with an assigned project manager may be much more detailed than this. Conversely, a RACI matrix for a small project may have just one task line.

Teams Involved

Aside from the specific groups identified in scoping and the RACI matrix, you'll likely have your own Rolodex of project contributors and collaborators.

I'm a huge fan of efficiency, and one of the best paths to efficiency is through communication. Determining who you need to communicate with, and when, is vital. Even for small projects it's helpful to identify collaborators early. Communicate early and often and you'll lessen the chance of having to do rework because a critical detail wasn't brought in until the end.

EVERY PROBLEM IS A COMMUNICATION PROBLEM

Nobel laureate and literary critic Bernard Shaw is credited with saying, "The single biggest problem in communication is the illusion that it has taken place."

One of the benefits of coming into a project as a contractor has been the ability to help facilitate communication within an organization, between peer teams as well as between IT professionals and non-technical leadership. Most technology problems are people problems, and most people problems are really just communication problems.

There are many reasons an IT pro may shy from certain communications, and rightfully so—it can introduce delays and complexity. But in the end, communication is critical to the success of projects, especially security architectures that cross multiple domains and disciplines.

Harkening back to the discussion of roles and responsibilities in Chapter 1, there are several teams or roles you'll likely need to connect with, as explained in the following sections.

CISO, Risk, or Compliance Officer

Regardless of the title, you'll want to have communication with the person who can best offer direction on security requirements for the organization.

Specifically, you'll want to understand regulatory and compliance mandates as well as any organizational policies related to your project. This person should also be able to provide details around any vendor management and supply chain security requirements.

REFERENCE **Revisit Chapter 1, "Introduction to Concepts and Relationships," for a refresher on the various people, teams, and roles, and how they relate to wireless architecture.**

Security Analyst or SOC

A security analyst or representative from the *security operations center* (SOC) can help identify requirements and processes for logging and alerting for the wireless infrastructure. This person or team can also collaborate on appropriate actions for escalations internally and define steps for incident response (IR) related to incidents on the wireless network.

Lastly, if there's a SOC they should play a role in the vulnerability management including scanning and reporting on the wireless system and track remediations around recommended security patches.

As one example, with WPA3-Personal networks especially, denial of service (DoS) monitoring and alerting is recommended to prevent DoS attacks through flooding that overwhelms the processor. Knowing this, the SOC should be monitoring and alerting on indicators of this type of attack.

Identity and Access Management Team

Networks based on 802.1X/EAP authentication or even MAC Authentication Bypass (MAB) will need to reference a user or device account. Your identity and access management (IAM) contact can help ensure the proper connection of the authentication server to the various user or device repositories and participate in any testing and validation.

This person may also need to be involved in help desk procedures that involve account lockouts or password changes.

The IAM team should also play a lead role in the management of your systems' shared credentials and secrets such as SSH keys and API keys (covered in Chapter 6, "Hardening the Wireless Infrastructure").

Network Architect and Network Operations Team

As you saw in the first two topics of Chapter 2, "Understanding Technical Elements," and in Chapter 4, "Understanding Domain and Wi-Fi Design Impacts," every wireless infrastructure relies on the wired infrastructure to varying degrees.

Whether the wired infrastructure is managing the data paths such as in bridged topologies, or simply serving up domain services, it will always play an integral role in wireless security and connectivity. As such, the network architect and network operations teams (if that's not you) should be your buddies throughout the planning stages.

Domain Administrators

Domain administrators (if not the same team as IAM) play an instrumental role in connecting or provisioning domain-based services such as DHCP and DNS as well as certificate and RADIUS services.

Help Desk

The help desk teams and processes are an integral part of supporting a secure wireless infrastructure. In addition, they're also your first line of defense (and offense) to help ensure a positive end user experience. Workarounds and shortcuts to bypass complex security controls lead to often overlooked lapses of security. The help desk can reduce friction with end users and participate in a feedback loop to you as the architect as part of the iterative *optimize* and *validate* phases.

Whether you're adding a network, updating captive portals, or planning a complete overhaul of the wireless, the help desk can provide valuable input into user behavior and ensure approved processes are followed as part of the security policy guidance.

Other System or Application Owners

If your project entails designing and securing a network for a specific application, or for access to/from a specific resource or data set, that application owner should be involved in planning and testing. This person can provide guidance on what the technical requirements are, offer insight on past challenges or issues, and detail the type of access required to and from the application.

He or she may also provide details on how the application is to be used, by whom/what, when, and how. For example, requirements for stationary IP-based temperature sensors are very different than those of mobile telemedicine carts.

Vendors, Integrators, and Other Contractors

Project contributors often overlooked are the contacts from your technology vendor, integrator, or other contractors. There's value in involving each. Having worked with hundreds of vendor representatives over the years, your mileage will vary.

Vendors can bring you product technical expertise, and if your specific account team can't help, they will have an escalation path to the support team, a product subject matter expert, or a product manager. A vendor sales engineer that's fully steeped in the industry can bring valuable insight combining both their deep product knowledge and industry knowledge.

Similarly, if your product reseller is a true value-added reseller (VAR) and integrator, they'll have expert technical resources that can help ensure you're considering all options—including those outside the vendor's portfolio. At the end of the day, the vendor team only gets paid if you buy their products, whereas a reseller often has a broader offering, and other consultants typically only offer services and not products, and therefore have no attachment to the products you purchase.

Aside from vendors, integrators, and consultants, you may also have third-party contractors such as a structured cabling company that may be physically mounting APs, or a penetration testing team, or compliance auditor. These are all great people to identify and communicate with early to reduce the chance of hitting a snag later.

SIZING IT UP: TEAMS, ROLES, AND RESPONSIBILITIES

As you saw in Chapter 1, not every organization will have clearly defined teams for the roles mentioned here. In small organizations, the person responsible for security may also be responsible for all aspects of IT, from networking and servers to help desk and end users.

In large organizations not only will there be teams for the roles described here, but they may be even more granular, with entire teams dedicated to managing routers, firewalls, or WAN infrastructure. In those larger environments, there's usually a cross-functional architect familiar with many aspects of the environment including the network and applications. Instead of wrangling an unmanageable group from possibly a dozen different teams, loop in the architect and work through them for as much as possible then supplement as needed.

Organizational Security Requirements

Obviously, this entire exercise of planning secure wireless architectures addresses the countless technical security requirements such as authentication and segmentation.

At a higher level, the organization's more holistic security temperament should also factor into your planning. Chapter 1, "Introduction to Concepts and Relationships," covered the quantitative and qualitative aspects of aligning wireless security with organizational risk.

The answers to organizational security requirements will shape your approach to designing robust and secure wireless and will define requirements for activities such as audits and testing. The following questions aim to reveal an organization's security requirements:

- *How would you describe the organization's security culture?*

  It's easy to gauge an organization's security culture; here are some questions to consider in support of understanding an organization's overall temperament about security.

  Has the organization made security a priority? Does it provide meaningful security awareness training to all employees? Does emphasis on security come from the top down (e.g., from the CEO or board)? Does the organization have written policies around security, and does it enforce them? Do managers tolerate attempts to bypass security?

  If the organization has a top-down security strategy that's communicated throughout the organization, then you're working in an environment where security is a priority.

- *What is the organization's overall risk tolerance?*

  Risk tolerance isn't the same as security culture. In a perfect world, there would be a correlation that organizations with a low risk tolerance (high security requirements) would have a strong culture of security, but that's not always the case.

  If your organization has a mature risk management program, then it probably has quantitative data around the risk tolerance and risk posture at any given time.

  Without that granular guidance, as demonstrated in Chapter 1, you can make some qualitative judgment calls using a high, medium, low scale.

  If the organization is highly regulated and/or very security conscious, it will have a low risk tolerance. Conversely, if the organization is less regulated and has a weak security culture, it probably has a higher risk tolerance and is willing to accept more risk in favor of ease of use or minimizing complexity.

- *Are there specific compliance requirements that need to be adhered to?*

  Any specific compliance requirements, auditable controls, or frameworks specified by the organization should be carefully considered during the planning of secure wireless.

Often these include requirements for access control, logging, authentication, approved cryptographic algorithms, and segmentation.

- *Has there been a defining security incident?*

  As mentioned earlier, security incidents within the organization, within a peer organization, or threats targeting an industry can all introduce a heightened level of fear, especially from non-technical stakeholders.

  If there has been a defining security event that's driving your project, spend time researching and analyzing it to ensure your architecture addresses that vulnerability.

Current Security Policies

The organization's current security policies (which may include high-level policies as well as more technical standards and procedures) are a requisite input into your planning.

The policies may address who can access systems, when, how, and from what devices. At a minimum, the organization should have guidance around any controls related to compliance requirements. Following are a few other topics that should have policies you'll want to review.

In addition to being an input, updates or additions to policy documents are also an expected output of your architecture. Standards and process documents should be updated to reflect the evolution of industry standards such as the WPA3 security suite, recommended EAP methods, and endpoint capabilities. The following are examples of security policy topics:

- Privileged access and account management
- Use of multi-factor authentication
- Use of personal devices—bring your own device (BYOD), as it's called—to access corporate data
- Guest access rights
- Minimum OS and patching levels for endpoints
- Vendor management and supply chain
- Remote access requirements
- Use of encryption
- Acceptable use policy (internal and guest networks)
- Security incident handling
- Event logging
- Configuration backups

Endpoints

The wireless network's purpose is to facilitate the connection of endpoints to networked resources, whether internal or on the Internet. It stands to reason that defining and characterizing the endpoints is an integral part of the design puzzle.

The endpoint capabilities (or lack thereof) will drive all aspects of the wireless design, from connectivity and RF to security. The following sections describe wireless endpoint considerations.

CAUTION **Endpoint capabilities can make or break a secure architecture. If your enterprise endpoints can't participate in technologies such as 802.1X and FT, your options will be limited. The same holds true for outdated endpoints that don't (and can't) support the WPA3 security suite and Enhanced Open. Wi-Fi Alliance certification of features ensures both conformance and interoperability for these vital protocols.**

Wireless Connection Type

Wireless encompasses a multitude of technologies that can vary from layer 1 and up. Security considerations for 802.11 WLANs (aka Wi-Fi) are different than (for example) wireless personal area networks (WPANs) based on 802.15.4.

Form Factor

The form factor of a device—whether it's a handheld, a tablet, or traditional laptop, for example—correlates to its onboard capabilities (such as processing and memory). Restricted devices are limited in what they can do in terms of computationally intensive processes such as certain cryptographic functions.

How mobile a device is (based on a combination of form factor and use case) also factors into requirements around roaming, among other things.

Lastly, often form factor indicates the bandwidth capabilities of a connected endpoint. A tiny IoT sensor can't possibly transmit the same bandwidth a laptop streaming a movie can, a side effect of onboard storage capabilities and radio capabilities that are often correlated to form factor.

Operating System

The device OS will further shape capabilities and factor into decisions such as supported authentication methods, whether a device can support 802.1X natively, whether a supplicant agent can be added, and which EAP methods are supported.

The OS of personal devices may also shape policies around BYOD and the ability of the organization to monitor and manage corporate data on a personal device. IoT devices are often running slimmed-down versions of OSs, which impacts the organization's ability to secure and interact with these endpoints.

Ownership

Who owns the endpoint is another factor for consideration. Whether a fully capable device such as a laptop or smartphone, or a less capable IoT sensor, the endpoint may be owned by the organization, by an employee (such as in BYOD), or even a third party.

Third-party ownership of endpoints is not as rare as one might think. It's common in organizations to have subsets of endpoints being managed by a third party such as in many healthcare biomedical devices, or for organizations that have managed print services where printers are more or less rented and maintained by a service provider.

Management

Ownership of the device doesn't necessarily correlate to management of the device. For example, in the realm of personal use devices at work there's not only BYOD but also a cornucopia of acronyms such as CYOD, COBO, and COPE for *choose your own device*, *corporate-owned business use only*, and *corporate-owned personally enabled*, respectively. The alphabet soup demonstrates the mixed ownership-management models prevalent in today's enterprises. These ownership models are detailed more in Chapter 8, "Emergent Trends and Non-Wi-Fi Wireless."

Also, as with third-party owned devices, there may be managed services models with corporate-owned externally managed devices.

Who owns and/or manages an endpoint will greatly impact the organization's ability to manage and change configurations such as moving an endpoint to a new SSID, or enforcing DHCP versus static IP addresses.

Location

Where are the devices located? Are they in an employee's home? At an office? For IoT-type devices, they could be distributed over large areas including outdoors.

User-Attached or Not

Whether an endpoint has a user attached or not is a factor in security, specifically how the device can be provisioned, connected, and maintained throughout its life cycle.

Headless devices not only are not user-attached, but may not have an interface for configuration input.

TIP Originally the descriptor of "headless" referred to endpoint devices that lacked a user interface. Later the term evolved and is now most often used to describe an endpoint that's non-user-attached, such as printers and the types of non-traditional endpoints often classified as LAN-based IoT.

Roaming Capabilities

As you saw in Chapter 4, fast secure roaming protocols play a major role in wireless security. Instead of falling back on passphrase-based networks, endpoints that support Fast BSS Transition (FT) can connect to 802.1X-secured networks without sacrificing speed and reliability.

Because of that, the endpoint's roaming capabilities remain a major factor in planning secure Wi-Fi networks.

Security Capabilities

Along with roaming, the WPA version of security an endpoint supports (WPA2 or WPA3 and Personal and/or Enterprise), plus the cryptographic algorithms, and suite of EAP methods determine its security capabilities. To be considered along with WPA3 is support for the newer encrypted Enhanced Open.

The endpoints' security capabilities will then determine how you must architect security profiles, and whether you'll need to use Transition Modes (to support legacy security protocols) or strict WPA3-only modes. Traditional OS-based endpoints receiving updates from the manufacturer can likely be updated to support newer roaming and security features.

REFERENCE See Chapter 2 for more on security profiles including the security considerations, recommendations, and warnings related to the use of Transition Modes.

As with other considerations in the inputs, the endpoint security and roaming capabilities can also be an output of the design, if the organization is in a position to specify requirements for procurement or prepared to upgrade incompatible endpoints.

Quantities

The volume of any class or group of endpoints will help inform requirements for IP address space, segmentation, and options for provisioning.

Classification or Group

Taking all of the above into consideration, you'll then classify or group similar devices together. This is your starting point for the planning templates and tables provided later in this chapter.

Users

For endpoints that have users attached, such as traditional OS-based devices that support user logon, several aspects of the user should be considered. Following are a few bullets to help uncover relevant user attributes.

- *User expectations*
  What expectations does the user have about the wireless connection? They may make assumptions about security, performance, uptime, and support, among other things.
- *Account information*
  How are the user accounts organized and managed? Where the user accounts reside, how they're managed, and referenced in network connection policies such as those in 802.1X-secured networks will dictate how granular your access control policies can be.
- *Relationship to the organization*
  Is the user an employee, a contractor, or guest? The answer to this question sets expectations about the ability of the organization to monitor or manage the user's endpoint, and what type of control the organization has over the user.

System Security Requirements

In this phase of information gathering, the architect will use information about the endpoints, users, and systems to begin documenting who needs access to what. Those elements are further characterized by the knowledge of security compliance requirements and risk tolerance to later determine appropriate security controls—segmentation, authentication, cryptography requirements, logging, etc.

In addition to the questions and information gathered so far, following are a few questions to direct this activity:

- Who (what users/devices) need access to what resources?
- Which networks should be segmented, and to what degree?

- Which security compliance requirements apply to which endpoints, users, or resources?
- What current security policies apply to which endpoints, users, or resources?

Applications

The application(s) to be supported on the wireless network will have requirements for connectivity, service level agreements (SLAs), quality of service (QoS) needs, bandwidth demands, and possibly roaming requirements. For IoT deployments, factors like power capabilities and duty cycle also come into play.

In addition, the application and its data may be in scope for compliance regulation such as PCI DSS, Health Insurance Portability and Accountability Act (HIPAA), the federal Cybersecurity Maturity Model Certification (CMMC) program, or in the U.S. even fall under the purview of the U.S. Food and Drug Administration (FDA) Center for Devices and Radiological Health (CDRH) division. The following are factors to consider in supported applications:

- Connectivity requirements that will help down-select wireless technology (e.g., distance, battery, and bandwidth requirements)
- SLAs and availability or uptime requirements that will dictate resiliency planning
- Quality of service requirements including factors that may impact options for fast secure roaming
- Regulatory compliance or internal policy requirements
- Ownership of the application, data, and endpoints accessing it (e.g., internal versus third party)

Process Constraints

The design may need to happen within the boundaries of certain system or process constraints, such as options for onboarding or provisioning devices.

Process constraints can apply to both greenfield and brownfield deployments but are naturally more prevalent in brownfield deployments where there will be limitations of existing infrastructure, applications, or workflows.

An example of a process constraint would be a requirement to provision secure remote AP access and onboard home users to the enterprise network without having access to the endpoint ahead of time—a common scenario in today's world with many organizations having moved to a 100 percent remote workforce.

Other process constraints may be related to provisioning options for headless or IoT devices, workflows within change management procedures, or an inability to change a directory structure for authentication purposes.

Wireless Management Architecture and Products

Last but certainly not least—in a brownfield deployment the wireless infrastructure may be already in place, without the option to upgrade, redesign, or replace those products. If that's the case, you may be working within product and management architecture constraints. Although not ideal, it's probably the most common scenario, especially for 802.11 WLANs.

And, as with a handful of other inputs, the wireless management architecture (e.g., controller versus cloud) and/or the wireless products may be a design output. In a greenfield deployment, or an upgrade scenario where budget is allocated, the architect may have the option to specify the product solution.

Your analysis during both the *discover* stage and the planning tasks coming up next should address any limitations of the current network infrastructure as it relates to the desired security configuration. There may be opportunities for incremental changes, or the organization may simply need to begin budgeting for refresh cycles.

Planning and Design Outputs (Design, Optimize, and Validate)

The information collected in the *discovery* tasks (*define* and *characterize*) will have informed what security controls are required for the network, connected endpoints, and accessed resources.

The outputs of the architecture planning are tied to the phases of *design*, *optimize*, and *validate*. As explained earlier, the *optimize* and *validate* phases are iterative and often intertwined.

The output will entail about a dozen different aspects, from the overall connectivity technology (e.g., 802.11 vs. cellular vs. Bluetooth), the types of wireless products including infrastructure and (if you're lucky) endpoints. The outputs of the design will also specify the requisite segmentation and data paths, wired infrastructure requirements, domain services needed, wireless networks/SSIDs, hardening, and possibly additional tools or software.

Wireless Connectivity Technology

Digital transformation projects and networks that are greenfield for novel use cases present the perfect opportunity to build a full secure wireless architecture holistically, including specifying the underlying connectivity protocol—802.11 WLAN, cellular LANs, Low Power WANs, Personal Area Network protocols, etc.

As part of the earlier work, the task of identifying which wireless protocols might work for the use case will have been completed as well and may further inform (or expand) the options around endpoints and endpoint capabilities.

For example, if you're tasked with researching and specifying a new network for point of care technology in a hospital, you might consider both 802.11 WLANs (Wi-Fi) as well as cellular LANs. In your research, you may find the cellular LAN solution is more cost effective, more secure, and more resilient than a comparable Wi-Fi solution in the same environment. In doing so, you'd then have the opportunity to specify many aspects of the wireless infrastructure, from the connectivity protocol up.

REFERENCE **Cellular LANs, IoT, and other non-802.11 technologies are covered more in Chapter 8.**

To summarize, here are a few of the dependencies that dictate which protocols and technologies can be considered.

The combination of requirements in these areas will determine the wireless technologies that can be considered:

- Distance or range of connectivity between the endpoint and access point
- Bandwidth of data transferred to or from the endpoint
- Power consumption and battery capabilities of the endpoint
- Duty cycle of the endpoint
- Security requirements
- Provisioning needs
- Encryption requirements

Endpoint Capability Requirements

The wireless network is only as secure as the endpoints attaching to it—and if you're lucky, you may get a rare opportunity to specify endpoint capabilities as part of the secure architecture. As described in the preceding scenarios, digital transformation programs, new use cases, and projects with a greenfield deployment are the perfect times to jump in and be prescriptive with endpoints and therefore endpoint capabilities.

In Wi-Fi networks, two key factors are support for the current version of WPA security (in this case WPA3) and support for fast secure roaming protocols (currently 802.11r being the reigning technology). Aside from that, endpoints should be Wi-Fi Alliance Certified to ensure functionality and interoperability (you'd be surprised how many aren't).

In other non-802.11 wireless technologies, endpoint capabilities to specify include of course the connectivity support, authentication and provisioning options, and encryption.

It was noted that greenfield deployments are a rare opportunity; in brownfield deployments options will be more constricted, as you'll have to work with what's there and make migration plans (e.g., from WPA2 to WPA3) accordingly. Having said that, when working on secure wireless enhancements in existing environments, there may be opportunity to upgrade existing clients to meet the latest standards for fast secure roaming and WPA, as follows:

- In Wi-Fi, when possible, upgrade or select endpoints that support the latest fast secure roaming and WPA versions (currently 802.11r Fast BSS Transition and WPA3)
- In greenfield deployments, select endpoints with capabilities that meet the connectivity requirements defined previously, and security requirements discovered in *Define* and *Characterize* phases
- In brownfield deployments, clearly document and communicate security limitations and risks associated with less-capable clients

Wireless Management Model and Products

The wireless management model (on-prem versus cloud) and products may be an input into your design, or the product of your design as an output. Again, greenfield deployments present unique opportunities but upgrades to a new product or new vendor solution during a refresh cycle present a similar opportunity.

Cloud-managed platforms offer organizations yet another distinctive opportunity during refreshes and expansion. The nature of cloud management makes it exceedingly easy to deploy a new product incrementally and in parallel with an existing solution. Meaning whatever exists in the environment currently, a network architect could deploy a second solution (from another vendor or a different platform from the same vendor) as part of an expansion or partial replacement of the existing solution.

Cloud solutions alleviate the hefty upfront expenses of on-prem infrastructure, plus offer a more linear cost model with AP-based subscriptions. It's easier now than ever to deploy a new solution and manage costing models associated with AP allocation to departments or locations.

However, cloud-managed solutions aren't for everyone. During your design, you may simply decide to modify an existing controller configuration for the desired effect, combining tunneled and bridged traffic to meet requirements, as follows:

- In a greenfield deployment, select a management architecture (controller, cloud) that suits your security and data path models
- For cloud-managed solutions, add a tunnel termination if needed to support client data tunnels (versus bridged traffic)
- In a brownfield deployment, modify existing controller-based solutions as-needed to meet requirements

RF Design and AP Placement

RF design and AP placement are outside the scope of this book, but a critical piece of planning any wireless network. As you learned in Chapter 4, fast secure roaming protocols (which enable the use of more secure standards like 802.1X versus passphrase-based networks) depend on a proper RF design. The following are reminders for proper RF design:

- Verify or specify an RF design that supports the coverage needed for fast secure roaming
- Verify or specify an RF design that has AP placement and redundancy measures to support your planned system availability and resiliency
- Verify or specify an RF design that meets requirements for coverage and quality for the least capable endpoint to be supported

Authentication

Authentication in secure wireless encompasses authentication of endpoints, users, as well as the infrastructure and the admin access (as in hardening). Chapter 6 is all about infrastructure hardening with a deeper dive into that topic.

Each of those decisions will be informed by the requirements identified during the *define* and *characterize* phases. Your decisions around authentication schemas should include the following considerations:

- Incorporate controls from policies or compliance requirements that specify rules for admin access to the wireless infrastructure, such as multi-factor authentication
- Address privileged remote access by internal and external teams that may be managing the system remotely
- Plan based on needs for endpoint device authentication requirements, and/or user authentication requirements, or a combination of both
- Remember 802.1X-secured networks are based on strong mutual authentication, whereas passphrases and MAC addresses are considered identification, not authentication
- Meet requirements for certificates in support of authentication schemas
- Design appropriate infrastructure authentication as part of hardening best practices
- Plan based on highest load expectations, such as factoring in the spikes during the start of a workday or other events
- Address timers including authentication timers and session timers

Data Paths

Decisions around data paths will be commensurate with the management model (cloud or on-prem with a controller or tunnel gateway) and security requirements.

For example, in a highly regulated environment it's common to prefer tunneling some or all client traffic to controllers or a tunnel gateway. At a minimum, restricted traffic such as guest networks and regulated traffic such as those falling under HIPAA and PCI DSS regulations should be tunneled for segmentation and the ability to extend encryption over the wired network.

REFERENCE **Revisit Chapter 2 for more on data paths, segmentation, and the differences with controller-managed APs, local cluster APs, cloud-managed APs, and cloud-managed APs with a tunnel gateway appliance.**

To plan data paths, consider the following best practices with the usual "it depends" caveat and an understanding that any recommendation for bridged traffic comes with the stipulation of "if policy allows" and any recommendation for tunneled traffic comes with the assumption the underlying network architecture supports it.

- Plan to tunnel restricted traffic such as guest and Internet-only networks
- Plan to tunnel data that's sensitive or regulated by a compliance policy
- Plan to tunnel traffic that can't otherwise be secured or segmented if bridged to the network at the edge
- Plan to bridge traffic (if policy allows) when the client and resources are co-located but the controller or tunnel gateway is elsewhere—for example if the wireless controller is hosted in a remote datacenter
- Plan to bridge restricted/sensitive traffic if tunneling is not desired but a network virtualization overlay is in use
- Remember the data path determines where and how segmentation is applied, on the wireless network versus wired
- See Chapter 2 for additional options around data path filtering and segmentation

Wired Infrastructure Requirements

The wireless architecture will dictate many aspects of the wired infrastructure, from security and segmentation to meeting resiliency and uptime requirements.

The edge switch port, edge switch configuration, router/routing switch configuration, and firewall all are impacted by the wireless network. Beyond the firewall, the WAN infrastructure may also be in the equation.

The edge port must support the power over Ethernet (PoE) requirements of the AP as well as bandwidth requirements, with a reminder that new Wi-Fi 6E will oversubscribe a standard 1 Gbps edge port and multi-gigabit Ethernet may be needed.

The edge switch will have to meet the overall PoE load across all ports, and ideally also support always-on PoE that would keep APs up during a switch reboot. The edge switches also play a role in the high-availability design. Of course, the edge switch is center stage in bridged deployments where the ports require configuration to support any bridged client networks.

The router or routing switch has an especially active role in bridged topologies, likely the point of access control through ACLs, and also DHCP proxy. Moving further north in the traffic path, the firewall will (at a minimum) be configured for inward-facing routes and firewall policies allowing traffic to and from the wireless networks from the Internet. Those firewall policies also define any security inspection of the traffic. A firewall may also be the point of traffic filtering for segmentation.

Beyond the firewall, the WAN infrastructure may need tweaking to support wireless traffic efficiently, specifically through proper routing of both client and management traffic. Where VLANs are in use, IEEE 802.1Q (aka QinQ tunneling) may need to be configured along with jumbo frame support. In fact, throughout the wired infrastructure, frame sizes are also a consideration—there's a delicate balance required to minimize roundtrips with larger frames, but not introduce latency to sensitive applications (like voice) by stacking or aggregating too much data at once.

Consider the following when planning wired infrastructure elements:

- Ensure edge switches support minimum PoE and bandwidth requirements (plan for multi-gig Ethernet ports on future switch purchases)
- When possible, select edge switches that support always-on PoE to prevent AP reboots and outages from switch restarts
- Always plan secure network designs with fully managed switches, including edge switches; fully managed means it can be configured entirely through CLI and/or APIs (versus those limited to web UI management or no management)
- Plan for ACLs and/or DHCP proxy (IP helper) for networks that are using a bridged data path
- Plan appropriately for configuration changes to the firewall to support connectivity and security of endpoints to the Internet
- Address WAN routing if required; use SD-WAN for smart dynamic policy based WAN routing when available

- Plan to configure proper frame size support throughout the infrastructure
- Where applicable, plan for proprietary vendor configurations such as access roles that may be extended from the wireless to wired infrastructures

Domain and Network Services

Domain and network services such as DHCP and DNS will be dependent on the data paths and topology. Different wireless networks will likely have different data paths and therefore require configuration of network services at different places.

The wireless architecture will also determine the authentication (RADIUS) server policies to be configured, the user directories referenced, and the certificate services required (for servers and optionally endpoints). Consider the following when planning domain and network services:

- In tunneled networks, plan to serve or proxy DHCP from the wireless infrastructure
- In bridged networks, plan to serve or proxy DHCP from the wired infrastructure using IP helper addresses at the routing switches
- Specify client DNS servers based on whether they require internal access or Internet-only, remembering that many captive portal deployments for guest access require resolving an internal domain name
- Plan appropriate time synchronization services for the wireless infrastructure and endpoints
- Revisit Chapter 4 for additional information on domain services
- Revisit Chapter 3, "Understanding Authentication and Authorization," for more on RADIUS, certificates, and related topics

Wireless Networks (SSIDs)

Ah, we're finally here! The place most IT pros like to start is actually the last step in planning—the wireless network, or SSID in Wi-Fi.

Looking over the preceding half dozen topics, each of those will need to be decided and planned before configuring the wireless networks.

The example and language used here will be 802.11 WLAN specific (Wi-Fi), but the same concepts apply to other wireless technologies as well.

Once you have the prior elements figured out, you'll configure the wireless network and designate the following:

- Network name (in Wi-Fi, what's broadcast for users to see)
- Security profile (WPA2/3-Personal, WPA2/3-Enterprise, Open or Enhanced Open)
- If WPA2/3-Enterprise, additional options around encryption, accounting, RADSEC, RADIUS proxy, and more.
- Data path of bridged or tunneled (most often specified per network/SSID, some products support more granular routing)
- Wireless over the air segmentation (multicast filtering and inter-station blocking)
- VLAN or dynamic VLAN assignment
- Roaming settings (ideally 802.11r if supported by endpoints)
- RF parameters (channel and power settings, and/or dynamic radio resource management)

HOW MANY SSIDs DO YOU NEED? SEVERAL. . .

For years, we've been told to collapse SSIDs to as few as possible in an effort to save airtime by eliminating additional beacon overhead. Well, those days are over! It's a new era.

In planning, you should start to think of SSIDs more like VLANs. Why? All devices on the SSID will receive the multicast/broadcast traffic from all the VLANs. This is a function of all devices on the same SSID sharing the same group temporal key (GTK).

Appendix C, "Sample Architectures," includes more guidance on this, but the short version is this: stop collapsing SSIDs and create as many as needed for proper security segmentation.

If your planning includes the following, consider dedicated SSIDs for each:

- SSID for managed endpoints authenticating with 802.1X
- SSID for managed endpoints connecting with MAB
- SSID for managed endpoints connecting with WPA2-Personal
- SSID for managed endpoints connecting with WPA3-Personal
- SSID for unmanaged or BYOD endpoints connecting with any method
- SSID for guest and Internet-only portal experiences including Open and Enhanced Open

This list may be more than your organization requires, but it exemplifies the recommendation to separate SSIDs as granularly as possible based on security classification, not just the security profile.

Endpoints with different classifications, or that are accessing differing levels of data or assets, should be segmented over the air just as you would with VLANs.

In today's guidance, six secure SSIDs are preferable over two or three collapsed SSIDs. The Wi-Fi technology has evolved and the beacon overhead on a properly designed network shouldn't noticeably impact performance.

System Availability

Availability still being one of the three security pillars, the design should address all aspects of high availability and resiliency from the AP to switch and controller. Consider the following when planning system availability:

- Address resiliency of RF coverage from APs if desired
- Optionally, plan for dual AP port uplinks (only if required and planned properly such as in support of Wi-Fi 6E deployments)
- Ensure edge switch power and control plane redundancy
- Specify dynamic routing protocols, if applicable
- Plan controller redundancy through active clusters, if supported, or active-passive failover

Additional Software or Tools

The design may call for security features not natively supported in the infrastructure you have. One very common example is the need to validate or authenticate both a user and a device on an 802.1X-secured network. In a Microsoft Server environment, the logic AND statement is not supported in Microsoft NPS RADIUS. The options are to use the EAP-FAST with chaining (most likely via Cisco ISE), find devices that support the standards-based EAP-TEAP, and/or purchase a different authentication server with advanced policy support (such as Cisco ISE, Aruba ClearPass Policy Manager, Fortinet FortiNAC, Forescout, or similar).

REFERENCE **See Chapter 3 for more recommendations on authentication, 802.1X, EAP, RADIUS, and options for using vendor-specific solutions and NAC products.**

Along the same lines, the design may call for an endpoint agent (software) to be added for feature support or feature expansion such as supporting additional EAP methods for authentication.

Throughout the process, known limitations and security vulnerabilities may have been identified and documented, and the design may call for additional tools for testing and validation, or for security monitoring and alerting. At a minimum, there may be an overlay testing tool for troubleshooting or for continuous testing and validation of application SLAs. Consider the following when planning additional software or tools:

- Plan additional products or software as-needed to address gaps in meeting defined security objectives; this may include advanced RADIUS servers or NAC products or client agents among others
- Plan tools for ongoing monitoring testing to validate operation, security, and segmentation
- Plan tools for troubleshooting to reduce downtime and proactively find issues in the network

REFERENCE **Chapter 7, "Monitoring and Maintenance of Wireless Networks," covers options for wireless testing and tools in more depth.**

Processes and Policy Updates

Whether greenfield or brownfield, the secure design should include updating or creating related processes and policy suites. It's the job no one wants to do—documentation—but it's essential to maintaining a secure wireless infrastructure. In fact, many security standards (such as NIST) include documentation as part of the scoped requirement.

If this book has demonstrated nothing else so far, it's shown that secure wireless infrastructures touch almost every aspect of an organization—from endpoints to servers and Internet and from compliance regulations to wired switch configurations. The processes are no exception; expect to address processes ranging from end-user support to secure device provisioning and ongoing security monitoring.

In addition, the documented policy suite (policies, standards, processes) will need to be updated. As an organization moves (for example) from support for legacy WPA and WPA2 protocols to WPA3, network configurations will need to be updated, additional segmentation may need to be enforced, and the list is a long domino of changes.

The processes and policies that will likely need to be reviewed and/or updated include those addressing topics, such as:

- Device provisioning processes (endpoints and infrastructure)
- Network monitoring processes and policies
- Help desk processes and end-user support
- Vulnerability scanning and software patching processes

- Security operations and incident response processes and policies
- Security testing processes and policies
- Other wireless standards and procedures

Infrastructure Hardening

Infrastructure hardening bolsters the integrity of the network and works toward a trusted infrastructure design. It involves disabling unused protocols, securing management access, removing unsecured protocols in favor of encrypted ones (such as disabling telnet and enabling SSH), and authenticating the infrastructure components (such as controllers and APs) to one another.

REFERENCE **Chapter 6 dives into this topic deeply. The depth of hardening required depends on the organization's requirements and industry.**

At a minimum, every organization, regardless of size, should plan for a few basic best practices:

- Disable unencrypted management protocols and enable encrypted ones (e.g., HTTPS vs. HTTP, SSH vs. Telnet, and SNMPv3 vs. SNMPv2c)
- Disable management protocols or access not in use
- Use user-based privileged access (don't use shared accounts such as admin or root, instead use RADIUS, TACACS+, or LDAP to authenticate to a user directory)
- Enforce multi-factor authentication for management if accessible remotely
- Limit admin access only to the people or teams that require it
- Secure API keys appropriately (for products that use API keys)
- Maintain physical security of the infrastructure components, ensuring unauthorized personnel, guests, and the public can't physically access the controller, APs, network cables, or jacks
- Configure rogue AP detection to be alerted if an unauthorized AP is connected to the network
- Enforce inter-station blocking over the air when possible
- Use certificates, allowlists, or manually authorize APs to the system

These are considered the bare minimum effort; many organizations have requirements that extend well beyond these tasks—topics covered in the next chapter. The federal government and U.S. Department of Defense (DoD) and peer organizations outside the U.S. all have relatively extensive guidelines for network and system hardening.

Correlating Inputs to Outputs

The next task is to correlate inputs from *define* and *characterize* phases and map to outputs of *define*, *optimize*, and *validate*. Working with the ecosystem of wireless technologies can be more of an art than a science, and your design will require iterative phases of tweaking and repeated validation—validation here meaning validation against the originally defined requirements.

For example, you may design an ideal 802.1X-secured network with FT for fast secure roaming to support a new cloud-hosted VoIP deployment, only to find out the desktop phones and some users' personal smartphones don't support FT. In which case, you may have to re-architect for PMK caching, opportunistic key caching (OKC), or even sacrifice security and move some devices to a passphrase-secured network to eliminate the latency of key distributions during roaming.

SACRIFICING SECURITY FOR PERFORMANCE

In the preceding scenario I cite sacrificing security as an option, which my astute technical editors (and likely you) have called out as undesirable.

It absolutely is. But at times, architects are faced with an impossible task and refusing to implement a network design because it's not perfect simply isn't an option. Telling the organization to replace tens of thousands or at times millions of dollars of endpoints also isn't always a choice. Telling a hospital staff "Sorry, your nurse paying system isn't working," also isn't an option. If you go that route, I might recommend you brush up your resume.

The purpose of the planning exercises here are to give you the best chance at gathering data and communicating requirements to the organization so they can support you in your quest for a secured network architecture.

But there are times when you'll be on the losing end of a battle and I want to make sure *all* use cases and scenarios are covered, not just the ideal ones.

Appendix C offers specific guidance on further segmenting networks in sample architectures, and Chapter 7 includes recommendations for monitoring and alerting when working with less-than-perfect security configurations.

To further assist you in your endeavors, this chapter includes "Notes for Technical and Executive Leadership," which summarizes the highest priority guidance for non-technical audiences.

Figure 5.3 offers a visualization of the relationship of inputs from the *discover* stage (*define* and *characterize* tasks) to the outputs of *architect* and *iterate* stages (*design*, *optimize*, and *validate* tasks). This example is a general-use map, and yours may look a little different. It's possible not all inputs influence the same design outputs, and it's very likely your project might include custom considerations not reflected here.

TIP Remember this and other resources can be downloaded from my blog at `http://securityuncorked.com/books`.

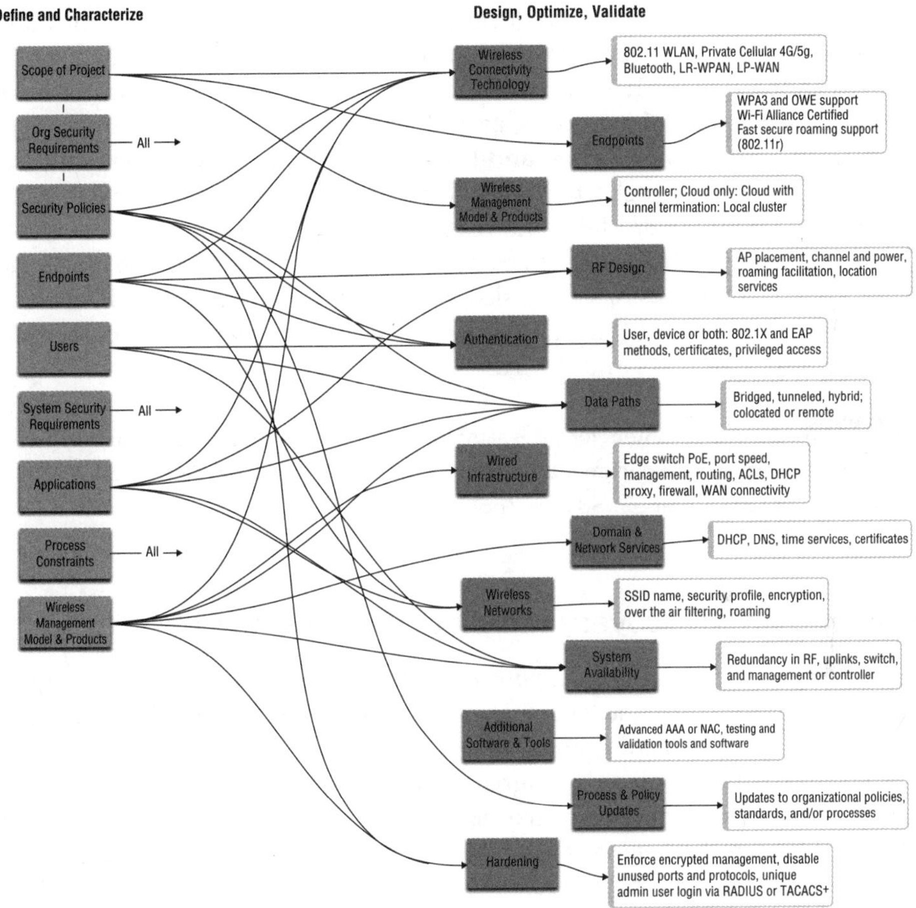

Figure 5.3: Visualization of the relationship of inputs to outputs

Take this mapping as a template to start with and adjust as needed for your projects. Once settled, the map will allow you and other team members to quickly identify dependencies in the architecture. If nothing else, the generic map here is a great tool to help other teams understand the depth of interdependencies of wireless network architecture.

In case you're interested, Figure 5.3 was created with a free tool found at `https://miro.com/templates/mind-map/`.

Planning Processes and Templates

Following are a few planning and design templates, modified slightly from the versions I've used in scoping and documenting client projects. As with the mind map in Figure 5.3, I encourage you to take these and make them your own.

This chapter thus far has summarized the considerations of the various elements to be considered as inputs, as well as how those affect the outputs of design. It's a lot of information and (in my opinion) too much to attempt to aggregate in one table or chart. Instead, there's a sequence of planning templates and guides that are interrelated but used at different phases.

The first template helps organize data and requirements from the *define* and *characterize* stages (*discovery* stage tasks). Following that is a sample SSID planner for Wi-Fi networks, and then a NAC Policy Matrix for planning access rights (authorization) for users and devices.

NOTE **Upcoming content includes sample forms with volumes of data that may be hard to read. Sample documents and additional resources will be maintained at** `http://securityuncorked.com/books`

Requirements Discovery Template (Define and Characterize)

Just as the header implies, the *Requirements Discovery Template* is a form designed to capture and organize the technical and business requirements for security of the system, users, and endpoints.

Referring to our borrowed Design for Six Sigma (DFSS) lingo, this document captures detailed scoping information such as types and counts of endpoints (*define*), and then characterizes their security requirements (*characterize*). At the end of the exercise, the deliverable is a well-organized description of security requirements.

The table in Figure 5.4 is meant to serve as an example of one way you might go about collecting and organizing data. The tables are populated with sample data for context.

As you're planning, here are a few considerations about this template example:

- This is not exhaustive and it's likely you'll need or want to add additional columns such as who the primary contact is internally or who the manufacturer is, etc.
- Documenting endpoint details with capabilities and access requirements in a table or template makes it easier to validate needs and aggregate

input from different teams. Departments that own or manage endpoints and applications (possibly outside the purview of IT) will have context to add around capabilities and access requirements.

- For the more arcane endpoints it's helpful to also capture links to datasheets or administrative guides if available online.
- Expect the rows to have some accordion behavior, meaning you may start with a basic grouping of laptops, printers, phones, and IoT-type devices, and further sift and separate those into more granular definitions. Conversely, you may find yourself being exceptionally granular to start and later rolling up and combining rows into larger groups. Of course, you can also add group headings and start organizing them that way; for example, corporate laptops, personal smartphones, facilities IoT, biomedical devices, or handheld scanners.
- This table for capturing inventory and capabilities does not necessarily translate into the final network design. For example, there may be endpoint groups listed as supporting 802.1X, but in your design those may ultimately end up on a passphrase-secured network.
- For purposes of preliminary scoping, a rough number for the endpoint quantity is sufficient. Don't worry whether it's 20 or 24; you want to know if it's 20 versus 200 or 2000.
- Capture and document the information that's meaningful to you and the teams collaborating with you—your template may be substantially different and that's OK.
- If you're evaluating your current architecture against security best practices, simply add columns to document where those endpoints are attached now, and how they're authenticated and secured.

Sample Enterprise Requirements Discovery Template

The table in Figure 5.4 is populated with generic but real-world data common in many enterprise environments. In these organizations, it's expected to have a mix of corporate-owned and personally owned devices and a blend of traditional OS-based systems along with non-user-attached endpoints like printers and VoIP phones, plus IoT-type devices for cameras, door entry systems, facilities monitoring, and screen casting.

In this exercise, don't worry about it being perfect or pretty. Just gather as much information as you can and organize it in a way you can sort, sift, and group later.

DEFINE			CHARACTERIZE				
Description	Form Factor	Est. Qty	Owner	Mgmt.	Access Req	Capability or Limitations	Security Notes
Sales team laptops	Laptop/Win	75	Corp	Corp	Internet only (all tools SaaS) + some users internal federal sales processing	802.1X with PEAP or TTLS natively supported	5 of team accessing CMMC CUI classified data, requires segmentation
General employee BYOD	Smartphone / varies	200	Personal	Corp via MDM	Internet only	Varies, 802.1X PEAP ant TLS supported on all	Can use MDM to onboard to secure network, segment and disable peer communication
Dev team laptops	Laptops/vary	15	Corp	Corp	Internal to Dev and Prod pods	802.1X with PEAP or TTLS natively supported	Team is accessing IP, Dev environments segmented from Prod
Printers	Printer	25	Corp	3rd party	Print server on prem, MSP management platform	Can do 802.1X PEAP/TLS	2 printers in conf rooms should allow guests to print
Airtame casting	Pluggable	5	Corp	Corp	Guest network; See additional notes on access for internal users	Can do 802.1X PEAP/TLS	Make sure casting devices are not serving local ad-hoc wireless. Define Wi-Fi filtering rules to allow casting

Figure 5.4: Sample requirements discovery template for endpoints and users in a basic enterprise environment

Figure 5.4 is a sample table, which captures the following information for each row (group or classification of endpoints):

- *Description*

  The description column is just to capture a commonly accepted label or descriptor of the group of endpoints.

- *Form factor*

  Form factor describes the endpoint such as a laptop, phone, tablet, printer, biomedical device, etc. Just as with description the form factor isn't a fixed set of options; use descriptors that are meaningful to you and as granular or broad as needed.

- *Estimated quantity*

  An estimated quantity is to capture a rough order of magnitude for each type of endpoint.

- *Owned by*

  Ownership describes whether the endpoint is owned by the organization, an employee (such as BYOD), a third party, or perhaps even a specific department in the company.

- *Managed by*

  Management of the device may also be by the organization, or personal/employee user, a third party, or as shown in the BYOD example it could be managed by the organization specifically with an MDM tool.

- *Access requirements*

  The column of access requirements is a rough estimation of the type of access the endpoint group needs. Some will require broad internal and Internet access, while specific subsets such as printers or handheld scanners may only require access to a specific server(s).

- *Capabilities or limitations*

  Capturing known capabilities or limitations of endpoints at this stage is helpful in planning network authentication and security profiles.

- *Security or other notes*

  Inevitably you'll have some notes-to-self, links, or reminders that can go in an ad hoc notes column.

Sample Healthcare Requirements Discovery Template

While best practices pervade across industries, there are some environments that require special attention when planning wireless security, and healthcare is one of those.

Figure 5.5 uses the same template as Figure 5.4, populated with a subset of data from a representative healthcare environment.

There are numerous special considerations for planning both wired and wireless network security in healthcare—specifically those with patient services and/or biomedical devices where both availability and security are equally important for ensuring services are available and meeting HIPAA requirements while keeping biomedical devices protected.

Along with healthcare, manufacturing, warehouse, and retail use cases bring similar complexity. For parallels, healthcare's nurse paging and biomedical devices can easily be substituted by manufacturing's needs to support handheld scanners and automated forklifts. Both have similar IoT-like qualities and demands for security and availability. Retail also includes heavy use of handheld scanners, has requirements for point of sale solutions, and may also be using location services.

DEFINE			CHARACTERIZE				
Description	Form Factor	Est. Qty	Owner	Mgmt.	Access Req	Capability or Limitations	Security Notes
Admin and intake laptops	Laptop/Win	20	Corp	Corp IT	Misc. internal and Internet services	802.1X with PEAP or TTLS natively supported	All systems access EMR/HIPAA data, see compliance requirements
EMR Mobile Cart/WOW	Laptop/Win	50	Corp	Corp IT + Clinical Engineering	Internet only (all tools SaaS) + some users internal federal sales processing	802.1X with PEAP or TTLS natively supported	5 of team accessing CMMC CUI classified data, requires segmentation
General employee BYOD and Vocera App	Smartphone / varies	40 (20 with app)	Personal	Corp via MDM	Providers and staff using Vocera app require access to comms server	Varies, 802.1X PEAP ant TLS supported on all, 11r for Vocera app required	Can use MDM to onboard to secure network, segment and disable peer communication; Vocera app users have addl. requirements
Vocera Badge for providers	Wearable	50	Corp	Corp IT + Clinical Engineering	Vocera comms and alerting servers	Can do 802.1X, required 11r for roaming	High availability and low latency requirements + security- see 1X/11r notes and testing; ask SE if CBRS is supported?
Rugged nursing tablets	Tablet/Win and Android	125	Corp	Corp IT	Misc. internal and Internet services	All can do 802.1X PEAP and TLS	Check roaming requirements
Biomed-Mobile telehealth carts	Laptop/Win	5	Corp	Clinical Engineering	Misc. internal and Internet services	802.1X with PEAP or TTLS natively supported	
Biomed-Vitals monitoring	Cart at bed	100	Corp	3rd party via Clinical Engineering	Capsule server hub on prem	700-series models support 802.1X PEAP and TLS; older 300-series PSK only	Some biomed have static IP addresses, check with clinical team
Biomed- IV pump PCUs	Biomed on cart	50	Corp	Clinical Engineering	Pump server, ports 443, 8443, 9100	Unsure, docs reference PSK	Check with mfr. on security capabilities

Figure 5.5: Sample requirements discovery template populated with sample healthcare endpoints

INDUSTRY INSIGHT: WIRELESS AND BIOMEDICAL DEVICES IN U.S. HEALTHCARE

In the U.S., the manufacturing and use of biomedical devices is regulated by the Food and Drug Administration (FDA) Center for Devices and Radiological Health (CDRH). Medical devices within their purview range from surgical mesh and prosthetics to wireless-connected implantable devices, IP-connected monitoring and testing equipment, and beyond. Devices are classified as Class I, II, or III, with oversight increasing with each class level.

Depending on the classification of the device, the manufacturer or vendor may have registered with the FDA for approval of the device via a 510(k) Premarket Notification, and changes to the device, its configuration, integrations, or use may require that it be resubmitted and approved by the FDA.

One point of friction when trying to enhance security for networks servicing biomedical devices is that some teams will counter any request for changes, citing the lengthy re-approval process involved in resubmitting a 510(k) or Special 510(k).

However, the FDA issued updated guidance specifically related to cybersecurity, noting that "...a change made solely to strengthen cybersecurity is not likely to require submission of a new 510(k). Cybersecurity updates are considered a subset of software changes that are implemented to strengthen the security of a system, protect information, and reduce disruption in service."

In general, if the clinical functionality of the device isn't impacted, it's unlikely the FDA requires a resubmit. Having said that, if a biomedical device was not originally approved for wireless connectivity, and that feature is added, it will likely require a resubmit and approval.

You won't make friends by being pushy on this topic but being informed and able to speak the language of clinical engineering teams may open doors to collaborate on increasing the organization's security.

For more fun reading, visit `www.fda.gov/medical-devices`.

Healthcare comes in many forms, ranging from hospitals to outpatient and ambulatory care facilities, labs and testing, and small primary care offices. Most of my experience has been with mid to large healthcare systems that include several hospitals as well as numerous outpatient clinics. When working in smaller doctor's offices or clinics, there's a bit less complexity with the biomedical devices, but it's still present.

Defining BYOD in Your Organization

This is a bit of a sidebar topic, but the timing is relevant while you're in the *define* and *characterize* phases. One important task you'll likely have is to work with various stakeholders to characterize what BYOD means in your organization; and it may mean different things to different people. Don't make any assumptions when working through the *discovery* phase.

I've found it helps to outline straightforward options and ask which scenario is intended.

Of course, there should be an organizational policy around BYOD that defines what users have access to from personal devices, what type of visibility or management the organization has over the device, and what happens in the event of an incident—will the phone be wiped, will the corporate container and accounts be wiped, what's the policy for using a personal account for business communication, what are expectations should the device be in scope of a legal investigation and e-discovery.

REFERENCE **The BYOD topic in Chapter 8 includes much more detail on planning BYOD. There you'll find a detailed explanation of the other ownership and management models including choose your own device (CYOD) and corporate-owned personally enabled (COPE) along with both legal and technical considerations and a summary of recommendations for securing BYOD.**

Legal and policy issues aside, there are a few technical descriptors for defining what BYOD really means, and ultimately, we're just aiming to understand if they're treated more like guests or corporate assets.

What does the user need access to when on their personal device? This answer varies depending on form factors, users, applications, management of the device, etc., but for our purposes at this stage you can categorize BYOD into two main headings.

Allow Personal Devices on the Network but with Internet-only Access If the organization intends to allow personal devices, but only grant Internet access, they usually want to reduce friction for users by not forcing them to re-register through a guest portal daily, or even multiple times a day. Personal devices may be smartphones, tablets, laptops, e-readers and the like, but should not include infrastructure devices such as Wi-Fi routers and switches.

In these cases, you'll want to tentatively plan for some method of device registration with a longer duration than a standard guest user. This can be accomplished with some vendors' Wi-Fi captive portals natively. Other solutions may involve a NAC or guest management solution with the option to register a device for an extended period of time.

The ideal scenario is to register the personal device to the corporate user and set it to expire after a predetermined time. In an enterprise, personal devices might be authorized for 90 days up to a year. In universities, it's more common to authorize them for the semester or year.

If a user-attached registration is not possible, a basic captive portal for this purpose with a longer account life is perfectly reasonable.

There's only one hard-and-fast rule here. You never want to allow corporate users to connect an unmanaged personal device to the secured internal networks using their domain credentials (specifically applicable to networks configured for 802.1X with a tunnel like PEAP and MSCHAPv2 as the inner authentication).

This is another opportunity to remind you that Microsoft NPS as a RADIUS server does not support restricting access based on both a user and endpoint—it's either or. To meet this recommendation requires an advanced authentication server or NAC product.

Allow Personal Devices on the Network with Some Level of Access to Internal or Protected Resources Alternatively, it may be the intention of the organization for users to use personal devices to access internal resources—internal meaning both on-prem but also protected IaaS and PaaS environments that may be hosted in the cloud.

If this is the scenario, be sure to review information on BYOD planning in Chapter 8 before getting too far along.

The truncated advice is that this should only be allowed if the organization has a well-defined executive-sponsored policy on the topic and has security controls in place to mitigate risk—specifically the organization should enforce use of a corporate-managed mobile device management (MDM) or other endpoint agent for visibility and control of the corporate data.

In addition, if personal devices are to have unfettered access to internal resources, they should all connect to the network securely; on wireless that means using 802.1X-secured SSIDs, which requires some onboarding or configuration of that device for the selected EAP method for authentication.

Summarizing the recommendations for planning in the preceding text:

- Internet-only BYOD is straightforward and can leverage a modified guest portal for longer-term access if desired.
- Internal access BYOD is more complicated, should be carefully considered, and should include MDM (or similar capability) and connectivity via 802.1X-secured networks.
- Looking forward toward zero trust strategies, even Internet-only BYOD models will have additional controls and likely a dissolvable or persistent agent to enforce security.

Sample Network Planning Template (SSID Planner)

It seems most organizations don't document their wireless networks unless forced to do so by regulation. It's really a missed opportunity to streamline planning, memorialize the configuration for others to reference, and for you to use in iterative updates, troubleshooting, and enhancements as new security features become available.

The sample forms here are very generic and capture the minimum information required for planning and documenting a wireless network (see Figure 5.6). Additional implementation- or product-specific details may be warranted, such as integrations with NAC or security monitoring tools.

Controller or Gateway Management Configurations

Item to configure	Input and notes	Instructions
Management IP		
Management IP mask		
Management IP gateway		*Default gateway IP for the controller*
Management VLAN		*Please indicate VLAN number and name*
DNS name for Controller		*DNS name for the controller. This entry will need to be added to your DNS server with the IP you provided above and should match any certificates*
Local domain		*e.g., acme.com*
Time server		*Internal time server suggested if available*
DNS Server		*Internal DNS server*
Physical location		*Building, MDF/rack*
AP authentication or control plane security		

AP Management Configurations

Item to configure	Input and notes	Instructions
Management IP Scope for APs		
DHCP/DNS		
Management IP mask		
Management IP VLAN		*Please indicate VLAN number and name*
AP naming convention		*Your preference for naming APs in the system, e.g., BLDG-FLOOR-ROOM or other*

Figure 5.6: Sample planner template for controller and AP management networks

The SSIDs in Wi-Fi networks should be planned and documented carefully and capture not only the various 802.11 and RF specifications, but also key elements of the security architecture such as the VLAN, DHCP server, default gateway, authentication, and secure roaming protocols (see Figure 5.7).

This format is designed for gathering planned configuration details with guidance. A different format such as a spreadsheet or script may be preferable for the implementation.

Sample Access Rights Planning Templates

The earlier template for cataloging endpoints, capabilities, and requirements included abstract information on what access might be required—Internet-only, certain internal networks or resources, or some combination of both.

The *Access Rights Planner* (something I refer to as the *Policy Matrix* when used for NAC planning) is a tool to further refine access rights based on multiple static and dynamic conditions. This form can be modified to be more VLAN or ACL-centric but in its current form is intended to specify authorizations at a higher level. How and where you apply the control or segmentation is implementation- and product-dependent.

Item to configure	Input and notes	Instructions
SSID Name to Advertise		
SSID Users description		*For example, "guests" "employee" "staff" "student"*
SSID Location Reach		*All locations/APs, or subset by type, building, locations*
SSID Client VLAN		*Please indicate VLAN of client IPs for this network, or enter "dynamic" for dynamic VLAN assignment*
SSID Client IP Address Scope		*The scope depends on the total number of endpoints you've planned for this network*
SSID Client IP DCHP Server		*The DHCP server could be the Wi-Fi controller, a network server, or other wired infrastructure device*
SSID Client IP Gateway		*Used for segmentation planning*
SSID Data Path	Bridged | Tunneled | other	*Select one: Bridged; Tunneled; other*
Fast Secure Roaming		*Select one or more: 802.11r (FT), OKC, or PMK Caching*
SSID Security Profile	Portal | Passphrase | 802.1X | none	*Select one Portal; Pre-shared key; 802.1X; none. If using 802.1X*
Security Generation	WPA2 | WPA3 only | WPA2 + WPA3 Transition Mode | Enhanced Open	*Note Transition Modes are not recommended*
IF 802.1X: **- RADIUS Server IP** **- RADIUS Shared secret** **- EAP methods allowed**		
IF passphrase, enter key(s)		*Mixed generation (WPA2 and WPA3) should not reuse the same keys. Your Wi-Fi product may also support personal PSKs or API-created dynamic PSKs.*
IF Portal describe registration type		*Types include Self-registration; sponsored registration, acceptable use only, BYOD registration, etc.*

Figure 5.7: Sample planner to be used per SSID

Two versions of the planner are provided: a full NAC-based planner that takes into account endpoint security posture, and a simplified planner for wireless connections only and without security posture variables.

If you're planning at this level, this should happen before the Network Planner (shown prior) to inform decisions around VLANs and other access control or role enforcement.

TIP **Remember, these planners and templates are meant to serve as an example and this methodology represents just one option for planning your wireless security architecture. Make it your own and by all means modify or rework anything in this chapter to fit your preferred planning methodology!**

Sample Access Rights Planner for NAC

This version of the planner would be used with advanced authentication services or products with NAC features (such as Aruba ClearPass Policy Manager, Fortinet FortiNAC, Cisco ISE, Forescout, and others).

As with the other templates, this is just one example of how to organize the information and serve as a starting point (see Figure 5.8).

Organization-Managed Standard	Device ownership	Device Type	User	Location/Group (all sites unless otherwise noted)	Access	Agent/Posture	Notes
	IN COMPLIANCE						
	Corporate-owned	Laptops	Domain User	LAN (wired or wireless)	Production (or VLAN X)	Persistent Agent in compliance	Can be persistent or dissolvable agent, or no posture assessment needed
	Corporate-owned	Laptops	Domain User	Remote VPN	Production (or VLAN X)	Persistent Agent in compliance	
	Corporate-owned	Desktops	Domain User	LAN (wired or wireless)	Production (or VLAN X)	Persistent Agent in compliance	
	Corporate-owned	Mobile Phones	Domain User	Wireless Only	Production (or VLAN X)	MDM (AirWatch, etc) in compliance	
	Corporate-owned	Printers	n/a	LAN (wired or wireless)	Printers (or VLAN W)	n/a	
	Corporate-owned	VOIP Phones	n/a	LAN (wired or wireless)	Voice (or VLAN Y)	n/a	
	Corporate-owned	Security/ Environmental	n/a	LAN (wired or wireless)	Security (or VLAN Z)	n/a	
	OUT OF COMPLIANCE						
	Corporate-owned	Laptops/Desktops	Domain User	LAN (wired or wireless)	Remediation Network	Persistent Agent OUT OF COMPLIANCE	
	Corporate-owned	Laptops/Desktops	Domain User	Remote VPN	No access	Persistent Agent OUT OF COMPLIANCE	
	Corporate-owned	Mobile Phones	Domain User	Wireless Only	Internet-only	Persistent Agent OUT OF COMPLIANCE	
	IMMEDIATE RISK						
	Corporate-owned	Any	Any	LAN (wired or wireless)	Quarantine Network	INFECTED OR POSES IMMEDIATE RISK	Quarantine network accessible only by/for IR team
Healthcare Managed Clinical	**Device ownership**	**Device Type**	**User**	**Location/Group (all sites unless otherwise noted)**	**Access**	**Agent/Posture**	**Notes**
	Corporate-owned	Infusion pumps	n/a	LAN (wired or wireless)	Clinical VRF/Network	CMDB lookup	
	Corporate-owned	Vitals monitor group 1	n/a	LAN (wired or wireless)	Clinical VRF/Network	CMDB lookup	
Employee Owned	**Device ownership**	**Device Type**	**User**	**Location/Group (all sites unless otherwise noted)**	**Access**	**Agent/Posture**	**Notes**
	IN COMPLIANCE						
	Employee Owned	Laptops	Domain User	LAN (wired or wireless)	TBD* Can be Internet-only, Production, or Other	n/a	
	Employee Owned	Laptops	Domain User	Remote VPN	TBD* Can be Internet-only, Production, or Other	Persistent Agent in compliance (or optionally dissolvable agent with posture checks but less security with dissolvable)	Note: Dissolvable agent only checks once at connection, vs Persistent Agent has continuous endpoint posture assessment/reporting
	Employee Owned	Smart phones	Domain User	Wireless Only	TBD* Can be Internet-only, Production, or Other	TBD or managed by MDM	
	Employee Owned	Smart phones	Domain User	Remote VPN	TBD* Can be Internet-only, Production, or Other	TBD	
	OUT OF COMPLIANCE						
	Employee Owned	Any	Any	Any	No Access	Agent or MDM	
Guest Owned							
	Guest	Any	Self-registration	Wireless Only	Guest/Internet Only	n/a	

Figure 5.8: An advanced access rights planner that factors ownership along with posture in authorization

In most cases, I group sections based on who owns the device—the organization, employee/personal, contractor, or guest. For each group, the table then outlines what access rights and segmentation is to be used depending on (in this example) the security posture of the device, the type of device, and the user logged on to the device.

- *Device ownership*

  Defines whether the device is owned by the organization, employee (personal/BYOD), a contractor, or guest. In some cases you may prefer to group by department ownership as well.

- *Device type*

  Describes the type, class, or form factor of the devices, such as laptop, printer, smartphone, VoIP phone, handheld scanner, biomedical device, etc. The language here is meant for ease of identifying and communicating internally so use what makes sense to your team.

- *User*

  Identifies the condition of what user is logged on to the device, such as a domain user, no user, or guest

- *Location, network, or group*

  Some access policies may only apply to specific locations or network segments. In the example planner there are different policies for wired versus Wi-Fi versus VPN remote access. They may also vary for roles in different countries or regions.

- *Access rights*

  Defines how segmentation will be imposed. This basic example set includes VLANs and VRF references. With some Wi-Fi products you may also be able to specify a role or policy with an associated set of access rules or ACLs already defined. Access rights should also address duration if the access is only for a predetermined amount of time.

- *Agent or security inspection*

  Devices accessing the production resources should have some level of visibility or control by the organization. For domain devices, that's usually accomplished with a software agent. For personal, contractor, and guest devices, options vary depending on touch but typically include MDM (for BYOD control) or no additional controls (as for guests).

Figure 5.8 is an advanced access rights planner for a typical enterprise organization broken down by organization-managed, employee-owned, and guest. Additionally, a sample healthcare clinical device is included to demonstrate the flexibility of the template.

Sample Access Rights Planner for NAC in Higher Education

Universities and other institutions of higher education bring a bit of a twist to network security planning since the bulk of their use models are BYOD—students with personal devices both in the classroom and in residential halls.

This sample form (Figure 5.9) outlines authorization for college-owned devices, student-owned devices, and guest devices under varying conditions. Note this environment plans for agents on school-owned devices, and calls out requirements for MDM for the school-owned smartphones that have access to internal resources.

College-Owned Devices	Device ownership	Device Type	User	Location/Group	Access	Agent/Posture	Notes
	IN COMPLIANCE						
	College-owned	Laptops	Domain User	LAN (wired or wireless)	Production (or VLAN X)	Persistent Agent in compliance	Can be persistent or dissolvable agent, or no
	College-owned	Desktops	Domain User	LAN (wired or wireless)	Production (or VLAN X)	Persistent Agent in compliance	
	College-owned	Mobile Phones	Domain User	Wireless Only	Production (or VLAN X)	MDM in compliance	
	College-owned	Printers	n/a	LAN (wired or wireless)	Printers (or VLAN W)	n/a	
	College-owned	VOIP Phones	n/a	LAN (wired or wireless)	Voice (or VLAN Y)	n/a	
	College-owned	Security/ Environmental	n/a	LAN (wired or wireless)	Security (or VLAN Z)	n/a	
	OUT OF COMPLIANCE						
	College-owned	Laptops/Desktops	Domain User	LAN (wired or wireless)	Remediation Network	Persistent Agent OUT OF COMPLIANCE	
	College-owned	Mobile Phones	Domain User	Wireless Only	Internet-only	Persistent Agent OUT OF COMPLIANCE	
	IMMEDIATE RISK						
	College-owned	Any	Any	LAN (wired or wireless)	Quarantine Network	INFECTED OR POSES IMMEDIATE RISK	Quarantine network accessible only by/for IR

BYOD or Student-Owned	Device ownership	Device Type	User	Location/Group	Access	Agent/Posture	Notes
	IN COMPLIANCE						
	Student Owned	Laptops	Domain User	LAN (wired or wireless)	TBD* Can be Internet-only, Production, or Other	n/a	
	Student Owned	Smart phones	Domain User	Wireless Only	TBD* Can be Internet-only, Production, or Other	TBD or managed by MDM	
	Student Owned	Game Consoles	Domain User	Wired or Wireless	TBD* Can be Internet-only, Production, or Other	TBD	
	OUT OF COMPLIANCE						
	Student Owned	Any	Any	Any	No Access	Agent or MDM	

Guest Owned	Device ownership	Device Type	User	Location/Group	Access	Agent/Posture	Notes
	Other/Guest	Any	Self-registration	Wireless Only	Guest/Internet Only	n/a	

Figure 5.9: An advanced access rights planner for higher education

Just as healthcare has some quirks when it comes to planning wireless security, so too does higher education where environments are predominantly glorified BYOD models.

Sample Simplified Access Rights Planner

If all of that feels overwhelming, or you simply don't need to plan for posturing and agents, you can simplify the model drastically.

In the sample form in Figure 5.10, the location column has been removed, along with any rows that didn't correlate to the wireless network. In addition, columns for the security agent posture and associated rows for endpoints that were out of compliance or at risk have also been eliminated.

This template can be further enhanced by adding context around the access—specifying the SSID or network name, along with the authentication schema to be used.

For more complex environments, an additional column could reference the directory to lookup the user or endpoint details—such as Active Directory, a NAC server, or other database or repository for MAC addresses.

Organization-Managed Standard	Device ownership	Device Type	User	Access
	Corporate-owned	Laptops	Domain User	Production (or VLAN X)
	Corporate-owned	Mobile Phones	Domain User	Production (or VLAN X)
	Corporate-owned	Printers	n/a	Printers (or VLAN W)
	Corporate-owned	VOIP Phones	n/a	Voice (or VLAN Y)
	Corporate-owned	Security/ Environmental	n/a	Security (or VLAN Z)

Healthcare Managed Clinical	Device ownership	Device Type	User	Access
	Corporate-owned	Infusion pumps	n/a	Clinical VRF/Network
	Corporate-owned	Vitals monitor group 1	n/a	Clinical VRF/Network

Employee Owned	Device ownership	Device Type	User	Access
	Employee Owned	Laptops	Domain User	TBD* Can be Internet-only, Production, or Other
	Employee Owned	Smart phones	Domain User	TBD* Can be Internet-only, Production, or Other

Guest Owned				
	Guest	Any	Self-registration	Guest/Internet Only

Figure 5.10: A simplified access rights form for wireless connections only, and no security posturing factors

Notes for Technical and Executive Leadership

This section summarizes key findings and recommendations appropriate for technical and non-technical executive leaders, offering fourteen bullet points organized into three overarching themes:

- Planning and Budgeting for Wireless Projects
- Selecting Wireless Products and Technologies
- Expectations for Wireless Security

It's impossible to summarize an entire book's worth of content into a few pages, but the selections here speak to the most common non-technical errors organizations make when planning, deploying, and managing secure wireless networks.

NOTE The following content addresses the main themes and questions I routinely field from CIOs, CTOs, CISOs, and non-technical leaders such as CEOs and board members. By no means is it exhaustive. My hope is to help ensure leadership is in alignment with the technologists' strategies for securing the enterprise wireless.

Planning and Budgeting for Wireless Projects

Managers and executive leadership have an exceptional opportunity to help the organization meet its security objectives in the earliest stages of planning and budgeting—before network architects are restrained by Wi-Fi and endpoint product selection—with these six key tips:

- Involve wireless architects early to save time and money
- Collaboration is king for zero trust and advanced security programs
- Stop planning 1:1 replacements of APs
- Penny pinching on AP quantities sacrifices security
- Always include annual budget for training and tools
- Consultants and third parties can be invaluable

Involve Wireless Architects Early to Save Time and Money

Secure wireless architectures permeate every aspect of IT, cyber security, and digital transformation initiatives. The considerations, decision factors, and inputs into a secure wireless design are innumerable. When the wireless architect is brought in after scoping, they can't always just "make it work."

Security in wireless has deep interdependencies on not only the wired infrastructure but also endpoint capabilities—an often overlooked and critical component. As such, you should involve all contributing architects early to collaborate at the onset of any ideas or projects. You'll save a lot of time and money if you don't pick a technology or product and don't purchase new endpoints or infrastructure until the entire team has been read in. If you don't have a qualified wireless architect in-house, use a consultant.

Collaboration Is King for Zero Trust and Advanced Security Programs

If there's one lesson I learned in fifteen years of leading advanced network security projects, it's that collaboration is critical. Whether you're planning a digital transformation program, a zero trust strategy, network access control, or just enhancing the wireless security, you need input from several teams or people.

The days of siloed networking, wireless, security, and identity management are over. Cross-functional teams will advance your mission, facilitate communication and collaboration, and build the trust required for these more complex projects.

Stop Planning 1:1 Replacements of APs

Most CXOs I work with have a gut instinct that newer technology equals greater range, but the opposite is true. Wireless standards are evolving rapidly. Sparing you the gory technical details, in general the faster speeds of Wi-Fi (due to modulation and encoding plus new spectrum) means the effective range of the Wi-Fi is much smaller.

That means the organization needs to stop trying to replace APs 1-for-1. The design you did nine years ago for Wi-Fi 4 (802.11n) simply won't cut it for today's Wi-Fi 6 (802.11ax) and tomorrow's Wi-Fi 6E deployments. Plus, the use cases and applications running on Wi-Fi a decade today pale in comparison to today's demands.

Proper RF design specifies AP counts and placements. A wireless architect or consultant can provide a brownfield design, reusing cabling drops and mounts, but you'll need to budget for additional APs. A completely unscientific but data-driven estimate is that you should allocate an additional 5–10 percent of AP count with each generation of technology. So if you're going from Wi-Fi 4 (802.11n) to Wi-Fi 6 (802.11ax) plan to increase the AP count by 10–20 percent.

In addition to procuring the proper count of APs based on a qualified design, it's smart to budget for at least an additional 5 percent overhead for spares or to fill in coverage or quality gaps.

Having said that, adding APs willy-nilly is also a poor strategy for wireless. Additional APs it not always the answer; if you have user or application issues, work with a qualified wireless architect to survey and recommend remediation.

Penny Pinching on AP Quantities Sacrifices Security

Choosing not to follow the design and instead reducing the AP count (or simply repeating a 1:1 upgrade) not only impacts coverage and quality, but also security.

Secure wireless relies on 802.1X networks that are authenticated and encrypted, with the downside being additional overhead and delay associated with the authentication and key exchanges. As users physically move, their device will roam from AP to AP. Wi-Fi networks support fast secure roaming to reduce latency, but if the APs are too far apart or otherwise not properly configured, the endpoint experiences a hard roam where the full authentication process is repeated.

A hard roam can easily take 10 to 20 times as long as a fast secure roam.

While the user may not notice, the applications will. Streaming media, video, voice, and online collaboration tools will be affected. That impact is devastating to any latency-sensitive application including mobile clinical devices and nurse paging technology as well as push-to-talk apps, handheld scanners, plus inventory and warehouse automation.

The unintended consequence of this all-to-common scenario is that wireless manufacturers instruct admins to put these critical devices on (unsecured) passphrase-based networks, which offer encryption but not authentication, and in current form of WPA2-Personal are very vulnerable to attack.

Poor RF design is also the leading cause of users plugging in rogue devices, which presents a litany of even more serious security concerns.

By getting a qualified RF design, following proper AP placement and radio settings, fast secure roaming can be supported, and more devices, users, and networks will be secured properly.

AP placement and RF design impacts:

- Basic wireless coverage and user experience
- Wireless quality including speed
- Roaming experience between APs
- Location services and asset tracking
- Security feature support

Always Include Annual Budget for Training and Tools

The worlds of wireless, networking, and IoT are moving at a fast pace, and the intricacies of the technology plus the vendor products and myriad monitoring and management tools are too much for even the most astute IT professional to keep up with. Wireless technology is its own beast, and even a seasoned network engineer will need ongoing training.

If security is a priority for your organization, then an annual budget for training and tools is vital.

Not sure where to start? I strongly suggest sending your primary architect to at least one in-person event or training a year and ad-hoc and online training throughout. If budget allows, two events or combination of event and training is ideal.

TIP **A good *starting* place is $3,500 USD for a multi-day training or event including travel. If you double that for a combination of two events and/or trainings, that's $7,000 USD.**

Mileage and pricing vary by region. For more junior engineers and admins, you may either need to spend more to get them up to speed, or alternatively rely more heavily on free online resources and allocate a smaller budget for those professionals.

If your organization really can't budget for professional development, or has temporary budget holds, there are free online resources and content from vendors and industry (vendor-neutral) events. Having said that, if an IT pro I'm mentoring repeatedly tells me their organization won't allocate budget for professional development, my recommendation to them is to find a new employer.

Wireless (including and especially Wi-Fi) is complex and multi-dimensional, and the technology evolves quickly, meaning proper and current tooling is critical. Wireless devices are dependent on the physical radio and your wireless architect simply cannot run a software upgrade on most RF tools to support the latest tech.

REFERENCE **Suggestions for wireless and security trainings, events, and tools are included in Chapter 7.**

Consultants and Third Parties Can Be Invaluable

If you're lucky enough to have a highly skilled wireless architect in-house, then you're leaps and bounds beyond most organizations. What's more common is a network architect or other system administrator fitted with multiple hats, wireless being one.

Whatever your team's expertise, consultants and third parties can be invaluable during the planning and budgeting phases or wireless projects.

And while, yes, I'm a consultant in this space, this isn't a sales pitch. You likely have resources available through your vendor or integrator. Not all systems engineers (SEs) are created equally, and you'll definitely want to seek out a professional who's steeped deeply in both wireless and security. Hey, maybe even someone reading this book!

Just remember that the vendor's team—sales or engineering—will have bias and possibly a very narrow scope to assist. Over many years, I can recall only a handful of wireless vendor sales engineers that were exceptionally knowledgeable both in their own product and in the industry technology.

Your reseller or integrator likely has a broader view and often isn't tied to one product or manufacturer, making them an ideal source for assistance.

If you're looking more for security best practices and answers to "can we" or "should we" then a more specialized advisor with knowledge of compliance requirements may be a better fit.

Selecting Wireless Products and Technologies

Often, the selection of wireless products and vendors occurs before the technical team is afforded any opportunity for input. Not only is it demoralizing to the technology professional, but it often results in poor product selection and an inadequate security architecture. Also, my hope is that leaders will embrace, or at least explore, novel solutions to meet business objectives.

I offer CXOs the following considerations when selecting products and technologies:

- Wi-Fi isn't the only wireless technology
- The product your peer organization uses may not work for you
- Don't buy into vendor or analyst hype
- Interoperability is more important now than ever

Wi-Fi Isn't the Only Wireless Technology

Wi-Fi is one of many wireless technologies that may be appropriate for your enterprise projects. Technology selection depends on the use case and varies based on:

- Endpoint types (traditional OS-based, IoT, etc.)
- Application use case and criticality (Wi-Fi may be less desirable for critical and latency-sensitive applications)
- Bandwidth requirements (how much data is being transmitted)
- Battery life and duty cycles of endpoints
- Distance or range of endpoint to AP

Details of these are covered in Chapter 8, and include:

- *802.11 WLANs* (Wi-Fi)

  Wi-Fi, for in-building and short-range outdoor applications
- *Private cellular 4G/5G LAN*

  New technology, for both in-building and long-range outdoor applications and support of use cases where Wi-Fi doesn't meet the connectivity and security requirements
- *Low Power WANs (LP-WANs)*

  Include LoRa, Sigfox, and modified cellular technology for long-range communication
- *Low Rate Wireless Personal Area Networks (LR-WPANs)*

  Including the suite of protocols based on 802.15.4 such as Zigbee Matter, ISA100.11a, WirelessHART, 6LoWPAN, Thread, plus BLE and others for shorter-range communication

The Product Your Peer Organization Uses May Not Work for You

Through my many years of security consulting for midmarket up through Fortune 5 clients, the question asked most often is "what are my peer organizations using/doing?"

There's value to running with the pack on certain matters, but when it comes to product selection, what your peer organization chose may not work for you.

One of the consistent roles I've had over the years is consulting with clients both ad-hoc and through structured readiness assessments for refreshes or disruptive and emergent technologies. Throughout those projects, there have been numerous considerations including a few that repeatedly appear as analysis inputs:

- Size, topology, and distribution of the networked environment
- Size and capabilities of internal teams for ongoing maintenance
- Related projects, technologies, and products within the organization
- Organization's long-term strategic vision
- Network infrastructure products in the environment currently
- Security and compliance requirements
- Budget and preferences of CapEx vs. OpEx spend models

Selecting the right product can make or break a project, but there's another side to this coin. With wireless technologies, there are many times the issue is not with the product but with the implementation. There have been scores of times I've been brought in to make suggestions to replace an inadequate wireless product only to discover the system was not properly designed and/or installed.

Others, such as NAC products, don't have the same feature parity and 70 percent of the time, a failed NAC project is due to improper product selection.

Don't Buy Into Vendor or Analyst Hype

Every time an executive selects a vendor based purely on where they fall in an analyst ranking, an angel loses its wings. Okay, maybe that's not true, but often the best and brightest solutions aren't necessarily in the Gartner Magic Quadrant Leaders' block. There's a place for the Visionaries, Challengers, and Niche solutions.

The world of industry analyst services is really bizarre:

Some firms rely heavily on academic researchers with no real-world experience; they just opine about the industry.

Others are defined by ego-driven pundits who are more influenced by the snacks you had at the ready than with your product features.

Plus, there is a pay-to-play aspect of many analyst reports; although a vendor can't pay for a ranking or position, it stands to reason an analyst will come to fully appreciate the depth of a vendor's offering by spending more time with the product. And there's a cost that accompanies an analyst spending time with the product.

NOTE **If you're interested in a juicy behind-the-scenes look at the world of analysts, check out *UP and to the RIGHT: Strategy and Tactics of Analyst Influence* by our friend Richard Stiennon. (IT-Harvest Press, 2012)**

It's a rare breed—the analyst with true industry expertise who can meld their understanding of real-world needs with a forward-looking strategy. Gartner has a lot of great industry information, especially with the Peer Insights, and sometimes they do get it right on the Quadrant. Other firms to watch are Forrester (`www.forrester.com`) with their Forrester Wave analysis. As an aside, if you compare Gartner's Magic Quadrants for datacenter and wireless against Forrester's Waves of the same segments in recent years, you'll notice the Gartner analysis trails the Forrester findings by one to two years.

There are also niche players like IANS Research (`www.iansresearch.com/`) who specializes in cyber security topics and uses practitioners instead of analysts as advisors. I have a relationship with IANS and serve as advisory faculty.

The vendors, too, bring a lot of hype to the process. The shiny new security features they promise may or may not work in your environment; may have dependencies they're not up front about and may not work nearly as well as advertised.

All of that to say—please involve your technologists and architects in planning and decisions before you or other executive leadership select a product. If you feel your technology professionals aren't equipped to participate in that decision, then involve a consultant.

Interoperability Is More Important Now than Ever

"I want one throat to choke" is a popular sentiment among CXOs. Yep, I hear you. But the reality is the next few years are defining moments for the technology industry. Zero trust strategies, digital transformation projects, and the convergence of IT, OT, and IoT along with a climate of significant security restrictions will strain even the most mature organizations.

The growing demands of our secure infrastructures require agility, flexibility, and scalability—all characteristics that rely on interoperability.

There are always exceptions, but these recommendations will help prepare your organization for whatever may come:

- Seek out products that allow for scalable, open integrations, such as those based on APIs and SAML
- Look for products that have been tested and certified for interoperability such as Wi-Fi Alliance-certified products
- In general, avoid technologies based on closed/licensed standards (as an example Z-Wave was licensed until recently versus Zigbee, which is an open standard)
- Avoid clunky and complex custom integrations that will have to be reworked after even minor updates

Expectations for Wireless Security

New security standards, new technologies, and new threats bring with them new recommendations. Expect to move away from PSK-secured networks and be prepared—the organization's relying on you to step up and help define BYOD security, privileged access management, and more.

Here are a few things to know:

- Consider PSK networks to be the "new WEP"
- You're not as secure as you think
- Get control of privileged access, especially remote
- Make sure you've addressed BYOD

Oh, and please help your technologists get the scheduled maintenance windows they need for critical patching and testing.

Consider PSK Networks to Be the "New WEP"

What does that mean? Networks secured with pre-shared keys (specifically networks that allow WPA2-Personal) are exceedingly vulnerable to a host of attacks including both online and offline dictionary attacks, among others.

Years ago, the mantra was "WEP is insecure, upgrade your networks." Well, today's mantra is "PSK is insecure, upgrade your networks."

Your organization should eliminate PSK-secured networks and instead move toward these options:

- WPA3-Personal secured with SAE (SAE is the replacement for PSK)
- WPA2- or WPA3-Enterprise secured with 802.1X/EAP (either is fine, but version 3 brings PMF and is preferred)

As an executive, there are a few key points to know:

- WPA3 is the new version of WPA security suite (replacing WPA2)
- Endpoints must be updated with software, upgraded, or replaced to support WPA3
- For Enterprise mode with 802.1X, both WPA2 and WPA3 are secure and okay to use but WPA3 is more secure
- Improperly configured transition modes that support WPA2-Personal and WPA3-Personal put your network at risk
- This book offers guidance to technical architects to properly design and secure transition-compatible networks

TIP **The technical guidance in this book prescribes expanding SSIDs—creating and adding more SSIDs to separate networks of different security postures or requirements. Over the air, each SSID is effectively a broadcast domain, just like a VLAN. Don't be surprised when your network architect proposes this.**

You're Not as Secure as You Think

More often than not, the security controls in a network are not implemented in the way a CISO or compliance officer believes them to be; or perhaps they're just not implemented in alignment with how security is described on paper through policies or compliance attestations. This book is one of the many ways I'm pursuing the lofty goal of better aligning security and networking, and my hope is that it arms technologists and executives with the information to make better informed security decisions.

This content is filled with concepts, examples, and guidance for building a secure wireless architecture. One representative topic is that of migrating from legacy WPA2 generation security to new WPA3. What the architect has learned is that the journey to WPA3 is not straightforward, and the opportunity for invalidating all the enhancements of WPA3 with a poor transition strategy is high. They've also learned the network security posture depends heavily on the organization's ability to update, upgrade, or replace endpoints that don't support the new standards.

Following are a few thoughts for consideration by executives as it relates to the overall security posture of the network:

- New Wi-Fi security standards demand precision and capable endpoints
- Common practices aren't necessarily best practices
- Just because you can't see or monitor something doesn't mean it's not a risk

- Wi-Fi and other wireless including Bluetooth and private cellular require tools for over-the-air and network-based security monitoring
- Gaps and lack of communication about network security policies introduce risk
- Regular assessments and/or penetration tests by third parties are exceptionally helpful in evaluating posture
- K.I.S.S. (keeping it simple) is still king for security; and most environments get so focused on advanced security they miss the simple, basic controls

TIP **New Wi-Fi security guidance deviates drastically from recommendations of the past decade. If your architect proposes something that seems foreign to you, please give them a chance to explain.**

Get Control of Privileged Access, Especially Remote

The practitioners out there may harangue me for this, but it's a topic that warrants discussion. Many organizations lack visibility and control of their IT teams exercising privileged access including remote access.

Network engineers, admins, and architects often have the keys to the kingdom when it comes to privileged access. Depending on the size of the organization, they likely have unrestricted management access to wired and wireless network devices including routers, domain and application servers, virtual hosts, user directories, and even firewalls.

In addition to their own access, they often extend management access to third parties including the vendor field and support teams, integrators, and other consultants.

Logon credentials are often shared among teams, stored in personal password managers, or even shared via spreadsheets or text documents without any encryption. Many technologists use their own applications for managing connectivity, such as personalized terminal services, which also have local (and unmanaged) credential storage.

While none of this behavior is malicious, it does put the business at risk. If your organization doesn't have policies around administrative access, storage and sharing of credentials, and remote privileged access, please consider leading that initiative. If the policies do exist, they should be communicated early and often to IT team members, with an opportunity for them to provide feedback. The goal is to enhance the security posture without inhibiting the professionals' work.

Some recommendations and points to consider include:

- Follow the principle of least privilege when possible
- Train IT teams at all levels/roles on the organization's policies around privileged access
- Provide easy-to-use secure credential vaulting and password management products
- Log and monitor privileged access and accounts
- Ensure devices are configured for user-based access and not shared credentials such as *root* or *admin* on shared accounts
- Include guidance for escalations and break glass emergency situations
- Ask for feedback from teams regularly to ensure the tools and policies are working and not being bypassed

Make Sure You've Addressed BYOD

Bring your own device (BYOD) and consumerization were hot topics ten years ago, but the pandemic has brought them front-and-center yet again.

With no warning or preparation, supporting a remote workforce was foisted upon companies across the globe. Decommissioned laptops were resurrected; VPN access expanded exponentially; employees were suddenly positioned to use personal devices for work. Security was sacrificed for availability in the name of business continuity.

Many organizations, even those with mature security programs, have found themselves now needing to retroactively address the use of personal devices for corporate access. BYOD is covered in depth in Chapter 8. The guidance for architects is that BYOD is complicated and policies around it aren't for IT to decide; it requires legal and executive guidance.

NOTE **As the executive, it's your role to lead the organization in this task. If there are no policies around BYOD or if they're stale, please make it a priority to create or update them and communicate changes to your network and security architects.**

Here are a few thoughts when considering BYOD policies:

- Along with BYOD there are other ownership-management models including CYOD, COPE, and COBO—choose your own device, corporate-owned personally enabled, and corporate-owned business use only
- Exceptions to BYOD or any network security policies should be approved by a board (security, change management, or other) not an individual person regardless of their status

- BYOD brings many legal implications and should be carefully considered to address data ownership, rights for wiping data, e-discovery, and liability of the organization for personal data
- In zero trust strategies, BYOD models that provide access only to Internet resources should still include protections and visibility related to the access of SaaS, PaaS, and IaaS
- Your BYOD policy on paper should have controls to enforce and/or monitor the behaviors described

REFERENCE **Chapter 8 includes BYOD topics that may be of interest to executive leadership, including legal considerations and best practices for when to use BYOD, and how to secure it.**

Summary

This chapter introduced a planning and design methodology loosely based on borrowed constructs from Design for Six Sigma. The planning entails five phases organized in three stages, which collectively comprise the inputs and outputs of the security architecture.

Discover Stage

- Phase 1: *Define* (scoping)
- Phase 2: *Characterize* (requirements mapping)

Architect Stage

- Phase 3: *Design* (functional mapping)

Iterate Stage

- Phase 4: *Optimize* (design adjustment)
- Phase 5: *Validate* (validate design against requirements)

It starts with the *define* phase to scope the project and collect requirements, followed by a *characterize* phase to refine and map requirements to the discrete design elements (such as client networks, infrastructure hardening) against known security requirements documented in policies or compliance mandates. Together, *define* and *characterize* compose the discovery tasks and serve as the inputs for the *architect* stage.

The output of the planning exercise was described through three additional phases—*design*, *optimize*, and *validate*, which comprise the *architect* and *iterate* stages.

The *optimize* and *validate* phases are entangled and non-linear, where *optimize* affords the architect the opportunity to further refine the design plans based on changing or additional inputs, while the *validate* phase serves as the opportunity to compare the design against the original requirements and validate the objectives are being met.

Instead of presenting a canned approach to planning, the content here continued with actionable templates and guidance for making decisions based on inputs and how those correlate to the outputs.

Sample templates, forms, and tables were provided along with representative data to illustrate the use of the templates.

Finally, the chapter concluded with pointed guidance written for an executive audience who wields the power to help make secure wireless more attainable.

REFERENCE **Find more detailed recommendations for planning in Appendix C, "Sample Architectures."**

CHAPTER

6

Hardening the Wireless Infrastructure

Infrastructure hardening has a special place in my heart; it's how I got involved with network security in the beginning. As part of architecting networks for clients and upgrading switches, routers, firewalls, and wireless, hacking into the devices (with the client's permission of course) was how we were able to access the systems in cases where a team inherited an environment or lost key personnel. Among the various projects, there was only one router we weren't able to access. Every other device or network offered a way in, either through misconfiguration or lack of hardening.

Secure architecture planning should address hardening the infrastructure, and many of these best practices should be in place before the wireless infrastructure is deployed, even in a test or proof of concept (PoC) deployment.

This chapter introduces concepts related to hardening the infrastructure including securing management access, implementing controls to guarantee integrity of the system, guidance for hardening client-facing services, and additional considerations and vendor-dependent features.

After you've worked through your planning tasks covered in Chapter 5, "Planning and Design for Secure Wireless," the next considerations to incorporate are the hardening aspects.

Hardening recommendations vary by organization based on risk tolerance, threats, and the resources available. It's important to note that not every organization will be able to implement all of the following controls. It may be an unpopular opinion from a security perspective, but security is a balancing act, and each organization should carefully consider the pros, cons, and trade-offs presented here.

For example, many of the hardening mechanisms presented will drastically reduce the organization's visibility of the environment—a trade-off that may not be acceptable or desirable. However, those limiting controls are appropriate for targeted environments such as those in federal agencies, national defense agencies, and financial organizations where additional resources and processes are in place to properly manage such an environment.

For that reason, this chapter includes tiered guidance to address recommendations for environments of all sizes, and organizations across multiple industries. All organizations should follow the minimum guidance for hardening, while additional controls are suggested for higher security tiers.

Content on hardening the infrastructure in this chapter includes the following:

- Securing Management Access
- Designing for Integrity of the Infrastructure
- Controlling Peer-to-Peer and Bridged Communications
- Best Practices for Tiered Hardening
- Additional Security Configurations

Securing Management Access

The first low-hanging-fruit task for hardening is to ensure all management protocols are configured to use encryption and proper authentication.

These tasks entail enabling encrypted protocols and disabling unencrypted protocols (such as enabling SSH and disabling Telnet for CLI access), removing all default passwords throughout the system components, tightening management access to enforce user-based logins (versus shared logins), and restricting the allowed sources of management traffic.

In this section we'll also address how to handle cases where shared logins are required, with proper use of credential vaulting for credentials, SSH, and API keys.

To wrap up the topic of securing management access, final considerations including privileged access management are addressed. Securing management access is broken down into these topic areas:

- Enforcing Encrypted Management Protocols
- Eliminating Default Credentials and Passwords
- Controlling Administrative Access and Authentication
- Securing Shared Credentials and Keys

- Addressing Privileged Access
- Additional Secure Management Considerations

Enforcing Encrypted Management Protocols

There are a few best practices for hardening that should be followed even in the laxer scenarios of labs and proofs of concept, and enforcing encrypted management protocols (along with removing default credentials) is one of them.

Even in a secured production environment with a dedicated management VLAN and restricted administrative access, configuring network devices for encrypted management protocols is a must. In doing so, you're ensuring critical and sensitive administrative traffic can't be exposed via eavesdropping—a scenario just as likely caused by security tools as it is a malicious user.

In most networks, some or all traffic is examined as it traverses the network for purposes of security inspection or populating baseline analytics. This could be via network-based security tools such as user and entity behavior analytics (UEBA) and security information and event management (SIEM) or through gateway security products performing advanced filtering, deep packet inspection (DPI), or SSL inspection of secured web traffic. Regardless of the source or reason, you never want management traffic to be passed in cleartext because you really never know where it may be exposed to eavesdropping and have its contents exposed or intercepted.

In addition to providing the confidentiality from encryption, secure management protocols add integrity through strong (ideally mutual) authentication of the administrative user to the system and also incorporate message integrity. These added integrity features ensure management traffic is not only confidential, but also has non-repudiation and is tamper-resistant.

Generating Keys and Certificates for Encrypted Management

Prior chapters addressed the use of certificates for purposes of authenticating RADIUS servers to endpoints, and optionally, endpoints to the server for 802.1X/EAP authentication. Discussion of certificates here is focused on securing management of the devices.

Using certificates for authentication and keys for encryption of course means that the certificates and keys have to exist on the managed device (in this case, some part of the wireless infrastructure). There's quite a bit of divergence among vendors as to how they handle certificate and key generation, but there's a general theme and common requirements underlying all products. You'll just have a bit of extra homework to do to research your vendors' latest documentation related to it.

Methods supported by vendors for certificate and key generation include:

- Self-signed certificates
- Certificates tied to secure unique hardware ID
- Certificates issued from third-party or public root CAs
- Certificates issued from internal domain CAs or PKI
- Keys generated on the device using key pairs

Certificates for secure device management are used in the following ways:

- For secure HTTPS web UI management
- For authenticating a controller to APs and vice versa

Key pairs for secure device management are used for:

- Secure Shell (SSH) authentication (as proof of an allowed certificate)
- Secure Shell (SSH) encryption
- Secure Copy Protocol (SCP), based on SSH

Some versions of wireless controllers also support SFTP for secure file transfers, which is not based on certificates or key pairs described here.

Although implementations and vendor support vary slightly, there's common guidance for creation and use of certificates and key pairs for secure management.

TIP **Know your bit strength. When it comes to crypto, key size matters. For certificates, consider RSA 3072-bit or ECC p-256 as minimums to meet today's standards and future-proof your deployment.**

Self-signed Certificates Self-signed certificates are supported by all wireless products but are not recommended. Instead, for a trusted infrastructure you should use either hardware-based certificates (if supported) or certificates issued from a root CA (either internal or third party).

The purpose of certificates is to provide integrity and confidentiality—specifically, authentication and encryption. By using a self-signed certificate, you're effectively invalidating the authentication portion of the mechanism since the client has no way to validate the server certificate against a known and trusted root CA.

In addition to the security risk, organizations will get penalized or flagged for using self-signed certificates and wildcard certificates in the environment during security assessments. The less secure certificate practices are never recommended, but they're especially bad practice when securing management of the infrastructure.

Certificates Tied to Secure Unique Hardware ID For more than a decade, the industry has been working toward cryptographic bindings tied to immutable secure hardware-based IDs. The value is that it provides devices (infrastructure and/or endpoint) with a strong integrated mechanism for interoperable authentication and encryption without having to manually provision certificates to the devices.

The IEEE 802.1AR standard specifies Secure Device Identifiers (DevIDs) to be used for authentication, provisioning, and management purposes, including 802.1X EAP authentication, using X.509 certificates that are bound to the device hardware at time of manufacturing.

In addition to the DevID, some products allow network administrators to create and add additional certificates with Local Device Identifiers (LDevIDs). The 802.1AR work builds on long-standing initiatives such as the Trusted Platform Module (TPM), which can be used with 802.1AR.

In more specific terms, this technology is found in Cisco's Secure Unique Device Identifier (SUDI), which is an 802.1AR-compliant secure device identity. Cisco's SUDI on the controller can be used for several functions including as the trusted certificate for AP join functions, the HTTPS certificate, SSH, and zero touch provisioning.

It's worth noting that Cisco wireless products produced between July 2005 to mid-2017 have manufacturer-installed certificates that expire after 10 years. That was addressed for 9800 series WLCs starting in 2019, but other models and products have an integrated certificate expiration of 2037. When the hardware certificate expires, any feature using the SUDI certificate will fail; see manufacturers' field notices for workarounds.

Other manufacturers are moving toward secure hardware-based IDs such as those based on 802.1AR and TPM. The 802.1AR device identities are hardware-specific and not applicable to cloud-hosted managers and virtualized appliances.

INDUSTRY INSIGHT: THE 802.1AR STANDARD

The IEEE 802.1AR-2018 standard document is available for free at `https://ieeexplore.ieee.org/document/8423794`. Source "IEEE Standard for Local and Metropolitan Area Networks - Secure Device Identity," in *IEEE Std 802.1AR-2018 (Revision of IEEE Std 802.1AR-2009)*, vol., no., pp.1–73, 2 Aug. 2018, doi: 10.1109/IEEESTD.2018.8423794.

The original standard was created in 2009 and updated in 2018 to add ECDSA (Elliptic Curve Digital Signature Algorithm) support.

The use of 802.1AR and hardware-based certificates extend beyond secure device management, as it can also be used for authenticating a device using 802.1X with EAP-TLS and identification of the device in the process of certificate enrollment with PKI infrastructures (such as through EST, described next) for local device IDs.

Certificates Issued from Third-Party or Public Root CAs Certificates for devices such as wireless controllers can be manually requested and installed, including from third-party and public root CAs. In these cases, you'll follow the request for creation of a certificate signing request (CSR), which is used to request a certificate from the issuing CA, and then you'll download and manually install the signed certificate(s).

Certificates Issued from Internal Domain CAs or PKI Manual certificate requests and installations are supported for both third-party and internal domain CAs, but enrollment can be more automated when using internal CAs.

In service of the enrollment of domain-issued certificates, network infrastructure devices such as wireless systems can utilize standards-based processes for requesting and installing certificates from the domain infrastructure. In most products, this is supported through the use of either Simple Certificate Enrollment Protocol (SCEP) or Enrollment over Secure Transport (EST). In both cases, the controller will proxy the requests for and installation of certificates on behalf of the APs.

Based on past experience and current vendor documentation, products will support one or the other protocol for certificate enrollment, so there's no need to choose; you'll simply use whatever the product supports. At time of writing, Cisco's IOS-XE code uses SCEP and EST and Aruba's Aruba AOS 8-code chain uses EST.

Keys Generated on the Device using Public/Private Key Pairs For protocols that rely on key pairs, such as SSH, simply follow the vendor's guidance for key creation.

As is always the case when using certificates, clock synchronization is critical and should be performed before generating or using certificates. Along with time synchronization settings, you'll also need to specify the device's hostname and domain so it has a FQDN when generating a certificate signing request (CSR) or a self-signed certificate.

Certificates come in several file formats; just like there are different graphics file formats (such as `.gif`, `.jpg`, and `.png`), certificate files used by vendors exist as X.509 PEM (encrypted or unencrypted), DER, PKCS#7, or PKCS#12, among others.

The most common are PEM (originally designated as privacy enhanced email) file formats, which are text ASCII-based and with readable headers such as `-----BEGIN CERTIFICATE-----`. A PEM file may contain a single certificate, a private key, or multiple certificates forming a chain of trust (to support the root CA and intermediaries), and are often of the file extensions `.crt`, `.cert`, and `.pem` (for certificates or chains) or `.key` (for private keys). DER and PKCS formats are binary formats, where DER and PKCS#12 can store certificates along with full trust chains and private keys like PEM, but PKCS#7 is a format for single certificates only.

For chained certificates used by APs to authenticate to the infrastructure, vendors will require the chained certificates to be in a specific order within the file, usually starting with the AP certificate, followed by the intermediate CA certificate, the root CA certificate, and then the private key.

The important point to note is that each vendor will support different file formats, and it's possible to convert to and from formats as needed. Consult your vendor product documentation or instructions from the issuing certificate authority for how-to details.

Figure 6.1 shows the standard readable header and footer of a PEM certificate with "Begin Certificate" and "End Certificate." DER certificates contain binary with no human-readable text as shown in Figure 6.2.


```
-----BEGIN CERTIFICATE-----
QIIH/TCCBeWgAwIBAgIMaBYE3/Q08XHYCnNSQcFBcjANBgkMhkiG9w0BAMsFADBy
QMswCMYDSMMGEwJSUzEOQAwGA1UECAwFSGS4YXQxEDAOBgNSBAcQB0hSdXN0b24x
ETAPBgNSBAoQCFNTTCBDb3JwQS4wLAYDSMMDDCSTU0wuY29tIESWIFNTTCBJbnRl
cQ11ZGlhdGUgM0EgU1NBIFIzQB4XDTIwQDMwQTAwNTgzQ1oXDTIxQDcxNjAwNTgz
...
-----END CERTIFICATE-----
```


Figure 6.1: Sample PEM certificate file format


```
3082 07fd 3082 05e5 a003 0201 0202 1068
1604 dff3 34f1 71d8 0a73 5599 c141 7230
0d06 092a 8648 86f7 0d01 010b 0500 3072
310b 3009 0603 5504 0613 0255 5331 0e30
0c06 0355 0408 0c05 5465 7861 7331 1030
0e06 0355 0407 0c07 486f 7573 746f 6e31
1130 0f06 0355 040a 0c08 5353 4c20 436f
7270 312e 302c 0603 5504 030c 2553 534c
2e63 6f6d 2045 5620 5353 4c20 496e 7465
726d 6564 6961 7465 2043 4120 5253 4120
5233 301e 170d 3230 3034 3031 3030 3538
...
```


Figure 6.2: Sample DER certificate file format

Enabling HTTPS vs. HTTP

Enabling HTTPS will require a certificate, which can be any of the certificate types just described—a self-signed (not recommended), a hardware-based certificate issued at time of manufacturing (such as those based on 802.1AR), or a certificate issued by a root CA either public or internal to the organization.

Infrastructure products will either create a self-signed certificate at initial boot or allow the generation of a self-signed certificate manually. Although self-signed certificates shouldn't be used in production, if the administrator is initially accessing the controller or device by web UI, then there will be a first-time use where unsecured HTTP or HTTPS using a self-signed certificate will occur.

Supported methods for the creation of, enrollment to, or installation of certificates are described earlier in the section "Generating Keys and Certificates for Encrypted Management."

Once a proper certificate is installed on the controller or device, access by HTTP should be immediately disabled, and instead enforce HTTPS, which is encrypted. If your organization doesn't manage devices by web UI, you can (and should) disable both HTTP and HTTPS.

Some products may support a setting to auto-direct from HTTP to HTTPS, which should be enabled if supported.

Lastly, many wireless products may require an additional setting to enable the management of the device over wireless.

Some products may also support authenticating admins via client certificates for HTTPS management, with or without the addition of username and password authentication. In these cases, only a client device with the installed certificate (designated as allowed in the controller) would be allowed management access to the device over HTTPS. With that configuration, access can require the client certificate by itself, or in combination with an administrator login.

CAUTION **The client certificate setting mentioned prior is one great example of why it's not a great idea to click buttons you don't understand. I've seen more than one client lock themselves out of the Wi-Fi management interface inadvertently with this exact setting.**

In most Wi-Fi products you can configure the use of HTTPS and disable HTTP in a web UI or CLI. The settings in a Cisco 9800 controller via the web UI are shown in Figure 6.3.

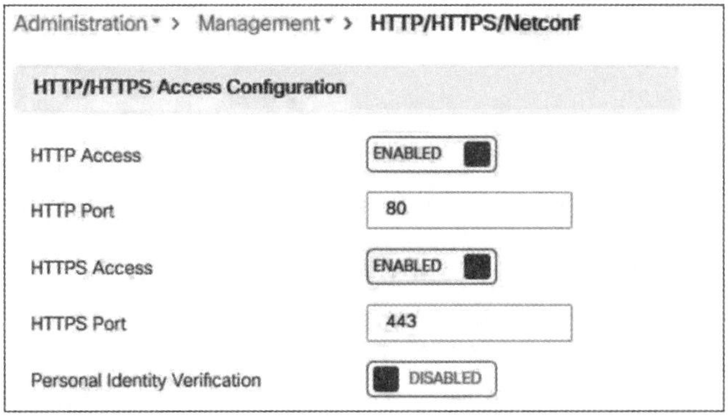

Figure 6.3: In most products you can enable HTTPS and disable HTTP in the web UI or CLI.

Enabling SSH vs. Telnet

Working with Secure Shell (SSH) for administrative access is two-fold—it can serve as both an encryption mechanism as well as authentication. First, SSH uses key pairs to enable encryption for CLI-based remote management access. Second, and often overlooked, SSH can use client keys as a management authentication mechanism (like certificates), and as described previously, this can be implemented with or without username and password credentials. Spoiler alert: in most organizations, I recommend always enforcing username-based logins since SSH keys are often mismanaged.

The first use case is straightforward. Just as with the other secure protocols, you'll need to complete the requisite steps to generate key pairs, and then enable SSH and disable its unencrypted counterpart Telnet. As with certificate creation, to generate the SSH key pairs, you'll first need to specify a hostname and domain for the device. SSHv2 should be used versus SSHv1, which has known vulnerabilities.

INDUSTRY INSIGHT: SSH IN FIPS MODE

Certain operational modes such as FIPS, enabled on wireless devices to meet the U.S. government Federal Information Processing Standard (FIPS) 140-2 standard for cryptographic modules require stronger cryptographic algorithms. When these modes are enabled, certain features may be disabled or auto-configured to meet the specified security parameters.

The second use case involves using SSH keys as a form of credential, either instead of or in addition to an authorized username-password credential.

When used for key-based authentication, an administrator or automated system generates a unique key pair for use by that user on that device (e.g., wireless controller), and the client key is then downloaded and used by an SSH terminal application (such as PUTTY) to authenticate the admin user to the wireless controller.

SSH public keys for authentication can be created using a common tool like PUTTYGEN (an add-in included with the PUTTY terminal application), or OpenSSL. See Figure 6.4. The private key should be saved and secured properly as part of a holistic SSH key management process. Then, in the wireless controller, you'll upload (or copy/paste) the public key of the pair and associate it with a user. Figure 6.5 shows an example of this operation in an Aruba Controller. The public key for the user is uploaded in the certificate store, and then associated with a user, as shown in this view.

TIP **You can configure SSH keys for authentication. It's not always easy to find this feature, but if you're interested, for Aruba Networks, search for "Public Key Authentication" in the user guide, and for Cisco look for similar commands around** `ip ssh pubkey-chain`.

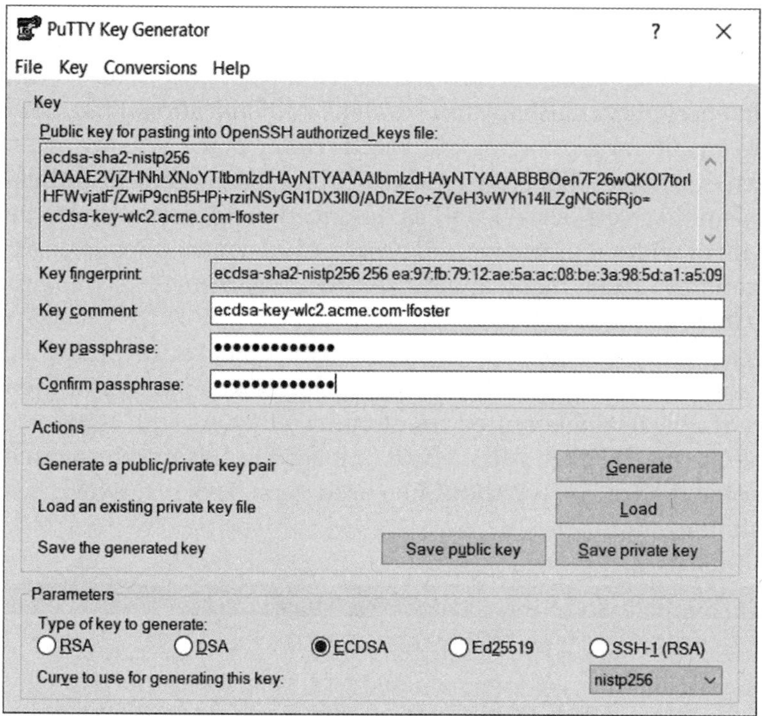

Figure 6.4: Creation of an ECDSA SSH key pair with PUTTYGEN

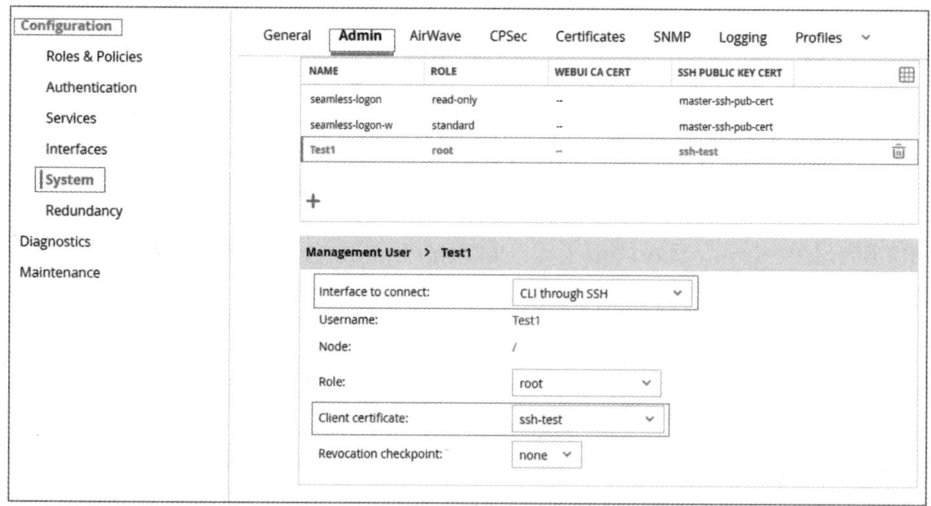

Figure 6.5: Many enterprise Wi-Fi products support public key authentication as shown in this Aruba controller example.

`https://the-ethernets.com/2020/10/ssh-aruba-mm-md-via-public-key-auth/`

Enabling Secure File Transfers

There are occasions where we need to move files around, to or from the wireless controller, and file transfer protocols should also be encrypted to protect the payload in transit.

Files that may get moved to or from the device include configuration files, software updates and patches, lists of APs, certificate files, MIB files, and CSV files such as a list of guest users or MAC addresses for an allowlist.

Files transferred using the web UI will be encrypted by the HTTPS protocol, but files transferred via the CLI should be secured with secure copy (SCP) or secure FTP (SFTP) to provide authentication and encryption.

SCP relies on SSH, and therefore configuring SSH at least for the encryption aspect (versus public key authentication) is a prerequisite.

Because SCP requires SSH, authentication, and authorization to have already been configured, many products default to unencrypted FTP (File Transfer Protocol) and TFTP (trivial FTP). While FTP is authenticated but not encrypted, TFTP is neither authenticated nor encrypted. Some products do also support SFTP (secure FTP), which also relies on SSH for encryption.

As with prior recommendations, enable SSH and then SCP and/or SFTP and disable TFTP and FTP, or configure management functions to each use SCP or SFTP if disabling TFTP and FTP is not a global setting. Table 6.1 provides an overview of the file transfer protocols and details, which include authentication and encryption.

Table 6.1: Overview of file transfer protocol security

PROTOCOL	AUTHENTICATED	ENCRYPTED
Secure Copy (SCP)	Yes	Yes
Secure FTP (SFTP)	Yes	Yes
File Transfer Protocol (FTP)	Yes	No
Trivial FTP (TFTP)	No	No

Enabling SNMPv3 vs. SNMPv2c

Simple Network Management Protocol (SNMP) is a standard protocol for monitoring and managing network devices such as switches, routers, and wireless products in addition to servers and endpoints such as laptops and printers.

SNMP data is organized in management information bases, or MIBs. Think of the MIB attributes as being a database with an element and then its corresponding configurations, such as "hostname" with the value of "`wlc02.acme.com`." There are standard MIB elements for most aspects of a network-connected device and they can be queried and/or modified. A query of "what is the hostname" would return the value "`wlc02.acme.com`" in this example, where as an SNMP management instruction to modify the MIB element of hostname may push a new value of "`wlc-east.acme.com`."

In addition to these examples of `get` and `set`, SNMP can be used for a trap function, which is a form of alerting to a logging system. Instead of a monitoring system polling the device for information, the system can gratuitously send a trap to a configured server. Traps can be enabled based on level or configured per group or type of trap.

The two versions of SNMP most widely used today are SNMPv2C and SNMPv3. There are a few devices in the world still relying on SNMPv1 but that has been deprecated for security reasons. Anecdotally, prior to SNMPv2c there was an SNMPv2, which offered encryption and other security but was not adopted due to its complexity.

SNMPv2c SNMPv2c relies on community strings versus true authentication, and it's not encrypted, which is troublesome. Community strings are simply strings of alphanumeric characters that act as a passphrase (much like a RADIUS shared secret). And as hinted, the community strings are sent in cleartext, meaning neither the management data nor the pseudo-authentication of community strings is protected.

SNMPv3 SNMPv3 was then created to address the lack of real authentication, encryption, and data integrity. With SNMPv3, users are created and specified with an authentication scheme (MD5, SHA) and an authentication password as well as a privacy protocol (AES, DES) and a corresponding privacy password. An SNMPv3 *user* is more of a machine account than an actual user account attached to a human. Users are created for each management or monitoring system.

SNMPv3, like any password-based mechanism, can be compromised by brute force and dictionary attacks but is leaps and bounds beyond better than SNMPv2c and any predecessors. In addition to the privacy and authentication controls, SNMPv3 includes security mechanisms to prevent spoofing and eliminates the use of cleartext credentials or data payloads.

If your infrastructure is managed or monitored with SNMP, use of SNMPv3 is always recommended over any prior versions.

INDUSTRY INSIGHT: SNMP AND NETCONF

For controller-based wireless systems including Cisco and Aruba, SNMP is used for monitoring and alerting only, and not for configuration of the controller or APs. Even in this use case, the encrypted SNMPv3 version is still recommended, and SNMPv2c should be disabled (community strings removed).

Cloud-managed solutions like ExtremeCloud (formerly Aerohive) also support SNMP including SNMPv2c and the secure SNMPv3.

Other cloud native solutions like Juniper's Mist wireless use secure APIs and do not use or support SNMP at all. Instead, these implementations use APIs (covered later) to push configurations, pull configs for monitoring, display configurations in dashboards, and to support other third-party product integrations.

With the entrance of cloud networking and expansion of network function virtualization (NFV), vendors are beginning to favor management options that involve much greater automation and integration such as NETCONF and YANG (both now IETF standards).

Together, NETCONF and YANG bring standard configuration protocols and data modeling that can standardize read and write functions across multiple vendors. Similar to SNMP MIBs, the YANG data model includes modules that are defined by standards. Unlike SNMP, CLI, or any traditional configuration methods, NETCONF is designed for intent-based networking, which allows the administrator to define the desired end state or behavior instead of configuring individual parameters.

YANG data is encoded in XML and transported by NETCONF between the console and infrastructure device via SSH. Configuration changes (and reads) can be scripted and executed from Python scripts. At time of writing, several vendors offer free tutorials and live sandbox environments along with Python scripts on GitHub for testing.

Cisco, Juniper, and HPE Comware devices are some of the most common that support NETCONF.

See Table 6.2 for a summary of secure management protocols.

Table 6.2: Summary of management protocols and recommended encrypted versions

USE CASE	UNSECURE PROTOCOL(S)	SECURED PROTOCOL(S)
Web UI Management	HTTP	HTTPS
CLI Terminal Management	Telnet	Secure Shell (SSH)
File Transfer	FTP, TFTP	Secure Copy (SCP) and Secure FTP (SFTP)
SNMP and API	SNMPv1, SNMPv2, SNMPv2c	SNMPv3 and APIs

Eliminating Default Credentials and Passwords

Our next hardening task is to seek out and eliminate all default credentials on the wireless infrastructure devices. As you'll see, this isn't always as clear cut as it sounds.

Changing Default Credentials on Wireless Management

For on-prem-based products, such as wireless controllers, tunnel termination gateways, and wireless management platforms, ensure the default users and credentials are removed. At a bare minimum, this means changing the default password, but a much better case is to change both the default usernames and passwords. This decreases the likelihood of a successful brute force attack because the combinations of passwords and usernames outstrip the attack tools.

For example, if the default credential is user "admin" with password "admin123", by changing the password to "somethingreallylonglikethis", the password length will make it challenging, but a malicious user can initiate a brute force attack using dictionaries or rainbow tables with the user "admin" and different passwords. If, however, the username was changed to "adminsys242" along with the longer password, the likelihood of a successful attack drops drastically. If proper monitoring tools are in place to alert on repeated failed login attempts, then you're in a great position.

PASSWORD LENGTH AND COMPLEXITY

In my security consulting circles, we're always collaborating and comparing notes on current best practices for password lengths. While some password lengths and rotations are prescribed in part by compliance requirements, the reality is many of those documents lag behind current attack tools, and some of the compliance frameworks are in opposition of one another when it comes to recommendations.

In today's world, length trumps complexity, and most security-conscious organizations have moved toward passphrases with no complexity enforcement but a minimum length. In addition, longer passphrases can alleviate the stricter password rotation schedules and as you'll see in the upcoming findings, many organizations opt for password rotation enforcement of once a year versus the prior standard of once every 90 days.

Currently these are recommendations from CISOs and security analysts of several major organizations, across many industries:

- User passwords: minimum of 14–16 characters, no complexity requirements, one year rotation, layered with MFA
- Privileged account passwords: minimum of 24–26 characters, no complexity requirements, varying rotations, layered with MFA

These recommendations may be shocking, and certainly not every organization is following this guidance. In addition to the privileged account length, most compliance requirements specify these management accounts should not be used for daily user activities, meaning Sally Sr. Network Engineer should have a sally@acme.com account for basic user access and some other account, let's say mgnets12@acme.com, for access to the management of critical systems.

The sally@acme.com account could be secured with the 14-character password, whereas the mgnets12@acme.com account would require a longer password and only be used for privileged management access.

Different products support different management protocols, as we saw in the preceding section, and it's not uncommon for products to come preconfigured with a set of default credentials, such as default console access, default CLI/

terminal credentials, and even a default web UI management account. These default accounts are becoming less common in products with more robust security features but are still prevalent.

Lastly, in addition to the normal management interfaces, highly integrated platforms may have credentials used for integrations with other products from the vendor or integrated third-party solutions.

The takeaway is to hunt down any and all default users and credentials, knowing there is likely more than one.

For cloud-managed products, there are usually no default credentials or users, except for access granted on the back end of the system by the vendor's support team (which can be disabled). The benefit is there's no default user; the downside is your management interface is exposed to the entirety of the Internet, making strong passwords and use of multi-factor authentication essential for securing management access.

Changing Default Credentials on APs

The management platforms and controllers aren't the only place default passwords are lurking. In every controller-based product, the APs are preconfigured (or configured through the controller) with default username and passwords. This can be configured and set centrally within the controller or management software.

Figure 6.6 shows an example of this setting on a Cisco 9800 web UI with options to change the default AP username and password from Cisco/Cisco to custom attributes.

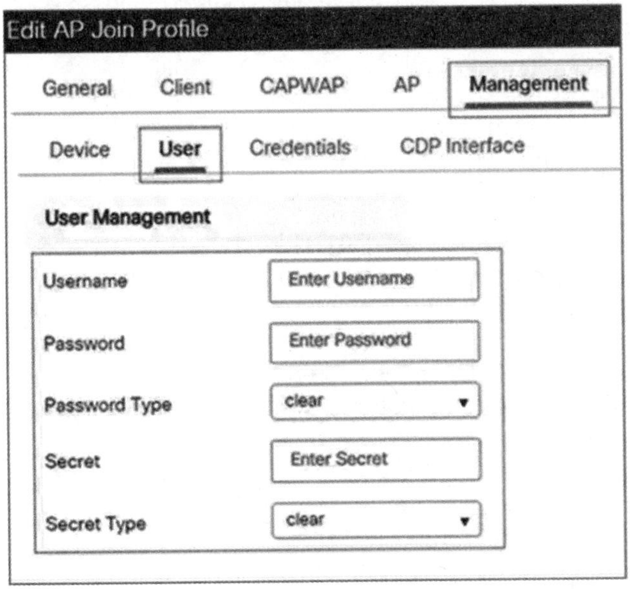

Figure 6.6: Where possible, always change the APs' default username and password.

Removing Default SNMP Strings

In the previous topics of enabling encrypted management protocols, we covered why SNMPv3 is greatly preferred over SNMPv2c or any prior version, even if used for read-only and monitoring purposes.

Historically it has been common for vendors to include default community strings for SNMPv2c, and also to have SNMPv2c enabled by default. That means, even if you've chosen to use the more secure SNMPv3 and configured it properly, it's possible the infrastructure is at risk due to the default settings from the vendor.

Always check, double-check, and recheck SNMPv2c settings and make sure it's not enabled (if you're using SNMPv3) or if it is enabled, make sure the community strings are custom. Specifically, avoid the use of "public," "private," and any common permutations of those two words. Because SNMPv2c community strings are sent in cleartext, do not configure strings that are real passwords for anything else. Your SNMP strings, if used, should be "somerandomstring" not "supersecretpassword". Even if SNMPv2 is not enabled, be sure any default strings are not configured.

This is probably the one setting you need to check regularly, as there have been cases where vendor code upgrades have changed (re-enabled) protocols such as SNMPv2c. Regular audits and monitoring of configuration changes is a topic covered in Chapter 7, "Monitoring and Maintenance of Wireless Networks."

SNMPv2c IN THE WILD

There are still many large deployments that rely heavily on the unencrypted and unauthenticated SNMPv2c flavor of SNMP. SNMPv3 has been around for 20 years and has been crowned a full Internet standard by IETF since 2004, the highest honor bestowed upon an RFC.

That means this protocol has been blessed for almost 20 years. Go forth and conquer SNMPv2c and try to eradicate it from the network.

Controlling Administrative Access and Authentication

The task of securing management access continues with controlling administrative access and authentication. This topic addresses the discrete considerations and features available for restricting management access only to the people (or non-person entities) that need it, from approved networks or sources, and ensuring that access is logged and auditable.

Enforcing User-Based Logons

An easy and exceptionally impactful management control is the enforcement of user-based logons versus using shared accounts and credentials such as *admin* or *root* accounts.

Aside from segmentation, one of the most pervasive requirements for any security controls or compliance framework is the restriction of shared and group account credentials. For example, NIST SP 800-53 specifies a suite of controls and baselines for restricting use of shared accounts, and additional controls for monitoring and life-cycle management in the event a shared account is required. The same control suite also specifies requirements for credential vaulting in the cases of shared accounts, something covered shortly.

Administrative access needs to be logged and auditable with attribution (to a specific person) in addition to being authenticated and encrypted. To support that objective, it's recommended that all organizations (regardless of size) rely on user-based management logons. In most products, this is supported through AAA (authentication, authorization, and accounting) protocols primarily via RADIUS or TACACS+, or alternatively through a connection to LDAP, which offers authentication but not accounting.

Options for user-based management authentication include:

- RADIUS
- TACACS+
- LDAP to on-prem
- LDAP to cloud with SAML (e.g., via Azure SSO services and ADFS)

In talking to clients over time, the perception they often have is this is a daunting and complicated task that requires a lot of time, additional infrastructure, and a TACACS+ server. The reality is, TACACS+ is just one option for user-based management authentication and many organizations opt to use RADIUS for administrative access authentication, allowing them to leverage the infrastructure already in place for user-based 802.1X/EAP authentication. Also, the configurations for authenticating using AAA (RADIUS or TACACS+) can be pushed to multiple devices (such as edge switches or APs) at once through scripts or APIs and can be easily configured in a traditional wireless controller.

SIZING IT UP: RADIUS VS. TACACS+ FOR MANAGEMENT ACCESS

While granular authorization policies are ideal, they're not required in all organizations and not specified in compliance requirements (other than the need to enforce principle of least privilege).

For small organizations, the granular policies afforded by TACACS+ simply aren't required, and RADIUS for authentication and accounting is adequate, and certainly better than nothing.

For large organizations with complex infrastructures, multiple tiers of architects, administrators, and operations users, the additional control and auditing of TACACS+ is desirable.

Although not discussed much throughout this book, there are SaaS-based cloud RADIUS platforms that support both end-user and management user authentication via the cloud. This is a great option for residential and personal users that require enterprise-grade protection but for businesses and even small organizations, most will have at least one server with domain services and if so it's preferable to simply enable RADIUS services locally if possible.

RADIUS and TACACS+ can both be used for management authentication and authorization but differ in granularity of authorization and security features (see Table 6.3). Most specifically, TACACS+ supports specific authorization policies, allowing the super user to specify which discrete commands an administrative user can issue on the device.

Granularity in the TACACS+ implementations vary, but usually follows one of two models:

- Mapping of privilege level to authorization rights
- Mapping of specific commands to authorization rights

As an example of mapping of privilege level to rights, for web UI access authenticated against TACACS+, Cisco's current code version on a 9800 maps users with privilege levels 1–10 to accessing the monitor tab features, and users with privilege level 15 have full administrative access. That's a web UI example; when implemented for SSH, per-command policies can be enforced versus the basic two tiers in the web UI.

As an example of specific commands, Aruba controllers support per-command authorization when used with ClearPass Policy Manager TACACS+ policies for management. In this example, a management user's rights could be restricted to only allow executions of specific commands, versus mapping to an entire privilege tier.

Table 6.3: Comparing TACACS+ and RADIUS for management

FEATURE	RADIUS	TACACS+
Industry Adoption	Open standard	Until September 2020, proprietary Cisco protocol that has to be licensed by third parties; new IETF RFC 8907
Authentication Security	Username may be sent in cleartext if not secured with other means or RADSEC	All authentication elements encrypted

FEATURE	RADIUS	TACACS+
Authorization Granularity	Authentication and authorization are tied, no granular authorization policies	Supports tiered privilege mapping or per-command authorization for granular policies
Logging and Accounting	Authentication	Authentication plus detailed authorization
Protocol	UDP, by design	TCP

Whether using RADIUS or TACACS+ for management authentication, the configurations are similar AAA configurations on the wireless infrastructure and will mimic the steps followed for 802.1X/EAP end-user authentication. For authentication against LDAP, a basic bind is configured along with the parameters for domain names, groups, and objects.

INDUSTRY INSIGHT: TACACS+ BY CISCO, ARUBA, EXTREME, JUNIPER MIST, FORTINET, AND OTHERS

Even though TACACS+ has been a Cisco proprietary protocol, other vendors do license it (just like CDP) and support it in their products.

As of September 2020, the IETF has a formalized RFC (`https://datatracker.ietf.org/doc/html/rfc8907`) based on TACACS+. With the opening of the protocol as a standard, more products may be supporting TACACS+, but already many other vendors licensed it and included support for TACACS+ server services in their authentication products. So far it appears the RFC for TACACS+ is stripped down, omitting many of the currently licensed features.

Aruba Networks supports TACACS+ in its ClearPass Policy Manager (CPPM) product. Verify your license types with your account team; Aruba has changed licensing models over the years and not all licenses include TACACS+, although that could change with the newly created RFC.

Fortinet FortiAuthenticator is a robust authentication product that has also included support for TACACS+.

FreeRADIUS now offers support for TACACS+ server services as well.

As for the cloud-managed products, Extreme supports RADIUS and TACACS+ from its legacy on-prem products and SAML2 integration from the cloud.

Juniper Mist is cloud-only and also supports single sign on (SSO) with Azure as well as SAML.

Creating a Management VLAN

Management VLANs are recommended as a best practice by vendors and also required by several cybersecurity controls frameworks.

A management VLAN can simply be a VLAN described for management of infrastructure devices, or it can be a more formal implementation by designation of a management VLAN within certain wired implementations. Formally designating a VLAN as a "management VLAN" in some wired products will automatically lock certain features and access to and from that network (which may or may not be desirable). Please consult your switch product documentation for guidance.

Formal designations aside, a VLAN that is simply defined as use for management of some or all of the infrastructure entails creating the VLAN, naming it, assigning an IP interface at the router(s), and configuring infrastructure devices (such as a wireless controller and APs) for that VLAN. Additional wired configurations for ACLs should be used to segment the management traffic from the client data traffic and other VLANs. Obviously in large environments, there will be multiple management VLANs, not one.

Depending on the size of the environment and organization, a single management VLAN could be used for several classes of infrastructure devices (such as switches, routers, servers, and wireless controllers), or the wireless infrastructure may need its own set of management VLANs. In this case, there may be one or more AP management VLANs, and one or more VLANs where on-prem controllers reside. Both scenarios are acceptable.

Management VLANs offer two primary benefits for security architecture:

- Segmentation of management traffic for integrity and confidentiality
- Segmentation of management traffic for availability

Segmenting for integrity and confidentiality helps protect management traffic from eavesdropping, injection, or modification from users or devices not authorized for management access.

Segmenting for availability offers the added benefit of reducing broadcast domains and protecting from denial-of-service (DoS) interruptions due to broadcast storms (often introduced via network loops), oversubscribed network segments (where management traffic is delayed due to other data along the same path), and malicious attacks.

If your wireless infrastructure devices aren't currently configured on a management VLAN, moving them (or moving other devices off their currently configured network) is highly recommended in all environments. Depending on the management architecture, it can be a tedious task, but the security benefits are well worth the effort.

WARNING **For environments adding a management VLAN, it is a best practice to not use the switch vendor's default VLAN for this purpose. If your edge switches come configured with a default VLAN ID of 1, do not use that for the management VLAN. If you do, any endpoint or user that connects to an unused port will be on (what should be) your most protected management network.**

Defining Allowed Management Networks

Network products, both wired and wireless, often support the specification of allowed management networks. This is a slight variation from a defined management VLAN described earlier in that allowed management networks enforce a basic access control list (ACL) for management traffic, which can be defined as allowed by one or more subnets. Management traffic received from other networks is simply ignored.

As an example, in many wireless products, you may have to explicitly allow management from clients connecting through the wireless interface.

Securing Shared Credentials and Keys

This topic, securing shared credentials, and the next one—addressing privileged access—go hand in hand, but are worthy of their own headings.

Even though best practices demand user-based management logins, just about every environment will have some shared credentials, and/or non-user-based accounts such as the admin and root accounts required for system provisioning. Even after user-based management authentication has been configured, there will still remain a certain volume of shared credentials, and those need to be addressed.

In the era of increasing security threats, remote work, third-party access to systems, and zero trust initiatives, credential vaulting is taking center stage as a critical network security tool.

Credential vaulting is akin to an enterprise password management tool on steroids. Mature vaulting products are designed to manage an assortment of secrets such as:

- Certificates
- Certificate private keys
- SSH keys
- API keys
- Shared and group passwords
- Service account credentials
- Other credentials for non-person entities (NPEs)

With vaulting tools, management of secrets is centralized and can therefore be controlled by the organization and integrated into identity and access management (IAM) security practices.

These tools also help organizations meet security compliance requirements, such as those mapping to NIST. For example, in NIST SP 800-53 revision 5, control item AC-2(9) mandates the restriction of shared and group credentials and control item IA-2(5) mandates that "When shared accounts or authenticators are employed, require users to be individually authenticated before granting access to the shared accounts or resources." (Source: `https://csrc.nist.gov/publications/detail/sp/800-53/rev-5/final`.) "AC" controls are Access Control, and "IA" controls are Identification and Authentication.

Credential vaulting solutions do just that and more. Let's say there's a scenario where our organization has user-based management authentication but for some reason the RADIUS server is unreachable, and the system must fall back to local authentication. Instead of logging in directly with the user "admin" and the associated password, my terminal application attaches to a credential vaulting tool, I log in to the tool as me, and the vault passes through the admin credentials to the controller. In that process, I never see the username "admin," nor the password associated with that user.

Most vaulting products have advanced credential life-cycle management and can update and rotate credentials on a predefined schedule. For example, if the organization rotates SSH keys or API keys every 90 days, the vaulting tool can automate those processes, or even update the password to a service account in the domain.

IT administrators and architects are often over-extended and putting out fires on a daily basis. In that daily grind and fray, it's common for sensitive credentials to be generated and stored insecurely. And, in fact, only the very large, very mature, or very regulated companies tend to have proper credential management.

If your organization doesn't have a credential vaulting solution now, please take this opportunity to make an appeal to your leadership to budget for that. It's a critical component to a holistic security strategy, and there are very affordable products, and probably even some free and open source options.

For context, popular credential vaulting applications include Azure Key Vault, Amazon AWS Key Management, and products like those from Beyond Trust (which can be deployed in the cloud or on-prem), all of which can be used to store not only SSH keys but certificates and other credentials.

SIZING IT UP: SSH KEY MANAGEMENT IN ENTERPRISE

In an earlier section, we covered how to generate and load SSH keys for the purpose of authentication, but SSH keys should be carefully managed and not generated and saved willy-nilly on IT administrators' computers.

One recent report cited that remote desktop (RDP) and Secure Shell (SSH) exposure increased by around 40 percent, likely due to the increase in remote working due to COVID-19. (Source: `https://info.edgescan.com/vulnerability-stats-report-2021`.)

In other study, over 90 percent of IT security pros reported they lacked a complete and accurate inventory of SSH keys. Additionally, nearly two out of three of the

cybersecurity professionals stated that they do not actively rotate SSH keys. (Source: `https://www.businesswire.com/news/home/20171017005080/en/Study-61-Percent-Organizations-Minimal-Control-SSH`.)

When used as a credential alone, SSH keys are truly the keys to the kingdom and should be protected as such. Mature and security-conscious organizations should have an SSH key management program that includes policies for SSH key life cycle and uses, inventory of SSH keys including occasional discovery and scanning, SSH key rotation, key vaulting and escrow, and verification that SSH versions and key sizes meet compliance regulations.

The National Institute of Science and Technology (NIST) provides in-depth guidance for SSH access management in its interagency report 7966 (NISTIR 7966) found at `https://csrc.nist.gov/publications/detail/nistir/7966/final`.

Addressing Privileged Access

Privileged accounts are defined as accounts with elevated privileges or access above and beyond what a standard user is authorized for. Traditionally, management of privileged accounts has focused on access rights for users and services accessing specific data sets or applications—typically those containing regulated records such as health records, financial records, or personally identifiable information (PII).

The trends of remote access, cloud-managed services (including SaaS, PaaS, and IaaS offerings), and rising security threats are causing the tides to turn a bit and privileged access management will be incorporating control of admin accounts for infrastructure components such as the wireless infrastructure.

Here, we'll hit the highlights of privileged access with the understanding that this is an initiative usually driven by the organization's identity and access management (IAM) or security compliance teams.

It's a relevant and timely topic for network and system administrators, though, and your architecture strategy should include processes and tools for securing administrative access to the infrastructure.

Securing Privileged Accounts and Credentials

Along with privileged accounts (used by humans) are privileged credentials that may include SSH keys as well as API keys and tokens used by services, applications, and microservices. Granting authorization to a thing rather than a person presents challenges and breaks many legacy authentication models (such as hardware tokens). To support these shifts in use cases, regulations are starting to address the security of access by what the industry is now calling non-person entities (NPEs).

The subject of non-person entities is most prevalent when discussing security of workloads such as containers and microservices as well as robotic process automation (RPA), but it's extending to the infrastructure as the use of APIs becomes more prevalent.

Examples of privileged accounts and privileged credentials as defined for compliance and security policy include:

- Local host admin accounts
- Super user accounts (including management access to wireless)
- Domain admin accounts
- Privileged business user
- SSH keys, API keys, and other secrets
- Service and application accounts
- Emergency accounts

It never makes sense to add complexity unless there's a need, but protecting privileged accounts will emerge as a central theme of network security architecture due to account misuse, abuse, and lost or stolen credentials.

In fact, privileged account management is such a prevailing factor in security breaches and incidents that Verizon's Data Breach Investigations Report (DBIR) now includes privileged access as its own risk classification in the report: `https://www.verizon.com/business/resources/reports/dbir`.

Figure 6.7 and Figure 6.8 demonstrate a few data points from the Verizon DBIR 2021 report.

While privilege misuse is in a downward trend, lost and stolen credentials and miscellaneous errors are both continuing and expected to grow with more cloud-connected and remote access models. The graphics represent data related to actual breaches, not just incidents, which are classified separately.

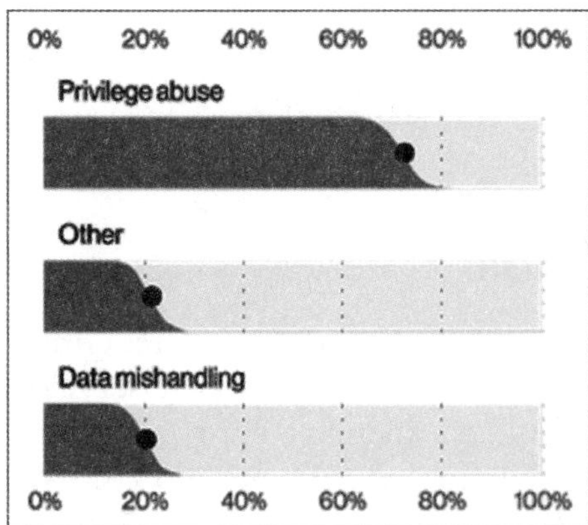

Figure 6.7: Figure 10 from the Verizon DBIR 2021 report, showing privilege abuse as the highest contributor in top misuses resulting in a breach. Command shell access is number 5 in the top hacking vectors in breaches.
`https://www.verizon.com/business/resources/reports/dbir/2021/`

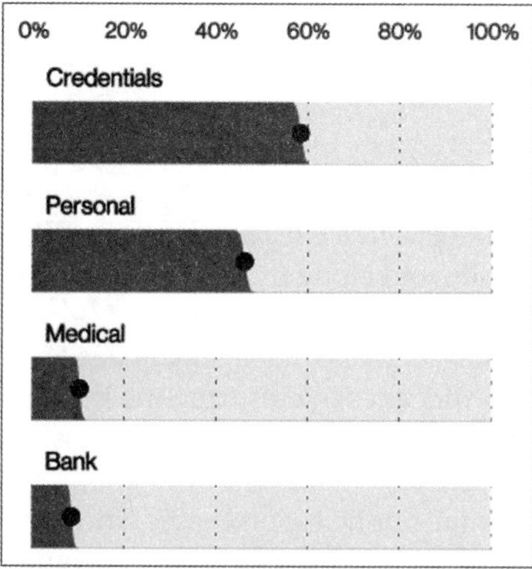

Figure 6.8: The report highlights credentials as top data varieties in breaches, above personal, medical, and banking data assets.

`https://www.verizon.com/business/resources/reports/dbir/2021/`

In a prior report, some notable findings demonstrated that:

- Weak or common passwords were the cause of 63 percent of all breaches
- 53 percent of the breaches were due to the misuse of privileged accounts

Privileged Access Management

The practice of managing privileged access is called *privileged access management* (PAM) and comes in the form of organizational policies, processes, and tools or applications.

PAM isn't just a best practice, it's also specifically called out as a mandated control in most regulations including Sarbanes-Oxley (SOX), the Federal and North American Energy Regulations Commission (FERC/NERC), HIPAA, and state-level regulations such as the California Information Practice Act.

International regulations for data privacy such as those outlined in the EU's General Data Protection Regulation (GDPR) may not specifically mandate PAM, but they do stipulate the organization must evaluate how it gathers and stores data as well as who has access to it.

Privileged access management processes and tools include:

- Privileged account and credential vaulting
- Monitoring and alerting of access
- Regular auditing of access and accounts
- Privileged account life-cycle management

- Proper password/passphrase security
- Enforcing the principle of least privilege

Privileged Remote Access

There's one more phrase to be familiar with—*privileged remote access*, or PRA, since we like acronyms so much. If you've followed along this far, there's likely no explanation needed.

Privileged remote access is a subset of privileged access management designed to address the unique aspects of accessing protected systems from the Internet or across other untrusted networks. It can also be used to manage access for employees and third parties, even if they're onsite.

Just like PAM, privileged remote access entails policies, processes, and tools but remote access tools are often different than standard PAM tools, meaning there may be a separate product used for this purpose, which is why I'm addressing it here.

Also, a recent forecast from IDG found 87 percent of enterprises expect their employees to continue working from home three or more days per week. This includes network architects, engineers, and sys admins who will rely greatly on privileged remote access moving forward. (Source: `https://www.idc.com/getdoc.jsp?containerId=prUS46809920`.)

With zero trust strategies surfacing and more and more users accessing systems remotely, the days of legacy VPN access where users are dropped on the network are coming to an end, and the dawn of granular PAM and PRA controls is coming.

Features for privileged remote access products vary, but in addition to the benefits of PAM, purpose-built remote access also addresses:

- Securely accessing systems remotely
- Maintaining detailed change logs for audit purposes
- Supporting various consoles and applications (e.g., for web UI or SSH access)
- Streamlining access by third parties including non-domain users (such as vendors' systems and support engineers)
- Enforcing multi-factor authentication

INDUSTRY INSIGHT: PRODUCTS FOR PAM AND PRIVILEGED REMOTE ACCESS

If you're interested in learning more about PAM and PRA products, some of the popular vendors include Beyond Trust (`www.beyondtrust.com`), CyberArk (`www.cyberark.com`), and Thycotic (`www.thycotic.com`). Each product solution has its own niche

and different offerings fit in different types and size of organizations, and products may include solutions on-prem, cloud, or both.

As a network engineer, I can say Beyond Trust has some especially compelling features like the ability to securely SSH to remote devices using your own local agent (e.g., your custom PUTTY) versus being forced through a weird proxy terminal. It also supports allowing third parties to join a session without having to share your own desktop—a must-have when asking the vendor's TAC engineer to remote in and help.

Oh, and if you do have a third party connect remotely, they can do that with or without you, through either pre-approval or on-demand request/approval. You can have the system auto record the entire session to view, replay, or audit any changes. And of course, it supports the full suite of secure access management, credential vaulting including securing SSH and API keys, which is equally important.

Other products likely have similar features. If you're advocating to leadership for a PAM or PRA, just kick the tires a bit and test a few options to see what integrates with your IAM and best serves your needs.

Additional Secure Management Considerations

There are a few fringe topics that don't warrant their own sections but are worth mentioning and considering in your security architecture.

Enforce Secure Passwords Follow the organization's guidance for enforcing strong passwords that are resilient to cryptographic attacks. Password length is the most important factor with complexity being a second factor. Password security should be applied to local administrative accounts as well as directory-based user logins such as those through TACACS+, RADIUS, or LDAP.

System Logon Banners Logon banners warn a user before attempting to log in to a restricted system, such as administrative access of the infrastructure.

Logon banners have legal implications and are recommended in most hardening guides as a best practice. The logon banner should include appropriate language warning an unauthorized user about attempts to access the system. It's worth noting that some vendors' features do not support the use of logon banners when enabled, so please consult your product documentation.

Limit Concurrent Administrative Sessions and Set Timeouts Network devices most often allow configuration of a maximum number of concurrent admin sessions through terminal VTY (virtual terminal) settings, with many products supporting configuration options ranging from one to an unlimited number of logins.

Best practice is to configure a reasonable maximum for concurrent admin sessions, taking into account not only human managers but security, management, and monitoring systems that may be using SSH.

Along with maximum sessions, there should be timeouts configured for all administrative access to the system.

Multi-factor Authentication (MFA) Although most compliance regulations define requirements for enabling MFA on administrative access of systems, it's still relatively rare as a direct control, meaning that box is checked and the requirement technically met via an intermediary like a jump box or through VPN access requiring MFA.

Regardless of how it's implemented, MFA is strongly suggested for all administrative access, whether local or remote. Consider it a crucial requirement (versus recommended control) for remote access and management access to cloud solutions (such as cloud-managed networking like Juniper Mist, Meraki, ExtremeCloud, and Aruba Central).

Logging and Monitoring of Administrative Access and Attempts Management access to systems including the wireless infrastructure and management should be logged and monitored appropriately. As with many of the topics here, this type of logging and alerting is specified as a requirement in most security regulations. This is a topic covered more in the next chapter. As part of your architecture planning, include monitoring for admin login password retries.

TIP When it comes to MFA, anything is better than nothing, but an authenticator app is better than SMS text. Attackers can use SIM swap techniques to trick a carrier into porting a number to a malicious user and gain access to text messages.

Designing for Integrity of the Infrastructure

Our hardening journey continues by designing for integrity of the infrastructure. At a high level, this involves authenticating the infrastructure to itself, ensuring the integrity of software and configurations, addressing secure backups, physically securing infrastructure components, and pruning unused protocols.

Infrastructure integrity is organized into the following topics:

- Managing Configurations, Change Management, and Backups
- Configuring Logging, Reporting, and Alerting
- Verifying Software Integrity for Upgrades and Patches

- Working with 802.11w Protected Management Frames
- Provisioning and Securing APs to Manager
- Adding Wired Infrastructure Integrity
- Planning Physical Security
- Disabling Unused Protocols

Managing Configurations, Change Management, and Backups

Configuration management is unfortunately often disregarded but remains a key element of infrastructure integrity. A common practice for many network admins is to save a known good configuration before a scheduled upgrade, but policies, processes, and activities related to managing configurations, change management, baselines, and backups go ignored in many cases.

If you're working in a regulated industry or conforming to NIST standards, your organization probably has a robust configuration management program. If you're not quite there yet, here are a few key items to incorporate into your security architecture.

Configuration Change Management

Change control processes address the life cycle of configuration maintenance and vary wildly. In small or unregulated environments, the change management process may be undocumented and a simple notification from the admin to a manager or request for a maintenance window is sufficient.

Large and regulated environments (including most federal government agencies and publicly traded companies) abide by strictly enforcing change management that involves some or all of the following:

- Formal request to implement a configuration change
- Analysis and rating of the change in terms of business impact, technical impact, and security risk
- Testing and validation of the planned change in a controlled test or lab environment
- Review by a change management board or approval committee
- Implementation of the change including notifications before, during, and after
- Testing and validation of the change in the production environment
- Exhaustive documentation and approvals at each step including rollback procedures

For professionals working in a more ad-hoc manner, it's recommended to incorporate tasks beyond simply scheduling a configuration change. These steps provide a minimal structure and documentation appropriate for organizations lacking mature change management:

1. Document the planned change.

   Capture the reason for the change and expected outcomes and known issues to resolve or monitor for. Also document the tests to perform validation of the configuration and any additional testing required to validate security controls.

   Ideally, document how the change should be implemented e.g., through web UI, SSH, and the commands or steps associated with the change plus the steps for a rollback if needed.

   If the organization doesn't have a platform for asset and configuration management but has an internal ticketing system, use that. At a bare minimum, create a text file or document with this information and start a file in a secure, shared storage area.

2. Review the planned change with at least one technical peer or manager.

   As a sanity check if your organization has no formal processes for change management reviews and approvals, review the planned change with at least one technical peer (which could be from another group or team) and/or your direct manager.

3. Optionally but ideally validate the planned change in a test or lab environment.

   This is highly recommended for numerous reasons, but some organizations simply don't have access to test or lab environments. If the environment you're in has a low tolerance for unplanned outages, it's likely worth the time and effort to make a request for lab equipment or even virtual appliances where applicable.

4. Schedule and document the change window.

   Before proceeding, back up the current configuration to a secured corporate-managed storage location. Schedule the maintenance window for the change to occur, and document the activities including the date and time stamps of when the changes were made or committed to the system.

   This is exceptionally helpful for later troubleshooting and technical support if there is an issue after an upgrade or configuration change.

5. Test and validate the change in production.

   For most changes, the testing and validation in production is a two-phased task. First immediate testing is performed at the time of the change, and then most often a window is defined to validate operation when the system is under full load.

Here's the process described using a real-world scenario:

1. The wireless architect is planning to change an existing WPA2-Personal SSID to use the more secure WPA3-Personal Only security mode.

   This is performed in the web UI under the template for the SSIDs using a radio button. The known issue is that endpoints not updated to support WPA3-Personal will not be able to connect; the organization is aware of this, has monitored the environment and is ready to proceed.

   Testing after the change includes connecting a WPA3-Personal-capable client to the SSID to ensure it can connect and attempting to connect a device only capable of WPA2-Personal and verifying it is not able to connect, which proves the configuration worked.

2. The architect has reviewed the planned changes with the network operations team, help desk team, and his/her manager and a maintenance window has been agreed upon.
3. The architect tests the process in the lab using a small physical controller running the same code as the production controllers.

   Both a WPA2-Personal and WPA3-Personal client are tested and results match with the expected outcomes defined in step 1.

4. The architect makes the change during the scheduled maintenance window, documents the date and time of the change, repeats the same client tests in the production environment, and confirms proper operation of the change.
5. The next day, the network operations team monitors the environment for failed connection requests on the updated WPA3-Personal network and the help desk fields calls from users with WPA2-Personal devices that aren't able to connect.

Although a possible real-world scenario, this example is drastically oversimplified and many change management tasks involve multiple changes, multiple testing sequences, and may have unknown or unintended consequences, such as a vendor bug.

In the absence of mature processes for change management, the best practice is to incorporate the due diligence tasks described here, ensure everything is well documented, and involve at least one other party or person in the process.

CHANGE MANAGEMENT IN NIST SP 800-53

The NIST Special Publication 800-53 *Security and Privacy Controls for Information Systems and Organizations* is a comprehensive framework referenced by many other compliance frameworks and regulations.

It includes detailed guidance for change management including baseline configurations (covered next) and control change processes in the Configuration Management (CM) suite.

Specifically, in revision 5 these are covered in sections CM-1 Policy and Procedures, CM-2 Baseline Configuration, and CM-3 Configuration Change Control. Additional supporting guidance is provided for impact analysis, monitoring, and other related tasks.

Configuration Baselines

Large and complex environments should have processes and practices around the creation and management of baseline configurations, which serve as the approved template for configurations that meet the organization's business and security requirements.

Baseline configurations describe the organization-certified state for implementing systems of a specific type. For example, a company with hundreds of locations and local wireless infrastructure at each site would create a standard configuration baseline to deploy secure wireless at each. Baselines can be detailed documents with specifications or well-documented configuration file templates or scripts.

There may also be variables within configuration baselines and elements dependent on the type of location or geography, for example.

Small environments may not have formal configuration baselines as the time to document them may be unrealistic for the value derived, and of course in many environments there may only be limited wireless infrastructure.

Having said that, for small organizations in some sectors and those required to meet specific guidance against NIST or ISO standards, it's possible there will be a requirement to document an approved secure configuration baseline.

Configuration Backups and Rollback Support

Configuration management encompasses ensuring configuration backups are retained and properly secured in an approved location.

And, an approved location, regardless of the organization's size, is not the wireless admin's laptop. An approved backup location should be a designated, secure repository accessible by other authorized administrators or systems. The repository could be a network management tool (such as Cisco Prime, Cisco DNA Center, HPE Intelligent Management Center, HPE OneView, SolarWinds, Aruba Fabric Composer, etc.) or it could be a designated place in cloud or network storage where backups can be sent via secure file transfer on an automated schedule (using SCP or SFTP).

Ideally, configuration backup management includes a platform that supports automating the configuration backup of the wireless infrastructure, added functionality to annotate configuration files to note inflection points or changes, a tool to perform a "diff" or comparison of two configuration files to highlight changes, methods to alert on unplanned configuration changes, and tools for validating security configurations against approved baselines.

Backups should be tiered and stored in two places, such as on-prem with a secondary sync to a cloud repository, or in two geographically diverse datacenters.

Rollback support is another consideration, and the backup plan should allow for an appropriate number of prior revisions and support retrieval for a restoration in the event a rollback is needed.

Monitoring and Alerting for Unauthorized Changes

Changes in the system configuration should be monitored and alerted upon. This could be implemented within the system itself, if supported, or via syslog or SNMP traps to a SIEM or logging server, or through network management products described earlier.

Logging and reporting are covered a bit more next, and examined in more depth in Chapter 7.

Configuring Logging, Reporting, Alerting, and Automated Responses

Hardening any infrastructure necessitates monitoring and alerting of attempts to bypass or breach security controls. For wireless, appropriate monitoring and alerting encompasses not only the management access but also client access and data security.

Chapter 7 deals with security monitoring in much more detail. In the context of hardening—logging, reporting, and alerting should (at a minimum) be designed to identify:

- Unauthorized configuration changes
- Attacks to the wireless (RF) medium including jamming

- Attacks to the wireless data infrastructure including spoofing and DoS attacks
- Attacks to the wired network initiated from the wireless network
- Administrative login attempts (failed and successful) and logging of usage
- Rogue and unauthorized devices (endpoints and infrastructure)
- Anomalous behavior

TIP **In addition to the upcoming chapter on monitoring and maintenance, for organizations aligned to NIST SP 800-53, consult the controls guidance in the System and Information Integrity (SI) topic domain for more information on monitoring.**

Verifying Software Integrity for Upgrades and Patches

Keeping the wireless infrastructure secure demands regular security patching and (optionally but recommended) steps to validate the software integrity.

Verifying Software Integrity

With everything so inter-connected these days, and masses of Internet resources, knowledge bases, blogs, articles, and file repositories, verifying software integrity is a little extra step worth considering.

With file or code integrity checking, the creator of the code (such as the wireless vendor) will use a standard hash algorithm and provide the hash output separately from the file. When you download that file, you can then verify the file hasn't been modified (through malicious tampering or inadvertent corruption) by using the same hash algorithm to verify the contents haven't changed.

UNDERSTANDING HASH FUNCTIONS

If you're not familiar with hash functions, they're one-way algorithms that take an input (e.g., binary from a data file) and perform a function that reduces it to a fixed-length alphanumeric value. Hashes are considered a cryptographic function, but unlike two-way encryption algorithms, there's no need (and no method) to reverse a hash, hence the one-way descriptor.

Popular hash algorithms include MD5, SHA-1, and SHA-256.

To get a feel for hashes, you can search for "hash generator" or "sha1 generator" for online tools that perform hash functions of raw text or files. There are also tools built into operating systems including Windows that perform the same functions.

As an example, let's say I'm going to share my prized award-winning, almost-famous recipe for Gluhwein (mulled wine). If I upload it on a file sharing platform for you to download, I can run a hash function—in this case SHA-1—and provide the output.

In fact, the SHA-1 hash of my Gluhwein recipe is `1DFE3F21AC012FE615CCF797DD439F436C378B7F`.

When you download the file, you want to verify that some evil person hasn't tampered with it and modified the recipe to include grasshopper legs. You also want to ensure you have the whole recipe and the second page didn't get corrupted in transit. To do that, you simply run a hash function on the file, and if all is well, you come up with the same value (the string ending in `8B7F`).

For the integrity checks to work, the hash function has to be provided in a way you know it's from the entity asserting itself (in this case me). Otherwise, a malicious user could simply post a bad recipe, named to match my file, with the hash of the fake/malicious file, and you'd never know the difference because the hashes would match.

When you download software and update files for systems, you have the option to perform the same validation.

Validation checks can happen a few ways—often it's supported in network management tools (like the ones mentioned in "Configuration Backups and Rollback Support"). You can also manually validate the hash using any online hash calculator.

HOW TO VALIDATE WIRELESS VENDOR FILES

Check with your vendor for specific instructions. As an example, Cisco document ID 211350 titled "How to Validate the Integrity of a Downloaded File" from `Cisco.com` includes instructions for this in an FAQ for one of its storage products. You can also look for "Integrity Verification" under Cisco DNA Center. Aruba Networks and other vendors have similar guidance where applicable.

Cloud-based management solutions have software platforms or microservices managed and maintained by the vendor. As such, the organization would have no need to download and install software, and therefore no need for this specific use case of integrity checks. The vendor will be performing these checks within its cloud management and AP ecosystem.

Upgrades and Security Patches

During the topic of system resiliency in Chapter 4, "Understanding Domain and Wi-Fi Design Impacts," it was noted that upgrading traditional on-prem wireless controllers was a gruesome and painful experience. The topic included a few anecdotes of upgrades gone wrong, and the general sentiment of Wi-Fi professionals that controller upgrades only be performed when deemed absolutely necessary (usually when there's a user-impacting issue that requires a code upgrade or a patch for resolution).

Even with my own personal intimate (and extensive) experiences with the headaches of controller upgrades, keeping systems up to date and patched is absolutely critical for maintaining a secure infrastructure. There are simply no exceptions to that—no matter how painful it may be.

In a perfect world, the realms of networking operations and security operations will come together and your newfound buddies in the security operations center (SOC) will be responsible for parts of the vulnerability management program, including scanning and alerting when patches are required. That doesn't alleviate the fact that you'll have to schedule and perform the upgrade, but it may help focus attention on the updates that are truly requisite for security purposes.

In most organizations, network admins are wearing many hats and it's untenable for them to read every line of every code release for every system they manage and make decisions about upgrades and patches. A patch may fix one problem but break three (or eight) others. And not all fixes and features are in use in every environment.

Knowing all that, and understanding the burden it imposes, regular updates for security patching is just not something to be sidestepped. Work it into your architecture planning. If you truly can't sustain the workload of updates, then discuss options for a cloud-managed platform that will alleviate much of that workload or approach your leadership about adding additional headcount or resources, and/or using a managed service.

It doesn't matter how the security patching happens; it just needs to be done.

Working with 802.11w Protected Management Frames

In 802.11 WLANs, there are management frames, control frames, and data frames. The data frames include the client data and are encrypted as defined in the SSID security profile, with a client that is authenticated. In the decades of 802.11 WLANs, the management frames have had no form of integrity—meaning, they've been neither authenticated nor encrypted.

The result of unprotected management frames is multifaceted. There are countless legitimate wireless operations that have exploited this gap to offer specific security-related features, such as an enterprise Wi-Fi system spoofing de-authentication packets to prevent a managed corporate endpoint from connecting to a rogue AP. In fact, many traditional wireless IPS (WIPS) functions have taken advantage of unprotected management frames.

Of greater concern of course is the reality that the exposed management frame model leaves the wireless infrastructure vulnerable to a host of malicious attacks, abuse, and misuse. To solve this gap and bolster the integrity of the wireless infrastructure over the air, the 802.11 standard was updated to incorporate additional protections.

I'm stretching the hardening topic a bit by covering PMF here, but there are components of the 802.11w standard for PMF that play a role in infrastructure integrity (including preventing AP spoofing). But also, Cisco has a proprietary feature called Infrastructure PMF, which is different than 802.11w PMF, and is a suite designed for infrastructure security. You'll see a sidebar on that at the end of this section.

The IEEE 802.11w standard for Management Frame Protection (aka PMF or MFP) has been available and supported since 2009 but only now required as part of the new WPA3 suite of security standards. In fact, it's the use of protected management frames that defines WPA3 (versus WPA2).

The benefits of PMF were introduced in Chapter 2, "Understanding Technical Elements":

- Prevent forgery of management frames (through SA query)
- Prevent spoofing of AP or endpoint
- Prevent replay attacks
- Hinder eavesdropping of certain management frames

Wi-Fi Management Frames

Wi-Fi management frames are used between APs and endpoints to find, join, and leave networks, and include:

- Beacons and probe requests and responses
- Association and re-association requests and responses
- Disassociation notifications and requests
- 802.11 authentication and de-authentication

Remember that the 802.11 authentication referenced here is the part of the association process all endpoints go through to join a network and is not the network authentication (such as 802.1X/EAP).

Unprotected Frame Types

The processes for endpoints to find and join a network requires that certain frames be accessible to any and all endpoints even before joining the network. In these instances, it's not feasible to encrypt or validate integrity of the parties. As an aside, there are some experts who are advocating for and proposing solutions to extend management frame protection to include frames prior to the 4-way handshake, as well as to control frames.

These unprotected management frames include:

- Beacon and probe request/responses before association
- Announcement traffic indication message (ATIM)
- 802.11 authentication
- Association request/response
- Spectrum management actions

Protected Frame Types

We've covered the 4-way handshake in several topics throughout the book, so it's probably no surprise that protected management frames are primarily tied to interactions that happen after the endpoint and AP have established a formal relationship.

Once an endpoint has joined the network (via the 4-way handshake), the AP and endpoint have exchanged information to derive encryption keys and are then able to prove their identity to one another as well as exchange encrypted packets.

Management frames that are protected include:

- Beacon and probe request/responses after association
- Disassociation (AP or endpoint terminating the session)
- 802.11 de-authentication (of endpoint to AP)
- Certain action frames such as block acknowledgments, QoS, spectrum management, and Fast BSS Transition (FT)
- Channel change announcements sent as broadcast
- Channel change announcement sent as unicast to the endpoint

The 802.11 de-authentication and disassociation frames are the two types of interactions most often spoofed (maliciously or intentionally from WIPS).

The new security association (SA) query function adds a lookup step, where the AP will check the association table before processing certain requests from an endpoint (such as a disassociation). If the association entry is present, the request will have to be sent via the protected frames or it will be discarded/ignored. This is part of the mechanism to prevent spoofing and forging of frames.

For unicast exchanges that are encrypted, existing cipher suites and keys are used, specifically the sender's pairwise transient key (PTK) is used for encryption of the unicast traffic.

For broadcast and multicast management frame protection, a new key is specified—the integrity group transient key (IGTK) along with a message integrity check (MIC) function. The IGTK is added to message 3 of the 4-way handshake, and after that 802.11w uses broadcast integrity protocol (BIP) to protect (integrity not encryption) broadcast and multicast management frames. The purpose of BIP is integrity and replay protection, not confidentiality.

Validated vs. Encrypted

With 802.11w enabled, some frames are not protected, others are protected with integrity by validating the sender, and a subset are also encrypted.

Which frames are encrypted versus which offer integrity only can be summarized as follows:

- After the 4-way handshake, unicast frames are encrypted (offering both confidentiality and integrity)
- After the 4-way handshake, broadcast and multicast frames are afforded integrity only (no encryption)

Encrypted unicast management frames include action frames, disassociation frames, and de-authentication frames. Contents are encrypted with the PTK, and while the payload is encrypted, the headers (including the endpoint MAC address) are not encrypted.

Multicast and broadcast traffic for PMF-secured endpoints use the BIP with IGTK just discussed to offer integrity and ensure the message is coming from a proven source and not spoofed.

TIP **One example of data encrypted with PMF are the disassociation and de-authentication reason codes. This will impact the ability of many Wi-Fi tools to watch and use reason codes and other data for troubleshooting purposes, something addressed further in Chapter 7.**

WPA3, Transition Modes, and 802.11w

To recap some of the material from Chapter 2, here are the scenarios that dictate the operation of 802.11w:

- WPA3 associations (including Personal and Enterprise) require the use of Protected Management Frames (PMF)
- WPA3 Only Modes support WPA3 clients only, and PMF is required for all clients to join
- WPA3 Transition Modes support both WPA2 and WPA3, where 802.11w/PMF is supported/optional and WPA3 clients will use PMF; WPA2 if supported/configured may use PMF (and will be classified as WPA3); WPA2 clients that don't support PMF will join without PMF

- WPA2 networks optionally support PMF, which may be enabled but is optional
- Enhanced Open with OWE (encrypted open networks) require the use of PFM

INDUSTRY INSIGHT: PMF VS. MFP VS. CISCO MFP

Let's take a stroll down to the netherworld of acronyms. The 802.11w standard is referenced by the Wi-Fi Alliance as Protected Management Frames (PMF) and referenced in the IEEE specification as Management Frame Protection (MFP).

Those are both the same thing (802.11w) but should not be confused with Cisco's vendor-specific feature suite also called Infrastructure Management Frame Protection (MFP), which is a collection of controls for infrastructure integrity.

Caveats and Considerations for 802.11w

PMF makes strides toward a high-integrity infrastructure but isn't a security silver bullet. A few of the limitations and security considerations include:

- Control frames are not protected including CTS and RTS (clear-to-send and request-to-send), which offers no protection against layer 2 DoS attacks.
- The 4-way handshake is not protected, meaning the endpoint is susceptible to man-in-the-middle attacks and evil twin attacks at first connection to the network.
- Slivers of time exist where de-authentication or disassociation packets could be spoofed in the small window of time between the 4-way handshake and establishment of PMF.
- On WPA-Personal networks, PMF does not protect against vulnerabilities of the passphrase to online or offline dictionary attacks.
- On WPA3 transition networks that support WPA2 and WPA3, PMF is not required and some clients will be vulnerable to attacks.
- The enhancements for broadcast and multicast traffic enhancements against endpoints not joined to the network, but offer no protection from malicious users or devices attached to the network and participating in broadcast integrity protocol (BIP).

INDUSTRY INSIGHT: USING 802.11w WITH 802.11r ROAMING ON WPA2 NETWORKS

Sheesh, Wi-Fi is a twisty topic. After noticing several vendors either didn't support or didn't recommend the combination of 802.11w protected management frames + 802.11r Fast Transition (FT) roaming on WPA2 networks, I went digging for answers. Stephen Orr was able to explain that at one time, WPA2 with 802.11w PMF only supported one SHA1 hash algorithm and FT and other authentication and key management (AKM) suites used a second. That meant enabling PMF broke things. The working group has since remediated that, and PMF should be supported on WPA2 networks as long as the Wi-Fi system and the endpoints have been updated. Remember, if an endpoint connects using PMF it will be classified as a WPA3 client. The difference between WPA2 and WPA3 is not defined by an information element as was WPA versus WPA2.

Provisioning and Securing APs to Manager

A trusted infrastructure relies on its composite components to be known, validated, and authorized to one another. Within the wireless infrastructure our goal is to control the environment and ensure only authorized APs are attached to our management system, that the management system and APs are mutually authenticated, and that we're being intentional about the AP authorizations and assignment to proper groups or roles. Optionally, we may want to encapsulate or even encrypt client traffic along the path.

The discipline of prescribing security for the rules that route and direct client traffic is described as *control plane security*, and it encompasses many aspects including infrastructure component authentication, approving devices into the infrastructure, and securing communication between the devices. In essence, control plane security introduces integrity and confidentiality to the infrastructure devices.

As you might imagine, control plane security is managed a bit differently in cloud-managed architectures versus on-prem controller architectures, but the overarching concepts and tasks remain the same.

Divergence comes when we ascribe responsibility to the tasks; in cloud-managed platforms, the onus is on the vendor to perform certain tasks such as ensuring the APs have a unique ID and a secure connection to the management platform.

Because the cloud-managed architecture blurs some of the lines between management and control planes, I've opted to not title this "control plane security" and am using the term "manager" to indicate a controller and/or a cloud management platform.

Execution of management and control plane security is covered in the following topics:

- Approving or Allowlisting APs
- Using Certificates for APs
- Enabling Secure Tunnels from APs to Controller or Tunnel Gateway
- Addressing Default AP Behavior

Approving or Allowlisting APs

Every controller or management platform will have a method (or several methods) to approve APs. This process can happen ahead of time, as in pre-authorizing APs by serial number or MAC address, or it can happen after an AP has discovered the controller or management system, but before it's fully adopted into the platform.

The settings may be called AP allowlisting (also known as whitelisting), AP adoption, AP approval, or AP authorization, among other things. Luckily the phrases are self-explanatory for the most part.

The general trend in the industry has been for vendors to support increased security postures by changing the default behavior to more secure configurations. For example, in the past a vendor may have had a default setting that automatically adopted APs that discovered the controller, authorized them, and put them in a default group. Over time, the default behavior changed and will now show the new APs in a pending state, force a network admin to manually approve them (or approve them through a predefined rule), and then assign them to a group.

Table 6.4 shows the most common options for ways the management platform may uniquely identify the AP; when the AP can be approved; and how manual or bulk approvals can be used. Methods may vary slightly between campus APs, remote APs, and mesh APs as vendor implementations differ.

Table 6.4: AP provisioning models supported

METHOD TO IDENTIFY AP	TIMES OF APPROVAL	BULK AND MANUAL APPROVAL
Vendor-issued certificate	Prior to AP discovery of controller	Serial numbers
Serial number	After AP discovery of controller	MAC address
MAC address	During AP installation via vendor app	IP address

Auto-discovery of the controller or management platform by APs is expected; that's not where we implement the controls. Once the AP has found the manager (or ahead of time if desired) we control the AP adoption from the manager (versus the AP).

As part of your hardening strategy, plan AP adoption carefully, and follow best practices including:

- Do not allow auto-adoption of APs to the manager
- Verify all expected APs are in the manager and none are missing
- For bulk adoption, use the most reasonable level of granularity such as the IP subnet of the AP management network(s)
- Audit inventory occasionally and alert on APs that unexpectedly leave or join the manager

SIZING IT UP: ENTERPRISE AP ACCOUNT ATTACHMENT IN CLOUD-MANAGED PRODUCTS

In cloud-managed products the authorization can get a bit wonky because sometimes the AP allocation is tied to the fulfillment supply chain process, meaning some vendor solutions will connect (although not necessarily authorize) APs to the organization via the distributor that processed the order. The distributor is the company behind the scenes, between the manufacturer and reseller, and they're usually deeply integrated with the manufacturer's systems and perform tasks such as these account provisioning functions.

What this means is that, with some vendors, the AP may be listed in your inventory and placed there by the distributor. At that point you then proceed as usual to authorize and assign the AP to a site or group (or not).

The downside of this is that in large organizations, ones with multiple locations, or departments that don't have central procurement, APs may end up in the wrong place.

I'll use a hypothetical example based on my own state's university system. Here in North Carolina, there's the University of North Carolina (UNC) school system. Within that system are several schools, most starting with "UNC." But there's also NC State University, which is part of the UNC system. That university is sometimes catalogued as "UNC-NCSU" or "NC State University" and even "NCSU" and "North Carolina State University." You get the picture. It's not only possible, but quite likely if NCSU were to order cloud-managed APs and the distributor wasn't consistent, they may end up with two, three, or even four accounts that would have to be consolidated. This scenario isn't limited to cloud deployments but it's more impactful if that's how the ownership is linked to the portal.

If APs you purchased are missing in action from your cloud product, contact your reseller; they will be able to work with the distributor to resolve it.

Using Certificates for APs

One way to authenticate APs to the manager is with certificates, and those certificates can then be used for encryption. This is another area where cloud and controller architectures differ.

Cloud-managed APs will have a certificate preinstalled at time of manufacturing. The AP phones home over the Internet to the cloud manager and has everything it needs to be uniquely identified and provisioned to the proper customer account.

Controller-managed APs could also have a certificate preinstalled, but it's more common (at least for now) for the controller to issue certificates to the APs when they register (and are approved) to the controller. Alternatively, most enterprise products support use of standard protocols to enroll APs (or other infrastructure) into an existing PKI infrastructure. As mentioned earlier in the section "Generating Keys and Certificates for Encrypted Management," Enrollment over Secure Transport (EST) and Simple Certificate Enrollment Protocol (SCEP) are two protocols supported by Wi-Fi vendors.

And while controller-managed APs may not ship with a pre-installed certificate, it's becoming more popular to see the controllers themselves pre-provisioned with certificates; specifically, vendors are beginning to make use of the TPM chips and IEEE 802.1AR (also discussed earlier).

With certificates in place, the administrator has the option to configure encryption between the AP and the manager.

Enabling Secure Tunnels from APs to Controller or Tunnel Gateway

It's been a bumpy road, but we've arrived at a time when enterprise-grade controller-based products ubiquitously offer the option to encrypt traffic from the AP to the controller, over the wire. Encrypted tunnels can be used to secure both management traffic and/or client traffic.

Encryption of data over the wire has been a staple of regulations for over a decade. Moving forward, zero trust initiatives will further the mission of pervasive encryption along the entire data path.

Just as with any encryption scenario, an encrypted tunnel requires two ends or terminations, mutual authentication, and encryption keys. Issuing certificates to the APs as described in the previous section sets you up for success here.

And, as described in the topic of data paths in Chapter 2, tunneling or bridging client data can be configured per-SSID (and in some cases more granularly than that).

From a secure management plane point of view, cloud-managed solutions have a bit of a leg up here since they're pre-provisioned with certificates and ready to connect securely to the cloud manager. Since the manager is on the

Internet, by definition traffic will be traversing an untrusted network and therefore demands encryption for the management plane.

For cloud-managed solutions that incorporate a local tunnel termination appliance (such as a Mist Edge appliance or an Aruba Gateway appliance in ArubaOS 10), encryption can be extended to the data plane for client traffic. And the configuration options are similar to that of a controller, but feature depth will vary per vendor.

It is important to note that enabling encryption between APs and a controller or tunnel termination gateway will incur processing and bandwidth overhead. There may be limits or recommended ranges from the vendors for a maximum number of APs on their hardware models. Please consult your vendor team or documentation to discuss limitations and other considerations.

ENABLING ENCRYPTION ON CISCO AND ARUBA

If you're looking for these settings in a Cisco controller architecture, look for AP authorization lists for AP adoption, AP join profiles, and CAPWAP with DTLS for encrypted tunnels.

If you're working with Aruba ArubaOS controller architecture, review its guidance for AP whitelists, provisioning, and CPSec control plane security.

Ideally your control plane security will meet or exceed your over-the-air encryption.

Addressing Default AP Behavior

On a related note, one last best practice in AP provisioning is to control the default behavior of a newly discovered AP. Whether you're working in a cloud or controller platform, this guidance is the same.

Along with the instruction to disallow auto-adoption, you also want to manage what happens to an AP after it's adopted or allowed, but before it's fully configured or assigned to a site or group.

If the platform includes a default group, never use that group for a live production network. Consider the default group a holding place, research to see if the default group is associated with any policies such as configured for broadcasting SSIDs, radio management profiles, etc.

The end goal is to exert full control over the AP as it joins the network. By whatever means needed—ensure that when an AP is adopted and allowed on the controller, whatever happens next is *not* that it starts broadcasting networks.

Adding Wired Infrastructure Integrity

Hardening may include adding integrity between the APs and the wired network. Chapter 4 included a few topics of wired infrastructure in discovery protocols, loop protection, and dynamic routing. There are hardening disciplines for every protocol used on networks—from routing to DNS, SNMP, and beyond—and there are volumes written on each.

For our purposes, this topic focuses on authenticating the AP to the wired infrastructure and managing the edge port VLANs that service APs.

As you'll see in the following content, these integrity controls are valid best practices, but reserved for organizations that have a need for a heightened level of security and which have the appropriate staffing resources to manage the additional operational overhead.

CONTROLLING AP PORTS IN PUBLICLY ACCESSIBLE AREAS

Along with heightened security requirements, environments with APs that are physically accessible by the public or by untrusted users should consider these controls along with detailed access control policies.

Specifically, for APs in tunnel mode, there should be prescriptive ACLs allowing access only from the AP management VLAN to the controller(s) and additional security monitoring. APs bridging traffic at the edge make this task more challenging, and in most cases if physical access to a bridged AP port or cable is a concern, strict AP authentication should be used, or you should consider tunneling instead of bridging traffic on those APs.

Authenticating APs to the Edge Switch

This topic comes with a caveat—it's often not manageable for organizations to perform this level of wired side hardening via AP authentication. However, there are organizations out there that require this level of integrity, so it's included here along with some how-tos and things to look out for. For most environments, there are many other hardening tasks that will give you a lot more bang for your buck.

Along with provisioning APs to the controller or manager, we can also authenticate or bind the APs to the wired infrastructure by using 802.1X or MAC address port security. Just as with client-based authentications, the APs too can use pure 802.1X/EAP, or 802.1X with MAB, or non-1X port security based on binding the MAC address to the port or switch. This section covers how to:

- Authenticate APs to ports with 802.1X/EAP
- Authenticate APs to ports with 802.1X and MAC Authentication Bypass (MAB)
- Bind APs to ports or switches with MAC address

Authenticating APs to ports with 802.1X/EAP

If you plan to authenticate APs to the edge switch using 802.1X, the requirements are that the edge switch will need to support AAA configurations—no problem, that's supported on every enterprise-grade switching product—and the AP will need to have an 802.1X supplicant and support for one or more EAP methods. Controller-based APs support full 802.1X authentication and most products can be configured with EAP-PEAP with MSCHAPv2, EAP-TTLS, or EAP-TLS.

> **FINDING 802.1X AP SETTINGS IN PRODUCTS**
>
> If you're looking for these settings on a Cisco controller, at time of writing, they're configured as part of the CAPWAP DTLS profile and within the `dot1x` configuration settings in the CLI.
>
> In the Aruba guides, look for enabling 802.1X supplicant support on an AP for instructions on configuring this. Both vendors support the use of factory-installed or controller-issued AP certificates.
>
> Cloud-managed APs are less likely to participate in 802.1X but check with your vendor for the most current feature support.

Authenticating APs to ports with 802.1X and MAC Authentication Bypass (MAB)

For environments that require hardening with APs that don't have an 802.1X supplicant, a secondary option is to authenticate them with MAC Authentication Bypass, or MAB. As covered in Chapter 3, "Understanding Authentication and Authorization," MAB comes with a few caveats and security concerns. Since it's a subfunction of the 802.1X protocol, the edge switch ports will require several lines of AAA commands as well as the added MAB commands.

The benefit of this model is that APs can be authenticated to the switch port (centrally by an authentication server) even if the AP doesn't support 802.1X natively.

In turn, that means you'll just need to configure 802.1X with MAB on the edge switch and ports servicing APs, and no additional configuration is required on the AP itself.

Binding APs to ports or switches with MAC address

To authorize the AP to the edge switch locally (versus via an authentication server) with the AP's MAC address, instead of configuring 802.1X (AAA) on the switch ports, you'll use a port security command. Consult your switch vendor's documentation but it will usually be found in the port security section of the configuration guides.

Most products support several options such as configuring the port for a fixed or static MAC address, a learned or dynamic MAC address, and then options such as sticky MAC (see Table 6.5). If you go this route, pay extra special attention to the maximum MAC address configurations. When used with an AP in bridged mode, the edge switch will see the client MAC addresses and you may find the switch has locked the port due to a MAC count violation.

Table 6.5: Comparison of AP-to-switch authentication options

METHOD	AP REQUIREMENT	SWITCH REQUIREMENT	VISIBILITY/ CONTROL
Authenticate APs to ports with 802.1X/EAP	802.1X supplicant required	802.1X AAA support	Centralized in authentication server
Authenticate APs to ports with 802.1X and MAB	None	802.1X AAA support	Centralized in authentication server
Bind APs to ports or switches with MAC address	None	MAC-based port security support	Distributed visibility and control; configured per switch

LOCAL VS. EXTERNAL SERVER AUTHENTICATION

As a personal preference for security and manageability, I avoid architectures that perform local authorization such as port MAC binding and prefer using 802.1X and/or MAB if this level of hardening is required.

In enterprise deployments the lack of central visibility and management adds operational overhead that becomes cumbersome quickly, and MAC binding is not considered secure, so it's a minimal security enhancement with a potentially hefty management overhead.

Wired 802.1X authentications also introduce a lot of overhead and should be reserved for organizations that have the resources to manage it, or focused in areas that need additional protection.

Just remember, the AP won't pass traffic or be able to perform any client services until it's authenticated to the network if any of these controls are enabled. In Figure 6.9, the switch port 2 is using port security with a local MAC lookup. Port 5 is configured for AAA/802.1X with MAC Authentication Bypass (MAB), which uses the AP's MAC address also. Port 14 is configured for AAA/802.1X with no bypass. The AP on this port will need an 802.1X supplicant and credentials—either a certificate or username/password—configured.

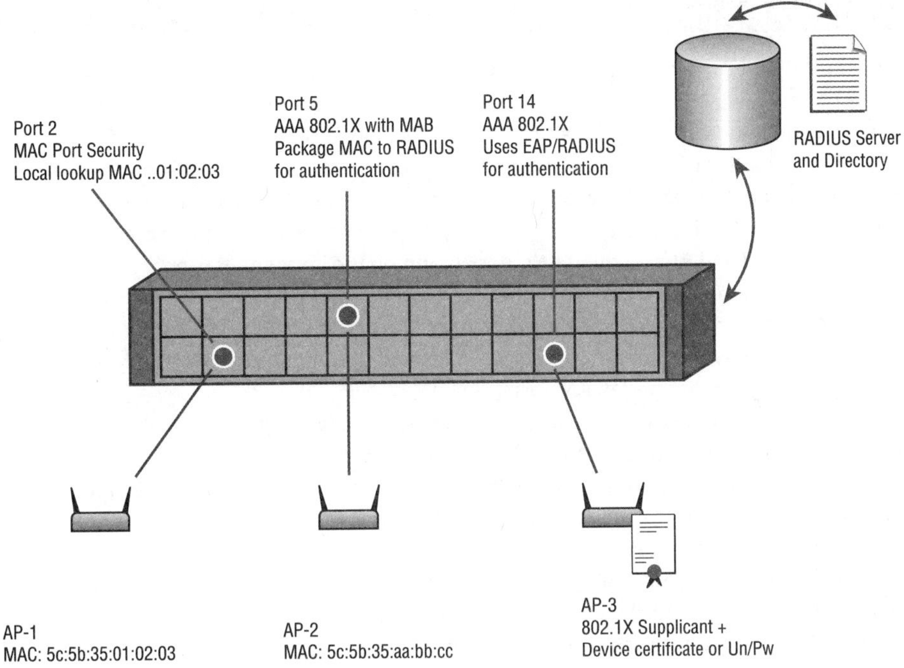

Figure 6.9: Demonstrating three AP to switch authentication options

Specifying Edge Port VLANs

A seemingly small task that's a best practice for any environment is to allow only the required VLANs on each edge port. This may entail pruning additional VLANs and/or specifying a native/untagged VLAN designated for AP management.

And, while this task sounds insignificant, it can be time consuming depending on the wired and wireless products. Some vendors, like Juniper's Mist products, use the APs to collect additional wired-side data and report things like VLAN mismatches (too many or too few) in the admin console. With most products, this has historically been a tedious, manual process.

This is another best practice that's dependent on the resources available. Due to the potentially heavy operational cost and the often dynamic nature of environments, this may not be the best use of time in organizations that aren't implementing change management processes.

Ideally though, you could perform this task once, and then ensure the proper VLANs (and only the proper VLANs) have been configured at the edge ports through those new change management controls and baseline configurations described earlier.

INDUSTRY INSIGHT: VLAN HOPPING

One topic that resurfaces every few years is the exploitation of VLAN configurations by VLAN hopping, which happens when VLAN trunk ports are configured insecurely, or when a default VLAN is in use, or a port has VLANs tagged that are not required.

In one scenario, an attacker can spoof a switch, connect to the network, and trick the edge switch into thinking the attacker is a peer switch. If automatic VLAN provisioning protocols are used, the enterprise switch will provision the attacker's port as a trunk with any assigned VLANs, or all VLANs, depending on the implementation.

Another scenario called *Double Tagging* can bypass layer 3 ACLs by double stacking headers in what appears to the first switch to be the native VLAN. This attack exploits ports configured with an untagged/native VLAN and does not require (actually doesn't even work on) trunk/tagged VLAN ports.

While these are legitimate attacks, they're considered less significant in the grand scheme of security concerns. Both require physical access to the environment and the Double Tagging attack is unidirectional (packets can only go one way) and requires the attacker to know the target IP, VLAN, and service, greatly limiting the risk of an attacker interacting with a target resource.

As with some of the other topics covered in hardening, penetration testers and malicious users have countless other easier ways to attack the network and while this is a best practice, it's not a task I suggest wasting resources on in the presence of other vulnerabilities.

If you're interested in learning more, vulnerabilities of VLANs (802.1Q) have been cataloged as Mitre CVE™ (CVE-2005-4440) since 2005. The following illustration shows the current VLAN related CVEs at Mitre. Visit `https://cve.mitre.org` and search the CVEs for "802.1q" for more.

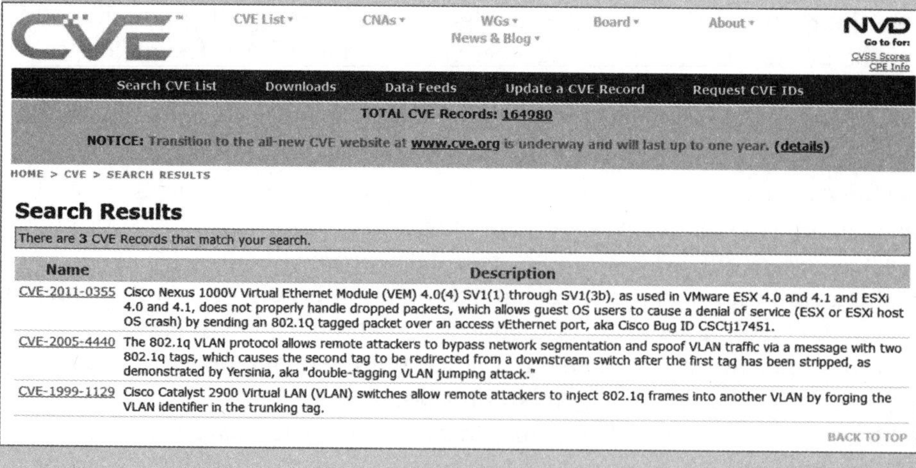

Name	Description
CVE-2011-0355	Cisco Nexus 1000V Virtual Ethernet Module (VEM) 4.0(4) SV1(1) through SV1(3b), as used in VMware ESX 4.0 and 4.1 and ESXi 4.0 and 4.1, does not properly handle dropped packets, which allows guest OS users to cause a denial of service (ESX or ESXi host OS crash) by sending an 802.1Q tagged packet over an access vEthernet port, aka Cisco Bug ID CSCtj17451.
CVE-2005-4440	The 802.1q VLAN protocol allows remote attackers to bypass network segmentation and spoof VLAN traffic via a message with two 802.1q tags, which causes the second tag to be redirected from a downstream switch after the first tag has been stripped, as demonstrated by Yersinia, aka "double-tagging VLAN jumping attack."
CVE-1999-1129	Cisco Catalyst 2900 Virtual LAN (VLAN) switches allow remote attackers to inject 802.1q frames into another VLAN by forging the VLAN identifier in the trunking tag.

Planning Physical Security

Many network attacks rely on physical access to the infrastructure by an ill-intentioned person. And many environments assume the assets enjoy a certain level of protection from physical access—the building access requires a key and/or an RFID badge; the server room requires additional access rights, etc. But in most cases, it's not fair to assume every network port is physically protected nor that a malicious user would not be able to access one.

Without a port security protocol or a network access control (NAC) product controlling the wired edge ports, many ports remain vulnerable to misuse.

In all cases, network security architecture should control points of entry to the network including edge switch ports. This means designing physical security controls at networking closets and edge ports, including ports that connect to APs, which may be in publicly accessible areas.

GETTING SWITCHES "SWEPT AWAY"

We've all seen it—the janitor's closet that's doubling as a network closet, the switch mounted haphazardly, inches away from the mops, cables strewn carelessly across the corner of the room and disappearing into the ceiling.

If you're gawking or rolling your eyes, then maybe you're one of the lucky few to have worked in environments with stricter access control and network provisioning. But over the years, I've personally seen hundreds of such closets and rooms. Not always the janitorial closet; sometimes it's the copy room, the library computer lab, a break room, storage closet, or simply beside someone's desk.

Organizations and IT teams do the best they can with what they have to work with. Although not ideal, sometimes it's how the cards are dealt.

The following series of hardening tasks are meant to address this and other vulnerable points in the network and offer alternatives when one or more best practices can't be followed.

Securing Access to Network Closets

Securing physical access to servers, storage, network devices (including wireless controllers), and networking or data closets is one of the most basic best practices for good network hygiene.

Ideally these assets should be located in places that enforce identity-based access, meaning not just a key but an RFID badge, biometric entry, or some way to correlate the access to a specific person. Earlier in this chapter we talked about enforcing user-based logins to manage network devices instead of using shared logins like "admin," and this control is simply an extension of the same concept. We want attribution along with access control.

If identity-based access control isn't possible, the next best option is some physical control such as a hardware lock. As my friend Jack Wiles would expound, be sure to select a lock type commensurate with what it's protecting. Locks have ratings like safes that indicate how hard they are to break or pick.

If you've never tried your hand at lockpicking, I suggest checking out a Lockpick Village the next time you visit an information security conference. It'll give you the chance to see exactly how vulnerable most locks are. Regardless of your lockpicking skills, consult a locksmith or consultant that specializes in physical security to choose the right lock types.

And while physically securing access to networking closets seems like a no-brainer, as noted earlier, sometimes that's not how the organization operates. For those cases (and for general added security) there are additional controls that can be employed. Not protecting physical access to network closets and datacenters may result in:

- Access by unauthorized parties including internal personnel
- Inadvertent reconfiguration e.g., through someone adding or connecting a network cable
- Denial of service through inadvertent introduction of a network loop, creating a broadcast storm
- Denial of service by taking a system component such as a switch or controller offline
- Subversion of visibility by taking a monitoring system offline, such as a management platform or security monitoring
- Outages due to improper system environment such as too high humidity or outside vendor–specified operating temperature ranges

Securing Access to APs and Edge Ports

It's important to maintain physical control of the entire data path, which means both ends and any terminations of network cabling. Outside of the networking closets and datacenters, protection should be extended to the network drops distributed throughout the environment including those servicing APs.

Ultimately the goal is to protect against an attacker removing an AP and using the AP's network connection to infiltrate the production network. Aside from the controls mentioned earlier—segmentation, AP authentication to the switch, port security, and monitoring—we can deter this behavior by making the APs difficult to access or difficult to remove. If an attacker gains access to an AP network port, the following actions could follow:

- Launch of denial-of-service (DoS) attack against the wireless infrastructure via the wired network
- Launch of denial-of-service (DoS) attack against other wired resources
- Lateral movement to endpoints and targets within the wired network
- Sniffing of network traffic
- Injection of malicious payloads
- Injection to switching and routing functions to disrupt or reroute packets

Wireless is full of dichotomies and AP mounting is no exception. Most Wi-Fi design professionals and vendors will direct clients to mount APs below a ceiling grid (where applicable) and generally otherwise not obstruct the RF signals by concealing an AP. So then, the APs are just hanging out, visible to all who pass.

If the AP is mounted on a wall 25 feet up and requires a lift to access, it's extremely unlikely to be accessed for ill deeds. But if that AP is mounted on a wall, or below a standard eight- or ten-foot ceiling, it's trivial for an assailant to get their hands on it.

As such, physically securing the AP (in addition to controlling the switch port security if appropriate) is key to protecting the network infrastructure.

There are several levels and methods of physically securing an AP, ranging from deterrent to dang near impossible to breach. In order of least to most secure, they are:

- Mounting hardware with tamper deterrence
- Mounting hardware and enclosures with latch locks
- Mounting hardware and enclosures with key locks

AP mounts come in an assortment of form factors and materials. Some are full enclosures (boxes), others designed to fit within existing ceiling grids, or attach to the underside of a ceiling grid. There are also surface mounts and right-angle mounts, along with concealed mounting such as bollards. Most are available with tamper deterrence and/or key locks. Many box enclosures may also have latch locks.

Figure 6.10 and Figure 6.11 show just a tiny sample of the thousands of mounting options available. Figure 6.10 is a polycarbonate enclosure manufactured by Ventev, designed for use in prisons and other harsh indoor and outdoor environments. Figure 6.11 shows a mounting enclosure with a key lock. This particular model is a suspended ceiling tile enclosure from Oberon.

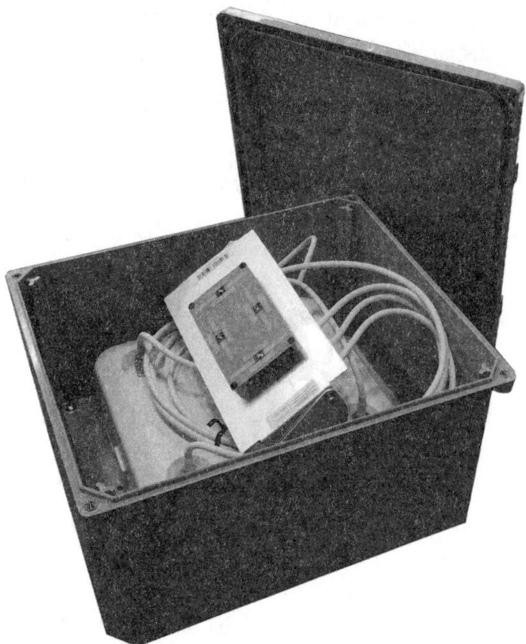

Figure 6.10: Example of a mounting enclosure with tamper-resistant screws (usually torx)
`https://ventevinfra.com/`

Figure 6.11: Example of a mounting enclosure with a key lock
`https://oberonwireless.com`

Locking Front Panel and Console Access on Infrastructure Devices

If controlling physical access to the wireless controller, APs, and other network devices isn't enough or isn't viable, hardening can include disabling or locking management functions available on the hardware itself.

Should a malicious user gain physical access to system devices without front panel and console security, they would be able to perform an array of attacks from denial of service to reconfiguration and locking out the legitimate system owners.

Without front panel and console security, the system is at risk of an attacker performing one or more of these actions:

- Unauthorized reboots (although this may be possible still by removing the power source)
- Factory reset (wipes the configuration)
- Password reset (some products include a button sequence to clear passwords)
- Reconfiguration (the system could be made insecure, unreachable, or otherwise compromised)
- Tamper with passwords (an attacker could change passwords and lock out legitimate management admins)
- Install malicious payloads (through software uploads via the USB or console)

For controllers, hardening front panel access involves controlling or locking ports on the hardware appliance, disabling the USB, disabling physical buttons to reset or reload services, and locking console access or any management features available to a person with physical access to the hardware (see Figure 6.12 and Figure 6.13). Vendors also offer pre-hardened configuration with FIPS-compliant products that will enforce front panel security (and other features).

Figure 6.12: An Aruba 7200 series Mobility Controller. Notice the console port, USB port, and function buttons on the front right panel.

`https://www.arubanetworks.com/products/wireless/gateways-and-controllers/7200-series/`

Similarly, most APs have console ports and USB ports, among other things. These additional inputs to management functions can be disabled via the controller or management system to prevent an attacker from compromising the infrastructure by penetrating the AP's configuration (see Figure 6.14).

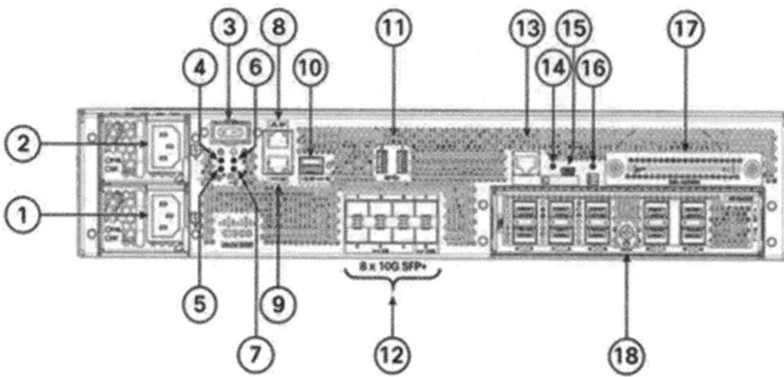

Figure 6.13: Diagram of the front panel components of a Cisco 9800-80 Controller. Labels 8, 9, 10, 11, 13, and 15 include various console and management ports including USB and Mini USB console ports.

`https://www.cisco.com/c/en/us/products/collateral/wireless/catalyst-9800-series-wireless-controllers/nb-06-cat9800-80-wirel-mod-data-sheet-ctp-en.html`

NOTE **Console ports aren't the only AP vulnerability. Joint Test Action Group (JTAG) ports, still used by some vendors during production, are available on printed circuit boards (PCBs) in APs. Used for testing, programming, and diagnostics, the JTAG ports can offer full access to the device and are rarely secured. An attacker would need to know what they're looking for but it's another reason to physically secure APs in sensitive environments.**

Figure 6.14: Image of an Aruba Networks 600 series AP with mini USB console

`https://www.arubanetworks.com/products/wireless/access-points/indoor-access-points/530-series/`

In addition, federal agencies and high-risk environments may want to take console access security a step further with additional controls:

- Tamper-evident labels (TELs) placed over console/management ports that are not in use
- Disable TAC support password recovery via console

Tamper-evident labels (see Figure 6.15) can ensure unused physical ports are only accessed in case of emergency, if permitted. The label is used to cover the port(s) and if it's removed to access the port, it will be evident. These can be custom printed for an organization and would be placed over management access ports, such as a console port, that should only be used in emergencies.

Most manufacturers support some form of password recovery as part of their technical assistance center (TAC) support processes. In these instances, someone authorized on the account would need physical console access to the device, at which time they can engage the vendor TAC for a one-time password with a short validity period. To prevent this bypass, the password recovery can be disabled.

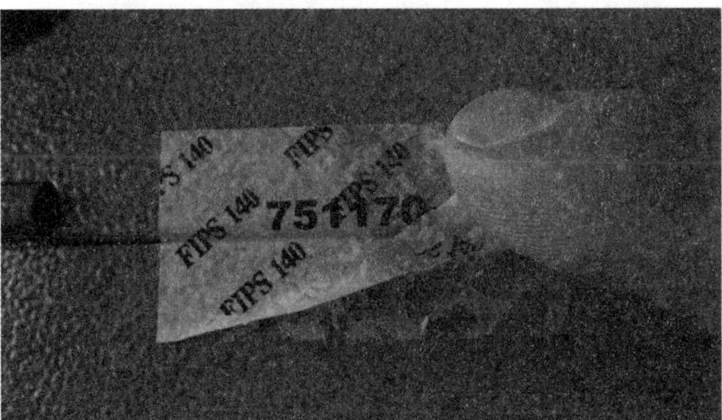

Figure 6.15: An example of a tamper-evident label
`www.digikey.com`

> **RELATED NIST SP 800-53 CONTROLS**
>
> **For organizations mapping controls to NIST, the grouping of Physical and Environmental Protection (PE) includes several controls related to physical security including Physical Access Control, Monitoring Physical Access, and Asset Monitoring and Tracking, among many others.**

Disabling Unused Protocols

In the same vein as disabling unused physical ports, hardening should absolutely incorporate disabling any and all unused protocols—not just the unencrypted management protocols presented at the beginning of the chapter.

Unused protocols are often ignored and become a blind spot for network and security operations. In an ideal world, the SOC (or someone) is running vulnerability scans internally as part of a vulnerability management program. These scans will certainly help identify at-risk devices using services with known vulnerabilities, but there may be known good protocols that still have no place in your network design.

The services, ports, or protocols you disable will be dependent on your infrastructure, but may include some or all of the following:

- Zero Touch Provisioning
- BootP Service
- IPv6
- UDP support of echo, discard, daytime, and chargen (small services)
- All unencrypted management protocols (HTTP, Telnet, FTP, TFTP, SNMPv2c, etc.)
- Encrypted management protocols not in use (e.g., HTTPS)

For the more security-conscious organizations and defense agencies, additional hardening is usually recommended by oversight bodies and can also include:

- ICMP
- Discovery protocols (such as LLDP and CDP)
- TCP and UPD support of echo, discard, daytime, and chargen (small services)
- IP source routing
- IP routing protocols not in use (specifically to/from a controller)
- ARP proxy
- Physical ports not in use

Disabling some protocols (such as ICMP, CDP/LLDP, Echo) may disrupt network monitoring and management tools. If you plan to follow more extreme hardening, ensure the organization has the proper processes and policies to support that, including detailed documentation, approvals, asset inventory, and strict change management.

TIP **Disabling unused protocols is part of the controls detailed in NIST SP 800-53 Configuration Management (CM) suite for least privilege.**

Controlling Peer-to-Peer and Bridged Communications

Restricting client traffic falls on the fringe of traditional infrastructure hardening but remains a relevant and timely topic.

Here, we'll look at a few specific methods to bolster integrity by eliminating or reducing risky traffic and behavior on the network. These hardening tasks include:

- Blocking Ad-Hoc Networks
- Blocking Wireless Bridging on Clients
- Filtering Inter-Station Traffic, Multicast, and mDNS

A Note on Consumer Products in the Enterprise

This seems like a logical time for a quick soapbox speech on consumerization; a topic I've had a long history with.

Consumerization is defined differently depending on who you ask, but I like this Gartner definition that says,

> **Consumerization is the specific impact that consumer-originated technologies can have on enterprises. It reflects how enterprises will be affected by, and can take advantage of, new technologies and models that originate and develop in the consumer space, rather than in the enterprise IT sector. Consumerization is not a strategy or something to be "adopted." Consumerization can be embraced, and it must be dealt with, but it cannot be stopped.**
>
> `https://www.gartner.com/en/information-technology/glossary/consumerization`

Enterprise environments have been dealing with consumerization for more than a decade, and frankly even though the conversation died out, the trend of bringing consumer-grade devices into the corporate network has only become more rampant.

These devices, designed for home networks and personal use, were simply never designed to be used within an enterprise environment, and certainly not within a secured network infrastructure. By definition, they're created to operate in the absence of formal network services such as DNS, DHCP, and authentication servers.

I remember the first time an enterprise client had me reconfigure their environment to support Bonjour. It was a highly regulated industry, I pushed back a bit, explaining the risks, but ultimately, I (and their CIO) was overruled because an executive demanded to be able to use his Apple iPad with his Apple printer, and wanted the ease of the services extended throughout the headquarters office.

If you've followed the book this far, I think the pattern that's evident is that the recommendations for best practices include a lot of conditional statements, "it depends," or "if, then, else" logic. What I'm about to say next will probably not be well received, but it has to be said:

Consumer and personal devices have no place inside a secured enterprise infrastructure.

Please take that sentence, highlight it, copy, paste, screenshot, share or do whatever you need to do and get it in front of your executive leadership.

However, I'm not a complete jerk, so if you've lost the battle and have to support these devices on the network, the following offers guidance to do it as thoughtfully and securely as possible. You *can* have consumer and personal devices on a network, it just won't be a very secure network.

Specifically, consumer and home devices that I personally do not recommend allowing include (but are not limited to) these:

- Home automation products including lighting, sensors, cameras not designed for commercial use
- Voice-assisted devices such as Amazon Echo suite, Google Home, and Apple HomePod, which introduce an excess of security and privacy risks

The ideal alternative is to seek out a commercial-grade version of a similar technology that meets the connectivity and security requirements of the organization. If these devices absolutely must be used for some reason within the environment, they should be on a dedicated wireless network, not with corporate devices.

Many professionals and non-technical executives are fearful of the risks from our hyperconnected ecosystem of IoT devices, but the scary truth is—consumer devices and protocols on the network are one of the biggest risks we're facing. They allow attackers easy access into the network, and unlike most IoT implementations, by nature of the use, these consumer devices are most often connecting to our smartphones, tablets, laptops, and printers. Meaning, isolating them from the endpoints they need to interact with is not possible (while retaining functionality).

Allowing an Apple HomePod because an employee finds it convenient or wants to listen to music is absolutely not a valid reason to expose the network to increased risk.

If your organization doesn't have a policy that addresses consumer and personal devices on the network, that should be a priority. Tools to identify, locate, and even block these devices and risky protocols are readily available in the Wi-Fi products.

Other devices that may use consumer-grade protocols (such as Bonjour) but have legitimate use cases in enterprise environments include:

- Digital TV and screen casting
- Certain wireless printing applications

Shortly, we'll look at specific protocols popular in consumer devices and discuss the behavior, security risks, and recommendations for each.

Blocking Ad-Hoc Networks

There are a few ways peer wireless devices can communicate to one other; two are:

- Peer communication through enterprise AP
- Peer communication in an ad-hoc network

Of concern here is the latter—ad-hoc networks. We'll address peer communication through an enterprise AP shortly.

With true ad-hoc networks, the endpoints are communicating directly, over the air, and through any part of the infrastructure you control or can even see. The lack of visibility introduces security vulnerabilities and prevents the organization from controlling (and securing) the data path appropriately.

Since ad-hoc networks aren't performed through the enterprise Wi-Fi system, they can't exactly be controlled that way. I say "exactly" because there are some mitigations available with wireless intrusion prevention systems (WIPS) we'll cover in Chapter 7.

Here are three actions you can take to identify and prevent ad-hoc networks:

- Disable wireless interfaces on devices that support ad-hoc networks (this may include certain features of screen casting applications, printers, etc.)
- Disable ad-hoc networking support on endpoints
- Configure the enterprise Wi-Fi to monitor the air for ad-hoc networks and alert on their presence

Figure 6.16 highlights the difference between peer client communication through an enterprise AP (shown in data path 1) and ad-hoc networks (shown in data paths 2 and 3). Data path 2 is an ad-hoc Bluetooth network for a presentation control to a laptop and data path 3 is an ad-hoc Wi-Fi network between a smart TV and a tablet for screen casting. In scenario 1, data can be filtered and secured by the AP or controller, whereas scenarios 2 and 3 are peer-to-peer connections that don't traverse the enterprise network. These may be 802.11 (Wi-Fi), Bluetooth, or other.

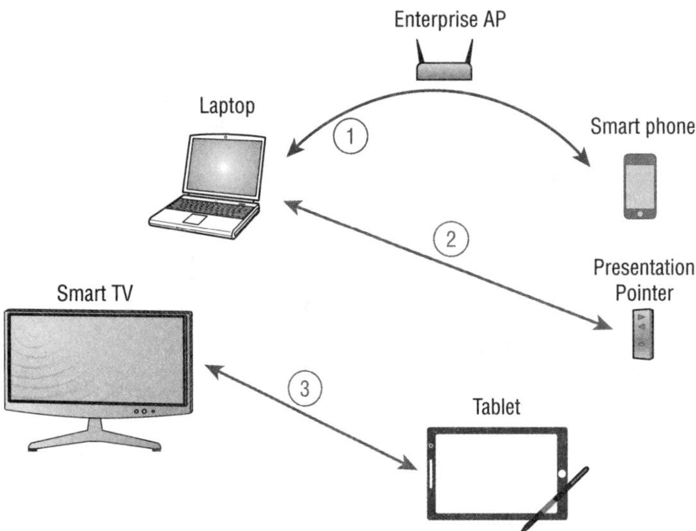

Figure 6.16: Peer communication through an enterprise AP vs. ad-hoc networks

Blocking Wireless Bridging on Clients

Another function that involves endpoint configuration and isn't controlled by the enterprise Wi-Fi is bridging of the wireless interface to other network interfaces (such as a wired interface).

As with ad-hoc networking, remediation actions are similar:

- Disable interface bridging on endpoints
- Configure endpoint security or endpoint detection response tools to alert if bridging is enabled
- Configure the enterprise Wi-Fi to monitor the networks for bridged clients and alert on their presence

Interface bridging can also be extended to service other endpoints with an ad-hoc network served by the sharing client. In Figure 6.17, the laptop is configuring its own Wi-Fi network and bridging peer traffic through its own interfaces. As soon as the user checks the box for Internet Sharing in the top window, the Mac will be broadcasting its own SSID and could be serving an enterprise network to unauthorized users or devices. This presents obvious security risks and can impact all Wi-Fi devices in the area by reducing the quality of the RF.

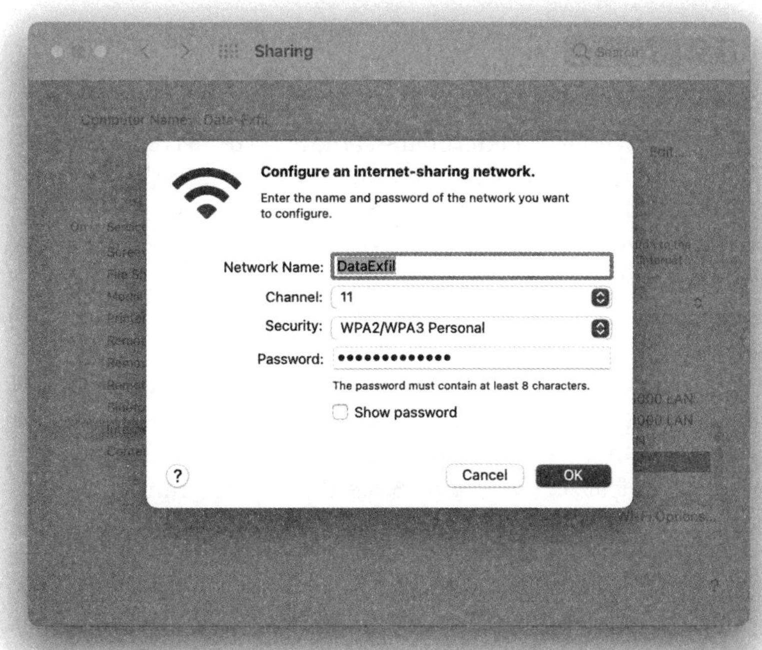

Figure 6.17: On most devices, it's trivial for a user to share network connections including sharing over Wi-Fi that isn't properly secured.

INDUSTRY INSIGHT: EXPLOITING A SECURE NETWORK WITH APPLE WIRELESS DIRECT LINK (AWDL)

If you're interested in some specific exploits, an article on Medium, "Escaping from a truly air gapped network via Apple AWDL," includes a detailed accounting of breaching an airgap network with IPv6 Node Information Query. See `https://medium.com/sensorfu/escaping-from-a-truly-air-gapped-network-via-apple-awdl-6cf6f9ea3499`.

For more on security topics related to Apple's wireless ecosystem, also check out the Open Wireless Link (OWL) project at `https://owlink.org/`.

OWL has published research on AWDL as well as privacy and security findings related to location services, device and user tracking, BLE, and DoS and MitM attacks.

Filtering Inter-Station Traffic, Multicast, and mDNS

This topic covers filtering and security controls for peer traffic that does traverse the enterprise Wi-Fi system. Client-to-client interactions over the Wi-Fi can occur through standard layer 2 adjacency (just like two endpoints on the same VLAN in a switch) and/or through peer-based multicast and broadcast protocols such as mDNS, Link-Local Multicast Name Resolution (LLMNR), and UPnP (covered shortly).

SSID Inter-Station Blocking

Inter-station blocking goes by different names, but in any vendor implementation it's the utility of blocking wireless clients on the same SSID from communicating directly with one another. Again, this is similar to blocking traffic between two wired endpoints within the same VLAN on a switch.

Inter-station blocking doesn't mean that endpoints can never talk to one another; it simply forces the clients to go through the AP or controller to be processed, where rules can be applied, and the traffic can be evaluated and allowed or blocked.

Long ago and far away, enabling inter-station blocking was a de facto state for any Wi-Fi deployment. Uncontrolled consumerization and the introduction of home-use protocols into the enterprise network caused the industry to flip and enable peer communication by default. Since then, more granular controls have been added.

Consumer technology aside, for the most part, secure enterprise environments should have no need for (and should not allow) direct peer client communication. If there is a situation that calls for it, an allow policy can be explicitly enforced within the Wi-Fi infrastructure.

This is an especially critical control in the heat of ransomware and zero trust initiatives, where implicit trust should be removed, and lateral movement should be limited.

Figure 6.18 shows options for inter-station blocking, multicast filtering, and Bonjour gateway on the Juniper Mist platform. Aruba controllers and Aruba Central cloud (shown in Figure 6.19) have two similar options. The Deny Intra VLAN Traffic option blocks both layer 2 and layer 3 traffic between endpoints on the same network. Deny Inter User Bridging blocks non-IP (layer 2) traffic between endpoints on the same network including ARP.

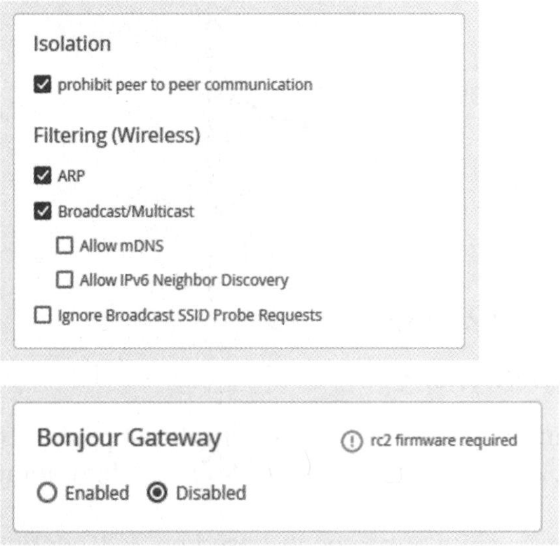

Figure 6.18: Juniper Mist options for inter-station blocking and multicast controls

INTER-STATION BLOCKING IN PRODUCTS

As you saw in Figure 6.18, Juniper Mist refers to this feature as *peer isolation*, and it's configured at the SSID.

In Cisco controller architecture it's configured as *peer-to-peer blocking* and applied per SSID.

Aruba Networks refers to the setting as *deny inter user traffic* and it can be applied globally within a controller, or per-SSID.

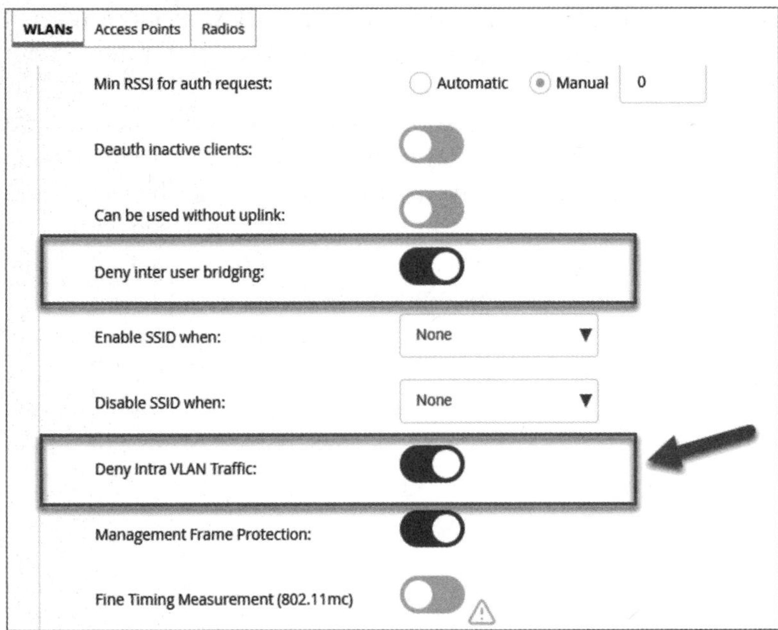

Figure 6.19: Aruba Networks Central Cloud SSID configuration

Peer-Based Zero Configuration Networking

And now back to the elephant in the room—peer-based multicast protocols such as Apple's Bonjour, mDNS, and other mechanisms of zero-configuration (zeroconf) networking.

Zeroconf networking describes a suite of protocols based on TCP/IP but designed to work in the absence of standard enterprise domain services such as DHCP and DNS. These are best thought of as peer-based self-organizing networks. Part of the zeroconf operations include:

- Link-local addressing, an address autoconfiguration (aka Automatic Private IP Addressing) that works with or replaces client IP addresses from DHCP
- Endpoint self-designation of a hostname (such as Joe-Smiths-iPad.local) instead of enterprise DNS entries
- Peer name service resolution such as mDNS, which uses multicast to all reachable zeroconf peers in the network, versus an authoritative DNS server
- DNS-based service discovery such as DNS-SD and SSDP in UPnP
- Universal Plug-n-Play (UPnP) protocols

Following is a sample of technologies that use Bonjour services, zeroconf, and mDNS for peer discovery and communication:

- Amazon TV
- Google ChromeCast
- Roku Media Player
- Misc. printing apps
- Phillips Hue Lights
- Airplay
- Apple TV
- AirServer Mirroring
- Apple AirTunes
- Apple AirPrint
- Apple FileShare
- Apple iTunes

The challenges supporting these and other zeroconf protocols on the enterprise network are two-fold:

- They introduce a very high level of risk to the network and endpoints with little to no options for reasonable mitigation.
- They are designed to work on layer 2 adjacent networks and not across IP subnets, as is a common (and recommended) architecture in enterprise networks.

In the recommendations later, Link-Local Multicast Name Resolution (LLMNR) will be referenced. This was a Windows protocol similar to mDNS, but mDNS was ratified as an IETF RFC, and Windows supports mDNS, so LLMNR is likely to be unused. If it exists in the environment, it should be managed the same as mDNS and Bonjour.

I'm not getting on the consumerization soapbox again, but these protocols were never designed to be used in enterprise networks, which you'll see in the following sections.

Disabling and Filtering Bonjour and mDNS Protocols

I'm going to level with you here; like others, I've been familiar with mDNS and related peer protocols but had never stopped to read the entire specification nor fully researched the vulnerabilities associated with it until recently. Capturing the result of that research, the sentiment went from "mDNS isn't ideal" to "holy smokes this is a ridiculously risky protocol."

Apple, the creator of Bonjour and mDNS, even states, "In a hostile environment other mechanisms must be used to ensure the cooperation of participants or to distinguish untrusted Multicast DNS messages." (Source: "Security implications of Bonjour protocol for developers and administrators," `https://support.apple.com/en-us/HT205195`.) "Hostile" in this context means a network comprised of endpoints that the user doesn't have full control over (e.g., a corporate or public network as contrasted with a home network).

As described earlier, these zeroconf protocols are peer-based and don't include any form of authentication or authoritative source. They bypass the enterprise services and are often not visible or reported on by monitoring tools.

The attacks on networks and endpoints through Bonjour and mDNS are countless. The security considerations around these technologies include:

- mDNS packets share open port information of targets, providing security details on par with vulnerability scanning and attack discovery tools.
- mDNS may also show full directory and file structures.
- mDNS and zeroconf protocols do not include any form of authentication or identity validation, making it easy to spoof a peer device or service and collect sensitive information or even deliver malicious payloads to an endpoint.
- Bonjour and mDNS bypass enterprise domain services including DNS, DHCP.
- Default Bonjour behavior usually shares (via multicast) the hostname as the device owner's first and last name, exposing the user's identity including information classified as PII (personally identifiable information) by many government regulations.
- Per the IETF RFC specification, fully securing the enterprise against mDNS requires extensive modification to the enterprise environment including moving to authenticated and encrypted DNS protocols within the organization.
- In some instances, an attacker can hijack or spoof legitimate traffic and redirect users to malicious websites and payloads.
- By virtue of its multicast nature, mDNS creates volumes of unnecessary traffic on large networks, flooding it with multicast traffic from hundreds or even thousands of devices.
- Bonjour introduces additional vulnerabilities by allowing peer communication between devices without visibility, security inspection, or control.
- mDNS automatically resolves any and all unconfigured and explicit local domain names, whether that's the desired behavior or not.

The information security community is in lockstep when it comes to Bonjour and mDNS—it's not appropriate for secured enterprise networks.

Hardening guides from almost every manufacturer including Apple recommend disabling Bonjour on endpoints and filtering it on the network. The NIST SP 800-179 explicitly states, "Bonjour multicast advertisements should be disabled in all environments except Standalone."

Even more shockingly, Trustwave's SpiderLabs security team demonstrated that mDNS multicasts (to all peers, unencrypted) the status of the endpoint's open ports, as shown in Figure 6.20.

In Figure 6.20, you'll notice the four lines at the bottom include the following lines, indicating Rodrigo's device has ports 22 (SSH) and 5900 (VNC, a desktop sharing and remote control application) open. This behavior is worse than allowing a user to run a port scanner; it's actively broadcasting open ports, in plain text, for all to see:

Hostname: Rodrigo.Lab.local with Port Listening: 22

Hostname: Rodrigo.Lab.local with Port Listening: 5900

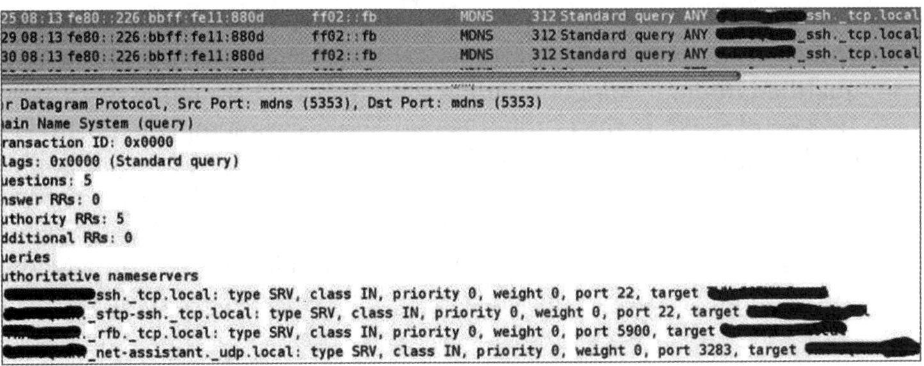

Figure 6.20: Excerpt from Trustwave's post on the security vulnerabilities of mDNS

`https://www.trustwave.com/en-us/resources/blogs/spiderlabs-blog/mdns-telling-the-world-about-you-and-your-device/`

Plus, in a 2020 article on poisoning attacks, a security consulting firm confirmed mDNS is still a major issue, citing that "mDNS can be abused via an attacker answering an mDNS request and impersonating a legitimate resource or computer on a network. Attackers can even use the same tool, the notorious 'Responder,' as published by Trustwave's SpiderLabs. The result is that the attacker can cause a device to send sensitive information directly to the attacker's machine, whether that be a print job for a document containing personal information or worse: a user's credentials." Source: `https://www.crowe.com/cybersecurity-watch/poisoning-attacks-round-2-beyond-netbios-llmnr`.

Excerpts of the security considerations included in the IETF RFC for mDNS include:

- "In an environment where there is a group of cooperating participants, but clients cannot be sure that there are no antagonistic hosts on the same physical link, the cooperating participants need to use IPsec signatures and/or DNSSEC [RFC4033] signatures so that they can distinguish Multicast DNS messages from trusted participants (which they process as usual) from Multicast DNS messages from untrusted participants (which they silently discard)."
- "If DNS queries for *global* DNS names are sent to the mDNS multicast address (during network outages which disrupt communication with the greater Internet) it is *especially* important to use DNSSEC, because the user may have the impression that he or she is communicating with some authentic host, when in fact he or she is really communicating with some local host that is merely masquerading as that name. This is less critical for names ending with '.local.', because the user should be aware that those names have only local significance, and no global authority is implied."
- "Most computer users neglect to type the trailing dot at the end of a fully qualified domain name, making it a relative domain name (e.g., "`www.example.com`"). In the event of network outage, attempts to positively resolve the name as entered will fail, resulting in application of the search list, including ".local.", if present. A malicious host could masquerade as "`www.example.com`." by answering the resulting Multicast DNS query for "`www.example.com.local`.". To avoid this, a host MUST NOT append the search suffix ".local.", if present, to any relative (partially qualified) host name containing two or more labels."

Disabling and Filtering UPnP Protocols

Universal Plug-n-Play (UPnP) is a vile, vile thing. Not only does the protocol perform some auto-discovery functions, but it takes the liberty of reconfiguring firewalls, poking inbound holes, and introducing extreme security risk. Read that sentence again.

And, in fact, it was UPnP that opened the doors that led to the massive Mirai botnet attack in 2016. Connecting to devices inside networks via holes opened by UPnP and using just a small list of default usernames and passwords for commercial devices, Mirai infected tens of thousands of IoT devices across the world and then used them to launch DDoS attacks to several targets, including the website of professional friend and reporter Brian Krebs.

Luckily UPnP performs this assault on consumer-grade routers, but there are plenty of small businesses, autonomous branch sites, or outposts that may walk into a retail technology store and pick up a consumer brand firewall or router for the office.

UPnP has been under fire since around 2001 when several government agencies issued security guidance to disable it within networks, but the vulnerabilities continue.

As recently as 2020, researchers updated reports on yet another flaw. A callback vulnerability named "CallStranger" allows an attacker to execute a variety of attacks including data exfiltration and DDoS attacks. It's not new, and a fix was issued years ago, but researchers expect CallStranger to remain a threat for years to come since many devices running UPnP aren't easily or often upgraded.

The recommendation for UPnP is simple—absolutely, under no conditions should it be enabled in networks. Not home networks, not small business networks, not enterprise networks. Specific actions include:

- In home and small offices using consumer Internet routers, or firewalls, access the management configuration and disable UPnP
- In enterprise environments, disallow consumer routers by policy, implement monitoring controls to identify them
- On IoT endpoint devices and gateway systems, change default passwords

TIP **If you want a bit more information and how-to, read Lifehacker's guidance on disabling UPnP in residential routers in their article "Disable UPnP on Your Wireless Router Already" at** `https://lifehacker.com/disable-upnp-on-your-wireless-router-already-1844012366`.

A Message on mDNS and Zeroconf from a Pen Tester

After reading the mDNS RFC, I reached out to one of my go-to pen tester teams and asked if and how mDNS and Bonjour played a role in their testing. I think their response sums up the risks nicely:

> **Both mDNS and zeroconf were never intended to be on corporate networks. They're literally designed for smaller networks that don't have an internal nameserver or IT folks capable of configuring network services. In the RFC for mDNS (**`https://datatracker.ietf.org/doc/html/rfc6762`**), the very first paragraph says that it's only intended for "the ability to operate with less configured infrastructure . . . in the absence of a conventional managed DNS server."**

The tool Responder [an mDNS tool] is one of the first things we run on internal pen tests. It listens for requests from mDNS, LLMNR, and NetBIOS names, and spoofs responses telling the requester that our attack machine is the system it's looking for. That forces the victim to connect through us, allowing us to man-in-the-middle the traffic.

I'd also point out that zeroconf is the exact opposite of zero trust which is becoming increasingly popular. The former is focused on trusting the network and accepting whatever anyone on it provides. The latter is designed on the assumption that you can't trust any network or anyone on it.

Nathan Sweaney, Principal Security Consultant at Secure Ideas (`www.secureideas.com`)

Recommendations for Securing Against Zeroconf Networking

In high security environments, zeroconf protocols should be completely disabled and strictly enforced. These tasks include:

- Disable all zeroconf protocol support on endpoints, including Bonjour, mDNS, LLMNR, and UPnP
- Monitor for and filter/block all mDNS and zeroconf traffic including Bonjour on the network
- Monitor for and restrict zeroconf protocols at the endpoint using endpoint security applications and/or endpoint detection and response (EDR)
- Create or update organizational and acceptable use policies to reflect this guidance
- Use assessments and penetration tests to validate posture

For organizations that are prepared to accept the risk of allowing zeroconf protocols on the network (against all hardening and security guidance), here are a few recommendations:

- Disable UPnP on devices and do not allow UPnP by policy; monitor and alert on UPnP
- Monitor for and filter/block UPnP traffic
- Do not allow organization-managed endpoints to participate in zeroconf networking, and disable Bonjour on these devices
- If permitted by policy, only allow Bonjour and mDNS on networks designated for Internet-only or personal device use (such as guest networks or residential halls) and properly segment those networks from organization-managed secured networks

- For mDNS, at a minimum, disable mDNS advertisements on all endpoints/devices
- Create or update organizational and acceptable use policies to reflect this guidance
- Use assessments and penetration tests to validate posture of the secured network(s)

The issues with mDNS and zeroconf networking protocols are well documented with numerous reports of exploits by security researchers and pen testers, guidance within the IETF RFC that these are not designed for enterprise use, and explicit guidance by technology manufacturers (including Apple) and NIST to disable these protocols in secure enterprise environments.

Knowing all this now, I personally consider it quite negligent for leaders to ignore the risks and continue permitting these protocols unchecked.

In my experience, the IT and networking teams within organizations are aware (at least in part) of the security risks, but unfortunately are often overruled by non-technical leaders who feel convenience trumps security. My hope is that the technical professionals are more heavily armed with this volume of data and can approach their leadership with solid evidence of the level of risk associated with Bonjour and mDNS.

Best Practices for Tiered Hardening

This chapter has covered a great breadth and depth of topics for hardening the wireless infrastructure. Promised earlier was tiered guidance to help you get started, wherever you are with the journey.

Some of the recommendations here are considered minimal-effort, must-do tasks, and then stringency increases from there based on the organization's risk tolerance and current security posture.

As with the rest of the best practices, there will always be exceptions and outliers, and while one could argue that a true best practice is applicable across the board, the reality is organizations have to work with the resources they have.

In the tables in the Figures to follow, recommendations are outlined by hardening topic and then organized by tasks for low security, medium security, and high security environments.

Low Security Environments Low security environments include minimum configuration even for labs and proof of concept (PoC) deployments. This tier is also appropriate for a small organization without many IT resources or a large footprint.

Medium Security Environments Medium security environment controls are what I consider baseline for mature and security-conscious organizations including moderately regulated industries such as healthcare and financial.

High Security Environments High security environments express requirements for an exceptionally locked-down environment. Many of these controls are specified in hardening requirements from federal governments and defense agencies.

Your organization or environment may fall between these three tiers and may also have unique circumstances; adjust accordingly.

Each of the tiers builds upon the prior, meaning the controls outline for low security environments should be followed or built upon by medium tier, and the same with medium to high. See Figure 6.21 for tiered guidance for securing management access, Figure 6.22 for tiered guidance for designing for integrity, and Figure 6.23 for guidance on restricting peer-to-peer communications and interface bridging.

Securing Management Access		
Low (minimum config)	Medium	High
• Enforce Encrypted Management Protocols • Eliminate Default Passwords • Enforce User-Based Logons • Enforce Secure Passwords • Enforce MFA for Remote Access	• Create Management VLAN • Secure Shared Credentials and Keys • Use Basic Privileged Access Management • Set Concurrent Session Timeouts • Monitor and Alert on Admin Access and Attempts	• Define Allowed Management Networks • Enable FIPS Mode • Use Advanced Privileged Access Management • Implement Logon Banners • Enforce MFA for All Admin Access

Figure 6.21: Tiered guidance for securing management access for low, medium, and high security environments

Additional Security Configurations

The direction offered thus far is, of course, nowhere near an exhaustive list. These highlight the activities that speak to meeting the most basic compliance frameworks and resolving the vulnerabilities penetration testers and malicious users exploit to gain access to networks.

In addition to the preceding lists, there are a few fringe considerations that warrant mentioning, including:

- Security Monitoring, Rogue Detection, and WIPS
- Considerations for Hiding or Cloaking SSIDs
- Requiring DHCP for Clients
- Addressing Client Credential Sharing and Porting

Designing for Integrity		
Low (minimum config)	**Medium**	**High**
• Backup Configs • Plan Regular Security Patching • Approve or Whitelist Organization Owned APs • Control Default Behavior of APs • Physically Secure APs in Public Areas • Use Port Security or MAB to Logically Secure AP Ports/Cables in Public Areas • Lock Network Closets and Datacenter Rooms • Disable IPv6, BootP (if not in use)	• Enforce Basic Change Management Processes • Document Configurations • Alert on Configuration Changes • Implement Vulnerability Management Program • Issue Certificates to APs for Authentication • Use 802.1X and/or MAB to Logically Secure AP Ports/Cables in Public Areas • Enforce Identity Based Access to Network Closets and Datacenters • Configure Front Panel Security on Controller • Disable Any Management Protocol Not in Use (e.g., HTTPS) • Assign Unused Ports to an Internet-only or Black Hole VLAN manually or with a NAC Product	• Enforce Advanced Change Management Processes • Establish Config Baselines • Verify Software Integrity • Disable Zero Touch Provisioning (ZTP) • Use Certificates on APs for Encryption Over the Wire • Use 802.1X to Logically Secure All AP Ports • Enforce Strict Access Control of All Facilities • Disable All Unused Management Access including Password Recovery and AP Console/USB Ports • Disable All Unused Protocols (e.g., ICMP, Echo, IP source routing, ARP proxy) plus disable physical ports not in use • Disable or Assign Unused Ports to a Black Hole VLAN

Figure 6.22: Tiered guidance for designing for integrity in low, medium, and high security environments

Controlling Peer to Peer and Bridged Communications		
Low (minimum config)	**Medium**	**High**
• Monitor Use of Consumer Devices in Network • Enable Inter-Station Blocking on all SSIDs • Disable mDNS Advertisements on Endpoints • Disable UPnP	• Restrict/Manage Use of Consumer Devices in Approved Networks • Disable Support for mDNS and Bonjour on Networks with Organization Owned Devices • Disable Support for mDNS and Bonjour on Organization Managed Endpoints • Disallow Devices with UPnP	• Deny Consumer Devices on Networks • Disable Interface Bridging on Endpoints • Follow Additional Endpoint and Networking Hardening Guidance

Figure 6.23: Tiered guidance for controlling peer-to-peer and bridged communications

Security Monitoring, Rogue Detection, and WIPS

Every secure architecture necessitates meaningful security monitoring, alerting, and reporting, and wireless networks are no exception. In fact, wireless probably demands a higher degree of monitoring than other systems just by virtue of the over-the-air access medium.

REFERENCE Coming up next in Chapter 7, we'll cover security monitoring, tools, testing, reporting, and how to navigate security audits and assessments.

Considerations for Hiding or Cloaking SSIDs

When it comes to hiding Wi-Fi SSIDs, there has been some poor advice doled out across the Internet by security enthusiasts who don't quite fully understand 802.11 WLAN protocols.

In residential and enterprise Wi-Fi products, the option often exists to "not broadcast SSID" or "hide SSID" or even "enable SSID cloaking."

All this setting does is tell the AP to only answer an 802.11 probe request when it contains the explicit name of the SSID. Otherwise, the default behavior is for the AP's beacons to include the SSID name. Functionally, the SSID name is what the user sees in the list of available networks. If the SSID is hidden, the beacon still exists, it just doesn't include the network name.

This adds zero additional security from an attacker, and in fact introduces additional vulnerabilities to mobile endpoints that may be used outside of the organization's walls.

When a legitimate client needs to connect to a hidden SSID, it's forced to announce to the world who it's looking for. It's another silly analogy but think of it like this—pretend you've been instructed to give $1,000 to Oscar. Oscar is some person you haven't met, but your contact assures you he's the guy you need and when you pay him, you'll get tickets to the sold out Six Nations Rugby match coming up. You walk down the street, holding the cash in your hand, calling out for Oscar. Finally, a nice young gentleman (noticing that wad of cash) approaches you, "I'm Oscar" he assures you. You hand him the cash. He said he was Oscar, but really his name is Jussi. At this point, he may run away, or he may lure you to a dark closet, tie you up, and steal your wallet.

This is demonstrated in Figure 6.24, where AP-1 is broadcasting the SSID names and Laptop-A is able to select and connect. However, AP-2 and AP-3 are both broadcasting all hidden SSIDs. Laptop-B has to announce it wants to connect to "Gizmo-Secure" SSID, after which AP-3 allows the connection. AP-2 is also shown to demonstrate it could be an enterprise AP and AP-3 may be a malicious user spoofing a corporate network.

That's what happens in the invisible world of Wi-Fi when we send our laptops and phones out into the world, and they're configured to look for a hidden SSID. The alternative would be going out onto the street and everyone having name tags. You walk until you find Oscar. It is possible Jussi could have stolen Oscar's name tag, but Jussi didn't know you were looking for Oscar, so that chance is low. And that's what it looks like when SSIDs are broadcast.

Figure 6.24: Conceptual comparison of connecting to a hidden SSID versus a broadcast SSID

Endpoints can be configured to connect to networks that are hidden. Additional configurations are required, as shown in Figure 6.25. The network name must be entered, and in most platforms it's case-sensitive. In the case of the Windows example here, the option to connect even if this network is not broadcasting is what tells the laptop this SSID may be hidden.

Add a new network

Network name

ACME-Secure-1X

Security type

WPA3-Enterprise AES

EAP method

Protected EAP (PEAP)

Authentication method

Secured password (EAP-MSCHAP v2)

☑ Connect automatically

☑ Connect even if this network is not broadcasting

Save Cancel

Figure 6.25: A Windows configuration option to connect to an SSID even if it's not broadcasting (meaning it's hidden)

Hiding the SSID has other impacts on availability and connectivity for clients, and the full list of considerations are as follows:

- Hiding the SSID only removes the network name from the beacons; an attacker can easily obtain the SSID name with basic attacks.
- It causes additional latency in the association process of many clients including Apple iOS devices.
- Other devices behave unreliably or can't connect at all to hidden SSIDs.
- Hiding SSIDs does not reduce the noise or traffic over the air.
- Additional risk is introduced by mobile clients actively probing for the enterprise SSID outside the organization.

There are some cases where it makes sense to hide one or more SSIDs. Primarily it makes sense in a complex environment for networks serving endpoints that remain within the physical boundaries of the organization. Therefore, the best practices for hiding SSID are as follows:

- Don't hide SSIDs for networks serving mobile clients that travel outside the physical boundary of the organization.
- Hide SSIDs for networks that are a) serving endpoints that remain physically within the boundaries of the organization and b) are preferred to be hidden to prevent attempted or accidental connection from users.
- Don't hide SSIDs as a security mechanism.

Two common use cases of hiding SSIDs include hospitals and manufacturing. Hospitals have one or more networks for clinical systems that remain in the building, and which aren't connected to by human intervention—meaning it's a pre-provisioned IoT-type device, not a user with a laptop that needs to select the network from a list. Warehouse, manufacturing, and retail all too have various handheld scanners, communications devices, and inventory management systems that would be hidden. See Figure 6.26 and Figure 6.27 for examples of settings to hide SSIDs on Juniper Mist and Aruba.

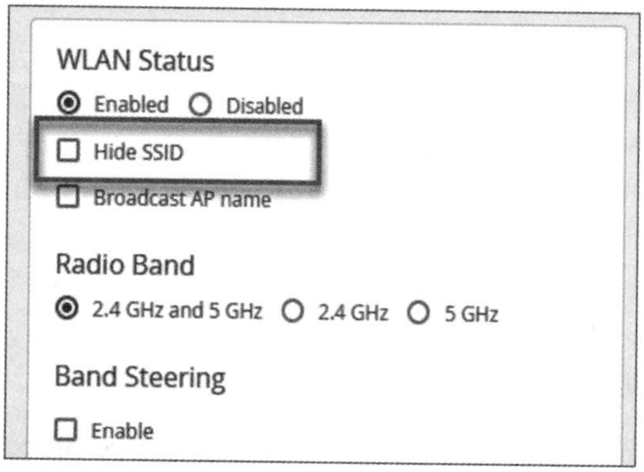

Figure 6.26: The Hide SSID option in Juniper Mist

NOTE Hiding SSIDs is not related to broadcasting AP names, which is also shown in Figure 6.26.

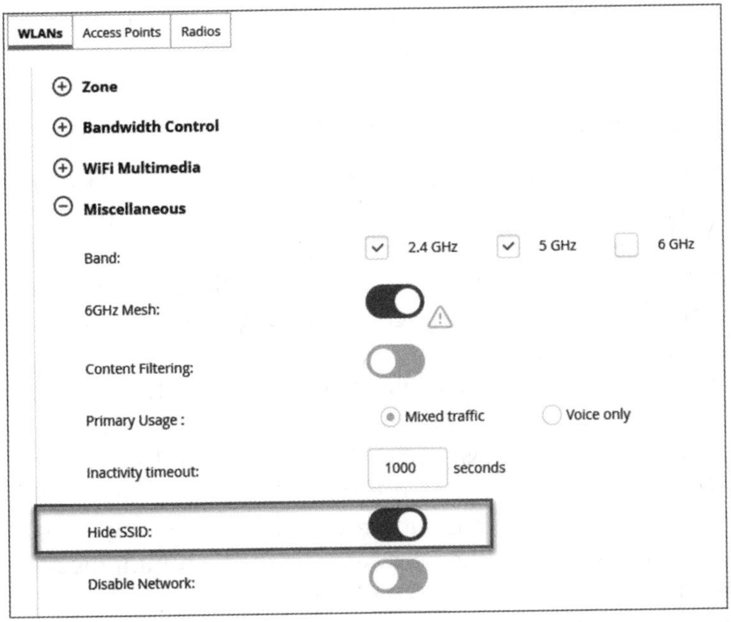

Figure 6.27: The Hide SSID option of Aruba Networks Central Cloud platform

NOTE The Disable Network setting will stop the SSID from being active, whereas the hide setting only removes the name from client view.

Requiring DHCP for Clients

One simple control to consider may be requiring DHCP for clients. These days, many network security and profiling tools rely on being able to see DHCP traffic (through IP helpers, network taps, or span/mirror ports) and conversely, subverting DHCP is one way an attacker can gain unauthorized access to networks.

Most Wi-Fi products offer a setting to force clients to use DHCP, and that's a great option for networks that aren't servicing intentionally statically configured clients. Requiring DHCP helps prevent access from misconfigured endpoints and ensures when a client connects it will have the appropriate network parameters (IP address, gateway, and DNS servers) to access resources.

The big caveat here is that there still exist many endpoints with statically configured IP addresses. This is less common on traditional endpoints, and more often seen on IoT devices or non-traditional IP devices that are being managed by a third party. For example, in healthcare networks there's pretty much always a subset of clinical devices with static IPs, and the hospital may not have purview or management access of the devices to change that.

When possible, all networks should rely on DHCP for addressing. If a static IP is needed, using a DHCP reservation in the server is the most appropriate approach.

Addressing Client Credential Sharing and Porting

Recently I put together a list of the top five vulnerabilities of Wi-Fi based on feedback from several penetration testers with different companies. After organizing and standardizing the responses, five topics bubbled up. We've addressed four of them so far in the book—including segmentation, Bonjour and other zeroconf protocols, porting credentials from clients, and lack of hardening.

The fifth one was a bit of a surprise to me; it involves attacks on credential sharing among endpoints, and several specifically cited the Apple Share Network Password function available in iPhone, iPad, and iPod devices.

With this option, any iOS user with a passphrase for a network (WPA2-Personal or WPA3-Personal) can easily port it to any other iOS user. It's as easy as the steps shown in Figure 6.29. From the infrastructure side, there's nothing you can do to prevent or even monitor for this; the peer devices may be communicating over Wi-Fi or Bluetooth, which is unlikely to be monitored.

This isn't an Apple-only problem. Many platforms incorporate some form of passphrase sharing. Android allows users to share passphrase secured networks via a QR code that can be scanned from the screen or emailed (see Figure 6.28). Microsoft also joined in when it introduced the hotly contested Wi-Fi Sense feature designed to share public hotspot passphrases. It removed the feature in 2016 according to Krebs on Security (Source: `https://krebsonsecurity.com/2016/05/microsoft-disables-wi-fi-sense-on-windows-10/`). Figure 6.29 shows the ease of sharing Wi-Fi passwords from an Apple iOS device.

Recommended workarounds are to not use passphrase networks where possible or to enable MAC address allowlists for authorized endpoints. New vendor privacy mechanisms such as MAC randomization further complicate any attempts to filter on MAC address or use an 802.1X secured network with MAC Authentication Bypass (MAB).

Options for controlling this behavior are limited and controls for a secure network would rely on not using passphrase-based (Personal) networks. With that being the only control, these are additional notes and considerations for protecting against credential sharing:

- Devices (such as Apple iOS) can share between devices of different owners
- Devices automatically sync network credentials to other devices by the same owner (via iCloud for Apple and other sync options for Android and Microsoft)

- If network passwords are shared between devices, the organization will have no knowledge of it
- There is no way to prevent network password sharing on unmanaged devices
- Network password sharing can be controlled on corporate-managed devices with an MDM through settings AirDrop for Apple devices
- The only protection from network password sharing is to not use passphrase-secured SSIDs with unmanaged devices

The last bullet is the real kicker since passphrase-secured networks are exactly what an organization will typically use for personal and BYOD devices.

Figure 6.28: The option to share a passphrase credential via QR code from an Android phone

How to share your Wi-Fi password

1. Make sure that your device (the one sharing the password) is unlocked and connected to the Wi-Fi network.
2. Select the Wi-Fi network on the device you want to connect.
3. On your device, tap Share Password, then tap Done.

Figure 6.29: Screenshot from Apple on sharing network passwords
`https://support.apple.com/en-om/HT209368`

Summary

This chapter has taken us through our hardening journey, from securing management access to building infrastructure integrity, controlling peer-to-peer and ad-hoc networks, and a few additional hardening considerations.

One important takeaway from this chapter is that overlay networks for personal- and residential-use devices introduce innumerable security risks to the enterprise networks. I hope this chapter has demonstrated the undeniable dangers of mDNS, Bonjour, UPnP, and other zeroconf networking protocols, and that your organization will take that into consideration when planning security enhancements. As Gartner cites in their definition, consumerization is something enterprises have to "deal with." Our hope is this chapter will help you deal with it securely.

The other takeaway is that hardening exercises are best taken in steps, and a sample of tiered guidance has been provided, starting with the lowest level appropriate for any devices under corporate ownership or touching any corporate owned networks (lab, production, or otherwise). The tiered guidance continues with a medium level security suite of recommendations, followed by a high-security tier with the most stringent controls. As a reminder, most environments will fall in between these three coarse tiers.

Some of the recommendations around hardening, mDNS, and SSID cloaking may be novel or unexpected. And the realization that virtually any consumer endpoint can easily share and port passphrases from WPA2-Personal and WPA3-Personal networks may be eye-opening. There's certainly a lot to consider in balancing usability with security these days.

Last, but not least, this chapter has probably demonstrated how twisty Wi-Fi technology can be, including with the numerous considerations for protected management frames (PMF), its interoperability with Fast Transition roaming, and WPA2/WPA3 secured networks.

One of the major tasks in hardening entails monitoring, alerting, and reporting of the security posture of the network, and we've dedicated an entire chapter to this topic next.

Part

III

Ongoing Maintenance and Beyond

In This Part

CHAPTER 7

Monitoring and Maintenance of Wireless Networks

Wireless, including (and especially) Wi-Fi is no longer set-and-forget. The systems need at least a bit of continuous care and feeding, along with basic monitoring and reporting to ensure integrity of the system is being maintained, and that the infrastructure is ready for tomorrow's demands.

This chapter includes the following topics:

- Security Testing and Assessments of Wireless Networks
- Security Monitoring and Tools for Wireless
- Logging, Alerting, and Reporting Best Practices
- Troubleshooting Wi-Fi Security
- Training and Other Resources

Security Testing and Assessments of Wireless Networks

First, let's talk about security testing and assessments, which specifically encompasses the types of services and testing that are frequently performed by a third party (exclusively or in addition to internal testing). These include audits (against some compliance framework), security assessments, vulnerability assessments, and of course penetration testing.

These four types of testing are often confused and conflated with one another, but in all cases in information security these are each quite unique with different methodologies, goals, and deliverables. Together, these practices along with ongoing internal monitoring comprise a robust security management program for wireless networks (or any network, for that matter.)

ADVICE ON TESTING: DON'T TAKE IT PERSONALLY

Before we get into these testing topics, let me start by saying all of these tasks are designed to evaluate and/or report on the security posture of the systems in scope. Whether tests or audits are performed by a third party, or another internal team within your organization, it's important to understand these are designed to help you maintain a secure system—not to attack or blame you for doing it wrong.

It's easy to get defensive and feel a bit like your work is being criticized, but that's not the intention at all.

We're all doing the best we can with the resources, knowledge, and time we have available. And usually, we're short on at least two of the three at any given time. If nothing else, I hope this book has demonstrated that network and especially wireless security is complex, intertwined, and ever-changing.

It's just not possible for any one human to know everything and protect against every possible threat out there. Keep that in mind as teams provide reports and feedback on the security posture of the system and just be thankful there are friendlies giving you the feedback instead of getting a call at 3:00a.m. because an adversary found their way into the network.

Oh, and, if they're good—they *will* find something. If they don't, then you may need a new pen tester.

Let's start with the boring parts first—audits—and then work our way down to the juicier topic of pen testing. In general, the order of presentation aligns with the level of touch starting with audits, which are basically checklists, and proceeding through to pen tests, which can be full-on assaults. The flow will look like this:

- Security Audits
- Vulnerability Assessments
- Security Assessments
- Penetration Testing
- Ongoing Monitoring and Testing

Security Audits

Security audits are one of those predominately boring but requisite tasks for environments that are under regulation, include internal audits as part of normal operations, or have requirements for cyber insurance. In Chapter 1, "Introduction to Concepts and Relationships," we covered compliance and regulatory requirements and took the nickel tour of regulations, frameworks, and audits. Not all compliance programs or regulations are formally audited, but at a minimum most include internal checklists or self-assessments.

Examples of compliance programs with formal audit processes:

- HIPAA (Health Insurance Portability and Accountability Act) in the U.S.
- PCI DSS (Payment Card Industry Data Security Standard)
- Payment Services II Directive (PSD2) and Regulatory Technical Standards for Secure Customer Authentication (RTS SCA) in Europe
- The Directive on Security of Network and Information Systems (NIS Directive) in Europe
- CMMC (Cybersecurity Maturity Model Certification) in the U.S.
- SOC Type 2 (System and Organization Controls) in the U.S.
- SOX (Sarbanes-Oxley Act) in the U.S.

If it is audited by a third party, then someone approved by the governing body will perform the audit and report on the posture against a specified checklist. The data may be entered manually, collected and reported on using automated tools, or (as is often the case) a combination of both.

Audits are most often performed at these times:

- At a specified cadence if part of the internal security program, such as quarterly, bi-annually, or annually
- At a prescribed cadence if part of a compliance requirement
- At a prescribed cadence if part of a cyber insurance plan
- Ad-hoc after a triggering event or concern of posture such as a system migration, upgrade, incident, or breach

An audit is evaluation of systems, processes, or people against a set of standards. The assessment is a checklist, and may include requirements of evidence, meaning the organization may need to provide some proof that they've met the checklist items.

Security audit items vary—the checklists for HIPAA are going to be different than those of PCI DSS for payment card processing. In general, there are some common themes and to offer some context, following are a few popular items within an audit scope:

- Evaluation of multi-factor authentication, especially for privileged and remote access (including management access to the Wi-Fi systems)
- Evaluation of the use of encryption for data at rest or in transit, based on requirements (including over-the-air encryption in Wi-Fi)

- Evaluation of vulnerability management of the infrastructure (including the Wi-Fi system components and endpoints)
- Evaluation of policies and processes, ensuring that processes or policies exist for topics such as *bring your own device* (BYOD), for example (which will influence your Wi-Fi network settings and policies)
- Evaluation of software and hardware configurations and maintenance such as patching, segmentation practices, and secure management (again, including the Wi-Fi systems)

For audits tied to compliance regulations, there may be penalties for non-compliance such as fines or de-authorization.

Vulnerability Assessments

Vulnerability assessments take place (ideally) regularly as part of an ongoing vulnerability management program. Vulnerability assessments consist of scanning infrastructure and endpoint devices for security posture and mapping that data against known vulnerabilities.

Security vulnerabilities are discovered daily, and there's a defined process for categorizing and disclosing them via a database of common vulnerabilities and exploits, or CVEs. For each CVE, additional data is added to offer more context including a severity rating and analysis of the exploitability and impact on integrity, confidentiality, and availability.

From 2016 through the end of 2021, vulnerabilities grew from around 6,000 to 20,000 CVEs a year. Meaning, entering 2022 and beyond, we expect an average of more than 50 new vulnerabilities each day. Of course, not every CVE applies to every environment, but that's a large and constant volume of new inputs to work with.

The goal of vulnerability scans is to identify which CVEs are applicable to the environment (which includes endpoints as well as the network infrastructure) and provide enough detail for the organization to prioritize patching and remediation.

Figure 7.1 shows sample data from a vulnerability scan of a lab environment with edge switching and Wi-Fi. This report shows several SSL/TLS vulnerabilities, a weak SSH key plus one device with Telnet enabled (an unencrypted management protocol that should be disabled per recommendations in Chapter 6, "Hardening the Wireless Infrastructure"). The second image from the same scan shows a vulnerability on two cloud-managed APs.

Other views of this data will show additional OS-specific guidance and reference the CVEs for each.

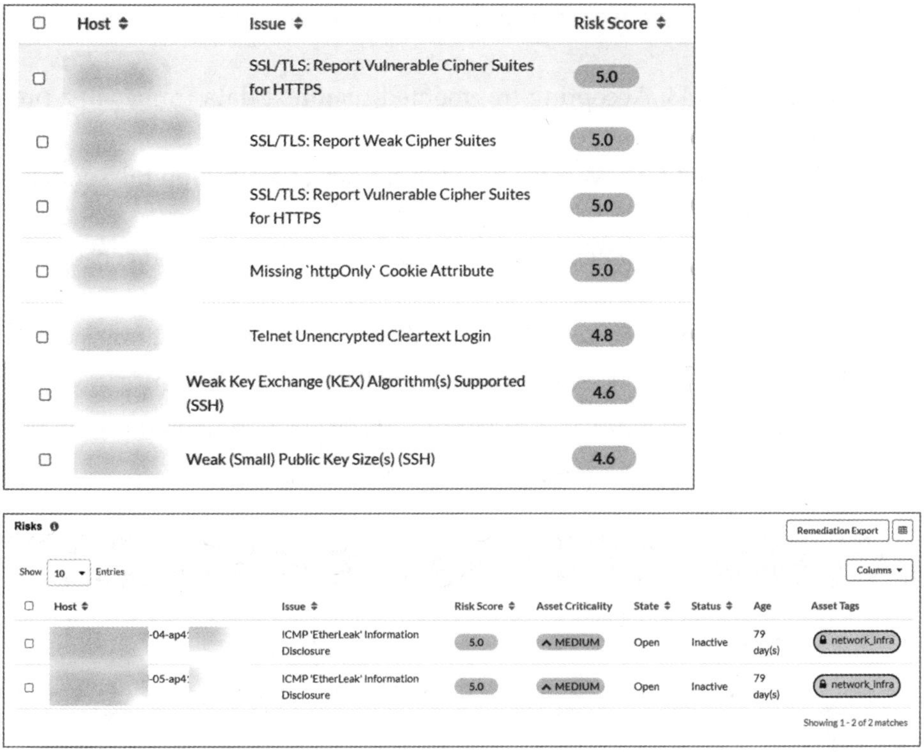

☐	Host ⬍	Issue ⬍	Risk Score ⬍
☐		SSL/TLS: Report Vulnerable Cipher Suites for HTTPS	5.0
☐		SSL/TLS: Report Weak Cipher Suites	5.0
☐		SSL/TLS: Report Vulnerable Cipher Suites for HTTPS	5.0
☐		Missing `httpOnly` Cookie Attribute	5.0
☐		Telnet Unencrypted Cleartext Login	4.8
☐		Weak Key Exchange (KEX) Algorithm(s) Supported (SSH)	4.6
☐		Weak (Small) Public Key Size(s) (SSH)	4.6

Risks

Remediation Export

Show 10 Entries

Columns

☐	Host ⬍	Issue ⬍	Risk Score ⬍	Asset Criticality	State ⬍	Status ⬍	Age	Asset Tags
☐	-04-ap4:	ICMP 'EtherLeak' Information Disclosure	5.0	MEDIUM	Open	Inactive	79 day(s)	network_infra
☐	-05-ap4:	ICMP 'EtherLeak' Information Disclosure	5.0	MEDIUM	Open	Inactive	79 day(s)	network_infra

Showing 1 - 2 of 2 matches

Figure 7.1: Sample data from vulnerability scanning in a lab environment

For organizations that have a vulnerability management program, or have requirements around vulnerability management, this process is an ongoing task, occurring weekly, daily, or even more often in some cases.

A vulnerability report will show the target of the scan (such as a wireless controller or wireless endpoint), open ports, and known vulnerabilities based on the operating system, system components, and patches.

Many attacks leverage multiple CVEs together. For example, the Wi-Fi Frag-Attack disclosed in 2020 exploited about a dozen different CVEs:

- CVE-2020-24588: Aggregation attack (accepting non-SPP A-MSDU frames)
- CVE-2020-24587: Mixed key attack (reassembling fragments encrypted under different keys)
- CVE-2020-24586: Fragment cache attack (not clearing fragments from memory when (reconnecting to a network)
- CVE-2020-26145: Accepting plaintext broadcast fragments as full frames (in an encrypted network)
- CVE-2020-26144: Accepting plaintext A-MSDU frames that start with an RFC1042 header with EtherType EAPOL (in an encrypted network)

- CVE-2020-26140: Accepting plaintext data frames in a protected network
- CVE-2020-26143: Accepting fragmented plaintext data frames in a protected network
- CVE-2020-26139: Forwarding EAPOL frames even though the sender is not yet authenticated (should only affect APs)
- CVE-2020-26146: Reassembling encrypted fragments with non-consecutive packet numbers
- CVE-2020-26147: Reassembling mixed encrypted/plaintext fragments
- CVE-2020-26142: Processing fragmented frames as full frames
- CVE-2020-26141: Not verifying the TKIP MIC of fragmented frames

Figure 7.2 shows data from the CVE-2020-26140 from `www.cve.org`.

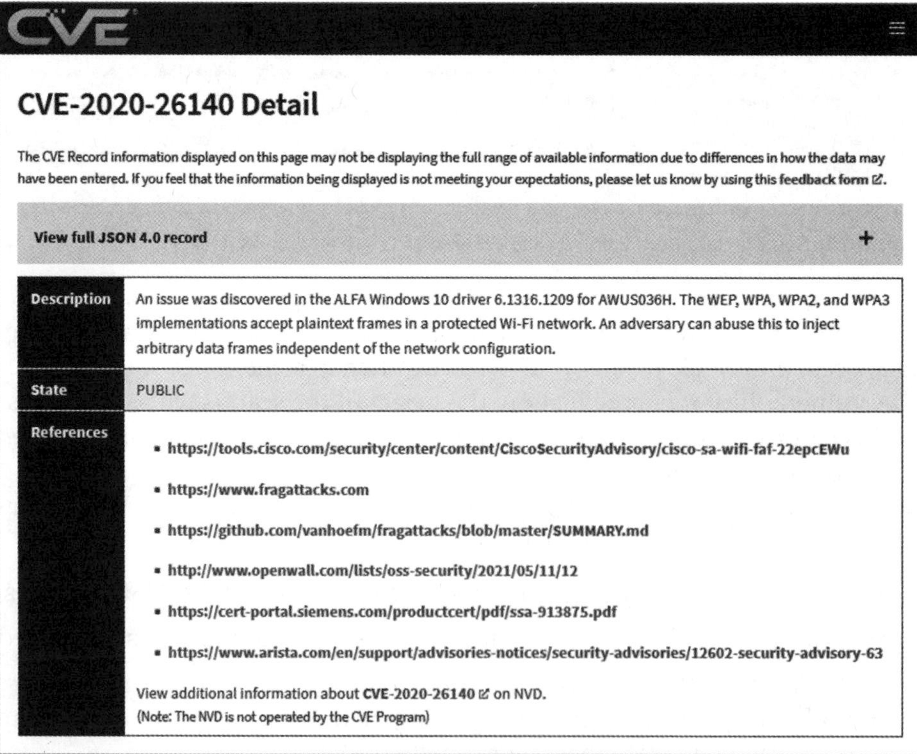

Figure 7.2: CVE-2020-26140 was one of many vulnerabilities exploited in FragAttack

Internal Vulnerability Assessment

These risks are usually evaluated from two lenses—internal and external. With an internal scan, the assessment is run based on how the target devices appear from the internal network. This is valuable since the internal network access will be unfiltered (not blocked by firewalls) as long as the scanning appliance has proper access.

Scanning for vulnerabilities from inside offers the entire picture of the device's posture, allowing the organization to fully secure it, including from internal (malicious or unintentional) threats.

External Vulnerability Assessment

For systems and networks that have Internet connectivity, an external assessment is also part of the vulnerability management program. The only difference here is that the scan is initiated externally from the Internet—offering a view of how the system appears to the outside world. The external assessment is critical since the Internet introduces access to millions of potentially malicious users, scripts, and bots.

As it relates to wireless systems, if your infrastructure supports remote sites with APs connected over the Internet, remote workers using remote APs, or remote management access to the controller, then the architecture will have necessitated inbound access from the Internet. In these cases, the external vulnerability scan will help ensure no gaps exist that may increase the risk of a successful Internet-based attack.

A vulnerability management program with ongoing vulnerability assessments is just one piece of an overall strategy for ensuring security of the environment. Outputs from these scans and summary reports may be included as part of the security audit and security assessments discussed in this chapter. However, security audits and assessments are more of a point-in-time static view of the environment, whereas an effective program will including continuous vulnerability scanning.

Security Assessments

The phrase "security assessment" can denote many things, and it's probably the one gray area that doesn't have an information security textbook definition.

Similar to the informal audits mentioned earlier, a security assessment is most often a checklist review of the security posture or configurations against known best practices or controls from a framework (such as NIST or ISO). The difference here is that this assessment is not a formal evaluation as is often the case in an audit. Also, some assessments are more hands-on in that the assessor may access the system being evaluated, versus only asking the questions about controls.

TIP **Think of a security assessment as a practice test, and an audit as a graded exam.**

There may be many types of security assessments used in the environment and depending on the use case and scope, they may be performed by internal teams within your organization or third parties. In fact, in practice, security assessments are a great substitute for or precursor to more in-depth penetration testing (covered next).

These assessments will vary in scope, depth, and breadth and follow no prescribed methodology. Each organization, consultant, or service provider will have their own process and documentation but the standards the systems are being assessed against are usually a subset of one of the common controls frameworks such as NIST, ISO, or the lighter controls of CIS Critical Security Controls (formerly the SANS top 20 list, now managed by the Center for Internet Safety).

For the purposes of wireless architecture, we use security assessment to evaluate the configuration and posture against what we know are vulnerabilities. Just as with audits and other scans, there are many low-hanging-fruit controls that should be in place before considering a pen test.

Examples of items that might be included in a wireless security assessment are:

- Review policies related to wireless use or access
- Review configured networks and segmentation
- Evaluate management access including use of user-based logins, strong passwords, timeouts, and encrypted protocols
- Assess encryption cipher suites in use and evaluate against industry best practices
- Analyze the configuration and posture of the wirelessly connected endpoints
- Scan the scoped components to evaluate vulnerabilities and posture

Notice the list to be action verbs here (as with audits) include "review" and "evaluate" versus "testing" or "exploitation" of the network. Figure 7.3 demonstrates the varying levels of touch across audits, assessments, and pen tests.

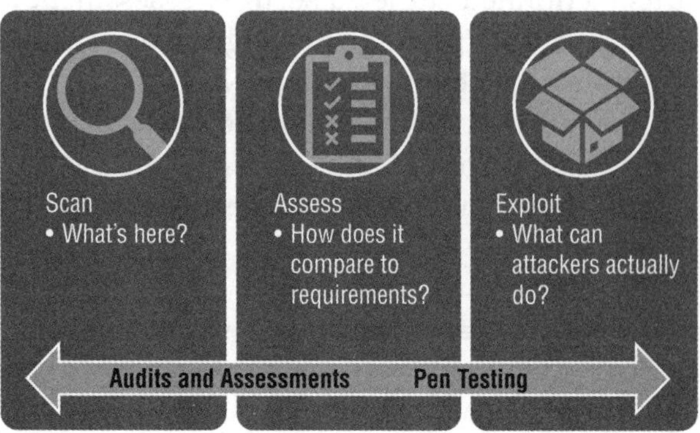

Figure 7.3: The level of touch varies from audits and assessments to penetration tests.

Penetration Testing

Penetration testing, or pen testing, goes a few steps beyond security assessments by actively exploiting and attempting to breach a target—in this case the wireless infrastructure.

I'll reiterate this again—there are countless ways to infiltrate a network and a skilled pen tester will absolutely find some way in; it's just a matter of how much time or effort is involved.

Think of pen tests like safe and fire ratings—there's a categorization of each. For safes, a rating of C Class is better than B Class and offers more protection. With safes, the Underwriters Laboratories (UL) has additional ratings that continue on to UL TL-15, TL-30, and the TL-30x6, which each describe how long the safe should be resistant to attack (here, that's 15 minutes, 30 minutes, and then 30 minutes from all six sides). Similarly, safes have fire ratings from UL which range, for example, from 30 to 120 minutes of viability under 1550° F direct heat.

Here's the bad news. The methodologies of testers vary by person, so there's not a comparable pen test rating that's industry standard; the safe rating is here merely to explain the concept that no network is impervious, and our goal is to reduce risk and increase time-to-breach, making it exceptionally hard for an attacker.

Before a pen tester is engaged, some activities and vetting must take place, and this is often arranged and managed by the CISO's office, a security operations team, or other technical leadership.

Just as there are internal and external vulnerability scans, so too are there *internal* and *external* pen tests. And there are different types of test models depending on the degree of information given to the tester about the environment ahead of time. In a *closed bo*x test, the testing team knowns nothing of the environment and works the process just as an external attacker would. Conversely, a *clear box* test model arms the testers with knowledge of the network, internal IPs, and systems.

Personally, for most situations I'm a fan of white box testing because, like a security assessment, having knowledge of the environment can be a leg up for the tester, and if it's secured against someone with knowledge, you're better protected. This methodology also offers better insight into protections against insider attacks.

INDUSTRY INSIGHT: USING CAUTION SELECTING PEN TESTERS

I'm ridiculously picky about the pen testers I work with and refer clients to. Why? Well, the skills vary greatly and there are thousands of self-proclaimed pen testers out there—network or systems admins who purchased an off-the-shelf or open source testing tool or went to a class or two. Those people and the quality of their work are far, far inferior to professionals whose vocation and career is dedicated to pen testing.

I've seen way too many pen test reports that are just outputs from common (often free) testing tools. Those are exceptionally unhelpful, they give organizations a false sense of security, and they're simply a waste of money. Quite literally, anyone could have downloaded a tool and run it; and I've seen organizations pay tens of thousands of dollars for garbage.

Skilled testers will use a certain set of standard tools to hit the low-hanging fruit, but these professionals have intimate knowledge of the systems, applications, and services they're attacking and layer various techniques against the target system.

Plus, not only are unqualified testers inefficient, using a person or company without the right credentials, experience, and insurance can be very dangerous. They may inadvertently cause damage to internal systems resulting in outages, at times they may use tools that introduce additional risk, and/or unknowingly initiate scans that are illegal. Here are a few pointers when selecting a pen test provider:

- Consider industry reputation of the tester and/or company, its longevity, and references (if available).
- Expect testers to have certifications such as Offensive Security Certified Professional (OSCP), Offensive Security Certified Expert (OSCE), or CREST (`www.crest-approved.org/`).
- Ask about the tester's process and look for it to include steps and deliverables beyond vulnerability scanning and to address your stated goals.
- Verify proof of insurance including professional liability and errors and omissions. This coverage is separate from standard business liability plans and even for the consulting services my company provides, I have this level of coverage.

Ongoing Monitoring and Testing

Along with periodic audits, assessments, and pen tests, your strategy should include ongoing monitoring and testing. Since this is a broad area with many tools and techniques, it's deserving of its own section coming up next.

Security Monitoring and Tools for Wireless

Monitoring and tools are another corner where wireless and security can intersect in wonderful ways. There are far more tools in the world than space to name them all; the content here is an attempt to describe the types of tools and their application within the world of wireless security.

In the upcoming section "Logging, Alerting, and Reporting Best Practices," we'll address the details of what exactly it is that we need to log, monitor, report, and alert on. The truth is, while many tools have discrete use cases, even more can be used in multiple ways, and therefore would introduce a lot of overlap if we combined the topics.

Along with the vital and obligatory topic of wireless IPS (WIPS), we'll address use of synthetic testing tools, securing logging, and a handful of wireless-specific tools with a bit broader scope and periphery to security.

- Wireless Intrusion Prevention Systems
- Synthetic Testing and Performance Monitoring
- Security Logging and Analysis
- Wireless-Specific Tools

Wireless Intrusion Prevention Systems

Wireless intrusion prevention systems (WIPS) are designed to detect and prevent attacks on wireless networks and endpoints, including (and especially) over-the-air attacks based on layers 1 and 2.

For the purposes of this book, I'm including several wireless-based security monitoring and alerting tools under the WIPS umbrella. Along with standard Wi-Fi and Bluetooth® WIPS there exist many technologies and products designed for securing and monitoring other types of radio transmissions including cellular, long range low frequency, and narrowband RF. These solutions incorporate powerful spectrum analyzers with software that performs various functions of decoding and analysis.

WIDS vs. WIPS vs. Wired IPS

First, let me go ahead and tackle the *WIDS* vs. *WIPS* conversation and get that out of the way. Just like its wired counterparts, wireless IDS is *intrusion detection*, and wireless IPS is *intrusion prevention*. Realistically there are very few use cases for a detection-only system. For obvious reasons most organizations are looking for the protection and to stop an attack.

If I left my laptop bag at a table with friends and said, "hey will you watch this for a minute?" and then a stranger came by and tried to steal my bag, my IDS friends would watch my bag get stolen and give me a detailed description of the thief along with the direction he went. My IPS friends would see the heist happening and grab my bag before the crook could abscond with it.

We all want those IPS friends. In wireless there are a few specific cases where we may be limited in the response and actions taken to mitigate certain over-the-air attacks, and I'll cover that shortly (and more than once).

NOTE **If you're wondering why anyone would want detection only and not prevention, it's because some entities may have a desire or need to prosecute on an action (in the legal sense). For example, if there's an illegal drug deal or a larceny, the authorities can't prosecute based on a criminal's foiled attempt. The activity is allowed to occur, and then they're busted afterwards. The same holds true for certain cybercrimes.**

The next explanation is the answer to a commonly asked question of "if we have IPS on the network, why do we need an WIPS?" It's a short and sweet answer. Wired IPS systems only have visibility into and signatures for wired-side attacks. They can't possibly see or parse an RF-based attack over the air. Hence, we require WIPS in addition to any IDS systems on the wired network.

WIPS sensors have visibility of layers 1 and 2 over-the-air, versus a wired IPS device, which will only see and evaluate wired traffic. Depending on the wired sensors, the wired IPS may secure traffic inbound from the Internet and/or internal LAN traffic, as shown in Figure 7.4.

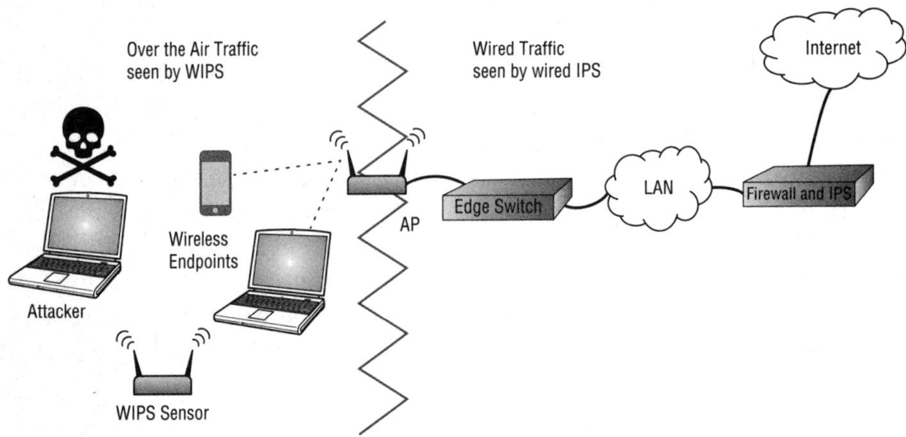

Figure 7.4: Visibility of WIPS vs. wired IPS

Requirements for WIPS

Because WIPS systems are monitoring for attacks over the air, there are a few specific requirements to consider when planning the architecture. The WIPS system will need to have sensors within the physical airspace of the assets to be protected, or area to be monitored; those sensors will need to have radios in the RF spectrum to be monitored; and the system (which by nature will have distributed sensors) should integrate with a centralized platform to ingest and alert on any incidents.

Radios, Modulation, and Encoding

Wireless IPS systems rely on radios, just like any other wireless device. The radios, modulation, encoding, and analysis in the WIPS solution has to mimic or match those of the systems it's protecting, meaning if the goal is to protect a Wi-Fi network, then the WIPS needs to be 802.11-capable. And not only does it need to be 802.11-capable, but the WIPS radios will need to be up to date on the latest 802.11 standards and chipsets in order to see and detect all attacks.

If we're monitoring Bluetooth or BLE, the WIPS system needs to have radios that support that RF and have the ability to perform Bluetooth packet analysis. For networks using Zigbee, Thread, or 6LoWPAN, WIPS will need to support 802.15.4. The same holds true for private cellular, which in the U.S. uses Citizens Broadband Radio Service (CBRS) frequencies and a different encoding and modulation than 802.11 and 802.15.4 technologies.

There are multi-band WIPS products designed to monitor multiple RFs along with various encoding and modulation types, but more commonly WIPS features are built into the infrastructure products offering those services. This means it's more common for an 802.11 WIPS solution to be integrated with a Wi-Fi solution than to have it as a separate product (at least for now). We'll investigate integrated versus overlay WIPS shortly.

Sensor Placement

In order to monitor the airspace, a WIPS system has to have physical proximity to the devices being protected. The coverage area of a WIPS sensor's radios can be drastically different than that of the wireless radios of the system it's monitoring. The footprint of the WIPS system should encompass the whole of the wireless service area to ensure full visibility, meaning WIPS sensors deployed too sparsely or without coverage at the periphery of the service area will result in gaps.

Central Monitoring and Alerting

The data and intelligence gathered by WIPS products are no help if the proper alerting isn't configured, making central monitoring and alerting a vital component of WIPS.

WIPS events can be processed within the WIPS platform, the Wi-Fi management platform (for integrated solutions), or via an external logging or monitoring platform such as a SIEM.

Regardless of where the alerts coalesce, it should be a centralized location and contain processing rules to get the right alerts triggered and sent to the right people. We'll cover more on which events should be logged, alerted, and reported in the next section of this chapter.

Integrated vs. Overlay vs. Dedicated

If I were writing this just a few years ago, I would tell you about integrated versus overlay systems. The truth is the industry went from three or four dedicated Wi-Fi WIPS vendors to approximately zero over the course of the past 10 years or so.

There are WIPS features in several of the special-purpose monitoring tools discussed with spectrum analyzers shortly. Those products have predominately been used in federal, military, and logistics operations with specialized needs and aren't mainstream in enterprise environments (yet).

INDUSTRY INSIGHT: THE BIRTH, DEATH, AND REBIRTH OF WIPS

Years ago, several solutions for Wi-Fi WIPS existed, including AirTight networks and AirMagnet Enterprise, among others. In 2016, AirTight became Mojo Networks (now part of Arista), changing from a WIPS product to a full cloud-managed Wi-Fi solution that incorporated WIPS as one of its features but no longer offered a stand-alone or overlay solution.

The other prominent solution was AirMagnet Enterprise, which was acquired by Fluke Networks in 2009, and later became part of NetScout. Around 2018, the AirMagnet line along with handheld testing tools was divested and spun off to what is now NetAlly. Unfortunately, through those transitions, the WIPS product was quietly sunsetted and had fizzled out by 2019.

The one exception (that I know of) in the world of Wi-Fi WIPS is Extreme's AirDefense WIPS solution, which was designed to also be deployed as an overlay to any third-party Wi-Fi products.

For many years, certain compliance requirements drove the WIPS market, going so far as to specify use of WIPS or other surveys to prove the non-existence of Wi-Fi within an environment. As Wi-Fi security increased, it was adopted more readily into security-conscious environments and the compliance mandates changed accordingly. Deploying an overlay WIPS requires additional cabling, mounting, and products—all of which add cost and effort that most organizations wouldn't pursue unless required to do so.

There is an emerging suite of security products that incorporate wireless sensors. In 2019, Outpost24 (`https://outpost24.com`) acquired ethical wireless hacking company Pwnie Express and incorporated those products into Outpost24's security assessment platform. You may have seen Pwnie Express products on the hit TV show *Mr. Robot*.

Wireless in the enterprise has converged and diverged over time and is again diverging into private cellular (4G and 5G) and 802.15.4 protocols including Bluetooth and BLE, in addition to 802.11 Wi-Fi. With IoT trends continuing upward and so many endpoints now wirelessly connected I believe we're on the edge of the rebirth of WIPS. It's possible a niche player may emerge, one of the synthetic QoS testing systems may add WIPS, or the enterprise community may adopt technologies currently used by federal governments and military.

So, instead, we'll talk about the differences between integrated and dedicated WIPS sensors, since the story is the same, but the characters are different.

Pretty much every enterprise Wi-Fi product has some form of integrated WIPS and security monitoring built into the product. If configured to do so, APs that are servicing endpoints can spend part of their time scanning and listening for signs of over-the-air attacks, anomalies, and rogue devices.

The benefit to allowing the APs to pull double-duty is that you've immediately satisfied the three requirements of WIPS—you have radios that support the RF bands of interest, they're physically distributed throughout the environment, and they're attached (or can be) to a central system, whether that's a management product, cloud manager, or controller.

The downside to the double-duty model is that the AP radios have limited time and bandwidth (as in RF bandwidth) to monitor. They can't really fully monitor while they're servicing clients—it's like the falsity of multitasking—they're not doing both at once, they're just switching quickly between tasks. As shown in Figure 7.5, if the radio needs to contain a threat, that radio will cease to serve clients. APs that have an additional dedicated radio can do both simultaneously by using the extra radio for scanning and containment.

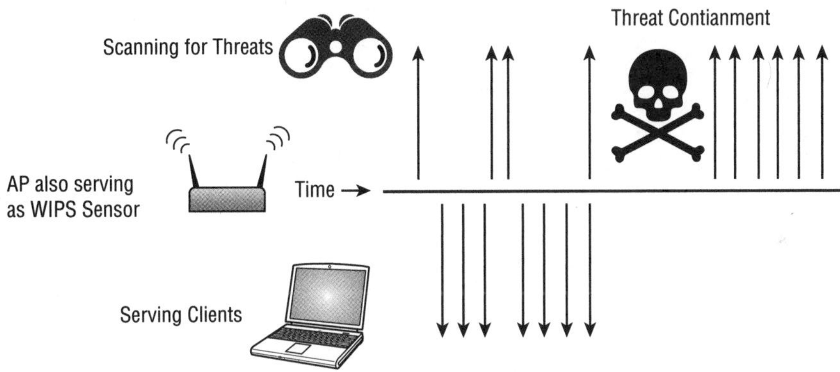

Figure 7.5: For any given radio, an AP also serving as a WIPS sensor will split its time between scanning for threats and servicing clients.

If there is an over-the-air threat, the only way to contain or mitigate it is also over the air (well, unless it's a rogue device you can unplug). In addition to the limitations of monitoring (listening), there are even more restrictions when an AP (or any radio) attempts to contain a threat. During the containment, that radio can't perform any other function, including servicing enterprise clients.

If you've seen *Ghostbusters*, the mental image of the particle thrower's stream while they're trapping ghosts is about as good as any to picture an AP "capturing" a bad actor on the wireless network.

If an AP has a dedicated security radio, it can perform both scanning and mitigation tasks without disrupting the client-serving radios.

The enterprise Wi-Fi products are usually configurable to a degree. The options change with code revisions but usually include the following:

- Set AP radio(s) to scan/monitor in downtime (when not actively servicing clients)
- Set AP radio(s) to scan/monitor a specified subset of channels or frequency (e.g., 2.4 GHz, 5 GHz, or specific channels)

- Set one or more AP radios as dedicated WIPS sensors (these radios will not service clients and this AP acts as an overlay/dedicated sensor)
- Set one or more AP radios as a dedicated spectrum analyzer (these radios will not service clients in this mode)

Granularity and options for WIPS vary by product. Juniper Mist has a limited WIPS configuration in the web UI, focusing on rogue and neighbor APs as well as honeypots, as shown in Figure 7.6. WIPS configurations in Aruba Central offer preset high, medium, low settings or the option to custom configure, shown in Figure 7.7. Alerts for WIPS events can also be configured in a different window.

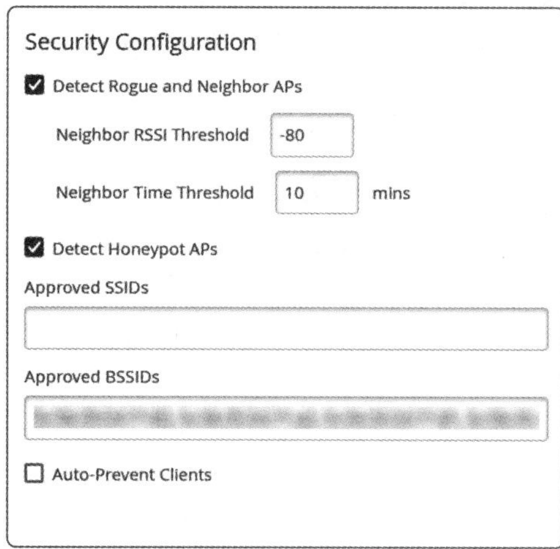

Figure 7.6: Juniper Mist has limited WIPS configuration in the web UI.

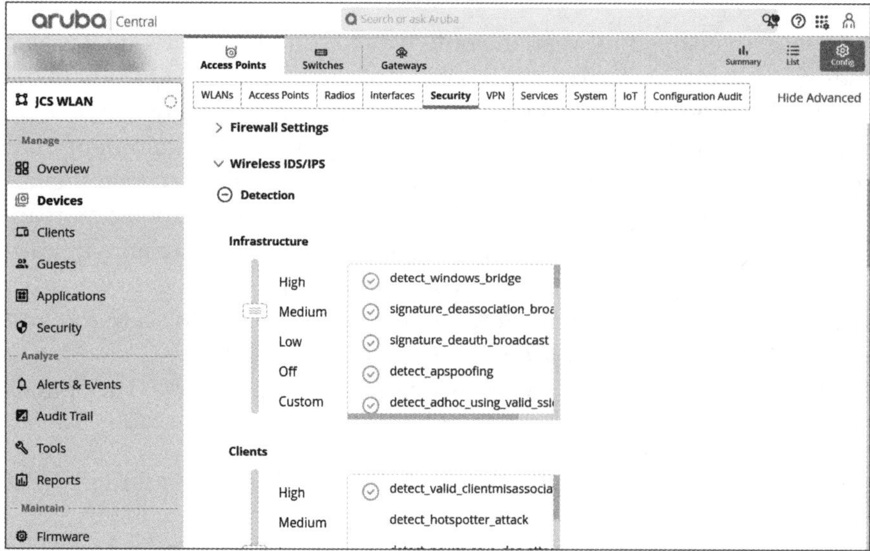

Figure 7.7: WIPS configuration options on Aruba Central cloud

What a WIPS sensor (or an AP moonlighting as a WIPS sensor) can do is dependent on what mode of operation it's in—integrated/dual function or dedicated.

INDUSTRY INSIGHT: WIPS ON CISCO, ARUBA, JUNIPER MIST, AND EXTREME

Features vary by code and platform so I'm only offering a nudge in the general direction for each of these vendors. It's pretty easy to find WIPS guidance for each. Cisco offers WIPS in its products and introduced Advanced WIPS (aWIPS) for the 9800 platform when connected to DNA Center. Prior generations require a dual-dashboard approach with the controller and Cisco Prime.

Aruba Networks' controller-based platforms offer two levels of WIPS, one with basic rogue detection and mitigation is included in the platform, and a lengthy additional set of signatures and actions is enabled with the RF Protect license. Note that Aruba refers to this feature as simply "WIP" and not "WIPS" if you're searching.

Juniper Mist offers a limited WIPS configuration in its web UI, and the feature is enabled at the site level. Most Wi-Fi 6 model APs have a dedicated third radio for threat scanning. Newer Wi-Fi 6E APs (that include a 6 GHz radio) now have a fourth radio added for threat scanning.

Extreme Networks includes a particularly robust suite of WIPS tools in its AirDefense platform. Some features are available as add-on licenses but include forensics, vulnerability scanning, WPA3 and Protected Management Frame (PMF) security, as well as Bluetooth monitoring. The WIPS solution is offered as stand-alone overlay hardware or an integrated AP.

For external antenna AP models that include a third radio for security scanning, don't forget to get the antennas! The following is a photo of an Extreme 460C tri-radio AP. The third radio is dedicated to threat scanning.

Attacks WIPS Can Detect and Prevent

Your WIPS mileage will vary depending on the deployment and products in use, but generally even the most basic WIPS will perform the most crucial tasks of identifying rogue devices and a subset of over-the-air attacks.

Attack signatures and methods have been collapsed for readability; Wi-Fi-capable WIPS often include these detection abilities:

Rogue AP By strict definition, a rogue AP is an unauthorized AP attached to the enterprise network. We'll cover rogue classification more in-depth next.

Rogue Client A rogue client could mean a couple of things depending on the vendor. Most often a rogue client is a client connected to an unauthorized or rogue AP. In other cases, a rogue (or unauthorized, not valid) client could mean any endpoint that is not associated with the enterprise Wi-Fi infrastructure.

Misconfigured AP This option is primarily designed for use when there are (for example) two separate enterprise Wi-Fi systems, and you want to instruct one system to check the parameters of the other.

Think of the prescribed parameters as being part of a secure baseline configuration—specifying the SSIDs to be served along with the corresponding security profile (e.g., WPA3-Enterprise or WPA2-Personal).

As an example, in a hospital setting, there may be a wireless controller with infrastructure APs, but there may be an overlay network by a third party in one area of the hospital using APs not attached to the controller.

The alert can also be used to report on misconfigured APs within a local cluster or using a management system with overrides (as in not controller-based).

Ad-hoc (Non-infrastructure) Wi-Fi Networks Ad-hoc networks occur when devices in the environment self-organize and establish wireless networks between themselves, bypassing the infrastructure Wi-Fi all together. Ad-hoc networks can be 802.11 WLANs, Bluetooth, or other networks.

Depending on the WIPS sensor capabilities, the system may only be able to see and report on ad-hoc Wi-Fi (blind to Bluetooth and BLE), which is especially dangerous.

Popular sources of ad-hoc networks in the enterprise include wirelessly enabled printers that advertise their own network, screen casting devices, smartphones in hotspot mode, and devices using services like AirDrop.

The infrastructure of WIPS sensors can see and monitor the RF, so even if the endpoints aren't talking to the infrastructure APs, their traffic can be heard over-the-air if the ad-hoc network is occurring in range of the APs.

AP or SSID Impersonation There's a subtle difference between impersonation and spoofing. AP impersonation occurs when an unauthorized AP assumes the identity of a legitimate AP. To do this, the malicious AP would be broadcasting the SSIDs of the infrastructure APs; this is often indication of an evil twin attack.

Aside from malicious use, it's also possible there is an enterprise-approved AP in the airspace and that AP may need to be manually accepted into the system as authorized. This is common if an organization has two different AP vendors in the same space, or within RF range of one another. For example, a main part of the building may be using a Cisco controller-based network, but in the manufacturing floor they've started migrating to Juniper Mist for the BLE location tracking. These two networks aren't designed for endpoints to roam between them, but they are both advertising the same SSIDs, and the Cisco APs can "hear" the Mist APs, and vice versa.

AP Spoofing Spoofing has to do with spoofing of a packet source, versus spoofing the identify of an AP. In this form of spoofing, an AP sends packets that look like they're from another AP. Figure 7.8 shows the difference between AP impersonation and AP spoofing.

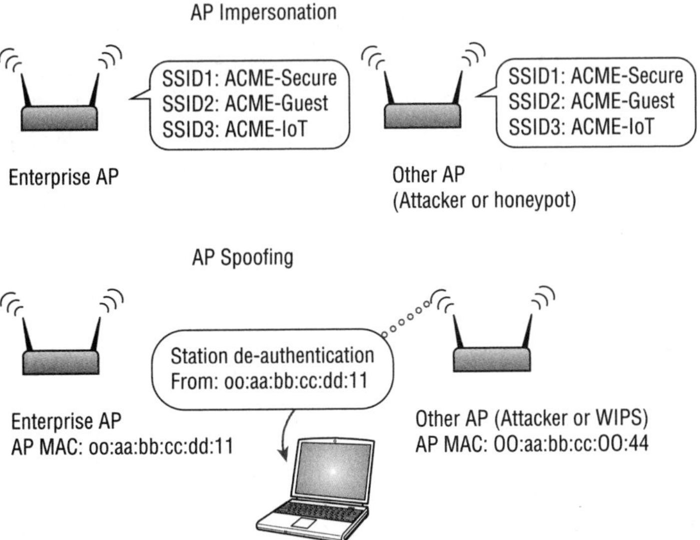

Figure 7.8: In AP impersonation (top), an AP is advertising the same networks as an enterprise AP, which is slightly different than AP spoofing (bottom) where one AP sends a packet that appears to be from another AP. The message is spoofed, not the entire AP.

I've avoided using the words "malicious" and "attack" here because spoofing has been used for many years as a legitimate mitigation technique for securing the enterprise environment.

In short, spoofing includes a forged frame that looks like it's from one AP but it's really from another. This can absolutely be one step of a malicious attack but it's also how current threat mitigations work (a topic covered shortly), when an infrastructure AP spoofs de-auth or disassociation packets from a rogue AP to keep valid clients from connecting to it.

After association (the 4-way handshake), endpoints and APs both participating in protected management frames (PMF) are protected from most AP spoofing attacks.

Client Spoofing Client spoofing behaves the same as AP spoofing, but in this case the packets are falsified to appear to be from a client (versus AP). The same model applies, and this could indicate an attack but could also be the result of a mitigation technique where an infrastructure AP again spoofs de-auth or disassociation packets, but they appear to come from the client to the rogue AP.

In addition to these two examples of spoofing, there's an assortment of related attacks with broadcast, floods, and fake beacon frames, all with the intent of introducing various denial-of-service (DoS) attacks.

Endpoints and APs both participating in protected management frames (PMF) are protected from client spoofing attacks.

Client with Invalid (Fake) MAC Address Every valid MAC address on a network interface indicates the manufacturer with the first 24-bits that define the organizationally unique identifier (OUI). For example, my laptop has an Intel® Wi-Fi NIC and it can be identified by the prefix of 60:F2:62 as registered to Intel Corporation. The OUIs are centrally registered with IEEE and are universally unique worldwide.

I always assume every network admin has extensive experience in hunting and searching down endpoints based on the OUI, but if not Wireshark's OUI Lookup Tool (`https://www.wireshark.org/tools/oui-lookup.html`) is very helpful.

NOTE **Randomized, or locally administered MAC addresses used for privacy are not considered fake MAC addresses.**

This event triggers when the OUI of an endpoint doesn't have a real (registered) OUI, indicating a fake MAC address. Randomized MAC addresses are not considered fake MAC addresses.

In recent years the use of MAC randomization has become prevalent in many endpoints including Apple phones and tablets, Android, and

Windows devices. The following image demonstrates how to identify a randomized MAC address.

The format of a random MAC address isn't really quite random; it's a locally administered MAC address as defined by IEEE so they're easily identified (see Figure 7.9)—meaning a WIPS solution should not trigger on IEEE-compliant random MAC addresses. If it is, the product may need to be updated.

- 32-28-6D-51-13-AF
- 56-EF-68-F6-0D-30
- 0A-13-A8-8E-B5-EF
- AE-83-37-55-A7-22

02-	32-	62-	92-	C2-	F2-
06-	36-	66-	96-	C6-	F6-
0A-	3A-	6A-	9A-	CA-	FA-
0E-	3E-	6E-	9E-	CE-	FE-
12-	42-	72-	A2-	D2-	
16-	46-	76-	A6-	D6-	
1A-	4A-	7A-	AA-	DA-	
1E-	4E-	7E-	AE-	DE-	
22-	52-	82-	B2-	E2-	
26-	56-	86-	B6-	E6-	
2A-	5A-	8A-	BA-	EA-	
2E-	5E-	8E-	BE-	EE-	

Figure 7.9: Image from a Cisco article detailing how to identify randomized MAC addresses. Any MAC address with a first octet that ends 2, 6, A, or E is a locally administered MAC address.

`https://community.cisco.com/t5/security-documents/random-mac-address-how-to-deal-with-it-using-ise/ta-p/4049321`

Broadcast De-authentication and Disassociation In normal Wi-Fi operations, de-authentication and disassociation frames are sent as unicast between an AP and client. Broadcast frames of these disconnect type frames are always suspicious.

It's possible some vendors may use a broadcast disconnect message to kick clients off an AP before an upgrade is initiated, but every other operation in 802.11 (e.g., channel change) has a specific type of message and it's not a broadcast disconnect.

Bulk disconnect messages are disruptive to the environment and indicative of a DoS attack.

Endpoints and APs both participating in protected management frames (PMF) are protected from spoofing attacks including forged de-auth and disassociation packets.

Anomalies That Impact Airtime and Quality Such as Those in CTS/RTS With Wi-Fi, airtime is a precious commodity as endpoints and APs take turns transmitting over the air; they do so using request-to-send (RTS) and clear-to-send (CTS) messages, among other collision avoidance coordination.

There are subsets of DoS attacks that abuse RTS and CTS and in doing so, they hog the airtime, making it impossible for legitimate clients and APs to transmit.

Vendors have different signatures for these, but in general they're looking for anomalies that indicate abuse (see Figure 7.10).

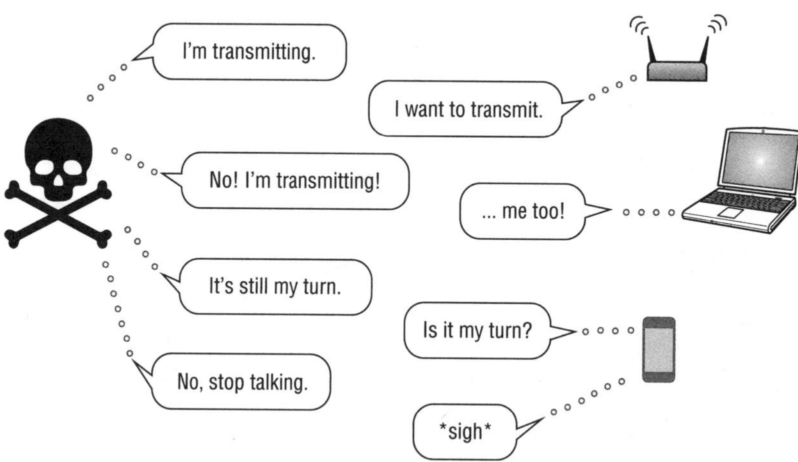

Figure 7.10: In request-to-send (RTS) and clear-to-send (CTS) attacks, airtime is used to interfere with normal Wi-Fi operations, impacting user experience and RF availability.

Malformed Packets and Fuzzing Fuzzing and malformed packets are a set of attacks that can be used as DoS attacks and/or to introduce malicious payloads (as in some fuzzing).

Fuzzing is especially nasty as data can be injected into legitimate packets (such as beacons), resulting in overflows, crashed drivers or OSs, and ultimately allow for remote code execution.

Not all vendors offer broad fuzzing signatures but they're useful in WIPS when available. In an ideal world, the endpoints' network interfaces would discard malformed or unrecognized frames instead of trying to process them.

Client with Interfaces Bridged Chapter 6 included details on why client bridging was undesirable and a security risk. Here, a bridge is defined as an endpoint that's connected two interfaces to pass traffic between them. Figure 7.11 and Figure 7.12 show how an end user can bridge interfaces in Windows.

Bridging interfaces is supported in just about every system platform with multiple network interfaces. Many WIPS products may only have signatures to detect this on Windows or a subset of OSs.

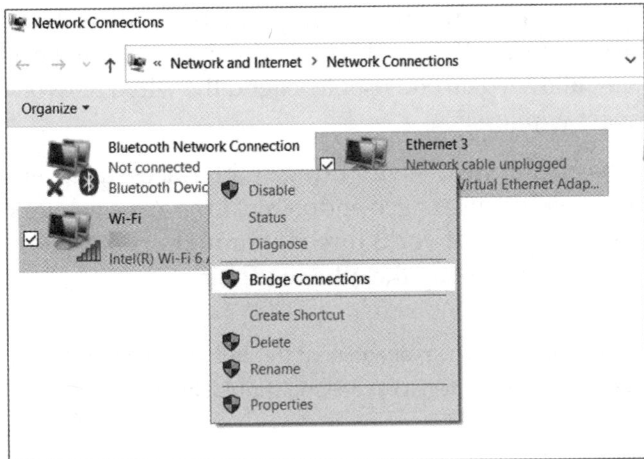

Figure 7.11: Screenshot from a Windows laptop showing the option for a user to bridge interfaces

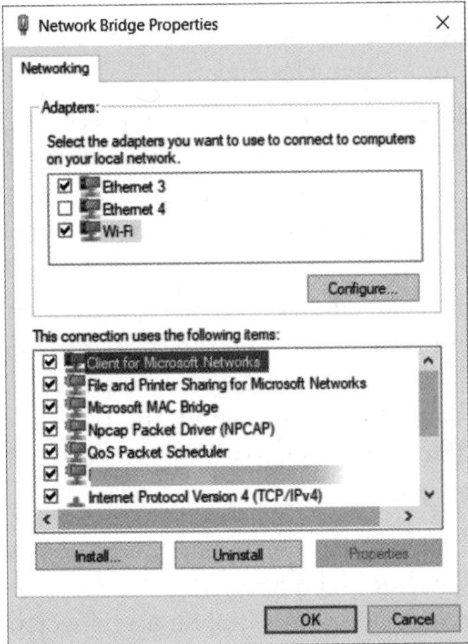

Figure 7.12: After configuring the bridged interface, an end user can specify the ways other endpoints can connect to and through their device.

If an endpoint is connected to the secure Wi-Fi (ACME-Secure SSID) and bridges the interface to its wired NIC, the laptop can offer access to the ACME-Secure SSID network to devices over the wired connection.

Note that this is not the same scenario as simply having a client that's connected both to the enterprise Wi-Fi and wired networks. Bridging can also work the other way, allowing an end user to extend the wired network to other users to connect wirelessly.

Just to make sure this is crystal clear—with bridged interfaces, the client device is allowing other (likely unmanaged endpoints) to access a network (possibly your secure corporate network) directly through its interfaces without authenticating or authorizing the other endpoints.

TIP For endpoints that are under enterprise management through domain policies or an MDM, options for end users to bridge interfaces can (and should) be disabled.

Wireless Bridges In this context, wireless bridges are connections between two or more APs designed to span a distance and bridge (for example) two buildings wirelessly. There are scenarios where an unauthorized wireless bridge may be introduced to the environment. APs acting only as bridges don't service clients, so in these cases the APs would not be classified as rogues, and the alert therefore requires its own classification.

Various DoS Attempts (Floods, Overflows) As you saw in the preceding examples, there are numerous types of attacks that may result in denial of service; these can happen starting at layer 1 (via RF jamming) up through the stack with the CTS/RTS abuse covered earlier, as well as various overflow, flood, and fuzzing attacks.

The signatures in various WIPS products are all over the place; in general I place higher value (or more risk) in events triggering the types of DoS attacks that may result in code execution or crashing devices, versus ones that will impact RF. While still an issue, if RF is impacted, the network admin will have several ways to see that in dashboards—and you can bet if they miss that the help desk will be getting calls.

Client Misassociation It's ideal to ensure enterprise-managed endpoints aren't connected to an AP or other unauthorized device. The client misassociation signatures will identify managed endpoints that have connected to rogue, external/neighbor, or honeypot APs (see Figure 7.13).

This prevents managed devices from accidentally communicating over non-encrypted networks, using unsecured or unauthorized resources, and mitigates opportunities for man-in-the-middle attacks.

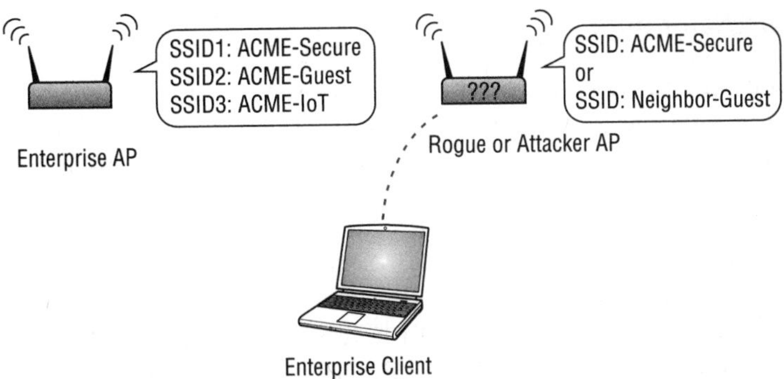

Figure 7.13: In a client misassociation, an enterprise client associates with a non-enterprise AP. It could be a rogue AP, honeypot, or an external AP.

Use of Apple AirDrop Another topic of Chapter 6 covered the security risks associated with zero configuration (zeroconf) networks including overlay protocols such as Bonjour and mDNS. Use of Apple's AirDrop protocol can absolutely introduce risks to the enterprise environment, as you saw in the last chapter. AirDrop facilitates direct file transfers between Apple devices; the discovery and connection is initiated over BLE, and the file transfer occurs over Wi-Fi. In 2021, researchers found vulnerabilities in the exchanges that occur over BLE, demonstrating it reveals the user's phone number and email address. That gap makes it trivial for an attacker to connect to a victim and deliver payloads. Assuming no attack in standard use, the Wi-Fi transfer is encrypted, obscuring any visibility into the payload; meaning the enterprise security tools can't inspect and protect an endpoint from malicious code (whether it was an intentional attack or malware).

For this reason, there are WIPS signatures designed to identify and alert on these peer-to-peer sharing instances when it occurs within range of the infrastructure APs or WIPS sensors.

Use of Attack Tools or Methods Most WIPS products will have signatures tuned to pick up on behavior indicative of common attack and cracking tools for Wi-Fi (and sometimes Bluetooth).

Specific tools you may see signatures for include Fake AP, AirPwn, Airsnarf, Chop, ASLEAP, Karma, cracking tools, Hotspotter, and Wellenreiter. Additionally other fragmentation attacks, DHCP attacks, unauthorized DHCP, and dictionary attacks can usually be detected.

Wireless Rogues and Neighbors

Wi-Fi and WIPS products use their own words for classifying devices they've discovered, and I don't feel they're always quite accurate in terms of determining risk level. Following are widely accepted definitions that more precisely reflect the risk and appropriate level of concern for each.

Authorized AP Authorized, valid, or approved APs are APs that your WIPS platform has identified as approved enterprise APs. These may be automatically discovered (especially with WIPS integrated into APs) or they may be manually identified and authorized in the system to prevent false positives.

Rogue AP Rogue APs are unauthorized APs that are connected to the enterprise wired network. Rogue APs introduce an extremely high level of risk to the network, since the unauthorized AP is likely broadcasting unfettered access to a live production wired network without proper security controls.

Rogues make their way onto the network in several ways. Most often rogues are introduced when employees or students decide the enterprise-provided Wi-Fi is not sufficient and they connect a personally owned AP (usually a consumer-grade wireless router) and configure it for their use.

As a quick aside, I've mentioned several times throughout the book that a quality RF design for Wi-Fi is crucial to security, and this scenario provides further evidence to that point.

NOTE Remember, vendors classify differently. As an example, Cisco considers any unknown AP as a rogue until it's classified by the network admin. Other vendors standardize on defining rogue devices as ones connected to the infrastructure without authorization.

Rogue APs can be prevented by enforcing edge switch port security, but due to a long list of complications, that is a rare scenario. And, while edge port security is the best way to protect the network from rogue devices, in the absence of that, detecting them over the air is second-best.

WIPS SENSORS VS. MOBILE HANDHELDS AND LAPTOPS

Remember that the theme here for over-the-air detection is that the APs or WIPS sensors have to be in the same airspace to see and hear the traffic they're inspecting. The best results are afforded when the entire space is monitored, meaning there should be WIPS sensors throughout the environment.

It also means that mobile handheld tools and laptops with software that are in the environment for short periods of time are not sufficient for full-time security monitoring including rogue detection; mobile tools are best used for troubleshooting.

Neighbor or External AP A neighbor or external AP is one that's in a close enough proximity to be heard over the air but is not determined to be connected to the enterprise network. That means an external/neighbor AP isn't a rogue, it's just something nearby, meaning it's not as high of a security risk, but can present complications and should be addressed.

To address a neighbor/external AP, simply label or flag it in the WIPS system as a known neighbor/external AP. Those likely won't move or change too much. The exception being many vehicles these days include mobile hotspots and those will come and go, creating a volume of alerts.

Other unexpected external APs may warrant further inspection. Specifically, there are two cases where external APs should be investigated.

First, if it's a user within your organization with an unapproved AP, even if it's not connected to the enterprise network, it is likely causing interference with the infrastructure Wi-Fi and should be removed. This includes mobile hotspots, MiFi devices, and anything else such as printers or casting devices broadcasting networks.

Second, an unexpected external AP could be a malicious user with either a hardware AP or a software-generated tool AP that is preparing for a Wi-Fi attack. The hope is that any attack launched would trigger one of the other signatures; this eliminates the need to attempt to monitor all external AP alerts (which is not reasonable in most organizations).

Honeypot AP Honeypots can be good or bad. Good honeypots are organization-owned and used in many forms for network security; they're designed to lure an attacker to a specific (fake) target. This allows forensics and incident response teams to study the attack methods and it can serve to lure attacks away from real production assets.

On the other hand, bad honeypots are used by an attacker to lure legitimate users to a malicious asset pretending to be an enterprise resource. In Wi-Fi applications, a malicious honeypot impersonates an enterprise AP and lures unwitting clients to it. By doing so, the malicious AP can attack the endpoint, deliver malicious payloads, or capture credentials, among other things.

If an organization has more than one Wi-Fi system or vendor in the environment, broadcasting the same networks, it will likely be categorized as a honeypot. This is a similar situation to the organizationally owned external AP described earlier.

Authorized APs are enterprise-managed or known APs in the environment. Rogue APs are unauthorized APs connected to the enterprise network. They may not be intentionally malicious but should be removed for security reasons. Honeypot APs impersonate enterprise APs in an attempt to lure clients to them, and neighbor or external APs are just innocuous APs that can be "heard" nearby. See Figure 7.14.

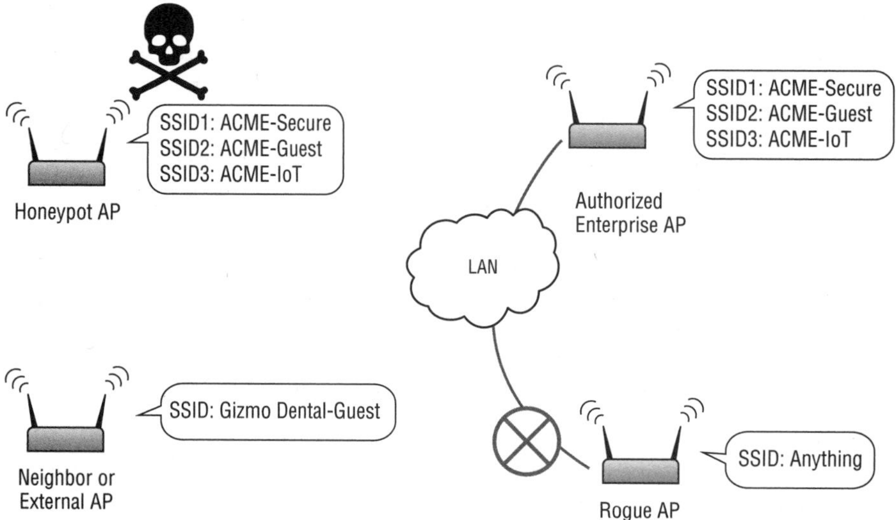

Figure 7.14: APs are commonly classified a few ways.

For detecting threats such as rogue APs, detection isn't always straightforward; it's not a binary rogue/not-rogue determination. Systems will classify APs as rogue but that may actually mean it's a suspected rogue, and for that reason many systems will layer inputs to build a confidence level.

For example, if several infrastructure APs see an external AP (an AP not within the management of the Wi-Fi system reporting it) then it may be a rogue. If the system then adds context based on seeing MAC addresses over-the-air and on the wired network from that AP, there's a much higher chance it's a real rogue.

Methods of rogue detection vary greatly from vendor to vendor; many of the methods are proprietary, and some are even patented. Also, these methods are continually evolving, which is why we're not delving to great depths on the classification methodology.

It is, however, important to understand the methods your WIPS vendor is using and ensure the environment is amicable to those. For example, products that add wired side visibility may need SNMP or other credentials to access switches and routers for polling. If the WIPS product is attempting to correlate broadcast traffic between wired and wireless, the sensor will need to have a presence in each VLAN being monitored, since broadcast traffic (unless specifically configured to do so) will not cross layer 3 boundaries. Keep in mind the VLANs you'll want to monitor aren't limited to the Wi-Fi designated VLANs; if you're looking for rogues, you want to watch all VLANs that have access to the internal assets.

The point is that each vendor will have a few conventional methods, and likely a few special sprinkles on top so do your homework and architect accordingly.

Table 7.1 offers a quick overview of the AP classifications. The *suspected rogue* category has been added to demonstrate the process most WIPS go through during classification. In these cases, the WIPS system has seen the unknown

AP over the air but doesn't have enough data correlation to determine whether it's connected to the wired network.

Table 7.1: Comparison of common rogue classifications

AP CLASSIFICATION	ENTERPRISE-OWNED	SEEN OVER-THE-AIR	SEEN OVER-THE-WIRE	NETWORKS
Authorized	Yes	N/A	N/A	Enterprise
Rogue	No	Yes	Maybe	Any
Suspected Rogue	No	Yes	No	Any
Neighbor or External	No	Yes	No	Any
Bad Honeypot	No	Yes	No	Impersonated Enterprise

FURTHER READING: WIPS FOR DUMMIES

Our friend, fellow author, and Wi-Fi guru David Coleman authored *Wireless Intrusion Prevention Systems (WIPS) for Dummies* (ISBN 978-1-119-80883-1), which is chock full of Wi-Fi WIPS knowledge.

For a more detailed look at WIPS in 802.11 networks, search for the title; complimentary downloads are provided by Extreme Networks.

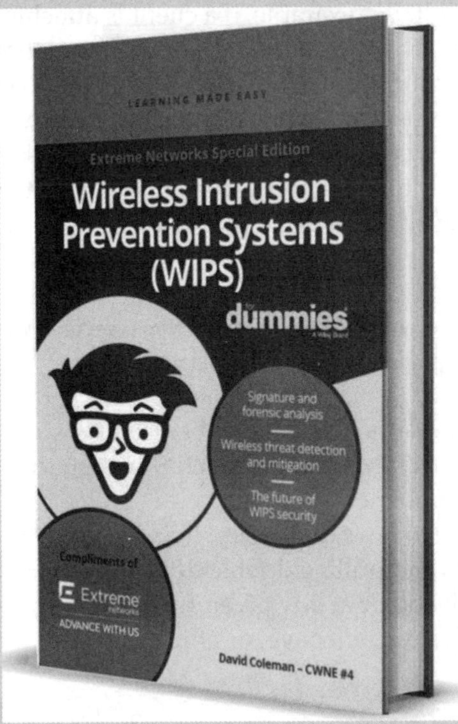

WIPS Mitigation and Containment

What happens after a WIPS identifies suspicious or malicious behavior? Depending on the product and options selected, the system may be able to perform certain containment and mitigation actions.

The WIPS system (which could be overlaid or integrated into the Wi-Fi APs) can always perform over-the-air mitigation, and some products also include a few wired network mitigations.

TIP As you'll see, rogue APs should be identified and physically removed from the network and not mitigated with WIPS.

For over-the-air mitigation, the process is simple—the WIPS will send spoofed de-authentication or disassociation packets to sever the connection between the targeted AP and client. The range of mechanisms include:

- Spoofed de-authentication packets as the AP (to the client)
- Spoofed disassociation packets as the AP (to the client)
- Spoofed disassociation packets as the client (to the AP)
- Spoofed channel change beacon as the AP (to the client, forcing the client to the wrong channel, effectively to a black hole or tarpit)

Vendors use combinations of these spoofed packets depending on the scenario and desired result. For example, if a client is attaching to a rogue AP and the mechanisms used are spoofed de-auth or disassociation packets, the client can (and will) simply continue reconnecting repeatedly to the rogue AP, and the WIPS would have to constantly work sending packets to sever the connection.

Instead, if the WIPS were to send a channel change packet, directing the client to a fake channel (really just a channel the rogue AP isn't using) then the client will continue carrying on as usual for some period of time instead of repeating the rejoin process.

TIP Remember, if the endpoint-to-AP connection is secured with PMF, spoofed frames, even from a legitimate WIPS system, will not work.

Combinations of the same packets can be used for varying forms of mitigation and containment. It's possible for a WIPS system to disable a rogue AP or rogue client in any number of ways.

Remember that due to laws of physics, in integrated systems where a Wi-Fi AP is servicing clients and pulling double duty as a WIPS sensor, the AP radios can only do one or the other at any given time. Relying on over-the-air remediation may not be desired, effective, or efficient.

In fact, in the cases of rogue APs, the action that should be taken is to identify the rogue AP, physically locate it (via wired and/or Wi-Fi locationing), and then have it removed from the network.

There should definitely be an organization-wide policy on the use of personal infrastructure devices within the enterprise environment, and that policy should address (and disallow) what would be classified as rogue APs.

Figure 7.15 depicts an over-the-air mitigation. In this example the WIPS is using both a broadcast de-authentication frame (to FF:FF:FF:FF:FF) as well as a unicast frame directed to the client (to 00:00:43:24:A4:C5). Both of the spoofed packets are sent from the WIPS but appear to be from the rogue AP.

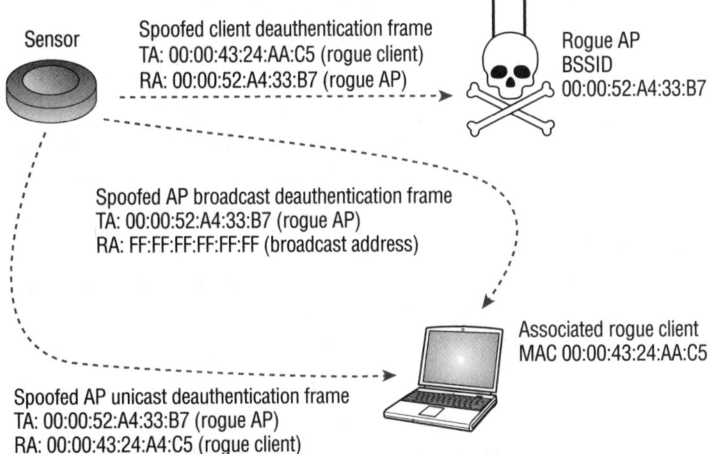

Figure 7.15: Over-the-air mitigation by a WIPS using spoofed broadcast and unicast de-authentication frames.

Wireless Intrusion Prevention Systems (WIPS) for Dummies; Wiley, 2021

The other option for mitigation is over the wired network. Wired mitigation is a feature not often found in the integrated Wi-Fi AP products but is common in overlay systems. The use of wired mitigation for rogue APs is yet another example of why holistic security architecture is so important. Our networked environments are an ecosystem and coordinating controls and mitigations across managed endpoints, the wired network, and Wi-Fi infrastructure offers a level of flexibility and security not attainable with a single security product.

In addition to WIPS, wired mitigation of rogue APs can be achieved with various network access control (NAC) products, which are able to disable edge switch ports via SNMP, CLI, or APIs. Of course, if the enterprise APs are being authenticated to the wired network as covered in Chapter 6, then edge ports using port security or 802.1X would have preemptive control over rogues. They simply wouldn't be allowed to connect and pass traffic through the wired network to start with.

Legal Considerations of Over-the-Air Mitigation

Over-the-air mitigations have legal ramifications every architect and organization should be aware of. The following example includes the Federal Communications Commission (FCC) in the U.S., but my understanding is this same model is true for most countries in Eastern Europe and North America.

Technically, blocking of wireless signals in any way is a violation of Section 333 of the U.S. Communications Act, which states that "No person shall willfully or maliciously interfere with or cause interference to any radio communications of any station licensed or authorized by or under this Act or operated by the United States Government." That statement encompasses any kind of jamming as well as the mitigation techniques described here with WIPS.

Although that clause has been in the Communications Act for over three decades, it had never been a problem for enterprises using WIPS including mitigation and containment. At least, not until there was an incident. Here's the short version of what happened in 2013 that sent the Wi-Fi industry spinning.

In 2013, Marriott used WIPS (specifically spoofed de-authentication packets) to mitigate (disable) the Wi-Fi hotspots of guests at a convention center in its Nashville, TN Gaylord Resort property. Regardless of the real reason, the perception and complaint by guests was that Marriott was attempting to force conference attendees to use (and pay for) their hotel-managed Wi-Fi at the venue, which was a non-trivial amount. Tsk, tsk. Oh, Marriott.

A complaint was officially made, a lawsuit was filed, and in 2014 the FCC formally fined Marriott $600,000 USD and issued an order summarizing the activities, the fines, and a multi-year probationary period of self-reporting. The front page of the order is shown in Figure 7.16. You can read the file by searching FCC records for file number EB-IHD-13-00011303.

In that order, the FCC cited Marriott as being in violation of Section 333 of the U.S. Communications Act. They underscored the expectation that airspace is not "owned" by any person or organization, meaning companies did not have the right to disrupt anyone else's use of that airspace, even within their own buildings and their own outdoor properties.

As you can probably imagine, the Wi-Fi and WIPS industries reeled after this finding. It's likely one of the main contributing factors to the demise of the overlay WIPS market. If organizations can't legally automate mitigation for security, offering an overlay solution doesn't add much value since the Wi-Fi APs could reasonably perform interim scanning and no enforcement means there's no requirement for multiple dedicated security radios.

However, there are legitimate use cases for wireless mitigation, and I firmly believe at some point the FCC (and possibly other regulatory bodies around the world) will need to reach an agreement with their citizens about what's appropriate. Here's why: Consider a hospital environment. Even now, they're rife with wirelessly connected systems that connect staff to electronic health records (EHR) on mobile carts and tablets, enable critical nurse calling systems, power telehealth, and facilitate lifesaving monitoring and alerting systems. Wirelessly connected devices are increasing, not decreasing, and the criticality of and reliance on these systems is also growing.

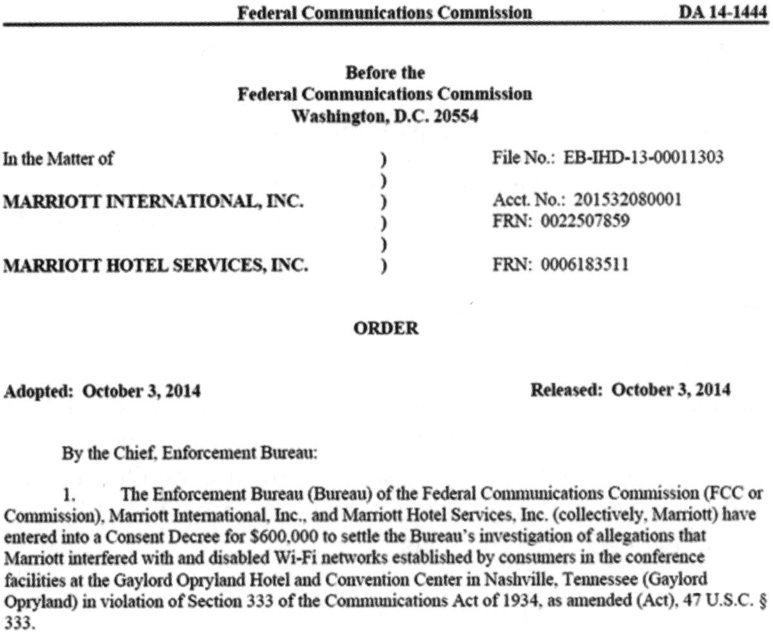

Federal Communications Commission **DA 14-1444**

Before the
Federal Communications Commission
Washington, D.C. 20554

In the Matter of	)	File No.: EB-IHD-13-00011303
	)	
MARRIOTT INTERNATIONAL, INC.	)	Acct. No.: 201532080001
	)	FRN: 0022507859
	)	
MARRIOTT HOTEL SERVICES, INC.	)	FRN: 0006183511

ORDER

Adopted: October 3, 2014 **Released: October 3, 2014**

By the Chief, Enforcement Bureau:

1. The Enforcement Bureau (Bureau) of the Federal Communications Commission (FCC or Commission), Marriott International, Inc., and Marriott Hotel Services, Inc. (collectively, Marriott) have entered into a Consent Decree for $600,000 to settle the Bureau's investigation of allegations that Marriott interfered with and disabled Wi-Fi networks established by consumers in the conference facilities at the Gaylord Opryland Hotel and Convention Center in Nashville, Tennessee (Gaylord Opryland) in violation of Section 333 of the Communications Act of 1934, as amended (Act), 47 U.S.C. § 333.

Figure 7.16: The cover of a nine-page order issued by the FCC to Marriott regarding their use of WIPS in the 2013 incident.
`https://docs.fcc.gov/public/attachments/DA-14-1444A1.pdf`

So then, in our made-up scenario, what happens in a few years from now when a malicious user walks into a hospital with an RF jammer, disrupting entire systems in the facility, and ultimately launches an attack that results in loss of life? Who's at fault here? Obviously, the attacker is at fault, but I imagine the families of the patients impacted will wonder why the hospital didn't protect itself, protect the patients, and ultimately protect human life—its most precious asset.

NOTE **On the flip side of that argument, consider the scenario that a nearby office inadvertently attacked the hospital's Wi-Fi through the business's WIPS containment. It really is a slippery slope and more complicated than checking a box to mitigate threats on a WIPS product.**

This is just my personal opinion, but if, as a society, we're saying we're ready to rely fully on wireless to connect our most critical assets, then we need to agree on ways to protect them accordingly.

De-authenticating users for monetary gain is not acceptable, but you can probably think of a thousand scenarios where mitigation is appropriate.

All of this is to say that organizations will have to make a cautious and informed decision about whether and how to use over-the-air mitigation. Like BYOD policies, this is not a topic to be decided by network or security architects or system administrators. It's a conversation about risk (legal and technical) to be had with the decision-makers (and lawyers).

Spectrum Analyzers and Special-Purpose Monitoring

Related to WIPS and security monitoring, there's a subset of technologies for monitoring layer 1 that deserve special attention before we address WIPS recommendations:

- Spectrum Analyzers
- Special-Purpose Monitoring

You'll find that many WIPS products include spectrum analyzer features, but capabilities are often limited by the type of radios built into the WIPS device.

In contrast, spectrum analyzers and other special-purpose monitoring tools are purpose-built products that can offer full visibility into a breadth of RF spectrum beyond integrated WIPS products.

Spectrum Analyzers

Spectrum analyzers are intuitively named; they use one or more radios to analyze the RF (the spectrum) and offer a visual representation of that data. Spectrum analyzers (or *spec-ans* for short) are great for looking beyond the data applications of wireless and seeing the entirety of the radio frequencies within a given band. They're often used in enterprises by Wi-Fi professionals to identify sources of non-Wi-Fi interferences (such as microwave ovens, heavy equipment, or other emitters).

Some WIPS systems include a spectrum analysis function, while others only monitor the slices of airspace in use for data transfer. If a WIPS is only concerned with layer 2 (non-RF) based attacks on Wi-Fi, it may scan through the specific Wi-Fi channels (such as 1, 6, and 11 on 2.4 GHz and 36, 40, etc. on 5 GHz) and report on security incidents within those ranges.

Other products may not only scan the known or in-use Wi-Fi channels but also watch the rest of the spectrum for interference and anomalies. Some products can be configured to do both; the extent to which a system can operate as a spectrum analyzer will depend on the radio configuration and is tied heavily to whether the sensor is integrated or overlaid.

A spectrum analyzer operates at layer 1 (RF for Wi-Fi, 802.11 PHY), a protocol analyzer at layer 2 (802.11 MAC), and packet analyzers at layer 3 and above, as shown mapped to the OSI stack in Figure 7.17.

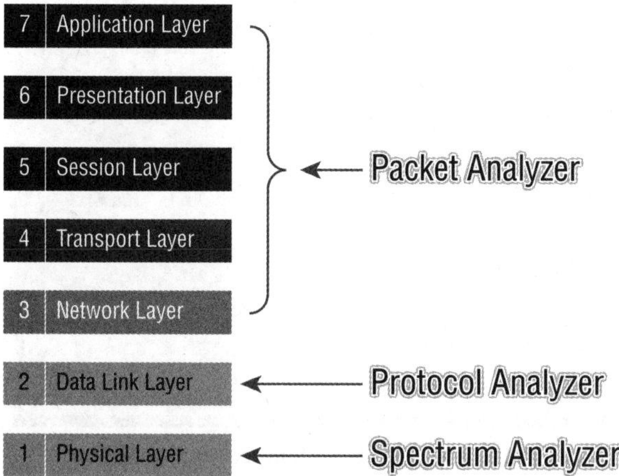

Figure 7.17: Spectrum analyzers, protocol analyzers, and packet analyzers operate at different layers of the OSI network stack.

From a security perspective, even without the decoding of 802.11 packets, a spectrum analyzer can provide insight into layer 1 RF-based events that may be impacting the availability of the enterprise's wireless networks.

Although I've mentioned 802.11 a lot here, the beautiful thing about spectrum analysis is it can happen on any radio frequency that the radio(s) supports. As you'll see next, spectrum analyzer hardware can be paired with accessories and software to create a powerful RF monitoring and security tool.

Many of the products in the niche monitoring space are built with software-defined radios (SDRs), which makes them easily reconfigurable to support new use cases and technologies.

As one example, the MetaGeek Wi-Spy hardware dongle radio is a spectrum analyzer. Combined with its Chanalyzer software, the spectrum (RF, layer 1) can be viewed along with the Wi-Fi networks (802.11 MAC, layer 2), as shown in Figure 7.18. Spectrum analyzers capture the RF utilization over a specified spectrum (based on radio capabilities) and display it as values over time using colors similar to a heatmap.

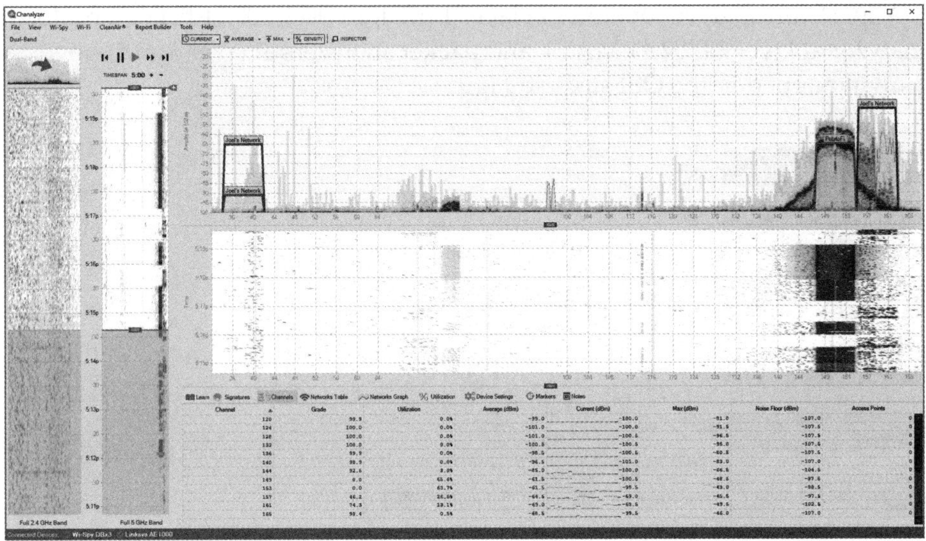

Figure 7.18: Visualization of spectrum analysis overlaid with Wi-Fi channel information on MetaGeek Chanalyzer

Some of the more common spectrum analysis tools in enterprise use today include:

- Spectrum analyzers built into Wi-Fi and WIPS sensors (many enterprise APs can be converted to operate as spectrum analyzers within their radio space with some limitations)
- NetAlly® AirMagnet® Spectrum XT, supports 802.11 and Bluetooth (`www.netally.com`)
- MetaGeek Wi-Spy + Chanalyzer, support 802.11 (`www.metageek.com/`)

The second and third bullets are representative of products that include software and a dongle (usually USB) with a radio. At time of writing most products still only support 2.4 GHz and 5 GHz radios with plans to release 6 GHz shortly. Products described for special-purpose monitoring (next) include products that extend well beyond Wi-Fi and Bluetooth.

NOTE **As with all vendor solutions mentioned in this book, the products and features will evolve and change so consider this list a starting point for research.**

NetAlly's AirMagnet Spectrum XT software offers many ways to view and analyze the RF spectrum (see Figure 7.19). Spectrum analysis software such as this one relies on access to a radio with capabilities in the range to be analyzed (see Figure 7.20). The adapter contains the radio required for spectrum analysis and can be used with a laptop or tablet and the Spectrum XT software.

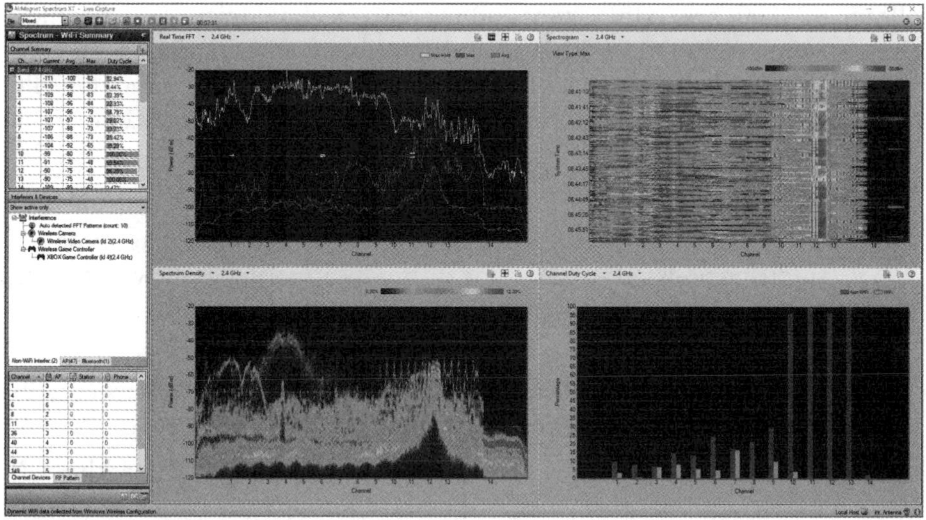

Figure 7.19: NetAlly's AirMagnet Spectrum XT software
www.netally.com

Figure 7.20: NetAlly dual band USB spectrum analyzer

Special-Purpose Monitoring

Some of the technology presented here may be new to many readers as they've primarily been reserved for federal and military applications. As Wi-Fi expands beyond 2.4 and 5 GHz into 6 GHz, novel 802.15.4 IoT use cases emerge, and private 5G and neutral host bring cellular technologies inside our walls, enterprises will find themselves in need of more robust monitoring than the typical 802.11 WIPS.

For that reason, I've chosen to include some of these more niche products for consideration as enterprise needs change and demand for over-the-air security grows.

The products and companies here are provided as an example of the types of technologies and solutions on the market today; inclusion here is not an endorsement, but this list has been complied with feedback from professional colleagues who work in this space and use these technologies. Also, the more robust monitoring for enterprise is an emerging and volatile market; some companies listed may have been acquired or changed direction throughout the life of this book, but you can use the terms for Internet searches to find comparable solutions.

Epiq Solutions Epiq offers wireless device detection and security alerting products that cover cellular, Bluetooth, and Wi-Fi. It also makes products for enforcing no-wireless zones and its products can integrate with security and incident management platforms such as SIEM and SOAR. (`https://epiqsolutions.com`)

CFRS CFRS provides military-grade RF monitoring for intelligence gathering and spectrum management and geolocation tools. (`www.crfs.com`)

Signal Hound Signal Hound provides spectrum analyzers that address RF from 1 Hz to 12 GHz frequencies along with its Spike software for decoding and analysis. (`https://signalhound.com`)

Recommendations for WIPS

Now that we've explored every aspect of WIPS including layer 1, we'll resume the WIPS discussion with recommendations. When it comes to planning WIPS in the security architecture, there are a lot of inputs to factor and additional research is always required. Implementations, depth of visibility with signatures, and options for mitigations vary by vendor. Also, the organization's security requirements and comfort with mitigation plays a major role in planning WIPS workflows.

Many compliance frameworks include requirements for some form of WIPS capabilities in the scoped environment—specifically alerting, not mitigation.

As you learned with the FCC ruling (and similar examples outside the U.S.), organizations may be limited by what they can do in terms of mitigation (legally) and that situation has made it a bit easier to rely on integrated AP WIPS sensors in years past.

Recommendations for WIPS can be summarized as follows:

- All organizations will benefit from visibility of WIPS, and it should be enabled (it's disabled by default in most Wi-Fi products).
- To start, use the visibility and alerting features only, and do not configure mitigation or containment unless authorized by the organization to do so.
- Like any IPS tool, WIPS will require maintenance, especially early in the project for tuning.

- Focus alerting on the most critical events, such as rogue APs to alleviate alert fatigue by removing or tuning down inconsequential or false positive WIPS events.
- Rogue APs should be addressed by physically removing them from the network, and there should be an organizational policy disallowing connection of personal APs to the enterprise network.
- Neighbor or external devices within your buildings but not attached to the network (such as smartphone hotspots, printers, and ad-hoc networks) can cause RF interference and impact availability. Reinforce desired behavior within your organization and ensure the enterprise Wi-Fi quality and coverage is sufficient to eliminate the need for users to supplement.
- Tune the WIPS appropriately for your environment; for example, you don't need to be emailed WEP weak IV alerts if you're not using WEP in the environment, and you may be less concerned about suspected but unconfirmed rogues if your infrastructure is authenticating APs to switch ports or uses NAC. As another example, in universities with a large personal device population, the decision may be made to ignore ad-hoc networks and AirDrop.
- Use mitigation and containment features only when approved formally by the organization and use them only for the most critical security concerns, remembering that integrated AP and WIPS products will be limited in containment capabilities.

There are additional security alerts we're interested in outside of the WIPS-detected events just covered, such as any incidents where admin access to the system is being brute-force attacked. The next major section of this chapter is "Logging, Alerting, and Reporting Best Practices," where I cover not only more recommendations for WIPS alerting, but also these other security events outside the purview of WIPS.

Synthetic Testing and Performance Monitoring

There exists a subset of wireless sensors that don't fall into the WIPS category but are worth mentioning for their usefulness in ensuring wireless availability and, in some cases, security.

These products fall into the categories of *synthetic testing* and *performance monitoring* products. They're wirelessly enabled devices that sit within the environment and are coordinated to test specific functions of the wireless network. Performance monitoring encompasses many aspects of Wi-Fi such as testing and monitoring for speed/latency, bandwidth, network services performance, and air quality.

The synthetic test devices are called such because they act as a client device would—connecting to the enterprise Wi-Fi and testing access, latency, and availability of predefined services and applications using fake or synthetically generated traffic.

The purpose of synthetic testing is to emulate the end-user experience and proactively alert on any issues that may result in a poor user experience. Products are most often hardware (sensors that look like APs) and occasionally offered as software agents that can be installed on a variety of endpoints including Android, Apple iOS, Windows, and Linux. See Figure 7.21.

Figure 7.21: Vendor 7Signal offers an app-based agent and hardware sensors including a permanent fixture model that looks like a small AP and a more portable device.
`www.7signal.com`

For example, synthetic testing and performance monitoring products can be used for:

- Monitoring and validation of layer 1 RF availability
- Monitoring various air quality metrics such as retries and airtime utilization
- Performance monitoring of network services such as DHCP and DNS
- Testing throughput to internal or Internet resources
- Testing access to secured networks (using a passphrase or 802.1X credentials)
- Testing access and latency to internal resources including applications with logins
- Testing access and latency to external and Internet resources
- Testing and validating SSID security settings
- Identifying a subset of WIPS events such as rogue APs

INDUSTRY INSIGHT: SYNTHETIC TESTING VENDORS

Currently there are a handful of vendors in the Wi-Fi performance and synthetic testing space. Each of the vendors offers a slightly different value proposition.

Wyebot (`https://wyebot.com`) is the newest product on the scene, offering performance monitoring, synthetic testing, and a subset of security monitoring features similar to a scaled-down WIPS system.

7Signal (`www.7signal.com`) has both hardware- and software-based solutions aimed at performance monitoring to help organizations meet strict service level agreements (SLAs) for Wi-Fi connected applications.

Aruba Networks (`www.arubanetworks.com`) acquired Cape Networks several years ago and incorporated their products into the Aruba User Experience Insights (UXI) product, which is focused on performance and synthetic testing, as shown in the following illustration.

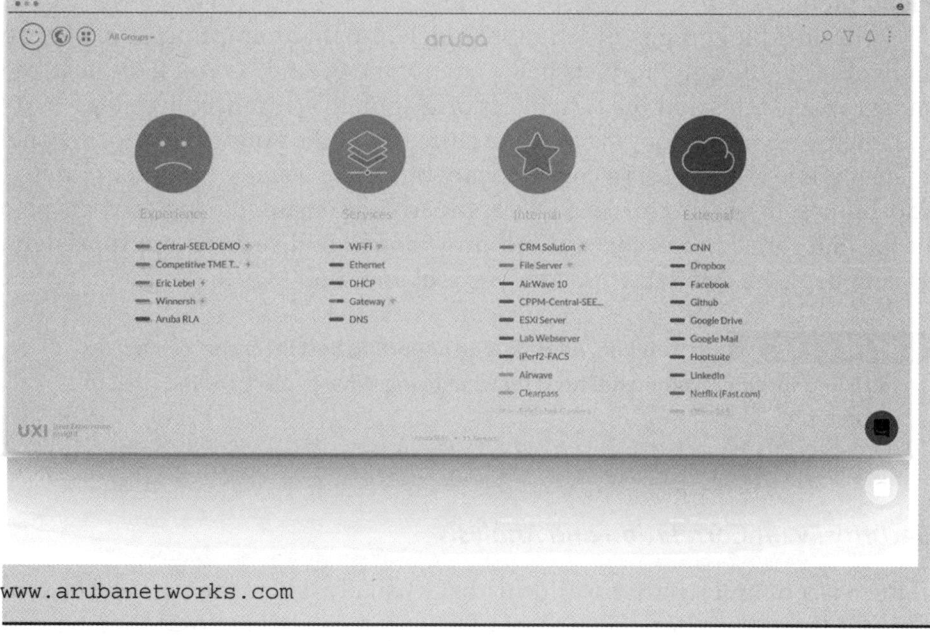

`www.arubanetworks.com`

Security Logging and Analysis

Congratulations. You have WIPS and other monitoring tools in place. Now, what do you do with all that data and alerting?

If your organization has a centralized logging and security analytics platform, it's ideal to integrate it with the Wi-Fi networks. There are meaningful differences between basic logging and security event analysis and correlation.

Security Event Logging

As with many topics in this book, centralized logging is a minimum requirement in most compliance frameworks.

Most enterprise Wi-Fi products support the option to send logs, alerts, or traps of various kinds to a log server. Event logging serves several purposes including:

- Capturing and memorializing significant changes to the infrastructure devices (such as configuration changes, software updates, and reboots)
- Capturing security events for further processing or alerting
- Logging administrative access for audit purposes
- Logging administrative access attempts for security alerting

Exceptions to products that support this type of logging include API-based cloud products with no SNMP or local logging.

Centralized logging is much more efficient than attempting to configure individual alerting on products in the environment. And, as you'll see next, the more capable tools add the benefit of correlation to centralized logging.

If there's no robust logging product currently, and no budget to procure one, there are free tools to serve as a stopgap, but these usually have very limited capabilities. If you do choose a DIY approach just ensure the log server is part of the enterprise infrastructure and provisioned and secured appropriately; meaning, please don't download a free tool and send logs to your laptop.

REFERENCE **In "Logging, Alerting, and Reporting Best Practices" coming up, you'll find specific recommendations for what to log, what to alert on, and what to report on.**

Security Event Correlation and Analysis

If there is a central security platform, that's usually where the logs will be sent for ingestion. This classification of products goes well beyond logging and includes event correlation, analysis, sometimes parsing, and workflows for incident investigation and response.

Tools that fall within the category of security event correlation and analysis include an alphabet soup of acronyms such as:

- SIEM (security information and event management)
- SOAR (security orchestration, automation, and response)
- XDR (extended detection and response)

By the time you're reading this book, there will likely be more new acronym bullets. I'll spare you the detailed nuances that differentiate these products but will offer a bit of explanation.

SIEM

SIEM is the next step up after basic logging. SIEM tools most often parse incoming raw data (such as alerts) and then correlate those alerts from various sources to identify possible security incidents and offer some context. SIEMs ingest alerts as well as raw text such as configuration files.

For example, a basic logging tool would indicate repeated failed logins for a domain admin. With the context of other events, a SIEM would be able to see that the repeated failed logins were on a database server with a "sensitive" data label, that event was followed by a successful logon, a configuration change in the database server, and all of that occurred minutes before an abnormally large volume of data left the network.

Whereas the logging tool could have alerted on a brute force attack of the original login attempts, the SIEM tool can correlate data and alert on an attack and data exfiltration with a high level of confidence. Once the SIEM has alerted security operations (SOC) teams of the event, they will investigate and begin containment and remediation.

SOAR

SOAR extends what SIEM offers by extending features beyond the incident alert. In our example scenario, a SOAR tool would continue the process of the investigation by walking the security analyst through the investigation with predefined playbooks, workflows, and automation.

In addition to the raw logs, alerts, and configuration files, SOAR tools integrate with many third-party systems; they ingest data flows such as threat intelligence feeds and connect to other parts of the infrastructure to facilitate automated remediation tasks.

XDR

XDR solutions fall into one of two buckets—*native* and *open*. The original value proposition of *native XDR* was that it would integrate different platforms within the same vendor ecosystem for a more cohesive and plug-and-play experience. Targeted for small organizations without full SOC teams and analysts, native XDR promises ease of use with a scaled-down SIEM/SOAR-type experience. The obvious benefit is that these XDR products would be easy to implement and maintain, with the downside of limited third-party integrations. The result is that native XDR may offer visibility into and protection from only a top tier of attacks.

Out of the original XDR scheme arose products designed to integrate with other vendors, and *open XDR* was born for organizations that wanted more flexibility. The open XDR products are designed to solve the vendor lock-in inherent with native XDR while taking advantage of the next generation of automated security response.

XDR is still a baby in the infosec market and at time of writing hasn't even reached the entry of the hype curve in adoption, but it's likely to continue to grow (or diverge into yet another acronym) in the coming years.

Wireless-Specific Tools

The wireless monitoring chapter wouldn't be complete without an honorary mention of the most popular wireless-specific tools.

The technologies covered here are not designed for security monitoring specifically, but they do play a role in maintaining a secure wireless infrastructure.

The tools can loosely be classified into four categories:

- Handheld Testers
- RF Design and Survey Software
- Network Protocol Analyzers
- Testing and Troubleshooting Applications

Here, I share specific vendors and products for context and because the Wi-Fi tools and testing market is not so volatile, so with few exceptions, these products or their next generation will be available for years to come.

Handheld Testers

Handheld testers are wonderful because they're small, portable, and very practical for sending onsite instead of having to dispatch a senior network admin for troubleshooting. By the way, I'm not covering spectrum analyzers here since they've been addressed earlier, but there are some handheld tools and applications that are purpose-built spec-ans.

Handheld tools with testing functions are designed to give technicians an easy way to test and validate Wi-Fi. Some of the checks and functions they perform include:

- Airtime availability and quality
- RF noise levels and RF interferers
- Network and network services availability (e.g., DNS, DHCP)

- Connectivity metrics such as throughput
- Security settings of the SSID
- Detection of possible rogue devices
- Verifying access rights and segmentation
- Path analysis to a specified target (internal or Internet)
- Over-the-air packet capture

Just as with spectrum analysis and WIPS, handheld wireless testing tools must have radios that support the RF spectrum to be analyzed. For Wi-Fi products, that's 2.4 GHz, 5 GHz, and now 6 GHz.

There are many handheld testers and different industries, especially in telecom, have specific brands and tools commonly used. For the rest of us, the products most often used for Wi-Fi testing come from NetAlly (`www.netally.com`). Earlier in this chapter when we covered WIPS, I provided a brief history of NetAlly. You may recognize some of its tools from being branded Fluke Networks or NetScout. NetAlly is well known and supported in the industry and has a strong community of Wi-Fi professionals.

NetAlly has a suite of wired and wireless testing tools, but the two prominent for wireless purposes are its AirCheck G2 and EtherScope nXG products (see Figure 7.22). In full disclosure, I'm particularly fond of the EtherScope nXG; it carries a hefty price tag but is, as they advertise about, as close as you can get to a "portable network expert."

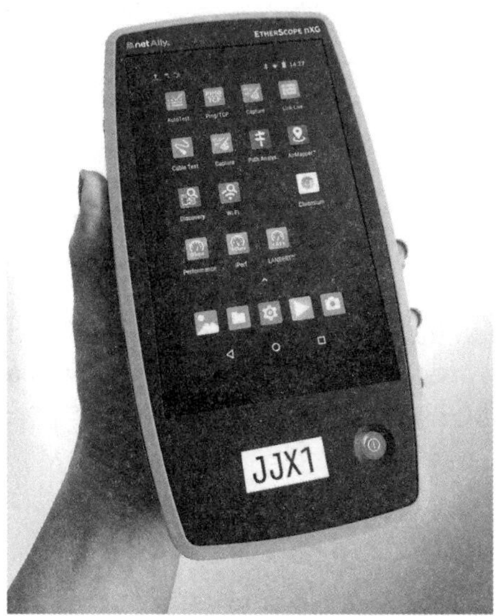

Figure 7.22: The NetAlly EtherScope nXG is a great handheld tool for testing both wireless and wired, and includes discoveries, tests, and reports that can be used for security analysis.

Among other things, these handheld testers are great for identifying security misconfigurations that result in access that should be blocked. In the testing sequences, you can configure it to report as pass or fail depending on the result. For example, one test parameter may be to verify a user can access an internal server, and if that's successful it's a "pass" result. A separate test may verify that that user cannot access another network or resource, and if the user is able to access it (via the test) then it can be flagged as a "fail."

INDUSTRY INSIGHT: AN ACCIDENTAL SECURITY ASSESSMENT WITH A NETALLY ETHERSCOPE NXG

I, like many others, have a lot of fun stories that start with "I was traveling and had my EtherScope with me. . ." In one such story, I was evaluating some of the reporting available from the EtherScope and had the device with me while out to dinner one night. To get sample data, I just connected to the city's downtown public Wi-Fi network, clicked on some auto-tests, and let the device run a generic discovery.

Just so we're clear—the discovery does not include any methods that involve attempting to access systems; it does not launch an attack, nor does it attempt to log in to anything. It's just looking.

So, you can imagine my surprise when I noticed the public Wi-Fi was configured in a way that it was bridged to the local municipal internal network, and in my very light touch discovery I was able to see wired desktops, laptops, servers, and printers at the courthouse, fire department, and one other municipal office. I could also see their internal switches, routers, and firewall as well as the entire public Wi-Fi infrastructure—including APs that were configured with default SNMP management.

It took a few days and a proxy introduction via one of my clients to get a response, but I was able to contact the network admin at that city and let him know what I stumbled on.

RF Design and Survey Software

There are two sides to the coin of RF planning—there's RF design and survey, which can be two separate vendors, or two products from the same vendor that are integrated.

RF design software estimates coverage and includes output such as predicted coverage heatmaps, while RF survey software takes live measurements with a radio and documents the actual coverage and quality in a space. *Design* is predictive and *survey* is actual coverage.

RF Design Software

RF design software uses a predictive modeling algorithm to estimate the RF coverage of a specified radio. In Chapter 4, "Understanding Domain and Wi-Fi

Design Impacts," and throughout other parts of this book, I espouse the importance of a proper RF design on wireless security. Because it's such a critical piece of wireless architecture, I'm covering it in more depth than other tools.

RF design accounts for not only basic coverage but also roaming, use cases, and endpoint capabilities, and a proper RF design addresses:

- Basic RF coverage of the space
- RF coverage requirements for seamless roaming
- Roaming and key exchange capabilities of the endpoints
- RF coverage requirements for location services (if in use)
- Requirements of the applications to be used
- Form factor, antenna, and radio capabilities of the endpoints
- Antenna and radio capabilities of the AP (analysis is specific to each model AP)
- Orientation and mounting height of the AP
- Building materials of all scoped spaces

This is only a partial list of design considerations. To use RF design software for predictive modeling requires an accurate and scaled floorplan, details of all building materials for the software to calculate attenuation, and knowledge of Wi-Fi design theory.

Many organizations make the mistake of purchasing a design tool but not sending network or wireless architects to any training on RF design theory. Even a cavewoman can drop APs on a floorplan and make the building look green. For that reason, in my experience roughly 80 percent of the time (or more) when an untrained professional creates a design plan, even for basic carpeted office space, it doesn't meet the actual requirements and there are problems later.

RF DESIGN SOFTWARE VENDORS

At time of writing there are three prominent and well-established RF design vendors. Ekahau's design tool (Ekahau Pro) and what's now NetAlly AirMagnet Planner focus on 802.11 Wi-Fi and 802.15.4 BLE design. A third vendor, iBwave, also has an option for cellular design. Here's a quick rundown of those three.

The benefit of AirMagnet Planner from NetAlly is that it's integrated with their cloud platform and handheld products and includes a free viewer, which makes it easy to share designs and heatmaps with team members and stakeholders without exhaustive exporting exercises.

Ekahau offers a suite of popular survey and design tools and is known for its training programs. The Ekahau Wi-Fi design course teaches not only the product, but deep RF design theory required for quality results.

iBwave currently offers two design products—iBwave Wi-Fi and iBwave Design. The latter includes design tools for both Wi-Fi and cellular wireless.

With the growth of cloud-managed Wi-Fi products and API-driven integrations, you can expect to see feature changes and new products emerge in this space.

The goal of RF design software is to ensure proper coverage and quality of the wireless signal against predefined requirements. An example is provided in the following illustration. Chapter 4 and Chapter 5, "Planning and Design for Secure Wireless," included additional details on why proper RF design is critical to wireless security architecture.

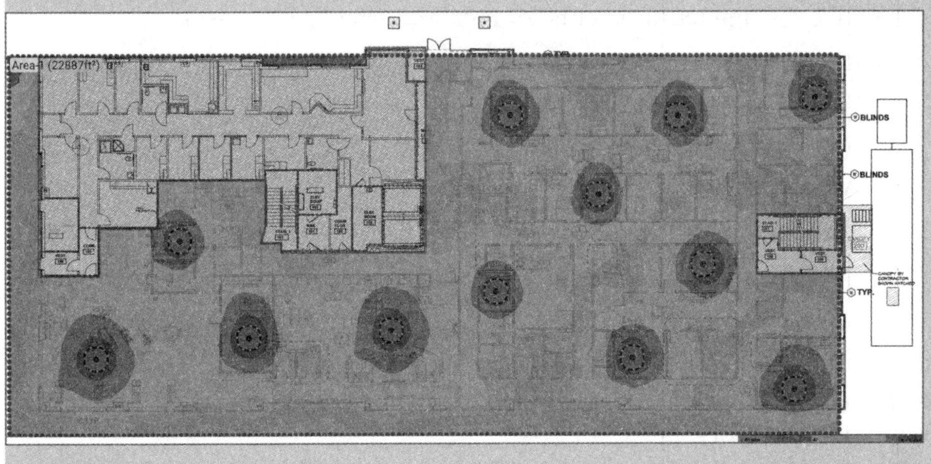

RF Survey Hardware and Software

RF survey tools are a set of hardware (antennas) and software that, used together, take readings from an environment, and visually represent the RF coverage and quality.

The same vendors of RF design software also have survey products that include either handheld tools (such as NetAlly) or antenna kits or purpose-built hardware (such as Ekahau's Sidekick Pro) for this purpose. Software runs on a laptop, tablet, handheld, or smartphone, which is connected to the hardware with the survey radios. Figure 7.23 is a heatmap from a live RF survey. The AP placements represented here are estimated based on location algorithms in the tool. The darkest gray areas designate spaces where the signal quality did not meet specifications.

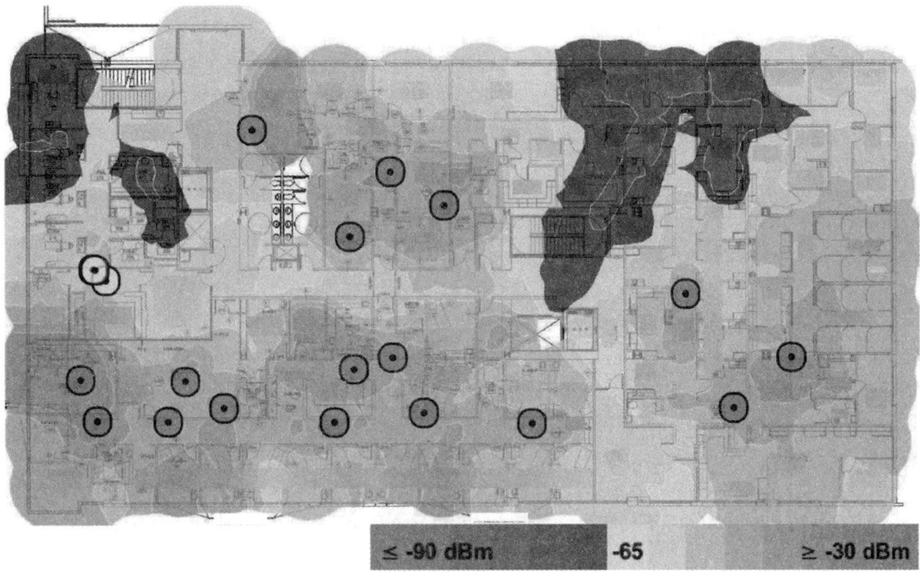

Figure 7.23: Heatmap output from a live RF survey

Network Protocol Analyzers

Network protocol analyzers offer unprecedented insight into the inner works of wireless, starting at layer 2. Again, as with any tool that's helping us over the air, use of a protocol analyzer relies on getting input from something that has radios capable of monitoring the RF in scope.

Depending on platform capabilities, packet captures can be accessed and viewed nearly real-time or offline as a static file.

The most common network protocol analyzer is Wireshark and there are plug-in color filters specifically for different wireless protocols to make life a bit easier (see Figure 7.24). Wireshark is supported by community volunteers and made available as a free download for Windows and Mac.

Cloud-hosted network vendors may also use cloud-hosted solutions for packet analysis, such as CloudShark. In fact, some cloud-hosted management platforms support auto-packet capture and direct integration with a cloud-hosted analysis platform.

Testing and Troubleshooting Applications

Along with the more formal tools and products mentioned thus far, there are scores of other testing and troubleshooting applications, many of which run on Apple iOS and Android smartphones or tablets. They come, they go, and there are too many to cover individually here, but many Wi-Fi and networking blogs and podcasts cover these supplemental tools.

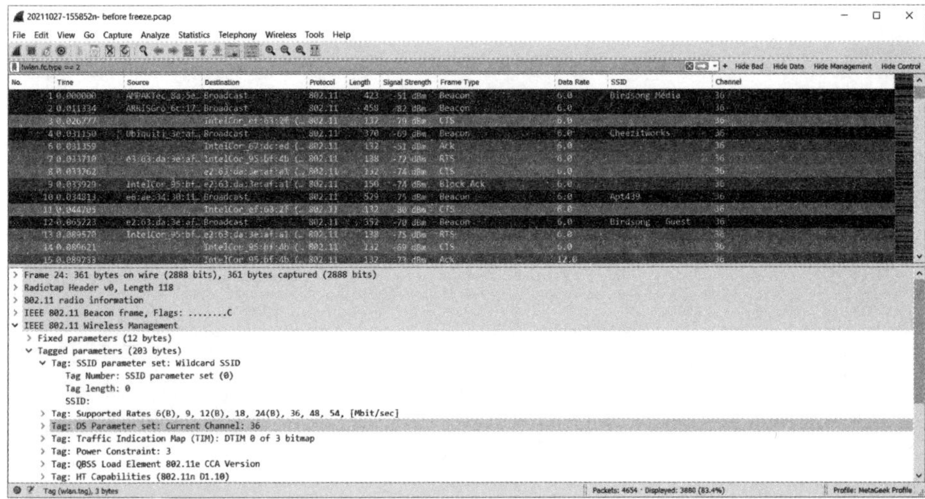

Figure 7.24: A Wireshark Wi-Fi packet capture can parse from layers 2 and higher to analyze protocols and packets.

At time of writing, a few popular tools include:

- MetaGeek (`www.metageek.com`)
- WLAN Pi (`www.wlanpi.com`)
- EAPTest (app available for Apple iOS)
- Wi-Fi Analyzer and miscellaneous apps available for Android)

Logging, Alerting, and Reporting Best Practices

Recommendations for logging, alerting, and reporting take the information covered in this chapter so far and combine it with Chapter 6 for a holistic approach to monitoring wireless for security.

REFERENCE **Following are suggested minimum events to log, alert, and report on; they're broken down by category following the topics of Chapter 6 with the added data from WIPS covered in this chapter.**

Peer-to-peer and bridged communication recommendations are included with the client security and other WIPS headings. Recommendations are laid out in the following order:

- Events to Log for Forensics or Correlation
- Events to Alert on for Immediate Action
- Events to Report on for Analysis and Trending

Table 7.2 provides a summary of the recommendations.

Table 7.2: Summary of recommended logging, alerting, and reporting by type

Events to log:	Events to alert on:	Events to report on:
SECURE MANAGEMENT ACCESS		
■ All successful management access ■ All failed management access attempts	■ Brute force login attempts ■ Anomalous logins	■ Presence of default users or credentials ■ Use of unencrypted management protocols
INFRASTRUCTURE INTEGRITY		
■ Configuration and software changes ■ Reboots of all infrastructure devices ■ All physical access ■ All other configured WIPS events	■ Rogue or honeypot APs ■ AP or controller offline ■ DoS attacks ■ Failed attempts of AP authentication to switch ■ Changes of an AP MAC or IP address ■ Unauthorized physical access or attempts ■ Expired certificates ■ Misconfigured AP ■ Devices that support UPnP	■ Configuration changes ■ Trending adoption of PMF ■ New APs or infrastructure devices ■ AP groups and configurations ■ Use of unused/disabled protocols ■ Expiring certificates
CLIENT SECURITY AND OTHER WIPS		
■ Airtime anomalies ■ Any client-related WIPS events ■ Allowed client traffic between zones ■ Attempts of clients accessing unauthorized resources	■ Ad-hoc networks ■ Client bridges ■ Malformed packets or fuzzing ■ Broadcast de-authentication or disassociation ■ Use of known attack tools	■ Trending adoption of WPA3 ■ Client mis-associations ■ Use of zeroconf protocols ■ Use of AirDrop and similar

Events to Log for Forensics or Correlation

The general rule of thumb, if you have central logging, is to log anything and everything short of informational traps. Ideally the logged events are ingested and correlated with security tools such as a SIEM, but in the absence of that, more log details increase the chance of success in forensics investigations following an incident.

Some logging platforms have costs based on the volume of data, so balance logging with budget as appropriate.

Secure Management Access

- Log all successful management access
- Log all failed management access attempts

Always log all management access attempts, both successful and failed. This supports event correlation and enables the system to alert on brute force attacks to the management of the system.

If your platform supports it, log discrete commands issued as well. This is possible with certain TACACS+ deployments and/or with privileged access management (PAM) tools. Both of these topics were covered in more detail in Chapter 6.

Infrastructure Integrity

- Log configuration and software changes
- Log reboots of all infrastructure devices
- Log all physical access
- Log all other configured WIPS events

Log configuration and software changes, which can be used to identify unauthorized configuration changes, and to trigger an event to match a configuration against the approved baseline.

Logging reboots of the infrastructure devices (wired and wireless) helps correlate outages and other user-impacting events.

Physical access to network closets and datacenters should be secured to the degree possible, and for systems that support identity-based access (such as card readers) that access should be logged for event correlation. In some platforms, physical location of users is an input for several triggers, including alerting if a user has physically entered one area of a building but logged in from another.

Also log all available events from the WIPS system, even if you may not be alerting on them all. This, too, helps with event correlation, and something that seemed innocuous may turn out to be activity preceding an attack.

Client Security and Other WIPS

- Log airtime anomalies
- Log any client-related WIPS events

- Log specially authorized endpoint traffic
- Log attempts of clients accessing unauthorized resources

WIPS-generated events related to airtime anomalies (such as CTS/RTS) may be false alarms. These events, when malicious, are specifically indicative of DoS attacks versus delivering payloads or infiltrating systems. For this reason, you may not want to alert on these events, but the data should be available for investigation later.

If your platform supports it, log attempts of endpoints accessing unauthorized resources along with logging access to specially authorized resources between security zones. For example, if traffic from a guest network to an internal network is not allowed, but there's a policy allowing access to specific assets (such as a printer or screen casting device) log the allowed traffic. Logging disallowed and allowed traffic can help during incident response and forensics.

Events to Alert on for Immediate Action

There are several events that may call for immediate investigation or action, and therefore warrant having an alert trigger. Alerting can come in the form of an email to a user or distribution group, automatic generation of an internal ticket, or be triggered via SOC workflows for immediate security analysis.

SIZING IT UP: LOG AND ALERT APPROPRIATELY FOR YOUR NEEDS

The volume of alerting a person or organization can handle is dependent on the resources available. If you're a one-man-band, use alerts sparingly and only for the most critical events such as rogue APs. Alerts don't do any good if they're ignored, and alert fatigue is a real thing.

For large environments with hefty resources, additional alerting can be managed through teams, a SOC, and/or tools that support automated response actions.

The events here are a recommended starting place; please adjust appropriately based on your capacity and environment.

Secure Management Access

- Alert on brute force login attempts
- Alert on anomalous logins

At a minimum, you want to be alerted if the system is experiencing any form of brute force attack, including (and especially) attacks against management logins. If the platform doesn't support this natively, this is one place to get a little digital duct tape and make it happen.

For more mature security programs with anomaly detection engines, it's ideal to also trigger immediate alerts on anomalous logins—for example a login from an IP address or network that deviates from the norm.

Infrastructure Integrity

The following alerts are loosely organized in order of precedence if your workflows and resources can't handle all alerts:

- Alert on rogue or honeypot APs
- Alert on AP or controller offline
- Alert on DoS attacks
- Alert on failed attempts of AP authentication to switch
- Alert on changes of an AP MAC or IP address
- Alert on unauthorized physical access or attempts
- Alert on expired certificates
- Alert on misconfigured AP
- Alert on devices that support UPnP

Many of these alerts will depend on a third-party tool of some sort—logging or SIEM. However, every Wi-Fi platform I know of supports basic alerting for rogue APs and you definitely want to enable that directly or via another tool. Along the same vein as strict rogues, alert on anything in the rogue family such as honeypots and wireless bridges (which may not appear as rogues because they're not beaconing).

I strongly suggest alerting on any infrastructure device that goes offline unexpectedly. There are many organizations and particularly universities that ignore offline APs because of the sheer volume in their network. Even knowing that, in large organizations with tens of thousands of APs, there are easy ways to flag maintenance windows and I advise security-conscious clients to alert on these events. It can be a precursor to something more, and even if it's not a security issue, it will impact users and should be addressed quickly.

If there's a WIPS function, it's good to alert on the DoS suite within WIPS (which will vary from a handful to dozens of signatures). These may require tuning to reduce false positives and they can be omitted in many cases if the tuning is not sufficient. If there's an outage, most environments will hear about it from the user population if nothing else.

The next two alerts speak to the risks of an unauthorized user accessing the network via the edge ports serving APs. If an AP is authenticating to the network, and repeatedly fails, that's a good indication someone has removed the AP and is attempting to connect to the network with its port. Similarly, in the absence of port security, if the AP suddenly has a different MAC address or IP address that also is an indication someone is attempting wired access through the AP's network drop. NAC products do a great job with profiling that adds a layer of protection against MAC spoofing for scenarios just like this. Again, it doesn't matter where the alert comes from—e.g., Wi-Fi product, log server, SIEM, or NAC.

Expired certificates can (and do) wreak havoc. In a perfect world, you'll be tracking infrastructure certificates and reporting on them proactively, but if somehow that got missed, an immediate alert is warranted before an entire subset of the wireless infrastructure is deemed unusable.

Alerts for misconfigured APs can be handled a few ways. The goal is to prevent any enterprise APs from operating in a manner less secure than designed. These alerts may be generated via WIPS (if supported) or can be configured through a SIEM that can compare current and prior configurations and alert on changes or deviations from a baseline.

Lastly, we've talked about the horrors of universal plug and play (UPnP), which reconfigures firewalls to allow unplanned inbound traffic. UPnP devices can be found by monitoring for Simple Service Discovery Protocol (SSDP) traffic. If your Wi-Fi or WIPS monitoring tools don't offer this, it's available in vulnerability assessment tools that offer network monitoring. SSDP is yet another zeroconf protocol designed for residential use.

You should also be alerting on added infrastructure devices—new DHCP servers, new routers, or any new device with OUIs from residential router manufacturers. Find some way to ensure these devices aren't on your network, ever.

Client Security and Other WIPS

- Alert on ad-hoc networks
- Alert on client bridges
- Alert on malformed packets or fuzzing
- Alert on broadcast de-authentication or disassociation
- Alert on use of known attack tools

Alerting on ad-hoc networks is appropriate for security-conscious organizations but isn't for everyone. Universities, retirement homes, certain healthcare facilities, and in general networks with residential areas are likely to see volumes of ad-hoc networks. In addition, printers and screen casting products within enterprise environments may also use ad-hoc networks; these are the types of events you want to alert on and investigate. It may take a bit to get them under control, but in an enterprise, direct Wi-Fi printing should not need to be enabled. The same goes for screen casting tools. Those should be investigated and monitored; it's possible to implement them securely without using ad-hoc networks, and if the decision is made to allow the ad-hoc connections, that should be monitored (perhaps not alerted on though). Also, remember that these extra (non-enterprise networks) use up precious airtime and affect RF quality for all Wi-Fi in the area.

I can't think of a legitimate use case to allow client bridging in an enterprise environment. The risks of this feature were detailed earlier in the chapter and client bridges should be alerted on and remediated immediately. Remediation can entail contacting the user and instructing them to cease; it could be auto-remediation by reconfiguring a managed endpoint, or it could be to denylist the device MAC address(es).

Malformed packets and fuzzing are attacks that may crash drivers, disrupt services, and allow an attacker to deliver a malicious payload to clients. Given the severity of these attacks, immediate alerting is warranted.

There's no valid use of broadcast de-authentication and disassociation frames in normal Wi-Fi operation. If your WIPS system sees this activity, it's worth investigating. It falls just below alerting on fuzzing since these attacks (by themselves) will simply result in DoS.

Lastly, it may be advisable to alert on any WIPS events tied to use of known Wi-Fi attack tools. Again, these may require a bit of tuning and some attack tools may not be effective against your infrastructure so pick and choose accordingly to alleviate alerts that don't apply.

Events to Report on for Analysis and Trending

We've got logging; we've got alerting. Next comes reporting, which is helpful for events that don't require immediate attention but warrant further analysis and to monitor trends over time.

Trending analysis is particularly helpful for organizations increasing security by migrating from WPA2 to WPA3 secured networks. Trending allows the admins to monitor the uptake of the new protocols including protected management frames (PMF, 802.11w) and follow capabilities of endpoints over time. This in turn informs the organization on when it may be appropriate to remove the legacy protocols from the environment.

In general, if there are events covered in the alerting section that you don't feel warrant immediate attention, you can configure reporting on these events and investigate them at some regular cadence, such as weekly or monthly.

Many of these items can be reported on in several ways, including through a network management platform (as opposed to a security platform). Enterprise network management tools often include built-in security audit reports designed to meet many of these reporting needs. As an example, the tools in this category would include several of the SolarWinds products, Cisco Prime and DNA Center, HPE Intelligent Management Center (IMC), and similar.

Secure Management Access

- Report on presence of default users or credentials
- Report on use of unencrypted management protocols

Maintenance of wireless security includes continuous reporting of the posture of the systems to ensure integrity is maintained. One such vital report should identify the presence of any default users or credentials, including (and especially) default SNMPv2 community strings (notably, "public" and "private") along with any default user accounts. Default accounts include management accounts and users such as "admin" or "root" along with vendor-provisioned accounts for help desk or other users.

Along with checking for default users, reports should identify whether unencrypted management protocols are in use—specifically looking for HTTP and Telnet as well as unencrypted file transfer protocols.

Infrastructure Integrity

- Report on configuration changes
- Report on trending adoption of PMF
- Report on new APs or infrastructure devices
- Report on AP groups and configurations
- Report on use of unused/disabled protocols
- Report on expiring certificates

As part of maintaining infrastructure integrity, configuration changes (planned or otherwise) should be reported on at a regular cadence. If configuration changes are reported on and investigated appropriately, many of the following reports won't be needed other than as an occasional audit. As referenced in the intro to this reporting topic, many products include a security audit feature with built-in report templates that cover most if not all of the items here. These reports may be run quarterly (or as needed) to ensure the posture of the infrastructure and any compliance mandates.

For trending analysis and migrations, adoption of protected management frames (PMF, 802.11w) should be tracked. The migration plans must also entail upgrading the endpoints to add this capability.

New infrastructure devices (planned or otherwise) should be reported on, including the presence of new APs. Ideally that report should be reviewed to ensure new APs are in the proper groups; if not, other means of reporting or alerting should be used.

Just as with the audits to ensure unencrypted management protocols are disabled, there should also be reports occasionally to confirm other unused protocols are in fact still disabled. This might include discovery protocols and certain routing protocols as an example.

Expired certificates usually cause quite a mess, and therefore reporting on expiring certificates early (before they're expired) is ideal.

Client Security and Other WIPS

- Report on trending adoption of WPA3
- Report on client misassociations
- Report on use of zeroconf protocols
- Report on use of AirDrop and similar

If your organization is migrating to WPA3, reporting on the adoption of WPA3 is helpful for trending analysis. At some point, decisions will need to be made around when to sunset WPA2-secured networks, or when to reduce the scope of WPA2 networks to the subset of devices unlikely to be upgraded.

If your organization doesn't have a specific plan, start with quarterly reports for migrations and use each report to target populations of endpoints that require driver updates or hardware refreshes to take advantage of the WPA3-secured networks and PMF.

Client misassociation indicates a managed endpoint that's connecting to unmanaged, external, or unknown networks while in the organization. This may be an indication that users are trying to bypass enterprise security or content filtering. Generally, this doesn't require the immediate action of an alert, but should be addressed for security hygiene.

Highly secured networks won't include zero configuration (zeroconf) protocols such as Bonjour and mDNS, and they won't allow ad-hoc and peer file sharing protocols like AirDrop. For those organizations, alerting on these may be in order but at a minimum for any organization it's preferred to at least report on them at some regular interval to understand how prevalent they are in the environment. Even if these protocols are allowed by policy, it would behoove an organization to understand if managed endpoints are participating, and to evaluate the risk regularly.

Troubleshooting Wi-Fi Security

Some days, it feels like we're living in that TV reality show "Wi-Fi Gone Wild." And, if it's not an RF issue, it's usually a security issue that requires troubleshooting.

Here, we focus on a few specific topics including:

- Troubleshooting 802.1X/EAP and RADIUS
- Troubleshooting MAC-based Authentication

- Troubleshooting Portals, Onboarding, and Registration
- Troubleshooting with Protected Management Frames (PMF) Enabled

Troubleshooting 802.1X/EAP and RADIUS

Troubleshooting Wi-Fi happens one of two ways—either through checking and validating logs and settings, or through packet captures. Packet captures are very powerful and are a staple for most Wi-Fi professionals. However, over the years I've encountered only a handful of network admins at client organizations that routinely used packet capture software (such as Wireshark). Also, as a consultant having to work in other people's environments often remotely or without easy access to take packet captures, sometimes it just wasn't an option.

I do highly recommend training (such as the CWNP CWAP discussed later in this chapter in "Training and Other Resources") to get familiar with packet captures if you're not already. Once you get past the initial hurdles of understanding the software and how to get and filter the captures, a little up-front effort will yield years of smooth sailing.

For those that just don't want to deal with packet captures, or who don't have access to captures, there are many easy ways to troubleshoot 802.1X, EAP, and RADIUS.

Things to Remember

We covered these protocols in depth earlier, but I realize Chapter 3, "Understanding Authentication and Authorization," was five chapters ago. Before you dive into troubleshooting, here are a few reminders about 802.1X, EAP, and RADIUS:

- 802.1X is an IEEE standard for port security on several media types including (but not limited to) 802.11 Wi-Fi.
- Extensible Authentication Protocol (EAP) is a standard for authentication and manages the key exchanges for encryption and is used with 802.1X for secured networks.
- 802.1X secured networks define mutual authentication, which means the server has to authenticate itself to the client (always with a certificate) in addition to the client authentication.
- A minimum configuration for an 802.1X secured wireless network involves the client, SSID, and an AAA server.
- AAA stands for authentication, authorization, and accounting (logging) and is most often a RADIUS server.
- WPA2-Enterprise and WPA3-Enterprise secured SSIDs use 802.1X with EAP.

- The EAP authentication happens between the client and the authentication server.
- Communication between the client and AP uses EAP protocol; the AP repackages them as RADIUS to communicate to the authentication server.

Things to Troubleshoot

There are twelve easy-to-check tasks for troubleshooting 802.1X, EAP, and RADIUS. The diagram enumerates where to check each item—at the RADIUS server, the AP/controller, wired network, or endpoint device.

Many of these requirements and additional details were covered in detailed in Chapter 3. Revisit that content for more detail.

Figure 7.25 is a holistic view of the 12 troubleshooting steps, which are divided into three groups. Some checks can be performed in the Wi-Fi infrastructure, some on the wired network, two on the endpoint, and seven can be checked on the RADIUS server itself. These tasks are listed ordinally, but you may not have admin access to all parts of the infrastructure included in the steps. If you only have access to the Wi-Fi system and the endpoint, then those are the first troubleshooting tasks for you, as indicated in Figure 7.25. The troubleshooting tasks are as follows.

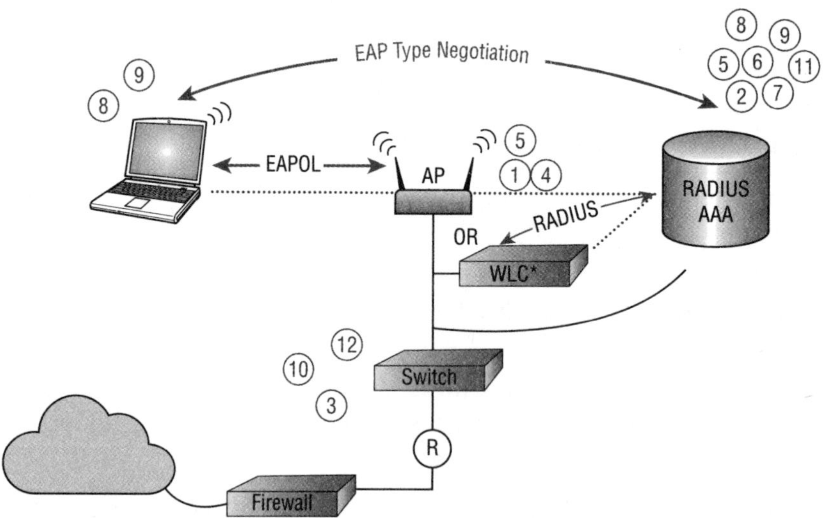

Figure 7.25: Connected components with numbers designating each troubleshooting step or location

If you're not seeing any activity at or from the RADIUS server, perform these five checks first:

1. Enable RADIUS accounting on the Wi-Fi system.

   Depending on the product, if accounting (logging) is not configured, you'll have nothing to check to investigate failure. Specifically, if you're using Microsoft NPS this is a manual setting within NPS.

2. Verify the Wi-Fi devices are allowed RADIUS clients.

   In the RADIUS server, ensure the Wi-Fi APs, controller, and/or AP network are configured as allowed RADIUS clients. If not, the system will not process requests from them. You should see this in the logs as well.

3. Test network layer 2/3 access between Wi-Fi devices and RADIUS server.

   If the Wi-Fi devices can't communicate with the RADIUS server, requests will not make it to or back from the server. The device(s) to check are the APs or controller acting as the 802.1X Authenticator. That will depend on your architecture and configurations. See Chapter 3 for more.

4. Verify the SSID is configured for 802.1X and the correct RADIUS server and ports.

5. Check and re-enter the RADIUS shared secret between RADIUS server and Wi-Fi devices.

   If logging is enabled, the RADIUS server will alert if there's an incoming request from an unknown RADIUS client.

If authentication requests are getting to the RADIUS server but the authentication is still failing, check these next four items:

6. Check the RADIUS server certificate.

   If the server certificate is not valid (expired), or the endpoint is configured (through group policy) to only allow specific certs, and that's not one, the authentication will fail, usually without a specific error indicating why. The same may happen if the RADIUS server was provisioned with a wildcard certificate.

7. Review the RADIUS server logs for failed reason code.

   Some RADIUS and NAC products will be very clear about the failure with a text explanation. If working with Microsoft NPS or similar, you may need to research the Microsoft reason code (available online).

   For example, NPS reason code 16 indicates authentication failed due to a user credentials mismatch. Reason code 22 means there's an EAP mismatch between the endpoint and the RADIUS server policy. Reason codes 48 and 49 mean the incoming request did not match an NPS network policy or NPS connection request policy, respectively.

8. Verify the configured EAP type of the endpoint matches what's allowed in the RADIUS server.

   The RADIUS policy may allow several EAP methods, such as PEAP with MSCHAPv2 as well as EAP-TLS or EAP-TTLS. The endpoint will probably be configured to use one; just ensure that is allowed in the RADIUS policy.

9. Check the client authentication user format.

   Most systems are pretty forgiving but if the logs indicate the user is unknown check the format, such as user@domain.com or `domain.com/user`. Also ensure the RADIUS server has connectivity to directory services it's relying on to verify the users.

The last group of checks are appropriate if the authentication was successful, but the endpoint still doesn't have network access:

10. Verify the endpoint's DHCP and DNS.

    If the endpoint wasn't able to get an IP address through DHCP, or it obtained an address but has a DNS server that's not reachable, it will not have network access after the authentication. If there's no IP address, retrace the path through VLAN and IP helpers to DHCP server to find the cause.

11. Check for a dynamic VLAN assignment in RADIUS server policy.

    If the RADIUS server is configured to return a dynamic VLAN (or any dynamic authorization such as downloadable ACL) it may not be correct; that network may not be fully configured; or there's a mismatch in the format of the VLAN with how the Wi-Fi expects to see it (e.g., VLAN ID versus name).

12. Check the endpoint's network access to the default gateway.

    If the endpoint has an IP address and reachable DNS server, check its access to the default gateway. If the gateway is not reachable, it's possible the IP address is in the wrong network, or another misconfiguration occurred on the wired side.

Troubleshooting MAC-based Authentication

MAC-based authentications can be used for endpoint authentication to Wi-Fi networks as well as for AP authentication to the switch. And MAC authentication, whether through a local port security policy or MAC Authentication Bypass (MAB) usually fails because of something simple. Following are three tips for troubleshooting MAC authentication.

MAC Address Formatting

Requirements for setting the MAC address format vary by product. Some just auto-magically work, while others require the delimiters to be specified to match the MAC addresses in the database (wherever that may be). The format specifies whether a MAC address is parsed as "aabbccddeeff" (no delimiter), "aa:bb:cc:dd:ee:ff" (multi colon), or any number of formats. The most common formats are outlined in Table 7.3.

Table 7.3: MAC address delimiter formats

DELIMITER FORMAT	EXAMPLE
No Delimiter	aabbccddeeff
Single Dash	aabbcc-ddeeff
Multi Dash	aa-bb-cc-dd-ee-ff
Multi Colon	aa:bb:cc:dd:ee:ff

NOTE **Most products are not case-sensitive, but a few do allow the delimited format to be specified along with a case (uppercase or lowercase).**

This is one quick and easy thing to check. The logs will show how the MAC address is coming in, and it's a trivial task to look at the database and see how they're represented there. Most vendors will include delimiter instructions, if required, in their configuration guides.

MAC Authentication Bypass AAA Settings

MAB is a fall-through sub-function of an 802.1X AAA configuration on a device (switch, AP, etc.). If MAB is not working properly and other parameters have been checked, it may be the AAA settings are not accurate.

Figure 7.26 highlights the complexity of port-level configurations of 802.1X with MAB on a Cisco switch. I'm using Cisco as an example, but all vendors suffer this challenge due to the nature of 802.1X and the fact that it's a port-level configuration.

Similar settings for an SSID are more streamlined but if you're troubleshooting APs not authenticating to a switch, this serves as a perfect example. This output shows an example of a port-level config for 802.1X with MAB, and includes instructions such as:

1. First, instruct the port to attempt to authenticate the endpoint with 802.1X.
2. Then, if 802.1X times out, attempt to authenticate with MAB:
   - Prefer 802.1X over MAB
   - Periodically reauthenticate

3. If the RADIUS server is unreachable, reinitialize to VLAN XX.
4. Reinitialize the voice VLAN on the port.

```
interface GigabitEthernet1/0/24
  switchport access vlan 200
  switchport mode access
  authentication event fail action next-method
  authentication event server dead action reinitialize vlan 200
  authentication event server dead action authorize voice
  authentication event server alive action reinitialize
  authentication host-mode multi-auth
  authentication open
  authentication order dot1x mab
  authentication priority dot1x mab
  authentication port-control auto
  authentication periodic
  authentication timer reauthenticate server
  authentication timer inactivity server
  authentication violation restrict
  mab
  dot1x pae authenticator
  dot1x timeout tx-period 10
```

Figure 7.26: Sample port-level configuration of a switch using 802.1X with MAB fall-through

In this example, the port configuration is taking advantage of multi-auth to allow multiple endpoints on the port (common with laptops daisy-chained with VoIP phones). With any vendor deployment there may also be temporary code to allow for a phased deployment.

In these cases, it's best to check with your vendor (switch or Wi-Fi) and consult their documentation or validated design guides. 802.1X gets especially persnickety on wired ports and behavior may be unpredictable as vendors (all vendors, not Cisco specifically) include operations that are off-standard. MAB itself is not a standard, nor is it part of the 802.1X specification, meaning each code revision or product may vary slightly in operation here.

Settings on the RADIUS and Directory Servers

Authenticating a device based on its MAC address as a username and password sounds far more straightforward than it is in practice.

By virtue of using a MAC address as both a username and password, a few basic security principles are violated and may require some tweaking. These include:

- Most policies disallow using a password that is equal to or similar to the username
- Most policies require the password to be rotated on a specified schedule (such as every 90-180 days)
- Most password policies require password length and complexity

These may require the directory or group policy to accommodate an exception for these devices.

Also, older products including prior versions of Microsoft IAS RADIUS required setting the account password to allow reversible encryption in order for MAC authentication to work.

Troubleshooting Portals, Onboarding, and Registration

Portals are used for guest registration, onboarding, and other device registration, such as BYOD.

While captive portals seem like a simple concept, the reality is they're usually constructed in house of cards manner with many interdependencies between redirects, DNS entries, certificates, and endpoint behavior.

For architectures with a captive portal built into the Wi-Fi product directly, the chaos is a bit more manageable. For solutions that integrate the Wi-Fi product with an external portal, it gets messy. Chapter 3 covered certificate requirements for captive portals and Chapter 4 touched on the use of DNS in captive portals.

There are scenarios with multi-tiered Wi-Fi infrastructure and external portals where the client may be required to have a certificate trust relationship with multiple different entities during the process.

To add complexity, the end-user experience with a portal depends heavily on the behavior or the endpoints—which changes regularly. Some endpoint devices won't allow connections to HTTP and will require HTTPS, and therefore, certificates. Some endpoints allow the user to bypass certificate warnings or to accept an unknown or invalid certificate; others won't. Some endpoints are okay with redirects, others aren't. Some are okay with DNS redirects but not HTTPS redirects, and vice versa. The list goes on.

Because of this complexity, the product- and implementation-dependent nature of portal requirements, and the ever-changing behavior of endpoint devices, I'm refraining from specific guidance other than to say portal issues are usually related to one of the following:

- Certificate on AP and/or controller
- Certificate on captive portal (if external)
- DNS entries related to the portal and/or controller
- Redirect methodology of the infrastructure
- Novel behavior within an endpoint device

Troubleshooting with Protected Management Frames Enabled

Working with protected management frames (PMF) will change troubleshooting substantially for some network admins and monitoring systems.

Unicast traffic exchanged between the AP and endpoint after the 4-way handshake will be encrypted, meaning third-party tools and passive sniffers won't have visibility into the exchanges. This may impact the ability to see (for example) disassociation and de-authentication reason codes or certain channel change messages.

For full visibility, the device or tool providing troubleshooting will need to have knowledge of, or access to, the encryption keys. The Wi-Fi infrastructure components (APs and/or controller), and packet captures created by the Wi-Fi infrastructure will be the best source of data for troubleshooting PMF-protected endpoints in many cases.

In addition, connections protected with PMF won't be susceptible to spoofed de-authentication and disassociation attacks. That also means over-the-air mitigation from WIPS will not work for PMF-protected endpoints and APs.

Training and Other Resources

Wireless is its own beast and own vocation on top of wired networking. And it changes constantly. As such, it's always recommended to keep the tools sharp with ongoing training and participation in the community.

I offer the following as suggested resources for additional learning beyond this book. Most of the courses and organizations here are long-standing and I expect they will persist for many years, but humbly ask for your forgiveness if these resources change in the coming years. If nothing else, they should serve as ideas and starters for searching. Resources here include the following categories:

- Technology Training Courses and Providers
- Vendor-Specific Training and Resources
- Network and Cyber Security Training
- Conferences and Community

Technology Training Courses and Providers

Aside from vendor-specific training for Wi-Fi products, there are a few offerings that are vendor-neutral, or contain a large volume of vendor-neutral content including:

- Wi-Fi Training and Certification
- IoT Wireless Training and Certification
- Network Security Training

The following training programs are offered only as a sample of what's out there. The commentary for each is designed to offer some context as to the content and how it relates to the topics throughout this book.

The beautiful thing about the technology industry is that whether you're interested in networking, wireless, or security, you'll find a ton of options and offerings including free content online in the forms of blogs, whitepapers, podcasts, and videos, in addition to paid training programs.

Also remember that many of these programs have both a training and certification component. If you're interested in education but not certification you can always take certification courses and use preparatory materials but elect to not take the exam.

NOTE **The training programs presented here are current as of time of writing and will change in the coming years. The main contents of this book will outlive its specific product and training info. As such, this list is meant to offer a starting point for research and ideas for continued learning in the topics of wireless and security.**

Wi-Fi Training and Certification

Certified Wireless Network Professionals (CWNP) is a training and certification organization that's part of Certitrek. CWNP has a suite of programs for vendor-neutral 802.11 WLAN (Wi-Fi) networking including:

- CWNA® - Certified Wireless Network Administrator
- CWSP® - Certified Wireless Security Professional
- CWDP® - Certified Wireless Design Professional
- CWAP® - Certified Wireless Analysis Professional

The CWNA is considered the entry-level certification and is required for recognition of any of the specializations (Design, Security, or Analysis). Although foundational, the CWNA content is by far the most extensive in the CWNP portfolio and is challenging due to the pure breadth of topics covered. Prep materials for CWNA teach all basic concepts of 802.11 including RF theory, modulation and encoding, antennas, network design and architecture, as well as troubleshooting.

The CWSP focuses on Wi-Fi security, but its content differs greatly from this book's. CWSP focuses more on security protocols and less on applied security concepts for architecture. It's a great supplement to this material for the Wi-Fi professional who wants to dig deeper into specific topics.

CWDP is the design course, and covers topics related to proper RF design, planning, and validation of 802.11 networks. You've heard throughout this book that proper RF design is vital to Wi-Fi security and the prep courses for CWDP are focused on that knowledge.

Last but not least, the CWAP (analysis professional) is regarded as the pinnacle of the specializations. It requires a deep understand of 802.11 protocols and packet structure and the courses for CWAP are heavy in packet analysis, frame formats, and troubleshooting.

Information on CWNP, its certification programs, training classes, and exam objectives can be found at `www.cwnp.com`.

IoT Wireless Training and Certification

At time of writing, the CWNP collection of IoT courses is the most prominent vendor-neutral offering targeted for Wi-Fi pros seeking competency in IoT technologies. There are industry- and product-specific trainings focused on specific technologies that readers are encouraged to research and seek.

CWNP doesn't only focus on Wi-Fi. In recent years, they've released a second track of certifications and training programs dedicated to non-802.11 wireless for IoT:

- CWISA® - Certified Wireless IoT Solutions Administrator
- CWICP - Certified Wireless IoT Connectivity Professional
- CWIIP - Certified Wireless IoT Integration Professional
- CWIDP - Certified Wireless IoT Design Professional

Like the CWNA program for Wi-Fi, the CWISA certification has a similar depth and breadth of content including RF technologies appropriate for Bluetooth and BLE, cellular, Zigbee, and others. Also as with its Wi-Fi counterpart, the CWISA is the foundational certification required to attain the specializations mentioned next.

The CWICP focuses on connectivity of IoT devices and takes a deeper look at the protocols in 802.15.4 networks (Zigbee, ISA, WirelessHART, 6lowPAN, and Thread) as well as long-range connectivity protocols such as Sigfox and LoRaWAN.

CWNP's CWIIP moves up the stack to topics of IoT integration, covering transport protocols like MQTT, CoAP, and AMQP. The integration content also covers use of APIs, programming, libraries, and Python programming fundamentals.

Lastly, in the IoT family, the CWIDP covers IoT design comparable to the CWDP for Wi-Fi design.

Network and Cyber Security Training

Security architecture is holistic, and as such it's a great idea to round out wireless knowledge with other network architecture and network security learning. $(ISC)^2$, CompTIA, and SANS offer training and certification in areas of network and cyber security that include but extend beyond wireless.

- $(ISC)^2$ Systems Security Certified Practitioner (SSCP)
- CompTIA Security+
- SANS various courses

$(ISC)^2$ SSCP is a vendor-neutral certification program designed for a breadth of IT professionals and system admins looking to integrate security with practical knowledge. It covers security operations, identity, authentication, and access controls, risk analysis and security monitoring, incident response, cryptography, PKI, network and domain security, and system security including MDM.

$(ISC)^2$ is unique in that most of its certifications are ANSI accredited, and that it's a not-for-profit membership-based organization. This means the SSCP (and other certifications) meet rigorous international standards for quality and process control of creating and administering certifications. Qualified professionals that pass an $(ISC)^2$ exam become part of the global membership. Information on SSCP can be found at `www.isc2.org`.

CompTIA Security+ is a vendor-neutral program for cyber security with a strong focus on networking technologies including Wi-Fi and other wireless. Many of the concepts overlap and extend topics of this book including risk and data privacy, identity and authentication, PKI, cryptographic concepts, hardening, types of attacks including Wi-Fi, security assessments and penetration testing, and secure network architectures. At time of writing the Security+ content includes WPA3. You can find the most up-to-date information at `www.comptia.org`.

SANS offers a breadth of security courses ranging from network security to pen testing, including ethical hacking, monitoring and detection, and specialized courses for automation scripting, cloud security, and forensics. Find their training portfolio at `www.sans.org`.

Vendor-Specific Training and Resources

Every Wi-Fi manufacturer will have vendor-specific resources available for users. Contact your field sales team for additional resources and assistance. Common resources available include:

- Fee-based vendor product training
- Free online self-paced product and technology training

- Knowledge base articles
- Validated design guides
- Product hardening guides
- Product installation and configuration guides
- Technical assistance centers (TAC, for support)

As an aside, I've done my best to represent today's most popular enterprise Wi-Fi vendors in this book, often painstakingly searching for hours to figure out what vendor X or Y calls a particular feature. The controller-based products (such as those from Cisco and Aruba) have basic configuration guides that are in the 1,200- to 2,000-page range and don't include hardening and certain other security features.

Many of those products also integrate with other vendor platforms for centralized management or monitoring, and those other products each have guides of similar length.

The point is—product training can go a long way and is almost always worth the time and money invested. Some of these are exceptionally complex systems with hundreds of feature interdependencies and "if, then, else" configuration requirements. A Wi-Fi architect can't scoot by watching the occasional 15-minute how-to or by listening to a 45-minute podcast. This field takes passion and a deep level of curiosity to be successful.

Conferences and Community

Last, but certainly not least—the networking, infosec, and Wi-Fi fields are full of industry conferences and amazing peer-based communities. Attending training and conferences is a great way to meet people, get questions answered, or share your knowledge. I've personally met many amazing professionals in-person and online, especially through Twitter where it's easy to search for key terms or get connected through crowdsourcing.

Just a few of the events and communities that include wireless are:

- CWNP hosts events, webinars, and online resources, `www.cwnp.com`
- Wireless LAN Professionals Conference (WLPC) hosts events in both the U.S. and Europe, `www.thewlpc.com/`
- Tech Field Day events, where vendors show off their new products to the technical community, `https://techfieldday.com/`
- Vendor annual conferences including Cisco Live, Aruba Atmosphere, Extreme Connect, and Juniper NXTWORK usually include technical training in addition to the conference

Summary

This chapter is the beginning of the end. As we enter Part III of the book, "Ongoing Maintenance and Beyond," this content draws on all prior concepts and requirements for a cohesive strategy for security monitoring.

The objective of this chapter is to offer real-world attainable suggestions for the continued maintenance of your secure wireless infrastructure such as expectations for assessments and testing and security monitoring tools and techniques.

The chapter took a pretty deep dive into the topic of WIPS and offered guidance that balances legal considerations with security concerns. You've also been introduced to a suite of tools and vendors in the wireless space.

We took a short tour through security event correlation, and how SIEM, SOAR, XDR, and/or network management tools may play a role in wireless logging, alerting, and reporting.

Within the section "Logging, Alerting, and Reporting Best Practices," I outlined where to focus your attention when it comes to monitoring the wireless environment, combining alerting for infrastructure integrity with the WIPS-enabled signatures for a holistic strategy. That same section provided detailed recommendations for which events warrant immediate attention, and which tasks may be best accomplished with which tools.

The fourth main area of this chapter highlighted troubleshooting tips and tricks for the most common issues in secure Wi-Fi. And, finally, we wrapped up with additional training and resources to continue your journey.

CHAPTER

8

Emergent Trends and Non-Wi-Fi Wireless

This chapter wraps up our content with a focus on some of the more fluid and transient aspects of wireless security.

Up until this point, the guidance and information in the book should persist for several years, with a few slight adjustments and evolutions here and there. Moving forward through this chapter, however, the content is less evergreen by design.

Instead of omitting emerging technologies that prove significant in wireless security, those topics that are more volatile are aggregated here, and presented with the caveat in large, flashing letters that everything you read moving forward is subject to change throughout the life of this book. Consider it a launching point for further investigation to update your knowledge.

This chapter's content is distilled down to two broad topic areas:

- Emergent Trends Impacting Wireless
- Enterprise IoT Technologies and Non-802.11 Wireless

This chapter concludes with final thoughts and key takeaways from throughout the book.

Emergent Trends Impacting Wireless

With so many connected devices, the explosive growth of IoT, and a cadre of digital transformation initiatives across all industries, just about every trend these days impacts wireless, and therefore security.

We're going to tackle a handful of the most pressing issues and the topics that address the most-asked questions by networking, wireless, and security professionals, including:

- Cloud-Managed Edge Architectures
- Remote Workforce
- Bring Your Own Device
- Zero Trust Strategies
- Internet of Things (IoT)

Cloud-Managed Edge Architectures

Over the last few years, the networking industry has begun its journey toward a converged edge architecture, and that journey is taking most of us to the cloud. Firewalls and gateways, SD-WAN appliances, and switches are being managed alongside Wi-Fi devices—and mostly in the cloud.

The trend is apparent when you look at Aruba Network's Central cloud, Juniper's Mist AI, ExtremeCloud IQ, and Fortinet's FortiLAN Cloud. These enterprise solutions followed suit after products like Meraki paved the way, proving the benefits of the cloud and ease of management in the SMB market.

The converged and converging edge and the move to cloud management brings a few considerations for security architecture and network operations in general, including:

- Modified privileged access management
- Use of APIs in favor of traditional network management
- Requirement of new skills
- Tighter vendor vetting, privacy, and security certifications

First, as soon as our entire network stack is accessible from the Internet, we have to rethink secure management access. For a refresher, visit Chapter 6, "Hardening the Wireless Infrastructure," and find the topic of "Addressing Privileged Access." The first thing I have clients do with cloud-managed platforms is enable 2FA. When the entire Internet becomes your management network, the architecture must change accordingly.

The next seemingly innocent shift is toward API-managed edge networks. With companies like Juniper moving away from SNMP and even CLI in favor of API-based management of APs and switches, tides are turning, and traditional networking tools and processes will likewise shift.

These two trends alone are early warning signs that network architects and administrators will need to retool themselves along with the network.

Lastly, managing devices in and from the cloud does present new security risks. Even if the vendor isn't collecting client traffic, the platform has metadata, user and endpoint location info, personal information including email addresses and usernames, RADIUS shared secrets, entire IP schemas—and all of that is in the cloud. For most organizations, that's okay if the vendor has properly secured their own environment. When you engage with a cloud provider in any way, the organization is effectively extending its trust boundary to include the provider.

Governmental requirements in many countries have driven vendors toward certain security practices and certifications, and the rest of enterprises should follow suit by better vetting vendors and demanding proof of ISO 27001-certified datacenters and SOC-2 compliance at a minimum.

Remote Workforce

The sudden work-from-anywhere model hit the world without warning in 2020, leaving companies still trying to adjust and address security gaps throughout 2021 and beyond.

In the U.S., several reports show we went from less than 5 percent of employees working from home three days a week or more to over 50 percent in 2021. Gartner reports similar spikes using slightly different criteria in its research. Figure 8.1 shows trending predictions of remote work by country from 2019 through 2025.

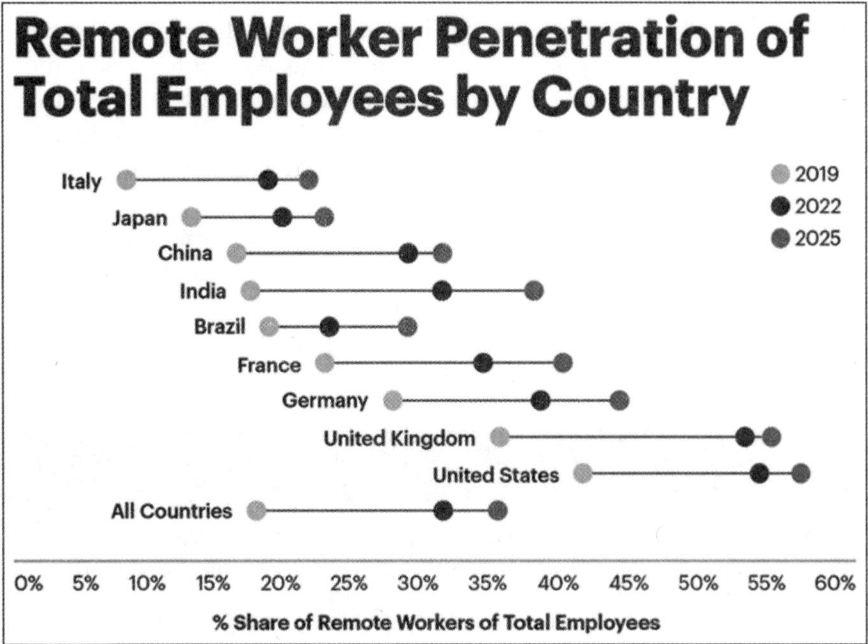

Figure 8.1: Gartner's predictive analysis shows a continued increase in remote workers across the world.

From a wireless and security architecture perspective, the impacts on our work have included:

- Addition of remote APs for home users
- Support for remote and home workers' devices
- Increase in use of personal devices
- Additional projects while offices are mostly unpopulated
- Talent shortage and staffing issues
- Architecture changes to support migrating applications and services to the cloud
- Changes in processes to accommodate zero touch and remote provisioning of infrastructure and endpoints
- Requirements for network admins' remote management access to systems

Challenges Supporting Work from Home and Remote Users

Supporting enterprise products in residential settings is challenging—the IT team has no control over the home user's Internet speed or quality; no visibility into the connectivity and physical location of devices; and APs in the home require power injectors and other accessories, along with cables.

In fact, Microsoft's report "The New Future of Work" (`https://www.microsoft.com/en-us/research/project/the-new-future-of-work/`) cites poor Internet performance as the top issue identified by IT professionals when asked about the biggest challenges for employees working from home. And enterprise remote APs or VPNs are only as good as the Internet connection at the remote site.

To add complexity, the use of personal devices for business applications has increased, forcing enterprise IT and help desk teams to support a much larger breadth of devices than before.

Technologically, the first obvious change for many has been the use of remote APs to support subsets of remote workers. Preferred over traditional VPNs by some users, remote APs extend the organization's Wi-Fi (and often, wired) networks to the user's home. Remote APs are described more in Chapter 1, "Introduction to Concepts and Relationships."

INDUSTRY INSIGHT: MICROSOFT'S "THE NEW FUTURE OF WORK" PROJECT

In 2021, Microsoft published "The New Future of Work: Research from Microsoft into the Pandemic's Impact on Work Practices," which aggregates the results from several studies.

In it, topics of connectivity challenges and security concerns are covered alongside personal productivity and well-being. This report and more can be found at Microsoft's The New Future of Work site at `https://www.microsoft.com/en-us/research/project/the-new-future-of-work/`.

It's a great resource for helping focus attention on the issues most impactful to users.

Balancing Additional Work and the Tech Talent Shortage

On the other end of the spectrum, work within the organization's traditional campus environments has increased for many professionals as teams take advantage of unoccupied and sparsely populated offices. Backlogged projects are being tackled while structured cabling and IT teams have easy access to workspaces and common areas, with little worry of disrupting employees.

The other side effect of vacant office buildings has been that many enterprises are downsizing, leaving, or remodeling offices to accommodate the new future of work. All of these lead to more workload on the IT and security teams, and there's little to no relief for many while the additional workload coincides with staffing shortages.

"The Great Resignation," as it's been dubbed, hit critical mass mid-2021, and it hit hardest in the IT sector. While 40 percent of remaining employees surveyed were considering quitting their job, that percentage was a whopping 72 percent for tech employees in a report from TalentLMS (Sources: `https://www.microsoft.com/en-us/worklab/work-trend-index` and `https://www.talentlms.com/tech-employees-great-resignation-statistics`).

IT and infosec roles were already posting record shortages before the pandemic stay-at-home orders were issued in 2020. The climate of added workloads, lack of social interactions with coworkers, and filling in for missing headcount is taking its toll—on personal well-being as well as enterprise security posture. When you're burning the candle at both ends, some tasks will have to fall off the list; things will get missed, and vulnerabilities will be introduced.

Process Changes to Address Remote Work

The to-do lists continue growing, as enterprises are forced to restructure applications to support a remote workforce—often moving servers and services to the cloud, some of which impacts the connectivity and security monitoring of networks.

Contributing to the resignation crisis among all employees, frustration with technology—laptops and remote network connections that don't work—are cited as a top reason people are leaving their jobs. The added friction of increased security requirements such as MFA aren't helping either. Confusion about what to do when technology doesn't work is also a leading problem.

Network services traditionally hosted on-prem are moving to hybrid and cloud models to help reduce user friction and ensure services are available without time-consuming access methods that further irritate users. These moves impact authentication and IAM processes, requiring integrations with APIs and a move toward SAML-based logins.

Along with changes in services and applications, the processes for provisioning infrastructure are changing—often moving toward automated and zero touch provisioning technologies, which may require additional upfront work and reconfiguration of existing network devices. The remote model necessitates modifications in how teams support end users and provision their devices as well.

And it's not just the end users that are remote. The IT teams, network architects, and help desk may also be remote, in whole or part and full time or part time. That means yet more changes to processes and architecture to ensure secure remote privileged access, account validation without in-person processes, and finding better ways to collaborate remotely.

Recommendations for Navigating a Remote Workforce

When it comes to supporting and securing remote workers, there are a lot of factors that are simply out of our control.

The most meaningful, and perhaps the most unexpected piece of advice I can offer fellow technologists and security professionals is to take care of yourself—physically, emotionally, mentally—whatever that means, and find ways to stay grounded, remain mindful, and build your own personal resilience.

The demands put on IT and infosec teams was overwhelming before the pandemic, and the speed of change in technology is dizzying without having new major projects thrown at us. Things may have to fall through the cracks, and some ends may be left loose. That, along with a few other tips and tricks for navigating the changing workforce include:

- Maintaining personal mindfulness and resilience to manage additional stress of changes
- Creating documentation, even if ad hoc, to track changes and notes
- Identifying repeatable processes and streamlining operations
- Offloading information-gathering tasks to vendors and resellers
- Creating processes to offload subsets of your work to more junior staff
- Leveraging zero touch deployments when possible

- Working on communicating clearly and often to avoid misunderstandings, friction, and rework
- Relying on peer groups within your industry or interest area to crowd-source resolutions to challenging issues

Bring Your Own Device

We're going to cover technical elements here, but if you take nothing else away from the topic of *bring your own device* (BYOD), what I hope sticks with you is the complexity of BYOD from the business and legal perspectives, and that BYOD policies are not something that should be left for IT teams to write or enforce single-handedly.

Also, there's only one hard and fast rule for BYOD: Access to any secured or sensitive resources should only be allowed from a device the organization has some level of visibility into or control over. You'll hear this repeated throughout this part of the book in various ways.

Retaining visibility and/or control can be accomplished in a few ways. The following subtopics will break down the models and best practices, but for context the goal can be reached in several ways, including:

- Personal device with corporate-managed mobile device management (MDM) agent to enforce device security and posture
- Personal device with NAC agent or agentless scanning to verify security posture
- Corporate-managed, personally enabled device under full enterprise management

Stats on BYOD and Policies

Even though around 60 percent of organizations have a BYOD program, more than half of them don't have or enforce a formal BYOD policy. And, regardless of the organization's official BYOD policy, two out of three employees use their devices at work—even if it's forbidden.

Without a policy for BYOD, organizations and employees are both left to be victims of circumstance.

Aside from the legal considerations, mobile devices continue to be the source of many breaches worldwide. At one point, 68 percent of healthcare data breaches were due to the loss or theft of mobile devices or files, as reported by Bitglass, and that number has continued to grow.

Policies and accompanying security controls can enforce meaningful protection of enterprise data on mobile and personal devices, offering mechanisms to encrypt data, enforce strong passwords and lockout timers, along with emergency remote wipe or GPS tracking to locate lost or stolen devices.

Worldwide, 97 percent of companies were affected by mobile threats, including reports that almost half of organizations (46 percent) reported at least one employee having downloaded malicious applications on a smartphone. It's probably no surprise then that nearly every organization surveyed reported at least one smartphone malware attack. (Source: Checkpoint Mobile Security Report 2021, `https://pages.checkpoint.com/mobile-security-report-2021.html`.)

SIZING IT UP: BYOD POLICIES

Creating a BYOD policy is a very personal thing for all involved—the organization and each employee. I almost cringe every time I'm requested for a consulting call and they ask, "What is everyone else doing for BYOD?"

It doesn't matter what everyone else is doing because their environment, user population, culture, intellectual property, platforms, and business model may be completely different than yours.

It's feasible to make a few sweeping generalizations and the following sections will walk through some of the inputs for a decision, but ultimately creating (and enforcing) a BYOD policy necessitates the organization's legal counsel, human resources, and executive leadership get involved.

Other Models for Ownership, Management, and Use

BYOD isn't the only way personal devices are used in the organization. There's yet another alphabet soup describing varying ownership, management, and use relationships, including the following:

Bring Your Own Device The traditional BYOD model describes a device that's wholly owned by the employee, and on which the enterprise may (or may not) request to install MDM or an agent to containerize corporate data for security purposes.

Challenges with traditional BYOD are that the organization's IT teams may find they need to support a breadth of device types, operating systems, and generations. App incompatibility, restrictions on MDM permissions, and the general overhead of help desk teams can be troublesome with BYOD.

Choose Your Own Device Choose your own device (CYOD) models were meant to address employees' desires to not be forced into a single type of device, while allowing the enterprise to focus on supporting a subset of devices with known compatibility.

In CYOD, the organization offers the employee their choice of a selection of supported devices. The ownership model can be corporate-owned or personally owned, but regardless, the organization will only allow and support use of the approved devices for business purposes.

Company-Owned, Personally Enabled Company-owned, personally enabled (COPE) devices tighten security one notch beyond CYOD. In COPE models, the organization owns and manages the device, but allows certain personal applications and uses such as voice calls, texts, and allowlisted applications.

COPE is emerging as a great option for security-conscious organizations that want to allow some workforce flexibility and, most notably, not force employees to carry two devices. COPE models should take advantage of built-in secure containerization or add-on software to separate personal and business data as much as possible.

Like CYOD, the COPE program may include a short list of devices for the employee to choose from, but unlike CYOD, the organization maintains ownership and control of the device, and pays for it.

Company-Owned, Business Use Only Company-owned, business use only (COBO) is just as it sounds—the device is locked completely, disallowing personal applications. COBO policies could be for any employee in a high-security environment such as financial, healthcare, or certain government agencies. It's also appropriate for shared device environments where one device is used by different employees at different times, such as for shift workers in a warehouse or field technicians.

Table 8.1 captures the various ownership-management models and the relative risk level associated with each.

Further defining BYOD in the organization can be completed now that you're familiar with the options for ownership and management of devices.

Table 8.1: Management and ownership models for devices

MODEL	DEVICE OWNERSHIP	DEVICE MANAGEMENT	USE	RISK LEVEL
BYOD (Unmanaged)	Employee	Employee	Personal + business	High
BYOD (Managed)	Employee	Employee, with company MDM	Personal + business	Medium-high
CYOD	Employee or Company	Company	Personal + business	Medium
COPE	Company	Company, allowing personal use	Business + personal	Medium-Low
COBO	Company	Company	Business only	Low

Further Defining BYOD in Your Organization

Chapter 5, "Planning and Design for Secure Wireless," outlined two basic starting points for defining BYOD in the organization:

- Allow personal devices on the network but with Internet-only access
- Allow personal devices on the network with some level of access to internal or protected resources

Our goal at that time was to gather enough information for a general planning table. Now diving deeper into BYOD, the objective is to further refine the definitions to begin drafting appropriate controls for securing BYOD (or one of the other management models). Considerations are summarized in Table 8.2 and include:

Form Factor Smartphone, tablet, laptop, etc.

Ownership and Management Model Unmanaged BYOD, managed BYOD, CYOD, COPE, or COBO.

Corporate Visibility and Control Based on the management model and the device type(s), what visibility or control does the organization have over the device to ensure security?

Access Requirements Internet-hosted SaaS applications, on-prem, or IaaS/PaaS resources, and the nature of the networks, endpoints, and data being accessed.

Access Source Where is the user accessing resources from—within the traditional perimeter in-office, or from home/remote? And if remote, what countries or regions?

Privacy and data protections vary in different parts of the world, as to approved encryption technologies. In addition, some industry segments will have additional restrictions for business conducted (and data transferred) to and from foreign countries.

Table 8.2: Table of BYOD access planning

FORM FACTOR	OWNERSHIP/ MANAGEMENT	ACCESS REQUIREMENTS	ACCESS SOURCE
■ Smartphone ■ Tablet ■ Laptop ■ Other (e.g., printer)	■ Unmanaged BYOD ■ Managed BYOD ■ CYOD ■ COPE ■ COBO ■ Plus, visibility and control based on management	■ Internet-based SaaS ■ On-prem resources ■ Cloud IaaS/PaaS ■ Plus, the nature and sensitivity of the networks, other endpoints, and data being accessed	■ Internal ■ Home or remote ■ In-country ■ From foreign country

The following sections will offer guidance through the decision-making processes, but the one combination to always avoid is any unmanaged device accessing anything other than Internet-based resources. And, in organizations beginning a zero-trust strategy, even that access should be metered and controlled in some way.

Legal Considerations for BYOD

The most critical input into any BYOD policy will be those of legal considerations. Organizations have a responsibility as stewards of their data, and they have a responsibility to ensure employee privacy is maintained—it's all a delicate balancing act. Legal obligations and options related to BYOD policies vary by region, and even by state within the U.S.

In a tragic 1977 plane crash, two members of the rock band Lynyrd Skynyrd were killed. The survivors and families made a blood oath that no one should ever again perform as "Lynyrd Skynyrd," and that oath was legally memorialized in a 1988 Consent Order.

What can Lynyrd Skynyrd possibly have to do with BYOD? Well, almost thirty years later, one of the few survivors, drummer Artimus Pyle, began consulting with a production company (Cleopatra Films) for a biopic about the band's 1977 plane crash. Other members from the original blood oath took legal action against Cleopatra Films, and won, blocking production of the movie. One large body of evidence was related to text messages exchanged between Pyle and the main writer (a contractor of Cleopatra Films)—the text messages (evidence in the case) were lost when the writer changed phones.

Shortening the story; the court found Cleopatra Films was guilty of "spoilation of evidence" by not having control over their contractor's text message as critical evidence in the case. The writer wasn't an employee of the film company, nor was he using a company-provided device. But there was also no policy in place outlining expectations of the use of personal devices, or ownership of data (including texts) on that device. The court awarded an adverse inference against Cleopatra Films, saying it was "common sense" the contactor's texts were in Cleopatra's control. There is a happy ending though; in 2018, the judgment was overturned and in 2020 *Street Survivors: The True Story of the Lynyrd Skynyrd Plane Crash* was finally released.

It's a fun story that brilliantly demonstrates the twisty legal issues organizations must navigate through BYOD policies.

And it's just one example in a long list of BYOD gone bad. In another case, audio manufacturer Klipsch won a case when it moved for e-discovery sanctions against a defendant accused of selling counterfeit product. The appeals court faulted the other company for "failing to have a software usage policy in place requiring its employees to segregate personal and business accounts or to otherwise ensure that professional communications sent through personal accounts could be preserved." The findings (and lack of policy and enforcement) cost the defendant $5M USD.

INDUSTRY INSIGHT: PRACTICAL LEGAL GUIDANCE ON BYOD

There's an exceptionally comprehensive document, "Commentary on BYOD: Principles and Guidance for Developing Policies and Meeting Discovery Obligations" by the Sedona Conference that outlines five principles for developing BYOD policies that meet e-discovery requirements.

Principle 1 Organizations should consider their business needs and objectives, their legal rights and obligations, and the rights and expectations of their employees when deciding whether to allow, or even require, BYOD.

Principle 2 An organization's BYOD program should help achieve its business objectives while also protecting both business and personal information from unauthorized access, disclosure, and use.

Principle 3 Employee-owned devices that contain unique, relevant ESI should be considered sources for discovery.

Principle 4 An organization's BYOD policy and practices should minimize the storage of—and facilitate the preservation and collection of—unique, relevant ESI from BYOD devices.

Principle 5 Employee-owned devices that do not contain unique, relevant ESI need not be considered sources for discovery.

The full 64-page document is available at The Sedona Conference website at `https://thesedonaconference.org/publication/Commentary_on_BYOD`.

The Sedona Conference (TSC) is a nonpartisan, nonprofit 501(c)(3) research and educational institute dedicated to the advanced study of law and policy in the areas of antitrust law, complex litigation, intellectual property rights, and data security and privacy law.

These are just a few of the many legal considerations that go into planning a BYOD policy:

- *E-discovery requirements*, defining what's in scope on personal devices, processes, and employee obligations during litigation holds
- *Data ownership and management*, including what obligations the organization has to protect its data, and what legal entitlements it has to wipe or back up data from personal devices

- *Data privacy*, including what the company is legally allowed to do to, or view from a personal device such as texts, calls, photos, GPS tracking, or browsing history
- *Employee lifecycle*, including what happens to the corporate data when an employee separates from the company
- *Illegal activity*, including what the organization is allowed or obligated to do in the discovery of illegal activity

Legal considerations don't stop at data protection. In 2021, Coca-Cola was slapped with a $21M USD judgment after one of its truck drivers hit a woman, reportedly distracted while chatting on the phone. Even though the company's policy prescribed the use of a hands-free device while driving, the plaintiff's lawyers convinced the jury that the policy was "vague and ambiguous."

Just to recap, we don't expect (or want) technologists or architects to be legal professionals. The point of this content is merely to demonstrate how bizarre and complex the laws can be, and the risk posed to an organization by ineffective or insufficient policies or controls.

Even if you stayed at a Holiday Inn Express last night, let the lawyers figure this part out.

Technical Considerations for Securing BYOD

Chapter 5 lightly touched on BYOD in the section "Planning Processes and Templates." In that content, the first exercise was to define what BYOD means in the organization. Picking up where that left off, the steps that follow are to work with executive leadership to define a BYOD policy and specify parameters around who can access what, from what device, when, and how. The technical planning for this was covered also in Chapter 5's "Sample Access Rights Planning Templates."

With any luck, BYOD policies will stipulate that the organization only allows access to secure resources from devices that are under some form of management by the organization. In that case, depending on the ownership and management model, solutions like MDM products may be in order.

MDM solutions don't stop at personal smartphones—they're the cure for all that ails us when we need control over otherwise unmanaged devices. For our purposes, I'm going to place anything that's remotely MDM into the "MDM" category as described with the following features. MDM products offer one or more of these technical controls:

- Enforcing passwords and password complexity
- Enforcing screen lock timeouts and incorrect login lockouts

- Containerizing corporate data away from personal data and applications
- Supporting e-discovery processes including holds
- Implementing a remote wipe sequence for corporate (and at times, personal) data and applications (if lost or stolen)
- Allowing remote lock of the device (if lost or stolen)
- Locating the device (if lost or stolen)
- Installing device certificates (including ones for Wi-Fi authentication)
- Installing server certificates (including RADIUS server certificates for 802.1X-secured Wi-Fi)
- Pushing enterprise applications to the device
- Pushing passphrases for WPA2- or WPA3-Personal networks
- Joining the endpoint to the enterprise domain
- Running security posture checks and enforcing posture

MDM-type products aren't just for phones; depending on the product and vendor, they may be used on any of the following:

- Virtual desktops
- Standard OS desktops and laptops
- Tablets
- Smartphones

One of the most challenging technical hurdles is that most MDM tools will work on (for example) iOS, Android, and Windows (along with virtual desktops) but the controls available per device OS will vary from one another and will vary over time. And the options and behavior vary depending on the MDM product used and the device's native containerization—for example, an Android device with Samsung Knox has security options beyond a standard Android device. Like so many of our other tools, MDM solutions may be integrated into our existing infrastructure (such as subsets of Microsoft Intune) or installed as a third-party solution.

Choosing an MDM tool is another choice very personal to each organization—the types of devices, operating systems, and the granularity of control varies by product, as does the integration with other tools.

Recommendations for Securing BYOD

At this point, I hope you're appropriately apprehensive of creating BYOD policies. It is also my hope you're able to motivate the organization to create or update BYOD policies if needed, but in the absence of that, there are a few best practices around BYOD that are generally safe for most environments.

Also, since the recommendations are dependent on the BYOD model (Internet-only access versus internal access), that's how they're organized All of these factors should be evaluated when considering an MDM platform.

REFERENCE **Revisiting the template from Chapter 5 titled "Sample Access Rights Planner for NAC" may be helpful. It includes several sample tables and access planners including examples that incorporate employee-owned devices.**

Recommendations for BYOD Internet-only Access

In Chapter 5 it was noted that planning for Internet-only BYOD is fairly straightforward. The one request that makes it more complicated than standard guest portals is that many organizations want their employees to be able to use the Internet-only connection without the friction and frustration of a daily captive portal experience.

This is a completely legitimate desire—however, to meet this objective, decision-makers will often instruct wireless architects to allow personal devices on the secured network—either by allowing users to join 802.1X-secured networks with their domain usernames and passwords, or by having personal devices on an internal passphrase-secured network (WPA2- or WPA3-Personal). Neither of these are advised because they introduce several security risks.

Instead, for personal devices requiring Internet-only access, with the goal of not repeating a daily captive portal, one of these methods should be used. These are presented in order of most to least secure, but all are valid options.

CAUTION **Remember as you're reading, these three options are appropriate for Internet-only BYOD use cases. These methods are not secured properly for access to secure internal networks.**

Registration-based BYOD Captive Portal This would allow employees to register their personal devices under their domain account using a portal page; the account can be authorized for a period of time or indefinitely.

With some products, the users can manage their list of approved personal devices, and a maximum number of devices per user can be specified. These features are most often available with a NAC product, but some Wi-Fi products are adding built-in support for this as well.

Notably, this option links the otherwise unknown personal device to a known user, but it does so safely, without exposing the user credential on the unmanaged device.

Basic BYOD Captive Portal with an Extended Cache Timer A basic portal would look just like a guest portal, but the timer would be set for a much longer duration; for example, instead of a 1-day guest account, the employee can join with their device and not have to revisit the portal for 3–6 months, or even a year. The portal can be a subfunction of an existing guest portal and of course should offer Internet-only access.

Passphrase-Secured BYOD Network Passphrase-secured networks for BYOD access is perfectly acceptable, *as long as the access is Internet-only*, but it's preferred to separate personal devices from any other corporate endpoints such as IoT devices.

This can be accomplished with separate SSIDs, or through dynamic VLANs (or roles, policies, etc.). If at all possible, use WPA3-Personal for this purpose—remember that WPA3-Personal networks are far superior in security to legacy WPA2-Personal networks.

Recommendations for BYOD with Internal Access

Circling back to the opening paragraph—the one hard-and-fast rule for secure BYOD is that access to any secured or sensitive resources should only be allowed from a device the organization has some level of visibility into or control over.

From a technical perspective, I'd encourage you to advocate for this model and/or to start with this access until better direction is offered by the organization.

Organizations should not allow personal devices on secure production Wi-Fi networks unless *all* of the following criteria are met:

- There's a business case outlining the requirements and there is no other way to meet the business objective
- There's a policy in place (approved by executive leadership) stating that personal devices are allowed, along with the conditions and access; and in the absence of a policy, request approval for personal devices in writing (email is fine) from your manager or other leader in the organization
- Each user participating in BYOD has read, acknowledged, and signed the BYOD end user policy
- The organization has some visibility into the security posture of the personal device such as through an MDM tool or NAC agent
- There is a mechanism to securely onboard the personal device to the network, such as through an MDM and certificates, or through a NAC product
- There's no onboarding mechanism that puts the user's domain credentials at risk; specifically, the users should not join an 802.1X-secured network from a personal device using domain credentials

In addition to the preceding requirements, there are a few additional recommended items that would offer additional security:

- Use granular access policies to allow access only to the required resources
- Preferably, do not co-mingle unmanaged personal devices with managed devices on the same SSID or VLAN
- Have a way to determine which devices on the network are personally owned, and also to identify the user/owner of the device
- If possible, use an MDM or agent for security posture scanning to ensure the personal devices are not infected, and meet minimum security requirements such as passwords and patching

The reasons for the recommendations here are covered throughout the book, and especially in Chapter 3, "Understanding Authentication and Authorization," and Chapter 6.

Zero Trust Strategies

Zero trust is one of today's (and tomorrow's) super-hyped topics and while I'm certainly not here to toss buzzwords around, zero trust is a timely and relevant topic for network and security architects. I don't plan to go deeply into the topic, but I work a lot in this space and feel there's value in clearing up a few misconceptions and points of confusion related to network security and zero trust.

The Current State of Zero Trust

Think of zero trust as a mindset, and not a product set. It's just like saying "layered defense." While the models and frameworks floating around seem to indicate zero trust is an all-or-nothing, super-granular architecture, the reality is that applying one-to-one access rules for every endpoint and user, to and from every service, server, and resource is impossible with current tools. It's neither practical nor desirable at this stage.

With today's zero trust strategies, the goal is to incrementally remove inherent trust and apply specific, context-based access rights. Moving away from traditional remote access VPN solutions is a perfect example of removing inherent trust—many VPN implementations today are an "all or nothing" access model; a remote user is either allowed on the network, or not. Believe it or not, relatively few VPN deployments apply granular access policies; instead, it's just as though the user were inside the office with the normal layer 2/3 access that accompanies it.

Instead, a zero trust strategy would direct us toward evaluating both the endpoint device and the user—perhaps performing a security posture scan and ensuring the user is requesting access from an approved region—and then the user is granted only the access needed, such as the ability to connect to a subset of servers or services in the organization.

Similarly, zero trust strategies will begin to inform stricter control for management access of systems, adding contextual information and enforcing greater security such as two-factor authentication (2FA). Zero trust isn't some imaginary mystical creature—it's really just the next evolution of network access control (NAC); one that incorporates cloud-routed technologies and Internet-based resources instead of (or in addition to) traditional on-prem models.

Moving toward a zero trust architecture is incremental and organizations will start identifying use cases and subsets of data or resources that are flagged as more sensitive than others, and that's where zero trust begins.

Zero Trust Language

Further proving the point that zero trust is just the next evolution of NAC, the terminology from the formal zero trust frameworks (such as NIST's) uses the same language from NAC and authentication models, such as:

Policy Decision Point The Policy Decision Point (PDP) is the brain of a zero trust ecosystem: it's the policy engine making decisions about access rights. Conceptually, the PDP takes inputs from various other systems (most likely via APIs) and then connects to policy enforcement points (PEPs), which will enforce the decision made. In reality, while a centralized PDP is idealistic, there is no one master product available to perform PDP functions for the breadth of use cases for zero trust.

Policy Enforcement Point Policy Enforcement Points (PEPs) enforce the access policies as directed and orchestrated by the PDP. PEPs can be any piece of software or hardware that can apply an access policy or control traffic flows; including (but not limited to) zero-trust enabled switches, routers, and firewalls. Software agents and endpoint software with built-in host firewalls can also be PEPs. From our earlier example, a VPN appliance would also be a PEP.

Subject The subject in zero trust lingo is just the device or user requesting access to a resource (aka target).

Target The target is then the resource being accessed. A target can be a specific server or device, or it could be as granular as a single database, service, or microservice running on a server or serverless infrastructure. Approaching it from a broader perspective, a target could be a set of resources, like an entire network segment—something referred to as an *enclave* model.

Action Action describes the action allowed or access to be granted. This could be allowing communication over a specific port or protocol, or it could be an action policy that allows all traffic to and from a network.

Condition If the action is only permissible with certain conditions, those are specified by the PDP. This may include dynamic conditions within the environment, subject, or target that determine access.

For example, a condition could be that the subject is making the access request from within a list of allowed geographies. The condition might also stipulate that the subject device has been scanned for security posture and passed. Or certain variables may be factored and require the user to elevate their privilege using 2FA to gain access to a secured resource.

Figure 8.2 shows a simplified diagram of a zero trust ecosystem with the labels for PDP, PEP, subject, and target. In the figure, a central policy engine is making decisions about access to the on-prem resources (in an enclave model) and to cloud-hosted resources, which may be granular per-service policies or also a group of resources. Although depicted with a user/device as subject, in zero trust models the subject could be a server or microservice.

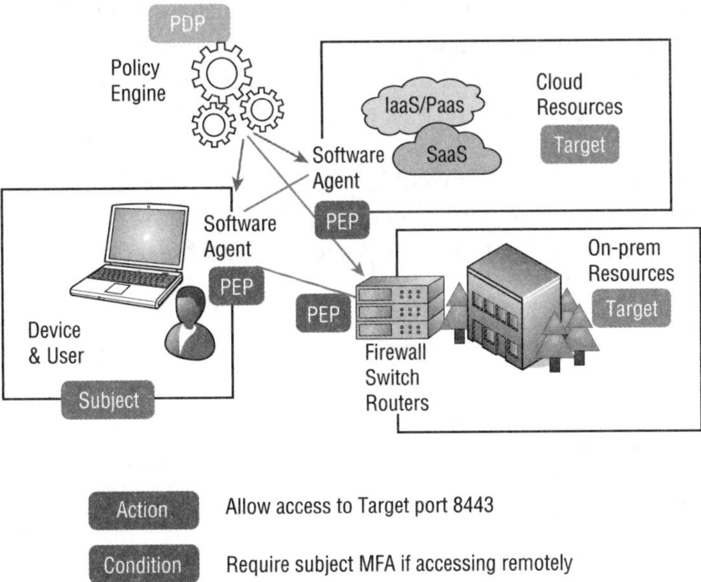

Figure 8.2: Zero trust terminology is just an upcycle of NAC

Types of Zero Trust Products

Common features of products that support a zero trust strategy may focus on one or more of the following attributes:

- Provide encryption of data in transit
- Manage the data path of connections

- Perform or enable security inspection of traffic
- Provide secure and granular remote access
- Provide secure and granular privileged access
- Enable concept of least privilege
- Ensure strong identity and authentication
- Use scalable integration methods such as APIs and SAML

I group the product sets into two main categories—cloud-routed and direct-routed. Currently, most direct-routed solutions are designed for on-prem use cases but that will evolve over time. The details of these are covered next.

At time of writing, these products are rarely converged, meaning there are virtually no products that are designed to (efficiently) enforce access control over the Internet and on-prem.

Segueing to the next topic, Figure 8.3 offers a simplified view of cloud-routed versus direct-routed models common in today's zero trust products. With cloud-routed solutions, not only is the policy engine typically in the cloud, but the cloud is also in the data path, and essentially "stitches" together access between two resources. Conversely, direct-routed models enforce connections directly between the user and the PEP, which may be a switch, firewall, or any of the aforementioned devices.

This model is simplified in that it's possible to use cloud-routed products even for localized on-prem campus deployments, and it's also possible to use direct-routed products for resources that are remote or cloud hosted.

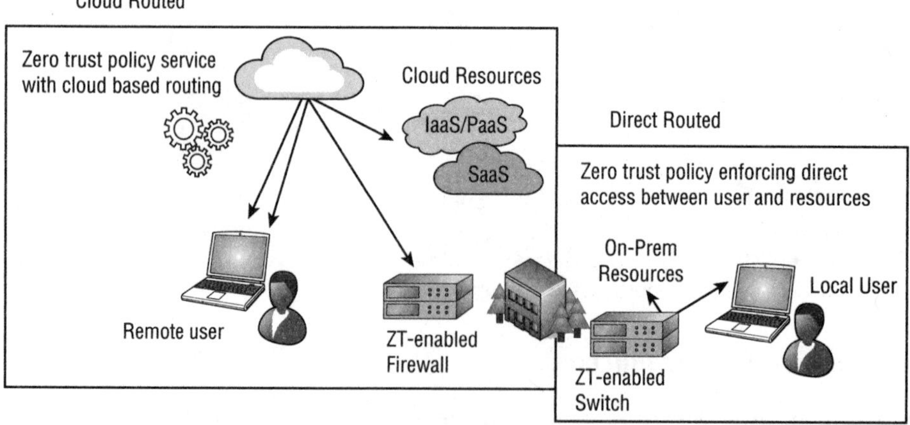

Figure 8.3: Simplified view of cloud-routed versus direct-routed zero trust product features

Cloud-Routed and Cloud Native Products

Cloud-routed and cloud native products are primarily focused on securing access to and from users that are not co-located with the resource because the data path involves the Internet. For example, if a user is working from home and accessing a resource on-prem or in the cloud, a cloud-routed product is well suited for the model.

Conversely, if the user is co-located on-prem with the resources, routing traffic out to and back from the Internet is not ideal. Depending on the network, data, and implementation, this hairpin routing out and back in from the Internet may be acceptable, but it's not how these products were envisioned.

These classifications of products, their features, and vendors change almost daily. These are all highly volatile markets, but for context, today's most common cloud-routed products include the following solutions:

- Zero Trust Network Access (ZTNA)
- Secure Access Service Edge (SASE)
- Cloud Access Security Broker (CASB)
- Secure Web Gateway (SWG)
- Software-defined WAN (SD-WAN)
- Workload and microsegmentation
- Software-Defined Perimeter (SDP)

ZTNA is an actual product and is not to be confused with zero trust as a strategy or architecture. Someone out there just wanted to further complicate and confuse us all with that name. SASE offerings are most often a ZTNA solution with additional services. The lines between the products listed here are blurry, at best.

TIP **ZTNA is based on the Cloud Security Alliance's model for Software-Defined Perimeter (SDP) and is also known as a dark cloud. The idea is that services and ports are not exposed at all. Only predefined brokered connections are allowed, and only after a user has been authenticated and authorized.**

Direct-Routed and On-Prem Products

The types of products designed to support zero trust strategies within a campus environment include:

- Network-based microsegmentation (next-gen NAC, SDN, VXLAN)
- Agent-based microsegmentation

- Network access control (NAC)
- Software-defined WAN (SD-WAN)
- On-prem components of Secure Access Service Edge (SASE)
- Zero trust enabled firewalls and gateways
- Purpose-built appliances

There are some crossover products and features between cloud-routed and on-prem products. There's also some re-use of language such as "microsegmentation," however that word means two different things. Originally a datacenter concept for controlling authorization between servers, services, and microservices—the term microsegmentation was then borrowed by the network security vendors to describe more granular access controls from their products.

The result is that a microsegmentation product for datacenter and cloud applications is not the same as a microsegmentation product for network security on-prem. They're completely different technologies, operate with different enforcement mechanisms (covered next), and therefore have no overlap nor any relationship to one another.

Segmentation Enforcement Models

Diving into the enforcement mechanisms for zero trust products could be its own book, so I'll try to distill this down to a few salient points.

Software-based PEPs

Software-based PEPs use agents on the requestor (subject) and/or the resource being accessed (target), and will most often use one or more of the following enforcement mechanisms to enforce an access policy:

- DNS
- Local host firewall
- Local routing control
- External routing control

The level of enforcement granularity may include one or more of the following, and the access policy may be unidirectional or bidirectional, meaning it may be initiated from the requestor (aka the subject), the resource (aka the target), or either. Some cloud-routed solutions offer limited granularity and may only be built to establish connections unidirectionally.

- Application signatures
- URL/URI

- TCP (any port)
- TCP (specific list of ports)
- UDP (any port)
- Microservices

Although software-based PEPs are viable on campus infrastructures with on-prem resources, they're more often used for access enforcement to and from Internet-based resources.

For this reason, most current products that are controlling access with campus environments are using on-prem products and the appliance-based PEPs described next.

Appliance-based PEPs

Appliance-based PEPs may be physical hardware or virtual appliances and include many of our traditional networking technologies such as firewalls, switches, routers, SD-WAN gateway, Wi-Fi controller or AP, or a purpose-built vendor device.

Just as with other access control technologies, the enforcement mechanisms in appliances most often include:

- Firewall policies or zones
- VLANs
- SDN or SD-WAN policies
- ACLs (static or dynamic)
- Routes (static or dynamic)
- VXLAN

The granularity of appliance-based PEP enforcement varies slightly from software PEPs, and includes the following:

- Application signatures
- Any TCP or UDP port or protocol
- TCP (specific list of ports)
- IP addresses or networks

Most of the appliance enforcement mechanisms will look familiar—they were covered as filtering and segmentation methods in Chapter 2, "Understanding Technical Elements."

Zero Trust Strategy's Impact on Wireless

The purpose of this content is to demonstrate that zero trust is not a mythical beast—it's built on principles (and even products) familiar to network and security architects. There's a lot of confusion in the market due to poor communication from vendors as to how their products work—specifically what enforcement mechanisms they're using, what level of granularity can be applied, and in which directions.

For now, the cloud-routed products are most appropriate for applying granular controls to and from users and resources that are not co-located. By inference then, for most readers of this book, you're interested in the on-prem infrastructure solutions for zero trust and the appliance-based PEPs. There will likely come a time in the future that the two may converge, but the overlap is so minimal currently it's negligible.

As it relates to this book's contents, zero trust initiatives may impact wireless architecture and security in one or more of the following ways:

- Prescribe stricter control of management access to systems including Wi-Fi controllers, and especially remote access
- Enforce more granular access rights for users with remote APs
- Define requirements for further segmentation of classes of endpoints such as separating IoT devices from other managed endpoints
- Define additional requirements for MFA/2FA for certain users or access and new SAML integrations for authentication
- Define additional requirements for device certificates along with user-based access
- Require new models for IoT connectivity and access control
- Involve new integration models including a broader use of APIs

ZERO TRUST ENDPOINT AGENTS AND WI-FI

There are two ways zero trust and Wi-Fi may converge sooner than later. First, these products often involve agents on the endpoints, which serve as enforcement and connection points for accessing resources. Depending on the vendor, the agent can control access to local resources and/or remote or Internet resources. When the agent is controlling access locally on-prem (such as a campus environment), it will impact the overall security architecture.

Second, as users continue working outside the corporate campus, zero trust solutions will continue adding tools to help enterprise IT teams support remote users. Vendors can use the agent already present on the endpoint for added visibility.

As an example, Zscaler offers a tool to help with root cause analysis for users with poor experience accessing applications or Internet resources. Among other things, the tool (currently named Zscaler Digital Experience, ZDX) can provide visibility into the

remote user's network connection and the path locally and through the Internet. ZDX also offers synthetic client testing similar to the Wi-Fi tools covered in Chapter 5.

Applications such as this from ZTNA and SASE vendors may become another tool in your toolbox. As they say, it's a "mean-time-to-innocence" solution and a great way to support remote workers on iffy home networks.

Internet of Things

Managing the Internet of Things (IoT) is certainly a pain point for network architects everywhere, and it's a force to be reckoned with. The following section, "Enterprise IoT Technologies and Non-802.11 Wireless," takes a deeper look at the models and wireless technologies prevalent in enterprise IoT, including (and especially) non-802.11-based wireless.

First, I want to present a high-level view of IoT models to serve as context because the term "IoT" tends to get thrown around, describing anything from a network printer to a structural strain gauge sensor the size of your pinky finger.

IoT models are based on connectivity, topology, and whether the protocol is natively locally routed, translated to IP, or routed to IP. These different models are grouped in the following way:

- LAN-based IoT
- Protocol-translated IoT
- Protocol-routed IoT

Figure 8.4 shows the three models side-by-side. LAN-based IoT models use standard IPv4 and/or IPv6 at the edge for LAN connectivity. Protocol-translated IoT has no IP protocol at the edge (instead uses Bluetooth, BLE, Zigbee, etc.) and requires a protocol gateway to translate the non-IP data to IP-based data to be then routed through networks to the Internet. Protocol-routed IoT seems similar to protocol-translated at first glance, with the important differentiator being the use of an adapted IPv6 protocol at the edge devices, which is then converted and routed.

LAN-based IoT

LAN-based IoT devices can be considered non-traditional endpoints that are connected to standard networks, wired, or Wi-Fi (802.11 WLAN). While it's common to throw non-traditional endpoints into the IoT bucket, strictly speaking (in the IoT communities) these don't quite qualify for an IoT designation.

However, since it's so commonly addressed along with IoT in networking, I'm including it as a model. It's certainly valid in that non-traditional endpoints take a bit more time and attention to identify and secure on any network.

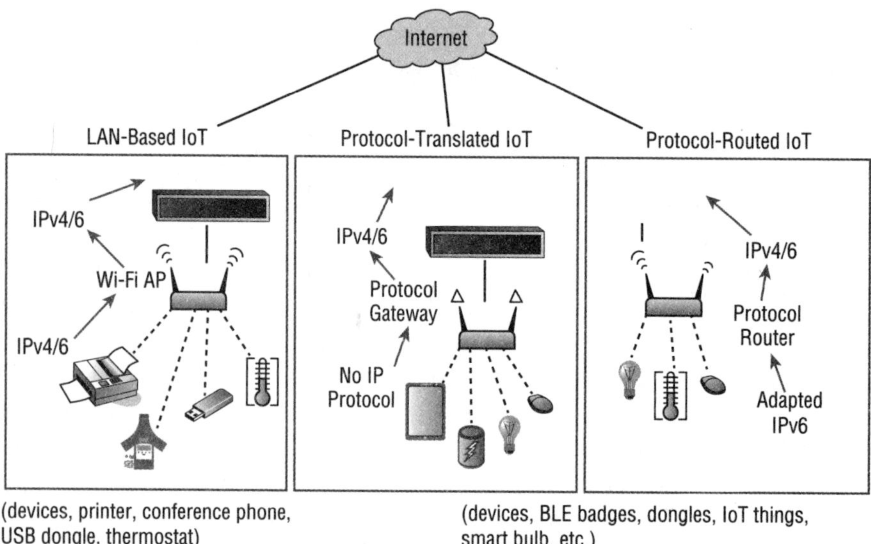

Figure 8.4: LAN-based IoT vs. protocol-translated IoT vs. protocol-routed IoT

Non-traditional endpoints on the LAN may include:

- Non-user-based devices such as printers
- VoIP and VoWiFi (voice over IP and over Wi-Fi) phones
- Screen casting devices
- Wi-Fi-connected facilities environmental control systems such as temperature sensors, lighting, and HVAC controls
- Wi-Fi-connected facilities security systems including door entry security and IP-based cameras
- Wi-Fi localized temperature sensors such as those used in pharmaceutical and food safety refrigeration
- Wi-Fi-connected vending machines, trash, and recycling

These endpoints have locally routed IP addresses (IPv4 or IPv6) and connectivity and data paths the same as other traditional LAN- or WLAN-connected devices such as laptops.

NOTE **LAN-based IoT devices are simply non-traditional endpoints and headless devices on the LAN that require local network access, and possibly Internet access for certain actions. Protocol-translated IoT and protocol-routed IoT are designated as true "Internet of Things" endpoints with a different use case and connectivity model than LAN-based IoT.**

The list here is provided only as an example; many IoT device manufacturers offer multiple connection models. For example, many kinds of sensors could be Wi-Fi connected but may also support deployment on RF in the sub 1 GHz range, and/or via direct cellular connection.

Also, remember that just because an endpoint is network connected doesn't mean it's Internet connected. Many devices described as LAN-based IoT are just local network-connected endpoints, not IoT.

Protocol-Translated IoT

Protocol-translated IoT devices aren't IP-addressed at the edge. Instead, there's a non-IP-based protocol in use along with a protocol gateway or protocol translator.

For context, common protocol-translated IoT technologies include Bluetooth and BLE and Zigbee. This IoT model isn't just for localized and personal LANs, though—it's also common in private WAN technologies (such as Sigfox and LoRaWAN) as well as some industrial automation technologies like WirelessHART. These other non-802.11-based IoT technologies are covered next in "Enterprise IoT Technologies and Non-802.11 Wireless."

In these models, the edge endpoints (which may or may not be in a mesh topology) communicate to a protocol gateway, which converts the data to IP for forwarding and routing to the LAN or Internet.

Protocol-Routed IoT

What's not included in protocol translated IoT are protocol-routed technologies such as Thread and others based on 6loWPAN.

These use IP addressing at the edge similar to LAN-based IoT but are not based on common Wi-Fi connectivity and may use an IP adaptation layer. This is explained more in depth in this chapter's next section.

Other protocol-routed IoT technologies include cellular IoT and private cellular.

For some technologies, an adaptation layer is used to compress (for example) IPv6 to a format that can be transferred over wireless technologies with tiny frame sizes such as 802.15.4 (the layer 1/2 RF standard that Zigbee and Thread use). This is exactly what 6loWPAN offers.

Enterprise IoT Technologies and Non-802.11 Wireless

A book on enterprise wireless wouldn't be complete without all the latest non-802.11 technologies. Those presented here are in varying states of flux and evolution, and so I've chosen to organize this in a non-traditional way and hit the highlights.

Security considerations covered here, instead of being protocol-specific (which change regularly) focus on less mutable characteristics such as the physical layer, topologies, connectivity models, and coverage areas. As these technologies grow over time, the guidance here should persist even through significant flux.

First is a high-level introduction of the various technologies to get familiar with the context, followed by characteristics that factor into connectivity and security architectures:

- IoT Considerations
- Technologies and Protocols by Use Case
- Features and Characteristics Impact on Security
- Other Considerations for Secure IoT Architecture

IoT Considerations

Among the countless connectivity and security considerations, IoT devices collectively are heavily influenced by a few IoT-specific characteristics, including the following features.

Battery Life Most edge IoT devices are not connected to AC power, nor do they have extensive battery or recharging capabilities. While there's been progress in the move toward energy-harvesting capabilities in IoT devices, batteries will remain a part of IoT life at least for the foreseeable future. If you're not familiar with the term, *energy harvesting* is a process to capture energy from renewable sources (such as movement, light, UV, RF, etc.) and convert it to usable electrical energy.

For the rest of the world's IoT devices, power is a precious commodity, and connectivity protocols for IoT are modified to preserve battery life. This means less frequent transmissions, shorter transmissions, antennas that use less power. and at times an inability to communicate with the IoT device between scheduled transmissions.

Form Factor The form factor of the device will dictate in large part the capabilities tied to onboard resources such as memory and processing power, in addition to battery. Although not a hard rule, and certainly not a linear relationship, generally the smaller an IoT device gets, the fewer capabilities it can accommodate. This translates into trade-offs for security and performance as vendor implementations streamline resources by reducing unique encryption keys in storage, bypassing provisioning security, or simply removing processing power required to perform computationally complex encryption.

Location Where the device is, physically, and how accessible it is—to you and to potential adversaries—is an important IoT consideration. When physical security can't be ensured, additional logical protections are needed. And, for unreachable devices deployed in remote areas, access for the purposes of software and security updates may not be viable.

Bandwidth The bandwidth a wirelessly connected IoT device can make use of will vary depending on the technology and deployment. In most cases, endpoints that are physically farther from the wireless base station and those that are extremely power-restricted will use radios that support very small transmission bandwidths. This reduction allows for a greater range in connectivity (by eliminating other interference with broader bands) and aids in preserving battery life. See Figure 8.5 for a high-level comparison of battery and range for several IoT technologies.

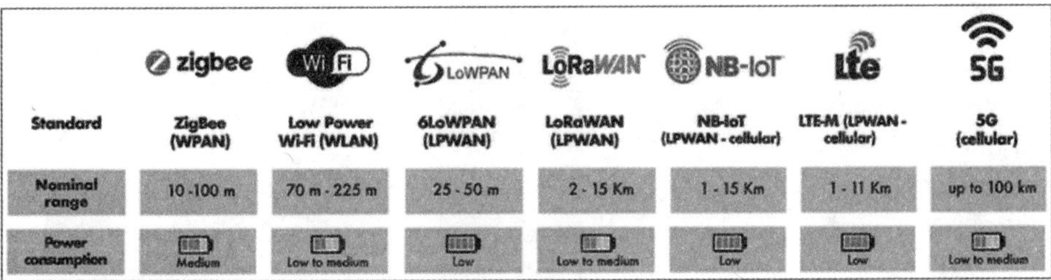

	zigbee	Wi Fi	6LoWPAN	LoRaWAN	NB-IoT	lte	5G
Standard	ZigBee (WPAN)	Low Power Wi-Fi (WLAN)	6LoWPAN (LPWAN)	LoRaWAN (LPWAN)	NB-IoT (LPWAN - cellular)	LTE-M (LPWAN - cellular)	5G (cellular)
Nominal range	10 -100 m	70 m - 225 m	25 - 50 m	2 - 15 Km	1 - 15 Km	1 - 11 Km	up to 100 km
Power consumption	Medium	Low to medium	Low	Low to medium	Low	Low	Low to medium

Figure 8.5: Comparison of power consumption and range for common IoT technologies

`https://www.saftbatteries.com/energizing-iot/impact-communication-technology-protocol-your-iot-application%E2%80%99s-power-consumption`

Technologies and Protocols by Use Case

There are a multitude IoT protocols, many of which are used primarily for home automation and hobbyist projects. The technologies here are of a scope limited to those appropriate for use in an enterprise environment. Other technologies and protocols not specified here follow similar models to the smart building and home automation suite.

We're going to take a brief tour of the following classes of IoT technologies:

- LAN-based IoT
- Bluetooth and BLE
- Smart Building and Home Automation
- Public Cellular for IoT
- Private Cellular and Cellular LANs
- Private WANs
- Industrial Automation

You can see the mapping of IoT technologies to edge IP protocol, connectivity model, coverage, physical layer, and whether it's mesh or not in Figure 8.6. Lighter connectivity lines indicate partial or limited applicability.

TIP This and other graphics and tables can be downloaded at `https://www.securityuncorked.com/books`.

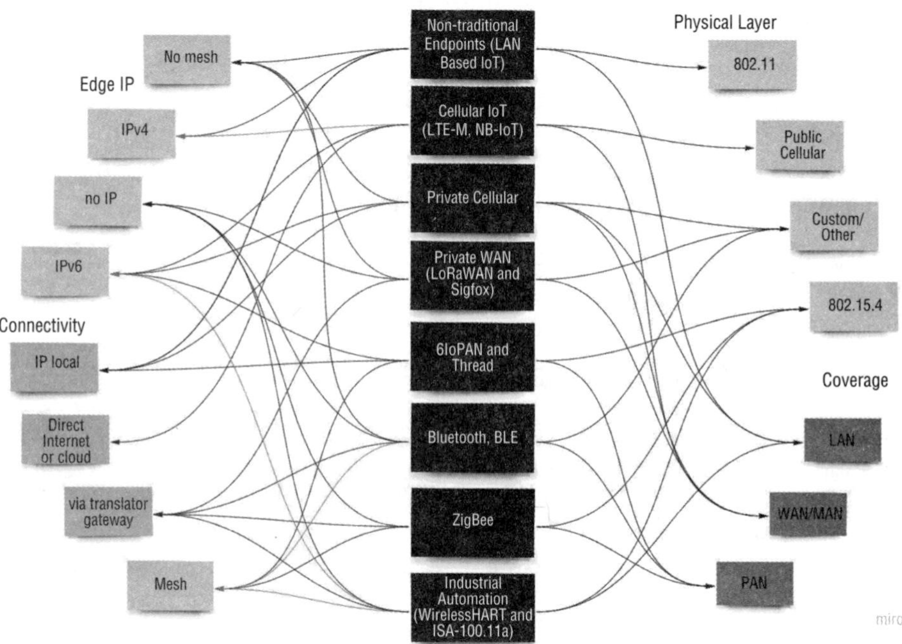

Figure 8.6: Visual map of IoT technologies, their topologies, coverage areas, physical layer, and edge IP model

LAN-based IoT

LAN-based IoT are the pseudo-IoT devices that don't quite earn the "IoT" title but are nevertheless connected devices we're forced to deal with and secure.

A large portion of this book has been dedicated to 802.11-based Wi-Fi security, and I'm not going to repeat all of those concepts here. There are, however, some unique considerations when dealing with the non-traditional endpoints on Wi-Fi networks, and those considerations are addressed here.

Securing LAN-based IoT includes the following challenges and considerations.

Device Identification

Most often, we have three ways to identify an endpoint on the LAN: it was manually provisioned, and the MAC address and other identifiers were documented;

it has a logged-on user that identifies the device; or it has a device certificate that identifies the device.

In the absence of one of those three more deterministic methods, processes for identifying a device is more of a guess. Profiling methods (common in NAC, UEBA, and Wi-Fi products) gain some context by looking at the NIC OUI to determine the manufacturer and by watching aspects of traffic passing through such as DHCP requests. Other profiling methods may layer additional criteria such as mapping open ports or parsing an HTTP header. In total there are about twenty common active profiling mechanisms.

Newer profiling methods incorporate data from watching mirrored traffic on the network to get even more context about the endpoint and what it's doing on the network. Each of these methods offers an increasing level of confidence about the identity of the device, but none guarantee 100 percent accuracy.

When it comes to device identify with non-traditional endpoints, this is by far the most time-consuming task in any project (including and especially NAC projects).

Device Authentication

Device authentication becomes another challenge with many LAN-based IoT devices. Even though connected with familiar Wi-Fi technology, the endpoint may not support authentication (remembering that passphrases aren't considered true authentication).

Most of the more standard network devices including printers and VoIP phones do support 802.1X, and therefore can be authenticated with either a device certificate or device domain credentials. Other endpoints support no authentication but could use MAC Authentication Bypass (MAB, which again isn't really authentication in the infosec sense.)

REFERENCE **Revisit Chapter 3 for a refresher on the types of authentications available for devices including tips for LAN-based IoT.**

Software Updates and Patching

Maintaining basic security hygiene with software updates and security patches is also challenging with many IoT devices (LAN-based and otherwise). It seems in many environments even printers and other readily accessible devices don't receive regular patches.

IoT endpoints without standard interfaces are even more cumbersome to update, often requiring additional management platforms.

Creation and Maintenance of Segmentation Policies

In a perfect world, each grouping of IoT devices would have granular access policies restricting traffic to and from only the requisite resources. In practice, many organizations have a single IoT-designated passphrase-secured network and throw everything there together, with little to no additional segmentation.

Wi-Fi, NAC, and authentication server products all support varying approaches to label and further segment IoT devices. Some offer integrations with external devices such as firewalls (which model traffic and apply labels to the endpoints); others rely on authentication policies with basic dynamic VLAN assignments.

Limited Options for Secure Wi-Fi Connections

Even when properly segmented, many IoT devices don't have robust support for secure Wi-Fi connectivity, making them vulnerable to over-the-air attacks. For example, Figure 8.7 shows a table from current product documentation for a vendor's industrial environmental monitoring systems. The wireless products referenced include sensors for temperature, pressure, humidity, sound, and cryogenic systems, among others, and are marketed to customers in healthcare, pharmaceutical, and food safety.

Network Defaults		Editable Y/N
WiFi Encryption	WPA2	Y
DHCP	Auto	Y
Protocol	HTTP	Y
DNS	8.8.8.8	Y
Destination Port	Port 80	N
Authentication	WIFI Authentication	N
Connection Time	Re-authentication not supported	N
Communication	FQDN: hat1.[illegible]monitoring.com	N

Figure 8.7: Excerpt table from a network connectivity datasheet for a Wi-Fi-enabled sensor product

In the network requirements, the table shows only WPA2-Personal is supported, and communication is by default HTTP over port 80. The port is not editable by the customer. This vendor also claims their system can "effectively reduce the potential for any type of network attack to zero" which is, of course, misleading.

Bluetooth and BLE

I'm a big fan of IEEE and standards, so I'm going to lead off with this. The Bluetooth specifications are no longer based on IEEE standards. Originally, they were part of IEEE 802.15.1 Working Group, but later spun off to the Bluetooth Special Interest Group (SIG, at `www.bluetooth.com`).

In addition, the bulk of Bluetooth devices have been created for personal and residential use—not commercial applications. The use cases have been changing in recent years with the growth of IoT in the enterprise, but there's a lot of legacy technology and the goal of most Bluetooth products is to simplify connectivity, reduce user friction, and reduce support calls.

As you'll see throughout the overview of Bluetooth and BLE, the default mode of operation is the least secure. To make these products appropriate for commercial use requires the product manufacturers to select secure implementations, incorporate those modules into the product, and to regulate allowing backward compatibility that may lead to less secure operating modes.

Bluetooth and BLE in the Enterprise

Classic Bluetooth and Bluetooth Low Energy (BLE) have been fixtures within most enterprise environments for some time. Classic Bluetooth has been well-suited for wireless accessories that require constant data transmission, such as wireless headsets and speakers.

But all that constant connection puts a drain on batteries, and battery life is of particular concern when it comes to IoT devices. The low energy (LE) specification of Bluetooth came along to help alleviate issues related to battery life and offer an alternate model with less power consumption (as well as some additional features).

Whereas Classic Bluetooth is still primarily used for connecting wireless peripherals, BLE expands on that connectivity and supports additional use cases including indoor location services. Just a few of the BLE applications include:

- Connectivity for power-sensitive IoT devices such as medical monitoring devices and industrial sensors
- Connectivity of personal wearables
- Mobile applications with BLE location points of interest
- Mobile applications with BLE door entry systems in hospitality

Bluetooth up through version 5.3 also supports a mechanism to switch to Wi-Fi-based connections. A museum with BLE-enabled points of interest and a mobile app may make use of this; the museum visitors can move throughout the museum and receive exhibit-specific information based on their location. The app might then push a document or video down to the app and could use Wi-Fi for that transfer to make use of higher bandwidths than what's available through Bluetooth (a feature referred to as *alternative MAC/PHY*, or AMP).

Bluetooth vs. BLE

Devices may support both classic Bluetooth and BLE operations simultaneously. Due to the differing battery requirements and use cases, whereas Bluetooth is always connection-based, BLE can be connection-less, meaning devices can interact with one another through listening to beacons instead of formally connecting through pairing.

From a layer 1 perspective, both Bluetooth and BLE use the industrial, scientific, and medical (ISM) portion of the 2.4 GHz spectrum (also used by 802.11 Wi-Fi), but Bluetooth and BLE both use different spectrum hopping methods and channel selection algorithms to avoid interference with other devices in the same airspace, including Wi-Fi.

Network Connectivity and Addressing

Considering network topology and connectivity, Bluetooth does not use IP addressing. Instead, it relies on 48-bit addresses (like a MAC address), the first half of which specify the organizationally unique identifier (OUI), just as in Wi-Fi devices.

Bluetooth creates small *piconets* (up to 8 devices due to 3-bit addressing) and multiple piconets can then connect to form what's called a *scatternet*. On the other hand, BLE uses 24-bit device addressing and could (theoretically) support piconets with millions of devices—although in practice they're much smaller—but this speaks to the expanded use cases of mesh IoT devices such as many sensors.

Provisioning and BLE Secure Connections

Bluetooth and BLE both use similar provisioning processes, which include:

- *Numeric comparison*, where one device displays a short code that can be matched to the other.
- *Passkey entry*, which is available for devices capable of input. The central device can generate a PIN that can be entered on a joining device, or the PIN is set during manufacturing and manually entered on both devices.
- *Just Works™*, which is a very lightweight pairing that uses a null value during the key exchange, effectively offering only connectivity and no security during the pairing process.
- *Out of band (OOB) pairing*, which allows Bluetooth devices to use a different wireless protocol to exchange keys and sensitive data not appropriate for Bluetooth (such as medical or payment card data). OOB could be used to connect capable devices over Wi-Fi or NFC, for example.

By far, the least secure mechanism is Just Works, yet it's the default mode of operation in Bluetooth devices. Other modes can be susceptible to eavesdropping and man-in-the-middle attacks, depending on the version and implementation. For more on why Just Works doesn't work, read the IEEE whitepaper on BLE

Just Works security vulnerabilities, "Bluetooth Low Energy Makes 'Just Works' Not Work," at `https://ieeexplore.ieee.org/document/9108931`.

BLE got a major bump in provisioning security with the BLE Secure Connections, introduced in version 4.2. With that, the devices use Elliptic Curve Diffie-Hellman (ECDH) key exchange, similar to many Wi-Fi operations. This addressed the problem of possible eavesdropping and man-in-the-middle attacks during the pairing process. The idea in legacy implementations was that the connection was only vulnerable during the pairing process, but the reality was that it introduced too great a risk for many of today's commercial and medical BLE applications.

An important note is that BLE 4.2 devices are compatible with older 4.0 and 4.1 versions, however the BLE Secure Connections operation is not—meaning if a more capable BLE 4.2 device is paired with a legacy device, it will not benefit from the increased security.

Speaking of encryption, the following modes are supported and in all cases AES-CCM mode is used after the devices are paired (through the secure ECDH or other mechanisms):

- Encryption disabled for all data
- Encryption enabled for unicast only
- Encryption enabled for all data

Bluetooth Exploits in the Wild

An Internet search will yield countless results for Bluetooth vulnerabilities. Here are a few that made recent headlines.

BrakTooth and SweynTooth Throughout 2020 and 2021, researchers released numerous new findings exploiting even the more secure BLE 4.2 modules. In fact, the BrakTooth suite of attacks in late 2021 resulted in disclosure of 16 new security vulnerabilities on top of exploiting 15 already known.

The result? Everything from nagging denial-of-service attacks to full code execution on the Bluetooth-connected endpoints. Many of the devices tested included boards used in commercial products and industrial automation and monitoring, in addition to LED light controls and gaming systems.

Other devices impacted by the vulnerabilities include smartphones, tablets, and laptops, as Intel, Qualcomm, and others used affected chips. The attacks were executed with equipment totaling less than $20 USD.

The same researchers also released SweynTooth in 2020, discovering 17 new vulnerabilities. In those attacks, researchers were able to trigger crashes, buffer overflows, device locks, and even completely bypass security on BLE devices. These attacks, like BrakTooth, were tested on several chips including those used in smart locks, Fitbits, TSA-compliant luggage locks, smart plugs, and other smart home products. They were also found to be successful against other wearables and medical devices including a blood glucose meter and a Bluetooth-enabled pacemaker shown in Figure 8.8.

The group's research including BrakTooth and SweynTooth can be found at `https://asset-group.github.io/`.

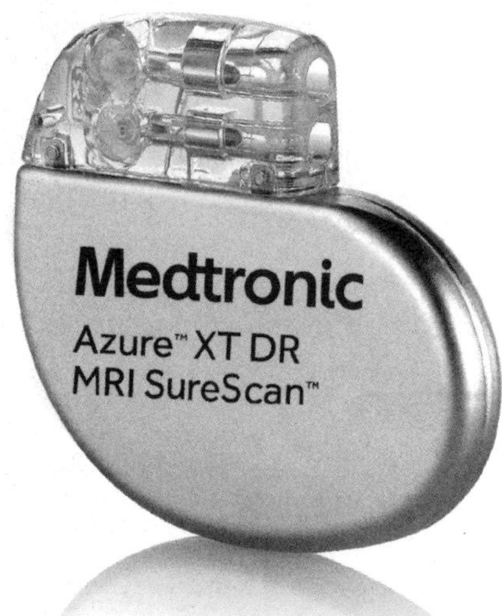

Figure 8.8: The Bluetooth-enabled Azure Medtronic pacemaker was just one of the many products vulnerable to the latest waves of Bluetooth and BLE attacks.
`www.medtronic.com`

Bluetooth Impersonation Attack Tested on more than 28 unique chips, the Bluetooth Impersonation Attack (BIAS) was successful on every device, including chips from Apple, Qualcomm, Intel, and Samsung. The vulnerability allows an attacker to impersonate any device that has been paired in the presence of the attacker.

Hacked e-Scooter In one incident researchers were able to abuse a misconfiguration in Xiaomi brand electric scooters' Bluetooth authentication, and control scooters from 100 meters away—sending unauthenticated commands to lock (stop) scooters, brake, accelerate, and even deploy malware.

You can read a recap of the research and accompanying video "Xiaomi Electric Scooters Vulnerable to Life-Threatening Remote Hacks" at `https://thehackernews.com/2019/02/xiaomi-electric-scooter-hack.html`.

Bluetooth and BLE attacks could go on for pages. My only reason for including these few examples is to demonstrate three points. First, even the newer versions of the BLE products are susceptible to attacks, and those attacks are non-trivial and can be life threatening in certain products. Second, the vulnerabilities are being disclosed almost daily. This proves the point that any specific guidance for Bluetooth and BLE specifically will be out of date quickly. Last, but not least, these attacks have been successful on popular wearables and smartphones (including Apple and Android), which further calls into question the use of peer-enabled protocols in a secure enterprise environment (a soapbox topic in Chapter 6).

I'm not suggesting organizations impose a moratorium on consumer devices, but they should be carefully considered, and the enterprise environment should be secured from these devices as best they can.

INDUSTRY INSIGHT: NIST GUIDANCE ON SECURING BLUETOOTH AND BLE

For Bluetooth Classic and BLE security, it is recommended to read and follow the latest security guide provided by the U.S. National Institute of Standards and Technology.

The current documentation is the "Guide to Bluetooth Security," NIST Special Publication (SP) 800-121 Rev. 2 available at `https://csrc.nist.gov/publications/detail/sp/800-121/rev-2/final`.

Smart Building and Home Automation

Aside from the ever-pervasive Bluetooth and BLE products, smart building and home automation products have evolved to use technologies based on the 802.15.4 physical layer such as 6loWPAN, Thread, and Zigbee—all of which differ from Bluetooth not only in the physical connectivity standard but also in coverage models.

Zigbee

Zigbee is the outlier here, being a non-IP-based protocol. Although a step up from Bluetooth, Zigbee has not gained the traction in large-scale deployments due to limitations of its mesh architecture and lack of IP stack. Because Zigbee (along with Bluetooth) doesn't use edge IP addressing, a translator is required to convert Zigbee protocol into IP protocol for forwarding or routing to and from IP-based networks.

Despite a few alliance agreements here and there, Zigbee has been fizzling out in recent years, replaced by newer, faster, and more scalable stacks like those based on 6loWPAN. Zigbee has been rebranded as part of the Connectivity Standards Alliance (`www.csa-iot.org`), along with Matter and other technologies.

Thread and 6loWPAN

Thread is one such communication stack based on 6loWPAN and therefore IPv6. All of these (Thread, 6loWPAN, and Zigbee) work on top of the 802.15.4 physical layer, which specifies tiny packet sizes of 127 bytes. The challenge is, IPv6 frames are about ten times that size at 1280 bytes. I'm about to switch from bytes to bits; one byte is eight bits. Also contributing to the length crisis, IPv6 addresses are long at 128-bits (for context, an IPv4 address is 32-bits).

6loWPAN is simply an adaptation layer that allows sending IPv6 over 802.15.4; it does this by compressing the IPv6 address and headers and handling fragmentation. Figure 8.9 offers a visual of the difference in ratios of IPv4 to IPv6 addressing and of the frame sizes of 802.15.4 versus IPv6. Effectively, 6loWPAN's task is to shrink IPv6 to about one-tenth its original size, in order to transfer it over 802.15.4.

```
0000 0000 0000 0000 0000 0000 0000 0000
0000 0000 0000 0000 0000 0000 0000 0000 0000 0000 0000 0000 0000 0000 0000 0000 0000 0000 0000 0000 0000 0000 0000 0000 0000 0000 0000 0000 0000 0000 0000 0000

8888 8888
8888 8888 8888 8888 8888 8888 8888 8888 8888 8888 8888 8888 8888 8888 8888 8888 8888 8888 8888 8888
```

Figure 8.9: Visualizing the ratio of IPv4 to IPv6 addresses (top) and 802.15.4 packet size versus IPv6 standard packet size (bottom)

Figure 8.10 shows the Thread communication stack and the placement of 6loWPAN as the adaptation between layers 2 and 3 of the traditional OSI stack. Layers 1 and 2 of Thread (and 6loWPAN) use 802.15.4. Unlike Zigbee, Thread stops short of the application layer, allowing for extensible uses

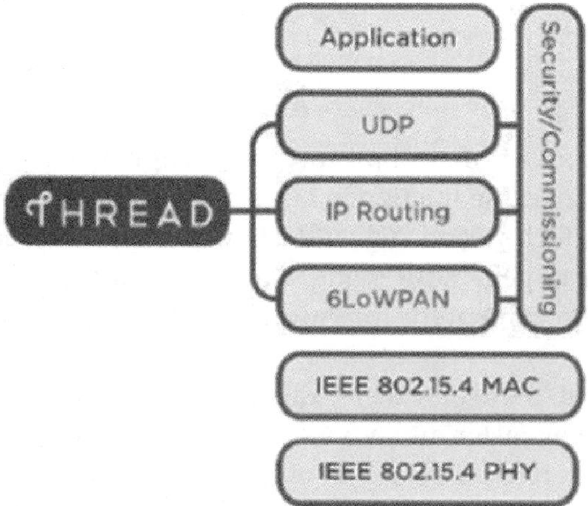

Figure 8.10: The Thread communication stack
www.threadgroup.org

The result of this stack and the IPv6 compression is that Thread (like many of its IoT cousins) requires a Thread-capable router to convert the local (compressed IPv6) addresses to something routable with full addressing.

I mention this because the language on the Thread Group site and documentation can be a little misleading and confusing, as it repeatedly notes there's "no need for a proprietary hub, gateway, or translator." That's true but it (and any 6loWPAN communication) does require a border router to convert the addressing and shortened headers.

The best way I've heard it described is like network address translation (NAT). If you contrast the private IP address space with publicly routable IP address spaces, the private address (such as 192.168.1.12) wouldn't be able to communicate directly with 4.2.2.2. Instead, it would go through a router or firewall that would convert the source and destination addresses so the publicly routed data can find its way back to the private IP address.

There is a major difference between 6loWPAN and non-IP technologies like Bluetooth and Zigbee. The border router is just that—it's routing the packets to and from the local (compressed) IPv6 addressed endpoints. Bluetooth and Zigbee aren't speaking IP at all, and their traffic must fully be translated to IP to interact with IP networks.

Public Cellular for IoT

As with everything else IoT-related, cellular connections needed a rework to be IoT-friendly. Standard cellular communication protocols are designed for high-bandwidth and low-latency data applications like video and voice, but IoT devices require much smaller bandwidth, and many want shorter communication time to preserve battery life.

Cellular IoT protocols address needs of:

- Low power utilization
- Low cost
- Reduced transmission frequency
- Extended connectivity ranges
- Low bandwidth

Public cellular (as opposed to private cellular, covered next) operates in licensed RF spectrums by mobile network operators (MNOs, or carriers). All current cellular technology is packet-switched, and IP-based for voice, data, and messaging.

Table 8.3 summarizes the recent generations of cellular technologies. Technically 4G-LTE is the marketing term for an enhanced version of 3G, and not full 4G. 4G-LTE, while much better than 3G, never met the specifications for

4G by the ITU Radiocommunication Sector (ITU-R), but it allowed older phone technologies to take advantage of faster speeds. The use of 4G LTE has become mainstream now since consumers tend to associate it with something better than just "4G." The standards are a little fuzzy in cellular all around. Today's 5G deployments are non-standalone (NSA) deployments that rely on a 4G network core, meaning just like 4G, our 5G isn't really 5G quite yet.

Table 8.3: Cellular generations and technologies

GENERATION	TECHNOLOGIES
3G (~2000)	UMTS, CDMA2000
4G (~2010)	LTE- Long Term Evolution
5G (~2020)	NR - New Radio

Cellular standards are managed by the 3GPP group (`www.3gpp.org`), which has a similar relationship to cellular standards as the IEEE does to 802.11. We hear about entire generations of cellular such as 3G, 4G (and 4G LTE), and 5G (NR), but in between these major families are smaller revisions and updates. In these releases, starting in 2008, 3GPP has introduced options to support IoT needs with direct cellular connectivity.

For context, the 5G in the wild now is release 15, which is considered phase 1. Phase 2 is attached to release 16, where the 5G benefits are to be fully realized based on the specifications. Release 17 will expand on 5G with industry-specific features; it was due for release late 2021 but 3GPP put the work on hold hoping to resume in-person meetings, and the new dates were mid-2022.

The technologies described here are all based on 4G/LTE, with the exception of Category NB1, which is 5G-ready as well. Within the realm of IoT-friendly cellular are those shown in Table 8.4.

Table 8.4: Excerpt of 3GPP releases and IoT-friendly classes

RELEASE	CATEGORY	MAXIMUM DOWN/UP	BANDWIDTH	FEATURES AND USE CASES
8	1	10 / 5 Mbps	1.4 - 20 MHz	Higher bandwidth, lower latency: use cases digital signage, audio/video
12	0	1 / 1 Mbps	1.4 - 20 MHz	First major IoT enhancement with power saving and reduced throughput

RELEASE	CATEGORY	MAXIMUM DOWN/UP	BANDWIDTH	FEATURES AND USE CASES
13	NB1 (NB-IoT)	0.68 / 1 Mbps	180 KHz	Support endpoints that don't roam; half duplex; very low bandwidth; low energy; use cases smart metering
13	M1 (eMTC)	1 / 1 Mbps	1.4 MHz	Supports endpoints that roam/move; enhanced power save modes; use cases outpatient monitoring, asset tracking

Cat-1 supports higher bandwidths and lower latency (great for IoT with voice or video) than the other IoT protocols but is based on standard LTE protocols and therefore doesn't offer the power and range benefits introduced later.

Among other issues, the challenges with cellular technologies for IoT is that they're designed for a certain level of interaction with the endpoints, which is not conducive to IoT devices that need long sleep times to preserve battery.

The remaining categories from the table (Cat-0, NB-1, and M1) all include significant changes to accommodate IoT, especially in the realm of power saving, with NB-1 supporting deployments with 10-year battery life.

Each of the standards come with pros and cons, and therefore are appropriate for different use cases.

SECURITY CONSIDERATIONS IN CELLULAR TECHNOLOGIES

Since 4G isn't really 4G, and 5G isn't really 5G, and security features are added with every revision, discussing security for cellular can get messy. Today's 5G deployments are limited to capabilities of a 4G packet core, but as hardware is replaced and new revisions released, the specs and implementations will change. With that understanding, detailed analysis of cellular security is outside the scope of this book.

But it's important to know that vulnerabilities exist in cellular, just as with Wi-Fi and every other wireless technology. Whether deployed as public or private cellular infrastructures, organizations should track risks in cellular as with any other networking technology.

And, while vulnerabilities exist in all cellular, public cellular and private cellular will have slightly different risk footprints when it comes to those. Attacks may be more applicable to one than the other.

For the most part these vulnerabilities have been disclosed in limited circles simply because the tools for attacking cellular have been less accessible than those for Wi-Fi, Bluetooth, and other IoT protocols. As 5G and private cellular are more widely adopted, that trend will change.

As the world transitions to 5G technology, network function virtualization (NFV) and software-defined networking (SDN) allow for much more flexibility for use case support. Combined, these support technologies such as network slicing, which allows 5G radios to service multiple devices with unique needs, simultaneously. If one device is a tiny IoT sensor that communicates infrequently and in short, small bursts and another device is using high bandwidth to support virtual reality (VR), 5G will be able to manage those needs independently with proper power, range, and bandwidth that's device-specific.

That brings a marked change to how IoT has been handled in 3G and 4G, and it opens up Pandora's box of IoT use cases. Figure 8.11 shows conceptual 5G network slices serving both a low-bandwidth IoT device and a high-bandwidth streaming media device. The cellular radio can adjust to meet the disparate power, bandwidth, and range demands of both at the same time.

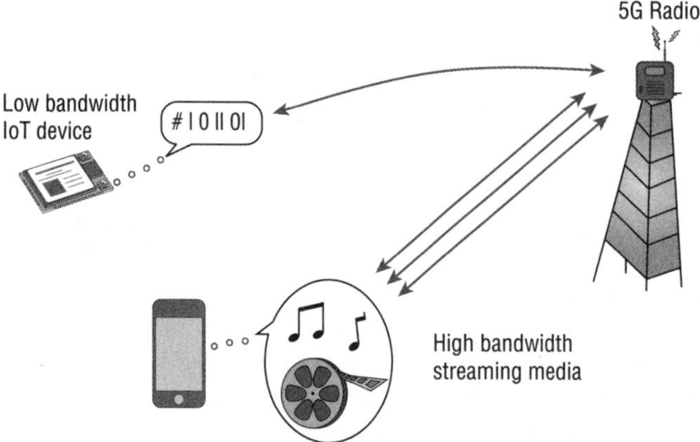

Figure 8.11: 5G network slices can divvy up airtime and bandwidth to meet varying needs.

INDUSTRY INSIGHT: MORE ON 5G AND CELLULAR TECHNOLOGIES

For a truly deep dive into cellular including security and IoT, check out *5G Second Phase Explained* (Wiley ISBN-10: 1119645506).

This book covers the 3GPP Release 16 enhancements and offers an authoritative and essential guide to the new functionalities while offering historical comparisons to prior generations of cellular technologies.

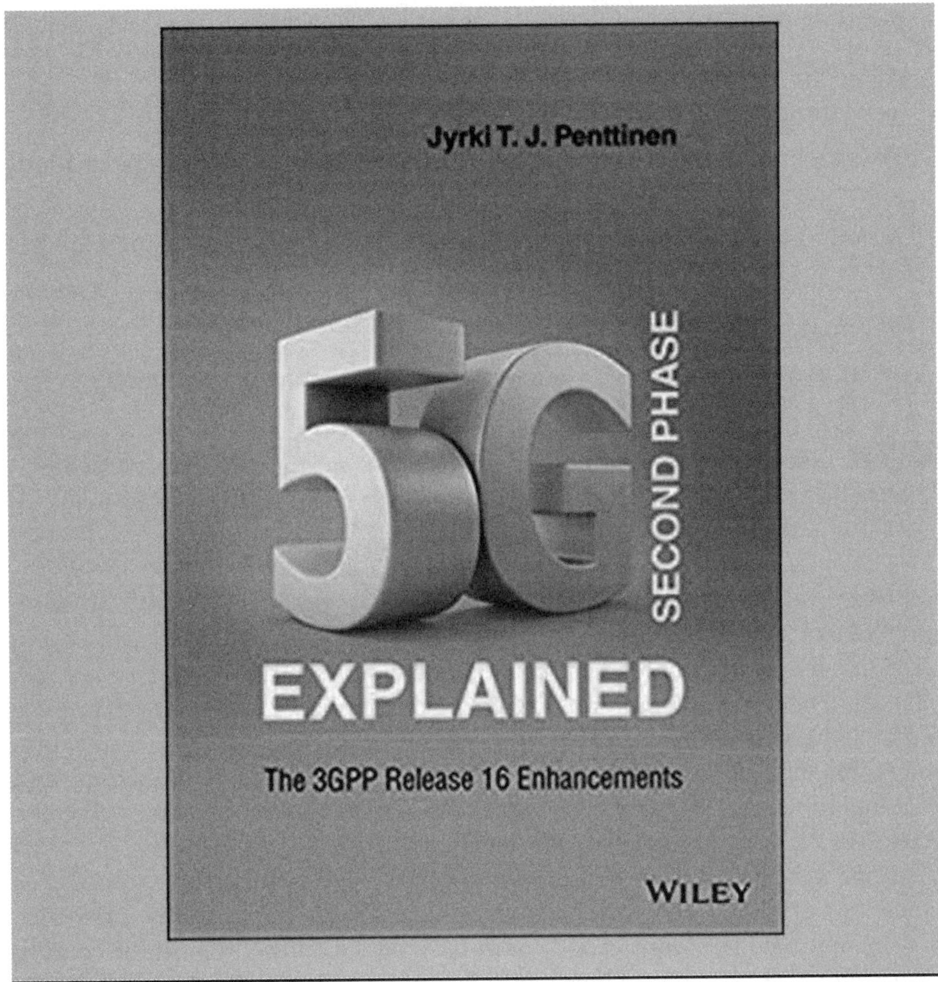

Private Cellular and Cellular LANs

We're switching gears here from IoT technologies supported by *public cellular* network carriers to the use of *private cellular* technology including the use of cellular within the enterprise LAN.

Private cellular solutions are designed for private use—meaning an enterprise organization has local infrastructure for its own connectivity. It's not shared with public users, and not meant for Random Randi Person walking through the building with their Vodaphone or Verizon device. Technically, there's one exception to this: an organization with a private cellular infrastructure could choose to deploy a neutral host network, but that's purely optional and outside the scope of this book.

Use cases for private cellular and cellular LANs include:

- Healthcare monitoring, collaboration, and paging
- Manufacturing autonomous guided vehicles
- Educational institutions' long-range backhauls for in-home student connectivity
- Retail point of sale and digital displays
- Push-to-talk applications in any vertical
- Smart building connectivity
- Security and facilities indoor, outdoor, and backhaul connectivity

TIP **Cellular LANs use cellular technology within the enterprise to deliver WLAN-like connectivity for specific use cases. Other private cellular models can be used for fixed wireless over longer distances.**

Private cellular LAN is the new kid on the wireless block, but it's growing quickly. Private cellular networks take cellular connectivity features and bring them into the organization for use as a local network, hence the term *cellular LAN*.

In some regions, a full private cellular deployment doesn't have to involve a mobile network operator (MNO), or carrier, at all. Instead, the organization may own and manage its own cellular LAN infrastructure, and the data isn't required to be passed over a carrier network. Effectively, it can be used just like an 802.11 WLAN but using cellular technology.

I'll spend a bit more time on the topic of cellular LANs than other technologies since it's very likely network and wireless architects across many industries will be involved in planning and/or deploying private cellular in the coming years. Figure 8.12 offers a simplified view of a cellular LAN. Not pictured is the cellular packet core (a bit like a souped-up Wi-Fi controller), which may be located on-prem, or in a public or private cloud.

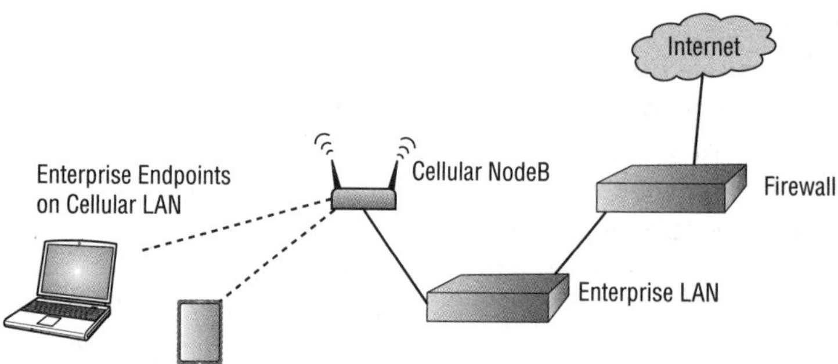

Figure 8.12: Cellular LANs are deployed much like Wi-Fi.

This overview of private cellular technology covers the following:

- Current State of Private Cellular by Country
- Deployment Models and Architectures
- Service Delivery Options
- Licensing and Spectrum Utilization
- Comparison of Cellular to Wi-Fi
- Considerations for Private Cellular

Current State of Private Cellular by Country

If you haven't heard of private cellular networks yet, it's probably because spectrum for this use case is just now being opened up around the globe.

The following is the current state as of early 2022, with the expectation that by the time you're reading this, the status for some countries will have changed.

In the U.S., the designated RF for private cellular is the Citizens Broadband Radio Service (CBRS) spectrum—in the 3.5 GHz band nestled neatly between 2.4 GHz and 5 GHz without interfering. On that spectrum, we can use both 4G and 5G technology.

Outside of the U.S.—Japan, Norway, Germany, the Netherlands, Poland, and Sweden—have all identified spectrum for private cellular use. Following a slightly different model, Australia, China, France, the Czech Republic, along with Hungary, Portugal, South Africa, and Spain offer leased spectrum for private cellular use. As of writing, spectrum allocations in Canada, Mexico, Brazil, and Turkey are to be determined.

Throughout Europe, many EU member countries have specified or are considering use of frequency bands n77 and n78 for 5G private networks. These operate in the 3.4-3.8GHz and 3.8-4.2GHz spectrums for n78 and n77, respectively. The UK has settled on n77 as well as band 40 and is exploring additional spectrum.

Deployment Models and Architectures

Cellular technologies have their own vernacular and architecture that doesn't correlate directly to Wi-Fi, aside from the concepts of radios in a cellular Radio Access Network providing connectivity conceptually as a Wi-Fi AP would. Behind the radios are a packet core in cellular, which offers a breadth of authentication and coordination functions. The packet core could be likened to a Wi-Fi controller but that's a grossly overly simplified correlation.

The two areas I want to cover are the deployment architectures and models that describe how private cellular is used, and how it may be integrated into the existing enterprise LAN:

- Deployment Architectures
- Deployment Models

Deployment Architectures Deployment architectures describe how, when, and where private cellular networks are used within an enterprise.

Private Cellular LAN for Endpoint Connectivity Private cellular can be deployed as a cellular LAN, offering endpoint connectivity both indoor and outdoor. This most closely mimics an 802.11 WLAN, with the notable exception that the RF technology used is not based on 802.11 standards.

Endpoints connecting directly to a cellular LAN require either a physical SIM card, a software-based eSIM, or an adapter/dongle.

The upcoming section "Comparison of Cellular to Wi-Fi" will offer more details about how cellular LAN compares to Wi-Fi.

NOTE While private cellular can connect phones, the use cases extend well beyond that. Supported endpoints include laptops, tablets, handheld scanners, and IoT devices along with CPE equipment.

Private Cellular Backhaul and Fixed Wireless Private cellular can be deployed in fixed wireless and backhaul architectures as well. In these deployments the goal is to offer point-to-point or point-to-multipoint connectivity acting as a wireless bridge, versus servicing endpoints directly.

One deployment option is to use private cellular over a distance to connect to remote gateways that can convert cellular to Wi-Fi. The use case of this architecture would be a municipality or school that may want to extend Internet access to employees or students in rural areas without reliable broadband access. Municipalities and various commercial organizations also use remote gateways to service endpoints in hard-to-reach locations where wired cabling is not possible or is cost prohibitive. Endpoints in this use case include video cameras, IoT sensors, and remote payment kiosks such as parking meters.

PRIVATE CELLULAR CASE STUDY: CITY OF TUCSON, AZ

In the U.S., private cellular has been the go-to solution for many municipalities and schools to extend network services and Internet to underserviced areas. The city of Tucson serves as a great case study for the use of private cellular to bring value to its citizens. Through 2021, the city deployed about 40 base stations, covering a 44-square-mile area. Private cellular was selected after the IT team determined it would take more than 7,000 Wi-Fi access points for the planned coverage area.

> The target coverage area included rural communities with no access to broadband. The project was initiated at the start of the COVID pandemic to ensure all citizens had equal access to the Internet.
>
> Since then, the city has extended the deployment use cases to include wireless on buses and is exploring IoT connectivity for municipal services such as traffic lights, fire services, and utilities.

Deployment Models Deployment models describe the level of integration into the enterprise LAN, and the option to extend services between the cellular and traditional LANs.

Private cellular LAN integration is vendor-dependent, as some offer only overlay solutions that are fully managed, others offer enterprise-managed LAN-integrated solutions, as well as everything in between.

The architecture and management model will also influence cost. For fully managed services there's obviously an ongoing subscription fee. With consumption-based models (like Amazon Web Services), customers pay based on the level of capacity and throughput.

Regardless of the pricing models, current solutions can be loosely categorized into parallel, overlay or integrated LAN solutions.

Parallel to LAN Many MNO-managed solutions are installed as a dedicated second network, parallel to the existing LAN. In this model, the cellular LAN infrastructure is deployed with a dedicated network infrastructure, cabling, and connectivity.

In this deployment model, there is no direct connection or integration between the private cellular network and the existing enterprise LAN. Traffic between the two networks would have to be routed externally. In this model, the cellular packet core is hosted and managed by the provider and is not located within the enterprise environment.

Within the realm of parallel solutions, the features and requirements may vary greatly. Some offerings are designed only for vendor-specific traffic (such as one offering from Motorola currently that prioritizes its voice and push-to-talk traffic only).

Considerations and variations in requirements for solutions deployed in parallel to the LAN include:

- Offerings may require dedicated cabling for cellular APs
- Some offerings require a full parallel infrastructure including switching, routing, and a separate broadband connection
- Offerings may be limited to specifying QoS for certain devices and applications

- Offerings with a remote packet core and no local handoff will introduce undesired hairpin routing and potential latency
- The provider will own and manage the packet core, meaning the enterprise does not have full control over the data or data paths for privacy

Overlay to LAN Private cellular deployments that rely in part on the existing enterprise infrastructure may be deployed as an overlay to the LAN. From an architecture perspective, the overlay model falls between the parallel and integrated deployment models.

Depending on the solution, the packet core may be located within the enterprise environment, or it may be offsite and managed by the provider. Most overlay solutions will provide the RAN but rely on the enterprise's existing switching, routing, and Internet connections.

Considerations for overlay solutions include many of those covered previously in "Parallel to LAN," and some items from the upcoming section on "Integrated to LAN" also apply.

Integrated to LAN Solutions that can be fully integrated to the enterprise LAN bring several benefits, and I'm partial to these solutions personally, whether delivered as fully managed solution or not.

In broad terms, with a cellular LAN solution, it may be integrated to the LAN in a few ways. And, more specifically, a fully featured solution will offer use of all integration options on a per-use case basis—similar in concept to how we're able to tunnel or bridge Wi-Fi traffic, except we're talking about the handoff between the cellular packet core and the LAN.

Currently available options to integrate the cellular LAN to the enterprise LAN include core handoffs that can:

- Bridge cellular traffic to the LAN
- Route cellular traffic to the LAN
- NAT cellular traffic to the LAN

One or more of these handoff techniques can be used based on the needs. Aside from the data paths, other integrations to the LAN and network services may include the following:

- DNS
- DHCP
- Specifying authorization from enterprise RADIUS/AAA
- Mapping endpoints to enterprise VLANs
- Applying ACLs to cellular LAN traffic to and from the enterprise LAN

For solutions that are integrated, the level of integration will vary by vendor. Some may offer basic local handoffs, while others offer deeper integrations with network and domain services including RADIUS.

Figure 8.13 shows how cellular LANs can be integrated into the enterprise LAN including support for routing to/from the LAN, bridging to VLANs, and NAT.

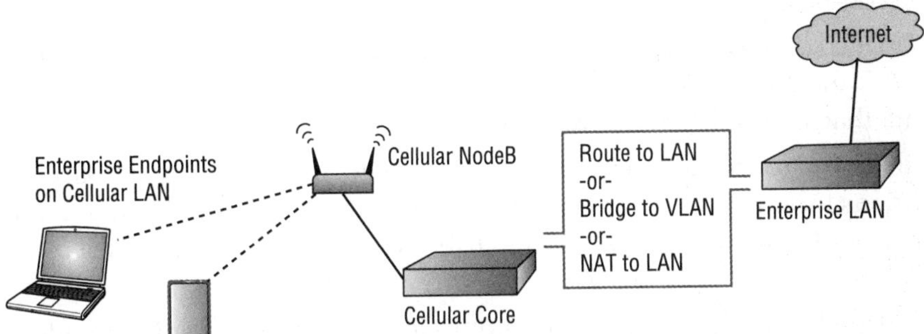

Figure 8.13: Cellular LANs can be integrated into the enterprise LAN several ways.

The deployment model (overlay or integrated) will dictate how tightly the services of the enterprise LAN and cellular LAN can be integrated. Obviously, in a pure overlay model with all traffic routed out to the Internet (and back into the enterprise if needed), the options for integrating services such as VLANs and RADIUS are limited.

Service Delivery Options

As with all topics in this chapter, the status of private cellular is in flux almost daily and varies by region. The statements about service delivery models are generalizations based on what's in the market today:

- Managed Service by a Mobile Network Operator
- Managed Service by a Third Party
- Enterprise-managed

Managed Service by a Mobile Network Operator Private cellular solutions may be offered by MNOs, and in some regions that may be the only option. A main consideration for MNO-provided private cellular is the privacy of the enterprise data. Just as with public cellular data, if an MNO is in control of, or has visibility into the enterprise data or metadata, there is opportunity for them to share or monetize that. It's also possible the contract may include clauses affording the MNO additional liberties. Reviewing the contract and privacy statements is always advised with any managed service, whether from an MNO or any third party. Most MNO-provided solutions follow the parallel to LAN deployment model.

Managed Service by a Third Party Private cellular solutions can also be provided as a fully managed service by non-MNOs and other third parties. In these instances, the infrastructure is managed by a third party, but the third party is not an MNO. This can assuage some of the data privacy and performance concerns of MNO-provided offerings, but as always, check the contracts carefully.

Managed deployments may be fixed-fee models or based on consumption. Check the billing structure carefully, as some services include billing for data that stays within the enterprise.

Fully managed deployments are vendor-specific, and some may impact the options for integration into the enterprise LAN, offering only an overlay option. At least today, backhaul and fixed wireless solutions are most often provided as managed services.

Managed solutions range in level of touch and control. Some providers may not allow enterprise IT teams to manage or co-manage the cellular LAN or endpoints at all, while others offer more flexibility.

Enterprise-managed In many regions, a private cellular solution can be deployed and/or managed by the enterprise IT team. This is especially easy with cellular LAN deployments where the solution can be managed very similarly to Wi-Fi.

For enterprises that wish to manage their own infrastructure, the requirements vary by region. Some require license and certification for specific use cases. For example, deploying CBRS in the U.S. requires involvement from a CBRS Certified Professional Installer (CPI). The level of touch depends on whether the solution is deployed indoor or outdoor.

TIP **For enterprises interested in managing their own cellular LAN, the private cellular vendors will often offer free services for assisting in planning and design.**

Licensing and Spectrum Utilization

Private cellular technologies can run in licensed spectrum, unlicensed spectrum, and shared spectrum with different models:

- Leased licensed spectrum
- Shared/coordinated spectrum
- Other dedicated spectrum
- Unlicensed spectrum (supported but not widely used)

As a comparison, *public cellular* (like your mobile phone) operates in licensed RF spectrums. Depending on the geographic region, *private cellular* may operate in licensed, unlicensed, or lightly licensed (shared) spectrum.

Leased Licensed Spectrum Today's cellular connectivity uses allocated spectrum that's licensed to various carriers. RF spectrum is managed per region or country. MNOs can lease portions of their licensed spectrum to an enterprise or entity for private cellular.

TIP **At time of writing, some countries rely on MNOs to provide private cellular solutions. In others, enterprises are free to provision and maintain their own private cellular infrastructure. Spectrum allocation is also regional, with some countries designating committed spectrum, many using leased spectrum, and others are to be determined.**

Shared/Coordinated Spectrum Private cellular in many regions including the U.S. uses a "shared" or lightly licensed spectrum that relies on coordination services, which are always subscription-based. In the U.S. the dedicated spectrum is CBRS, or LTE band 48. As you'll see in "Comparison of Cellular to Wi-Fi" this sits around the 3.5 GHz spectrum.

Specifically for CBRS, there are three tiers of access prioritization: incumbents, priority, and general. Incumbents are afforded the highest priority and include fixed satellite services, radar, and defense. Priority access licenses (PALs) are available for a portion of the dedicated spectrum. PALs are bid on regionally and those entities are given higher priority than general access (for that spectrum). General access is the level most used for localized cellular LANs and is everything else that isn't formally registered as incumbent or priority.

Since it's coordinated spectrum, to enforce the three access tiers CBRS includes a central coordination system called Spectrum Access System (SAS).

Other Dedicated Spectrum In some regions, private cellular uses dedicated spectrum that's not necessarily shared or coordinated but is also not a slice of an existing MNO spectrum. This varies by country and is in flux. For example, in the UK, spectrum may be applied for and granted for a small fee. The model is similar to coordinated spectrum but is static in nature through the manual application and allocation process.

Unlicensed Spectrum (Supported but Not Widely Used) It is also possible to deploy private cellular in unlicensed spectrums (as Wi-Fi does). This is supported by technology such as MulteFire but is not widely used. Using cellular technology on unlicensed spectrum forces it to behave more like Wi-Fi, which negates many of the RF benefits of cellular.

NOTE A full introduction to private cellular technology is outside the scope of this book. I'm focusing on the features and integration that are most applicable to wireless architecture, and specifically aiming to provide enough context of private cellular to help you determine whether it should be considered for specific use cases.

Comparison of Cellular to Wi-Fi

In many cases, cellular technology can offer specific advantages over other wireless. If you reflect for a minute on the different experiences you've had with mobile phones versus Wi-Fi, you'll begin to understand some of the advantages cellular connectivity might bring.

From an RF perspective, private cellular can offer more resilience and coverage than Wi-Fi. It can also operate at higher density, longer distances, and in harsher (RF) noise environments, making it ideal for certain applications. Plus, cellular has guaranteed QoS and SLA due to scheduling versus Wi-Fi's contention-based operation. This feature alone has a significant impact on the reliability of the connection and support for latency-sensitive applications.

TIP Every type of wireless technology has pros and cons, and the trick is to understand those along with the use case and project requirements. Private cellular is just one option but is exceptionally viable for many enterprise use cases, as you'll see here.

Table 8.5 touches on a few technology comparisons of private cellular to 802.11 Wi-Fi technology. As always, there are many other factors to consider when deciding what wireless connectivity is best for a given use case. This information is selected to demonstrate the drivers behind moving some enterprise applications from Wi-Fi to private cellular or cellular LAN connectivity.

The details following Table 8.5 focus on comparisons of:

- Distance/Coverage
- RF Resiliency, Spectral Efficiency, and Interference
- Device Identity and Authentication
- Transmission Scheduling, QoS, and Roaming
- Access Hardware
- Supporting Clients
- Traffic/Data Path

Table 8.5: Private cellular vs. Wi-Fi

FEATURE	PRIVATE CELLULAR	TRADITIONAL WI-FI
Coverage distance	Long, Kms (~25k sq ft / 2k sq m indoors; 1M sq ft outdoor)	Short, meters (~2k sq ft /180 sq m indoors and varies outdoor)
Resiliency on RF	High resiliency	Medium-low resiliency, depends on network design
Device identity and authentication	Inherent and mutual / SIM-based	Configured with services, profiling, authentication server, user login, passphrases, or certificates
Device density	High	Low
RF spectrum	Varies by region (dedicated 3.5 GHz CBRS band in U S ; shared/ coordinated); some countries use public cellular licensed RF	2.4 GHz, 5 GHz, 6 GHz
RF interference impact	Low/limited, depends on spectrum	Medium-high; all Wi-Fi devices and others on spectrum above, including microwaves, Bluetooth
Transmission scheduling	Coordinated and guaranteed	Contention-based using CSMA/CA
Security	Native and integrated	Overlay, varies
QoS	Integrated and configurable	Configurable, varies by Wi-Fi vendor and deployment
Roaming & latency	Seamless, very low latency	Varies greatly by deployment and endpoint capability
Access hardware	NodeB cellular AP	802.11 Wi-Fi AP
Supporting clients	Most phones, tablets, many laptops, growing number of handhelds and IoT devices, all tested for conformance and interoperability	All 802.11 endpoints (ideally Wi-Fi Alliance certified but remains optional)
Traffic/data path	Depends on deployment and region but can be LAN -> WAN and/or ISP	LAN -> WAN and/or ISP

There's a lot to unravel here, so let's look at a few of these comparisons and offer additional context. For readability, some topics in Table 8.5 will be consolidated and detailed together, and others are omitted from further explanation. The highlights are as follows:

Distance/Coverage Layers 1 and 2 of cellular are substantially different than Wi-Fi. The different modulation and encoding techniques mean cellular signals are viable well beyond the threshold of Wi-Fi.

We describe how well a wireless signal can be heard over the ambient noise floor in terms of decibels (dB), as a relative signal strength. The absolute power is defined in units of dBm (decibels relative to a milliwatt), which is a logarithmic (not linear) scale. Without boring you to death with RF math, in Wi-Fi we usually design for something in the neighborhood of –65 to –75 dBm, which gives us a net positive of around 15 dB signal above the noise floor.

Cellular technologies use small subcarriers, which means the endpoint can still receive and decode data with much weaker signals. By comparison, 4G/LTE coverage may be designed in the –95 to –110 dBm range. Both Wi-Fi 6 and LTE are using orthogonal frequency-division multiple access (OFDMA) modulation but applied differently. In addition, cellular clients can transmit at a higher power than Wi-Fi, meaning they can communicate farther distances from the radio.

The result is that a single 4G/LTE eNodeB (cellular AP) can cover three to four times the area as Wi-Fi indoors, and ten times the area outdoors. That's significant.

As an example, Figure 8.14 shows a vendor's budgetary network planner for a 500,000 sq ft convention center. This tool offers a rough estimate and doesn't take the place of a proper design, but the output estimates just 22 cellular APs, versus what would easily require over 125 Wi-Fi APs (and that's being very generous to Wi-Fi).

RF Resiliency, Spectral Efficiency, and Interference Proper Wi-Fi design can get complicated quickly. The 2.4 and 5 GHz RF spectrums used by Wi-Fi are crowded, and the transmission is contention-based (something covered in "Transmission Scheduling, QoS, and Roaming" shortly). In addition, in Wi-Fi, it's the endpoints that make decisions about roaming and connectivity. There are numerous roaming protocols, and not all endpoints support all protocols, further adding complexity.

The combination of the spectrum interference, contention mechanisms (endpoints having to fight for airtime), and the often unpredictable behavior of endpoints render the resilience of Wi-Fi over the air quite volatile.

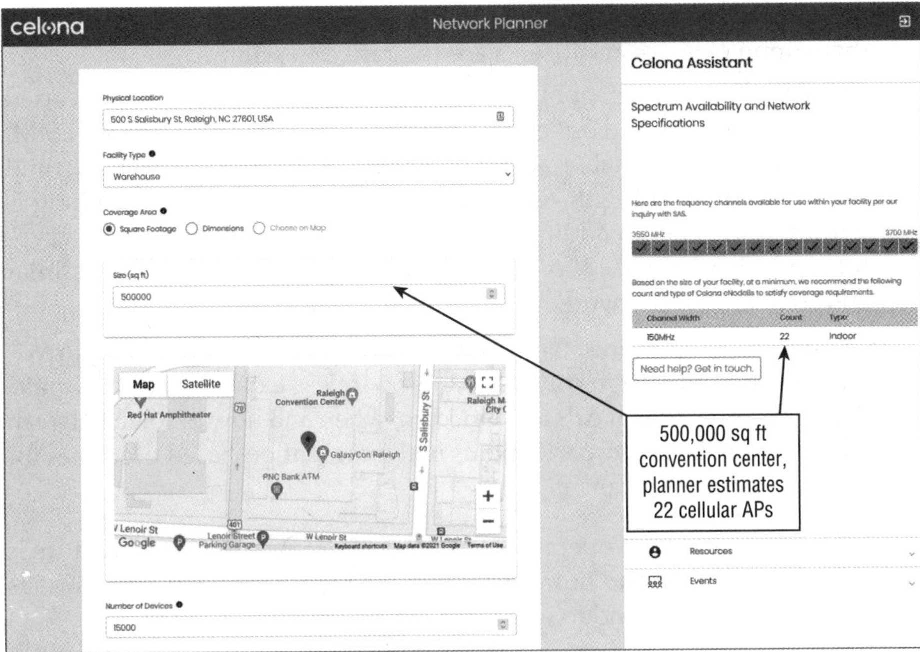

Figure 8.14: As an example of the coverage capabilities, this vendor's rough estimate for a 500,000 sq ft facility is just 22 cellular APs.
www.celona.io

Conversely, private cellular technologies work in a less cluttered (at least for now) and a more strictly managed spectrum, include guaranteed scheduling of transmissions, enforce system-coordinated roaming, and (as described in the prior section) can operate effectively with much weaker signal strength. The system coordination extends to spectral efficiency as well; in cellular, the coordination by the cellular APs allows for spectrum re-use and very granular scheduling. This brings the added benefit of allowing for much denser endpoint populations in the same airspace and even on the same radio.

Device Identity and Authentication As someone who's worked with network access control (NAC) technologies for many years, device identity and authentication is always a pain point. Earlier in this chapter, challenges with endpoint identity and authentication were captured in "LAN-Based IoT."

In Wi-Fi, if an endpoint doesn't have a user attached, we don't always know what it is—an especially daunting task with the insurgence of IoT in the enterprise. Simply identifying a device becomes a complex dance of profiling mechanisms and parsing network traffic from DHCP requests and/or mirrored ports. The level of assurance varies, and at best the IoT endpoints have an assertion of identity (most often based on MAC address), and rarely, if ever, true authentication. IoT endpoints that support device certificates are the rare exception.

Cellular technologies, public and private, use unique identifiers coded into the hardware and linked to a physical SIM or software-based eSIM. The identity is immutable, and private cellular endpoints are known and provisioned, meaning the enterprise knows exactly what something is without guessing. In addition, the unique device IDs along with cryptographic keys are used for mutual authentication between the cellular endpoint and the network.

Transmission Scheduling, QoS, and Roaming As hinted earlier, in Wi-Fi, the endpoints are in control of their wireless experience. They make decisions about which APs to connect to, when and how to roam between APs, and they're active participants in the contention-based processes for securing airtime to transmit.

Cellular technologies operate the other way around; the infrastructure is in control of when and how the endpoint connects, when it transmits, as well as when, how, and to where it roams.

That coordination is made possible by the mandatory RF feedback loop in cellular technology using channel quality indicator (CQI). These metrics let the cellular infrastructure know the RF conditions as experienced by the endpoint. That data allows the infrastructure to direct the endpoint to use the best client modulation through adaptive modulation and coding (AMC). The process can be compared conceptually to 802.11k and 802.11v in Wi-Fi with the understanding that client support of Wi-Fi protocols is optional and the experience inconsistent.

TIP **As an analogy to contention-based versus traffic scheduling, consider this: When driving a vehicle through an intersection, Wi-Fi would be a four-way stop and cellular would be a traffic light.**

Because of that one fundamental differentiator, cellular is able to guarantee time slots and create an environment of predictable performance. It's a good fit for applications that are latency-sensitive or require strict service level agreements (SLAs) and QoS. In fact, not only does cellular have an absolute command of the scheduling, but 5G technologies introduce network slicing and support QoS over the air to a great degree.

As you've probably learned through your journey with this book, Wi-Fi can get pretty dang complicated. Even on its best days, Wi-Fi networks will have collisions, congestion, and large variations in performance due to 802.11 contention mechanisms. Clients do what they want and often operate off-standard. Toss in the occasional monkey wrench of poor RF design, hidden node, and misbehaving clients, and it's a mess.

When it comes to roaming, the variability with Wi-Fi endpoints is just as troublesome. There are numerous standards and variations in endpoint support for Wi-Fi. Certifying devices (infrastructure and endpoints) with the Wi-Fi Alliance is purely optional, and even for vendors that pursue it, they may opt for the lowest requirements and not test and certify critical features such as Fast BSS Transition for secure roaming and WPA-Enterprise operation to support 802.1X.

In Chapter 4, in "Understanding Wi-Fi Design Impacts on Security," we took a deep look at the roaming technologies in Wi-Fi and the supplemental protocols required to facilitate fast and secure roaming between APs on encrypted networks. That same section discussed at length the challenges with supporting both security and speed and the resulting common use of less secure networks to support low-latency applications.

Cellular has been designed to support voice and has inherent mechanisms for ensuring both QoS and seamless, low-latency roaming.

NOTE **Already, LTE networks are in the sub-20ms handoff, and 5G promises speeds around 2ms. Compare that to the Wi-Fi handoffs that may take hundreds of milliseconds.**

Figure 8.15 demonstrates the roaming decisions in Wi-Fi (by the endpoint) versus roaming in cellular where the packet core makes the decision about endpoint roaming.

Figure 8.15: Comparison of roaming decisions on Wi-Fi (left) versus cellular and cellular LANs (right)

TECHNOLOGY THROWBACK: CELLULAR CONCEPTS IN WI-FI

In the mid-2000s, Meru Networks (since acquired by Fortinet) had a novel Wi-Fi offering that allowed for single channel deployments (e.g., all APs could be on the same channel). The technology also introduced virtual cell and virtual port, which delivered (among other things) a virtually zero-handoff roaming experience.

Does any of that sound familiar? If so, it's because the Meru technology borrowed several concepts from cellular. The entire Wi-Fi system was carefully orchestrated, and the infrastructure used patented secret sauce to trick endpoints into relinquishing control of transmission and roaming decisions.

Unfortunately, the product was a bit *too* flexible in configuration options and introduced a high degree of complexity. Plus, the single channel architecture didn't scale well in high-density deployments. After the 2015 acquisition by Fortinet, the legacy Meru technology was slowly phased out.

Access Hardware Where Wi-Fi has APs, cellular access radios are referred to as *NodeBs*, specifically eNodeBs in 4G, and gNodeB in 5G (for Evolved NodeB and Next Generation NodeB, respectively).

You'll also see references in cellular terminology to a *packet core* or *mobile core*. In private cellular LANs, this is a virtual or physical appliance that acts much like a Wi-Fi controller. It handles many functions like managing the cellular APs, authentication, routing, and applying QoS policies and access control policies.

Figure 8.16 offers a view of both an indoor and outdoor cellular AP. Very similar to Wi-Fi APs, most cellular nodes (or cellular APs) run on PoE and standard cabling. Unlike Wi-Fi the reach of private cellular is much greater, especially outdoors.

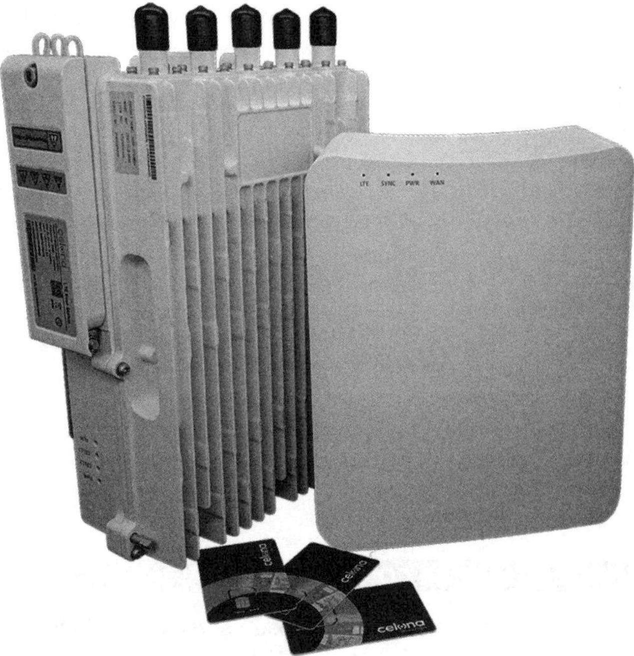

Figure 8.16: Cellular nodes (or cellular APs) look and operate very much like Wi-Fi APs. Pictured here are two models from Celona, one outdoor and one indoor, along with SIM cards. www.celona.io

Supporting Clients One area Wi-Fi has a huge leg up on cellular is with the breadth of endpoints that support Wi-Fi, but that's been changing since 2020. It's obvious that mobile phones and many tablets have cellular support, but what's less obvious is that many of today's laptops support it as well.

Endpoint adoption of the CBRS (band 48) in the U.S. has jumped significantly, and other regions of the world are following suit. To date, just some of the devices that support CBRS include Apple iPads; Dell, HP, and Lenovo laptops; Microsoft Surface devices; Panasonic rugged laptops; Zebra and other handheld scanners, and chipsets from Qualcomm.

For devices that don't natively support SIM, eSIM, or have private cellular radios, there are adapters that can be used with most IP devices to convert them to cellular endpoints.

TIP For U.S. implementations, a list of devices supporting CBRS is maintained by the OnGo Alliance at `https://ongoalliance.org/certification/fcc-authorized-end-user`.

Traffic/Data Path With cellular LANs, the data path can be identical to an enterprise Wi-Fi data path. And, depending on the vendor, it can be far more flexible than that of traditional controller-based APs. Cellular's use of NFV and microservices means the core duties could be distributed and virtualized on-prem or in a public or private cloud.

Security and Privacy in Private Cellular

Cellular brings several advantages in terms of security. Its inherent device identity offers the foundation required for data confidentiality and integrity. Looking back to the IAC (integrity, availability, and confidentiality) triad of infosec, cellular can help in all three aspects, and can do it natively, without requiring external products or services.

In the prior topics I've shared several bullets and explanations of the possible security benefits of cellular technologies. There are a few additional points and overarching themes to call out for private cellular LANs specifically.

Identity, Authentication, and Encryption Cellular has inherent strong device identification and authentication using SIM/eSIMs, eliminating the need to "guess" what mystery endpoints are for specifying access control. This is also used for strong encryption over the air and extended through to the packet core and beyond.

Cellular LAN products can come pre-hardened, and most can offer added integrity through authentication of the system components to one another, natively—meaning the endpoints authenticate to the cellular APs (or

NodeBs), which in turn authenticate to the core, which authenticates to a central management platform, if applicable.

Most private cellular deployments support the same AES-128 bit encryption comparable to WPA3-secured networks, with AES-256 on the near horizon—encryption comparable to the WPA3-Enterprise 192-bit, CNSA suite. In the Wi-Fi ecosystem, to get the same level of integrity and confidentiality requires the addition of NAC-type products, and/or a PKI infrastructure with device certificates.

Segmentation Also, cellular (especially 5G) has intrinsic segmentation over the air and private cellular systems can extend this end-to-end throughout the LAN infrastructure. Some products can slice and encrypt over the air at the granularity of a single flow.

Segmentation can be extended to or through the enterprise LAN, depending on the product and implementation. With integrated solutions, the cellular slice or segment can be carried through to the wired side and handed off to the LAN just like any other wired or Wi-Fi traffic, and segmentation can be preserved, changed, or modified at that point, based on policy.

Data Privacy Integrated cellular LANs have the potential to keep enterprise data in control of the organization and eliminates privacy concerns related to passing data through carriers/MNOs or other managed service providers.

The privacy statements are vendor- and deployment-dependent. Solutions offered by MNOs and/or as a fully managed service by a third party may have visibility into certain aspects of the data or metadata.

As with all hosted services, read and vet the provider's privacy statement that details how your organization's data privacy will be maintained, and whether the provider may be giving or selling data to third parties.

Centralized Security and Risk Management Cellular LANs that can be integrated to the enterprise LAN aid in centralizing visibility and control, allowing organizations to maintain security compliance and incorporate IoT connectivity into existing risk management models and zero trust strategies.

REFERENCE The earlier sidebar "Security Considerations in Cellular Technologies" applies to private cellular infrastructures as well. Cellular offers some enhancements in security over other technologies but it's not impervious to attacks.

A SIDEBAR ON NEUTRAL HOST NETWORKS

Neutral host networks (NHN) are outside the topic of private cellular, and beyond the scope of this book but warrant mentioning.

The NFV and SDN technologies applied in 5G architectures will, for the first time, divorce the cellular architecture and data from specific hardware, meaning cellular services can be virtualized and delivered in many ways.

One of those ways will be through non-MNO infrastructures, and that's what NHNs will offer. Organizations can deploy private cellular infrastructure and then offer virtualized carrier services through their hardware. The organization plays no role in authorization or authentication—all NHN traffic is tunneled from the cellular radio to the MNO.

With NHN, the organization can use its private cellular infrastructure to service internal private cellular endpoints in parallel with public cellular services for guests. The cellular APs segment and tunnel connections securely to each appropriate packet core.

It's being considered a much more viable option for indoor coverage than DAS and other small cells including LAA and LTE-U, which have proven to interfere with Wi-Fi.

Private WANs

There are many types of IoT WAN technologies, and most fall into the low power WAN (LP-WAN) category. Cellular IoT technologies also fall in this bucket, but the private WAN technologies differ in their use of proprietary, non-standard, or unlicensed connectivity over long distances.

Private WAN technologies, like other LP-WANs, offer connectivity over coverage areas ranging from municipal- to nation-wide. I'm omitting the spectrums for these since they vary per region.

These and other LP-WANs are used to connect IoT devices including over longer distances, and specific use cases include:

- *Agricultural*, for monitoring weather conditions, silo and tank levels, soil data, and temperature
- *Smart buildings*, including smoke alarms, security alarms, monitoring for rodent infestation, and garbage collection
- *Smart cities*, for monitoring water facilities, management of street lighting, structural integrity sensors, parking management, and connected digital signage
- *Logistics*, for tracking vehicles, warehouse security, and food safety monitoring
- *Utilities*, including consumption meters, remote facility monitoring, supply monitoring, and surveillance

Two private WAN technologies are Sigfox and LoRaWAN. I'm only providing a brief overview of these technologies because they (and the next topic of industrial automation) are a bit more of a niche than the other technologies here.

Sigfox Sigfox wholly owns the connectivity technology from the cloud services to the endpoint software, offering the solution much like a carrier model. The endpoint hardware, though, is open market.

Sigfox enjoys a much larger client base in Europe than the U.S. and other parts of the world (see Figure 8.17). Among other challenges, in the U.S., the designated 900 MHz frequency is a bit crowded, and the FCC limits the time-on-air. Also, only one Sigfox network can be deployed in any area, which is limiting. Originally designed for unidirectional sensor communication, it has limited bidirectional support.

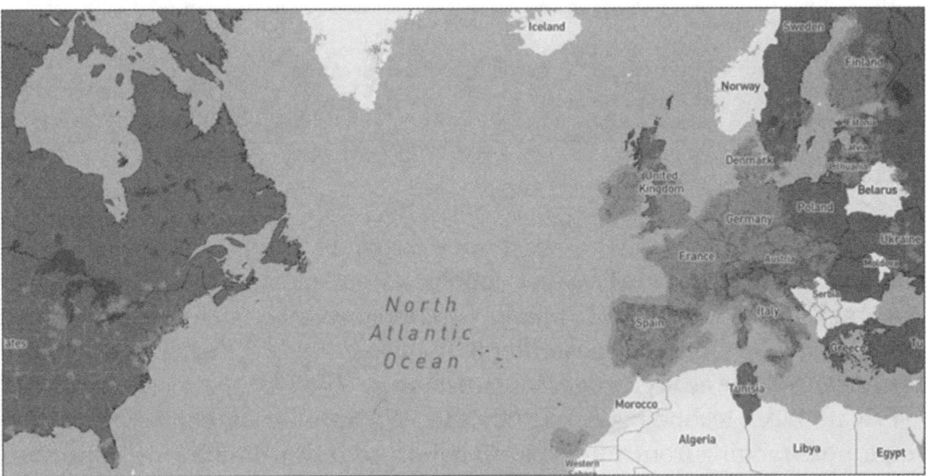

Figure 8.17: Sigfox adoption is focused in about a dozen European countries and a few smaller Asian countries, shown in the medium grey tones such as France and Germany. `https://www.sigfox.com/en/coverage`

Although Sigfox touts a presence in 75 countries, implementations outside Europe, Japan, South Korea, and Thailand are spotty, at best. In early 2022, Sigfox announced its intentions to file bankruptcy and seek a buyer, which will likely influence the future of this technology.

More information on Sigfox can be found at `www.sigfox.com`.

LoRaWAN LoRaWAN has similar use cases to Sigfox, but a business model that's opposite. The LoRa specifications are available for all to use through the LoRa Alliance, but there is only one hardware vendor for the LoRa radios—Semtech. For coverage context, the entire country of Belgium (12k sq mi/30k sq km) is blanketed with just seven LoRaWAN gateways. The LoRa Alliance can be found at `https://lora-alliance.org`.

Recapping, Sigfox is a service offering and a platform that's wholly owned, and the radio hardware is open source. Sigfox manages the network as a service (just like a carrier) and charges based on an OpEx model. LoRaWAN, on the other hand, offers a fully open ecosystem but only one hardware vendor, and unlike Sigfox, enterprises can build and operate their own LoRa network.

There are many other technical differentiators such as the modulation, bandwidth, encryption, and uni- versus bi-directional communication models.

LoRaWAN specifies encryption using AES-128 and offers some enhancements such as separation of authentication and encryption keys. On the other hand, Sigfox doesn't encrypt messages anywhere in its stack—something left to the customer to address.

Industrial Automation

Ready for more acronyms? Industrial automation technologies covered here fall into the low-rate wireless personal area networks (LR-WPAN) category—not to be confused with the LP-WANs just covered. And these are also based on the 802.15.4 layers 1 and 2, along with 6loWPAN, Thread, and Zigbee covered earlier in the chapter.

While all of these collectively fall into the industrial-IoT (IIoT) heading, the two mentioned here—WirelessHART and ISA100.11a—are particularly well-suited for industrial automation, the convergence of operational technology (OT) with wireless IT.

Industrial automation technologies are used for many industry 4.0 applications in a few primary categories, including:

- Safety
- Monitoring
- Control

NOTE **Wireless connectivity for industrial automation is focused on the second bullet, monitoring. With today's technologies, safety mechanisms are too critical to rely on wireless connectivity, and wirelessly connected control mechanisms are reserved for specific use cases and architectures.**

WirelessHART and ISA100.11a WirelessHART and ISA100.11a are the two main industrial automation connectivity protocols. Both used for similar applications, the decision of which protocol to use comes down to vendor selection.

WirelessHART is the protocol of choice for vendors such as Emerson, Siemens, and ABB, among others. ISA100.11a is mostly supported by Honeywell and Yakogawa.

Because of that, it's unlikely the wireless or network architect will be in the position to specify the protocol, but here are a few key differentiators to consider when securing these networks.

Network Topology and Security Both based on the 802.15.4 standard, ISA100.11a was primarily designed as a star topology, while WirelessHART uses mesh extensively. From a network management perspective, ISA100.11a supports administrative configuration of how devices and routers are allowed to communicate.

ISA100.11a doesn't define the full application stack and can therefore be extended to support Fieldbus, Profibus, HART, and other legacy protocols.

Another key differentiator is the use of IP protocols, with WirelessHART not using IP at edge devices, and ISA100.11a (like Thread) built on 6loWPAN and therefore IPv6 at the edge. And, although based on 802.15.4, both WirelessHART and ISA100.11a modified the layer 2 operations, meaning they're not strictly implementing the 802.15.4 standard.

Provisioning options are similar between WirelessHART and ISA100.11a, and both support a suite of security keys including join keys, network keys, and session keys. WirelessHART requires all keys, but ISA100.11a specifies join and session keys as optional.

The systems will ship with default keys to get started, but best practices entail creating new keys and exchanging them out-of-band to prevent eavesdropping and exposure of the keying information.

See Figure 8.18 and Figure 8.19 for WirelessHART and ISA100.11a architectures, respectively.

Features and Characteristics Impact on Security

All of the IoT technologies mentioned so far have varying (but often overlapping) features and characteristics, such as the RF spectrum in use, the topology, whether they're IP-based or not, etc.

While the protocols, security features, and implementations of those technologies may evolve over time, their more static characteristics offer some long-standing guidance on considerations for securing these networks.

Following is a brief overview of the considerations for IoT and non-802.11 wireless as it relates to:

- Physical Layer and RF Spectrums
- Coverage
- Edge IP Protocols
- Topology and Connectivity

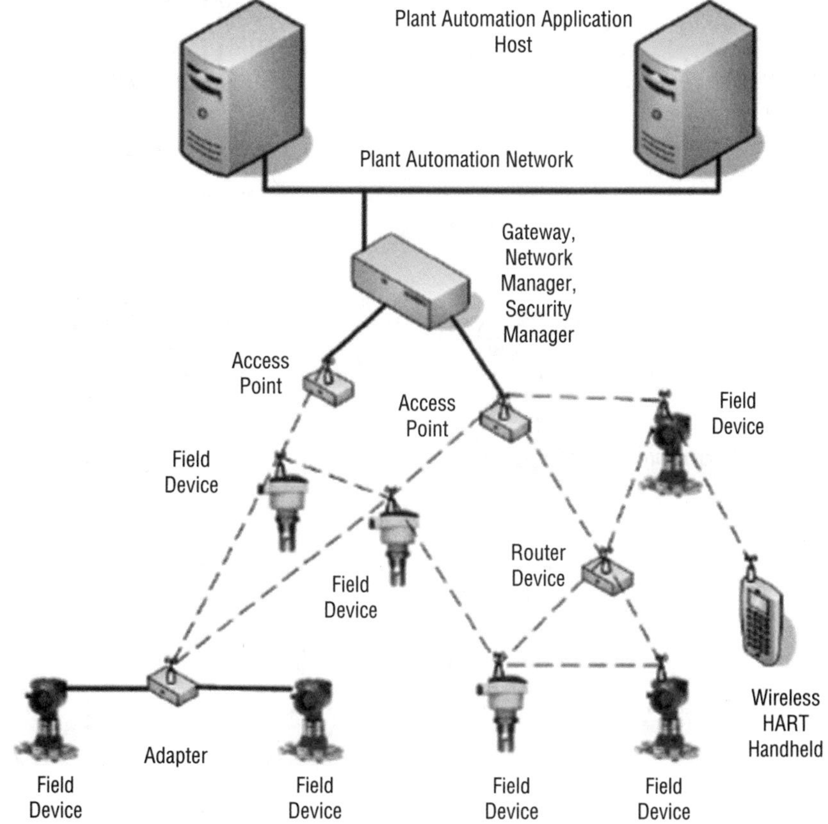

Figure 8.18: Diagram of WirelessHART architecture
`www.emerson.com`

Physical Layer and RF Spectrums

The rainbow of wireless connectivity standards ranges from sub-1 GHz up through 50 GHz and beyond. Most of the technologies covered here operate in the 800–900 MHz (~400 MHz in some parts of Asia) to 2.4 GHz (802.11, 802.15.4, and Bluetooth), 3.5 GHz (CBRS and other private cellular), 5 and 6 GHz (802.11), and other cellular spectrums with 5G.

And of course, these spectrums range from licensed to unlicensed (such as 802.11 and 802.15), and everything in between with the shared spectrum (lightly licensed) CBRS space in the U.S.

The same principles covered in Chapter 7 in the section "Security Monitoring and Tools for Wireless" apply here. For security monitoring over the air, the tools must have visibility into that spectrum. Whether a narrow or wide band sensor is used, it needs to cover the spectrum(s) in scope, and of course, multiple sensors can be used as appropriate for varying location and radio needs.

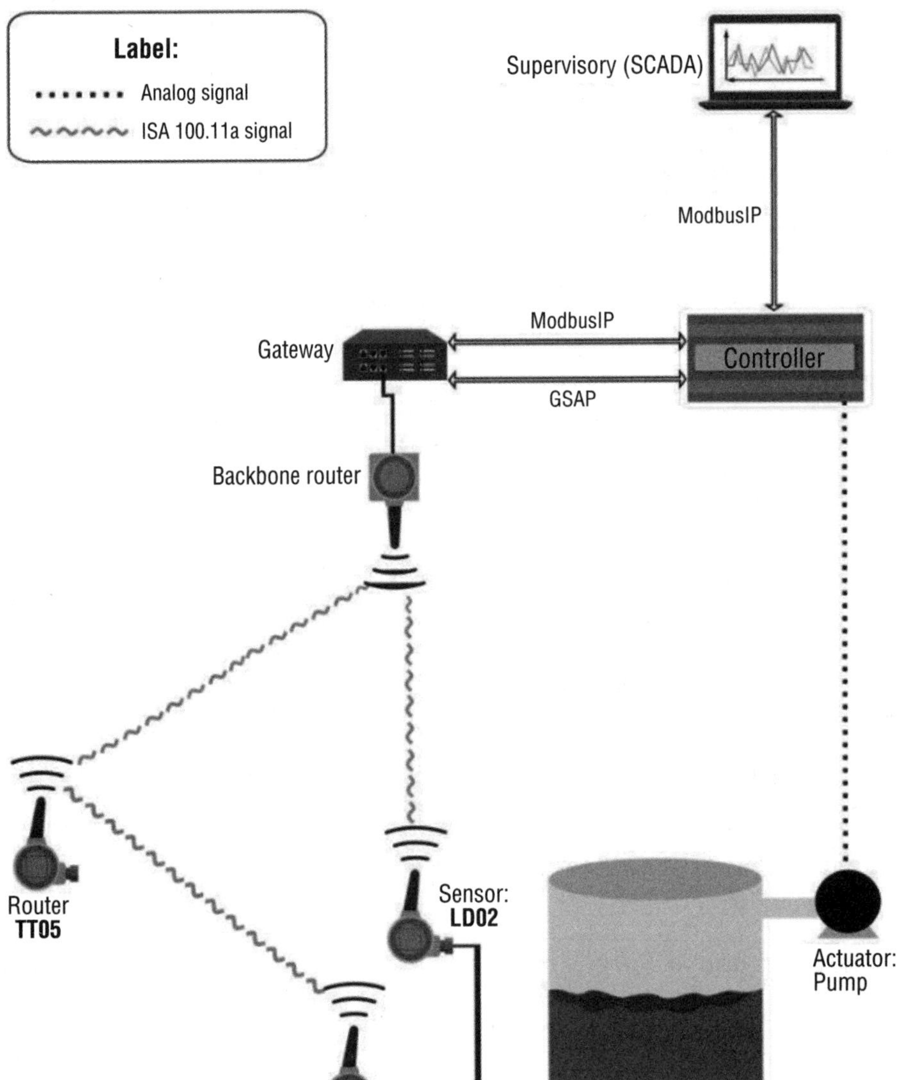

Figure 8.19: ISA100.11a sample architecture
`https://www.ncbi.nlm.nih.gov/pmc/articles/PMC7570649/`

Coverage

Connectivity models here range from very localized personal area networks (PANs) to local area networks (LANs), and metropolitan and wide area networks (MANs and WANs). Coverage models for IoT devices include:

- *Personal area networks (PANs)*, including Bluetooth, 802.15.4-based technologies, and localized industrial automation

- *Local area networks (LANs)*, including 802.11, in-building private cellular, and broader coverage industrial automation
- *Metropolitan and wide area networks (MANs/WANs)*, including cellular IoT, outdoor and fixed private cellular, and private WANs

PANs and LANs are easily monitored and managed due to their localized nature, while MANs and WANs present a bit more of a challenge.

Security considerations for coverage models include the impact on:

- Over-the-air monitoring ability
- Accessibility
- Physical security of endpoint

As always, over-the-air monitoring means security sensors must be in proximity to the assets to be monitored (in addition to requiring appropriate radios). For PANs and LANs in the enterprise environment, it's easy because if you have Wi-Fi APs, you have a tool for at least some basic monitoring. For those longer-range MAN and WAN deployments, depending on the nature and criticality of the network, security monitoring may be desirable; in those cases, the tools in Chapter 7's topic of "Spectrum Analyzers and Special-Purpose Monitoring" offer a few options.

For provisioning, updates, and general manageability of the endpoints, the accessibility of those devices is also a factor. Again, PANs and LANs are going to be easier than, for example, a sensor that's buried a meter under soil 10Km away, or a pressure sensor embedded in a bridge. Consider the accessibility in planning ongoing maintenance, and if endpoints aren't accessible after deployment, plan for contingencies up front.

Aside from your ability to access an endpoint for maintenance, consider accessibility of the device by an adversary. If someone were to get his or her hands on one or more endpoints, consider what risk that poses, and plan for mitigations prior to deployment.

Edge IP Protocols

Across the half dozen or so groups of technologies presented, some use IPv4 addressing at the edge, others IPv6, and still others use none at all:

- *IPv4 addressing*, used in 802.11 and older cellular
- *IPv6 addressing*, used in all 6loWPAN, Thread, cellular, some 802.11, and ISA-100.11a
- *No IP addressing*, used in Bluetooth, BLE, Zigbee, Sigfox, and WirelessHART

Whether endpoints are IP-capable or not usually speaks to their other capabilities in terms of memory and processing—features that impact their ability to perform computations for encryption and decryption, and to store multiple keys. The result is that, while a specification may define options for multiple levels of keys, the implementation may reduce or eliminate keys, or downgrade to a lesser protocol during provisioning/pairing or encrypting. Thus, the IP stack (or lack of) will often influence security decisions related to:

- Endpoint capabilities
- Requirement for translator gateways
- Protocol vulnerability and mitigation
- Enterprise visibility and security controls

Technologies that don't have IP addressing at the endpoints will require a translator gateway device to convert data to IP-routable packets for connection beyond the local network. By virtue of them being the Internet of Things, IoT devices will at some point require IP connectivity. Translator gateways can introduce both a point of failure and a point of vulnerability.

The protocol(s) in use will, of course, impact the specific vulnerabilities and threats for the devices as well. While many translate, IPv4 attacks vary from IPv6, and non-IP-based attacks use different tools and tactics. Conversely, for security monitoring tools, visibility into IoT devices is usually easier with IP-based communications.

NOTE **IEEE published the article "Roadmap of Security threats between IPv4/IPv6" capturing the IPv4 vs IPv6 threats to IoT, available for purchase at** `https://ieeexplore.ieee.org/document/9422653`**.**

Topology and Connectivity

The network topology is another security consideration. Networks described here are sometimes IP local connected, directly connected to the Internet, connected via a translator gateway, and/or mesh—with the note that mesh is not mutually inclusive or exclusive of the other three. And some technologies support more than one topology.

- *IP local connectivity*, used in 802.11, private cellular LANs, 6loWPAN, Thread, and ISA-100.11a
- *Direct routing to Internet*, available in cellular IoT

- *Connectivity via a translator gateway*, required with Bluetooth/BLE, Zigbee, LoRaWAN, Sigfox, and WirelessHART
- *Mesh topology*, used to varying degrees by BLE, 6loWPAN, Thread, Zigbee, and WirelessHART

The data path of any communication will, by necessity, impact the security architecture. As with prior bullets, security monitoring and policies are well equipped to monitor and control local IP-based traffic. It's a trickier task for endpoints that directly connect to the Internet, and for technologies that communicate on non-IP protocols and require translation. The provisioning processes and security of the devices is also influenced by the connectivity method.

Non-IP networks that require a translator gateway, as mentioned earlier, present both a point of failure and point of vulnerability. Additional monitoring may be required for these infrastructures.

In addition, mesh topologies introduce additional considerations for both system resiliency and availability and security. Mesh uses the endpoint devices as routing hops, and the exact functions vary by protocol and implementation. Where, how, and when data is pushed through the mesh is a factor in securing mesh—how the data is encrypted, by what keys, and when is a critical piece of information. The time-to-live or hop maximums also factor into the equation for availability and throughput. How trust and peering in a mesh is established is another consideration.

Topology impacts security considerations primarily related to:

- Data path
- Vulnerability of translator gateways
- Mesh protocols

Other Considerations for Secure IoT Architecture

The four preceding topics have captured most of the factors impacting decisions around securing IoT devices, and specifically non-802.11 devices.

Those are summarized here in brief:

- Over-the-air monitoring capabilities
- Accessibility of IoT devices and gateways by admins and adversaries
- Endpoint capabilities including memory, processing, battery
- Networks requiring translator gateways
- Vulnerabilities specific to protocols

- Monitoring and mitigations specific to protocols
- Data path based on topology
- Use of mesh routing

In addition, there are just a few additional observations of current IoT security operations not included in the preceding bullets:

- Integrity of device provisioning and updates is something to be examined closely. The examples of the constant volume of Bluetooth and BLE vulnerabilities related to pairing and joining are only to serve as an example for all IoT-connected devices.
- Not all IoT protocols specify or require encryption. One example was Sigfox, where the burden is on the customer to ensure security.
- Even protocols that specify encryption and integrity don't enforce it; it's an option. Those technologies can often be forced into a downgrade attack even if a more secure mode is in use. For example, several BLE attacks even with the newer security modules were able to force pairing to Just Works, with no authentication.
- Going one step further, even if the implementation does include encryption, the key exchanges, management, and hierarchy of keys may be flawed or insecure. For example, several define options for network (group) keys along with pairwise (unicast) keys. However, depending on the vendor and implementation, the system may use only group keys and shared keys, creating security vulnerabilities.
- The security capabilities, requirements, and implementations are dependent on intertwined relationship of numerous factors including ones mentioned here and beyond.
- In general, consumer-grade equipment is configured for the easiest plug-and-play operation. Many commercial products follow suit, but some take security seriously and offer hardened protocols and tested products appropriate for enterprise use.
- The security of your IoT devices should be aligned with the criticality of the system, the sensitivity of the data, and the connectivity to other assets and networks. Don't spend $1M defending a $20 asset. And don't render a $1M asset insecure with a $20 IoT device.

Final Thoughts from the Book

Well, we made it! Here we are at the very end of the last chapter of the book.

There's been a lot of content delivered in these eight chapters, and I truly hope you've found at least a few nuggets of useful information. Realizing this book has delivered an entire tour of information security concepts, roles, risk and compliance, technical controls, planning guidance, and a journey through IoT and other emergent trends, I'm taking this opportunity to recap the most significant points I hope you'll take away:

- Wireless (including and especially Wi-Fi) is a complex topic.
- RF designs greatly impact Wi-Fi security.
- There are several right ways to design secure wireless.
- Wireless security is described in levels, not absolutes.
- A wireless infrastructure is only as secure as its weakest link.
- WPA3 security is transformative but only as good as your migration plan.

And as for the recommendations for designing and maintaining secure wireless networks, here are takeaways by chapter.

Chapter 1, "Introduction to Concepts and Relationships"

- There may be many people and roles involved in architecting secure wireless; meet them and communicate often.
- You can justify many wireless design needs in the name of IAC (integrity, availability, and confidentiality).
- Strictly speaking in infosec and compliance terms, MAC authentication and passphrase-secured networks are considered identity, not authentication.

Chapter 2, "Understanding Technical Elements"

- Data paths, segmentation, Wi-Fi architecture, and network configurations are all intertwined.
- WPA2-Personal is "the new WEP": remove it from your network as quickly as possible.
- Contrary to prior guidance, secure networks will require separating SSIDs to a greater degree, instead of collapsing them.
- Using "Transition Mode" networks that support WPA2 and WPA3 together is not advised unless absolutely necessary.

Chapter 3, "Understanding Authentication and Authorization"

- Dynamic VLANs (or other authorization) can be easily configured as a standard RADIUS attribute using any product.
- EAP methods go beyond just PEAP and TLS.
- MAC Authentication Bypass is a good option, but most implementations are done insecurely; don't comingle MAB (or any MAC auth) with full 802.1X endpoints.

Chapter 4, "Understanding Domain and Wi-Fi Design Impacts"

- RF design and coverage impacts the function of key exchanges for secure roaming, and therefore security.

Chapter 5, "Planning and Design for Secure Wireless"

- There are dozens of interrelated inputs and outputs involved in designing secure wireless.
- There are three main topic areas technical and non-technical executive leaders should know including budgeting, selecting products, and expectations for security.

Chapter 6, "Hardening the Wireless Infrastructure"

- Hardening guidance is tiered, and strict hardening isn't recommended for all environments.
- Protected Management Frames (PMF) will add security but also impact current monitoring tools.
- Consumer technologies can have a place on enterprise networks, but it won't be a secure network.
- Ease-of-use zero configuration protocols like mDNS and Bonjour introduce extensive vulnerabilities to enterprise networks and are a primary entry point for pen testers and attackers.

Chapter 7, "Monitoring and Maintenance of Wireless Networks"

- Pen testers and security assessors are your friends; they're going to find problems, be glad it's a friend not a foe breaking in.
- All wireless networks should include some amount of WIPS monitoring and rogue APs should be addressed and removed.
- As non-802.11 wireless is introduced to the enterprise, security will necessitate additional monitoring tools.
- Events that present immediate security risk should be alerted and acted on; other events should be logged and/or reported for future analysis.

Chapter 8, "Emergent Trends and Non-Wi-Fi Wireless"

- BYOD policies are legal documents and shouldn't be created by IT teams without legal and human resources' input.
- The only hard rule of securing BYOD is to not allow unmanaged devices access to secured resources.

- Zero trust strategies are real, and they're attainable.
- There are numerous non-Wi-Fi IoT technologies appropriate for enterprise use.

As always you can find me online at `www.securityuncorked.com` and `www.viszensecurity.com`.

Until the next book—so long, and thanks for all the fish.

APPENDIX

A

Notes on Configuring 802.1X with Microsoft NPS

When learning 802.1X (wired or wireless) in the lab, I recommend starting with Windows NPS—it can be easily enabled on any Microsoft server platform at no additional charge, uses vendor-neutral standard RADIUS attributes by default, and doesn't have a ton of other vendor-specific NAC features you need to sift through such as profiling, posturing, and third-party integrations.

Most products, including NPS, will have an option to configure 802.1X policies using a wizard or walk-through, which I highly recommend on all platforms. The information here and in Chapter 3 will guide you through adjustments to the default policies to meet your needs.

The section "Planning Enterprise (802.1X) Secured SSIDs" in Chapter 2, "Understanding Technical Elements," includes a short list of six (plus some optional) components required for 802.1X in Wi-Fi. This content expands on that with explanations of the discrete configurations for each of the components. That was your list of raw materials needed and here's your instruction booklet for how to build your secure network with those pieces.

Wi-Fi Infrastructure That Supports Enterprise (802.1X) SSID Security Profiles

The first and most obvious requirement is a Wi-Fi product that supports the WPA Enterprise (802.1X) security profiles for SSIDs. This isn't a problem for any enterprise-grade products; they'll all support WPA2- and/or WPA3-Enterprise security, but it's unlikely to find this support in consumer-grade and residential ISP-provided wireless routers.

To access the latest features including WPA3-Enterprise, you may need to upgrade the product firmware. Ultimately you should see an option for WPA2-Enterprise, WPA3-Enterprise, and some representation of a WPA3-Enterprise Transition Mode, which would support both WPA2- and WPA3-Enterprise.

Endpoints That Support 802.1X/EAP

Long ago in the early days of 802.1X, there was no native support in operating systems, and we had to install a separate piece of software called a supplicant. Now, every user-based operating system (such as Windows, Apple, Android) has an 802.1X supplicant natively built in and the outliers may be the non-user-based devices such as printers, VoIP phones, and IoT-type devices.

For laptops, tablets, and smartphones, you shouldn't need to install or make any major changes for the endpoint to support 802.1X. Having said that, you may need to update the endpoints' software and/or NIC drivers to support the latest Wi-Fi standards such as WPA3-Enterprirse. As you'll see next, those endpoints may require some configuration to connect to the network with the correct parameters for authentication.

For those other devices, believe it or not, most printers, VoIP phones, and similar enterprise devices do support 802.1X. They just may require a bit more touch to get updated and configured. See the upcoming sidebar "Issuing Device Certificates to Endpoints."

VoIP phones frequently require software upgrades for the phones, pushed from the phone controller, as well as additional configuration for 802.1X support within the phone controller. Although outside the scope of this book, many VoIP phone systems were initially configured with a dual-boot DHCP mode whereby they make an initial request on the untagged/native VLAN and then receive a DHCP response that directs the phone to the tagged voice VLAN. When switching to 802.1X-authenticated VoIP phones, you'll likely want to omit the dual boot function and use LLDP-MED with 802.1X for a more streamlined and consistent operation.

Figure A.1 shows an example of a common office HP brand multi-function printer that supports 802.1X authentication to the network with either EAP-PEAP (user credentials via MSChapV2) or EAP-TLS (device certificate).

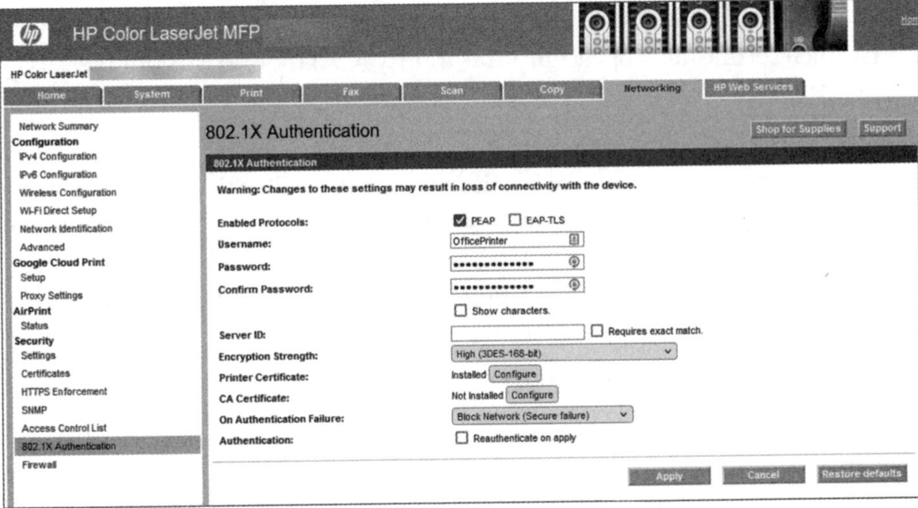

Figure A.1: Most non-standard endpoints such as network printers and VoIP phones support 802.1X.

A Way to Configure the Endpoints for the Specified Connectivity

For endpoints connecting to 802.1X-secured SSIDs, you'll want to consider the best options for pushing the settings in bulk where possible. While this can all be done manually, it certainly doesn't scale if a technician or admin has to touch each endpoint to configure it. And, possibly even more of a frustration—the settings for 802.1X configurations can vary even from version to version of the same platform.

Options for pushing configurations to common enterprise endpoints include:

- Active Directory Group Policy (or similar directory policy)
- Mobile device management (MDM, for laptops, tablets, smartphones)
- VoIP controller (for VoIP phones)
- Printer fleet management tools

The exact configuration needs will depend on the EAP methods and other parameters for authentication, but in general on the endpoint you'll need to specify:

- EAP inner and outer method (such as EAP-PEAP with MSCHAPv2 or EAP-TLS)

- The level of encryption to use
- Which certificate(s) or certificate authority (CA) roots to trust for the server certificate(s)
- Whether to connect automatically
- Selection of credentials or certificate; for EAP-PEAP, whether to pass-through Windows credentials or prompt the user and for EAP-TLS, which certificate to use
- Whether or not to validate the RADIUS server certificate (this should always be "yes" and is becoming mandatory)
- Whether to enable fast reconnect

The EAP methods and authentication methods will have to be negotiated between the endpoint and the authentication server, so you'll know without a doubt if that's not configured correctly.

PEAP Fast Reconnect settings on the endpoint shouldn't make or break anything. The Fast Reconnect setting simply bypasses the inner authentication method to facilitate a faster Wi-Fi roaming experience. Impacts of 802.1X on roaming were covered in Chapter 4. Most Wi-Fi manufacturers recommend enabling fast reconnect, so that's a good place to start and adjust later if it's causing issues.

NOTE As a reminder, everything related to 802.1X authentication and encryption is negotiated between the endpoint and the authentication (RADIUS) server. The wireless infrastructure is simply repackaging and ferrying packets around—with the one exception of WPA3-Enterprise certificate validation.

Lastly, different client operating systems handle 802.1X inputs differently, and some (especially smartphones) tend to be better at auto-sensing and auto-configuring parts of the endpoint's 802.1X configuration we'd otherwise have to set, whether manually or through a policy.

ISSUING DEVICE CERTIFICATES TO ENDPOINTS

For endpoints authenticating with device certificates, there are many ways to automate enrollment and issuance of certificates. One such service for headless devices (such as printers and network infrastructure) is Microsoft Network Device Enrollment Service (NDES) with SCEP. NDES is a service within AD Certificate Services.

- Domain-managed computers and servers can be auto-enrolled through AD Certificate Services (without the add-on NDES).
- Non-domain-managed endpoints (including BYOD, Apple, Chromebooks, etc.) can be managed through MDM products and/or vendor-specific onboarding products.

An Authentication Server That Supports RADIUS

The next component needed is an authentication server that supports the RADIUS protocol. Products that offer this include Microsoft Network Policy Server (NPS) services (which can be enabled on any Microsoft server), Cisco Identity Services Engine (ISE), Aruba ClearPass Policy Manager, FreeRADIUS, Pulse Secure Steel-Belted RADIUS, and the list goes on.

While authentication servers often support local accounts, by far the most common implementation is to connect the RADIUS server to an existing repository of users such as Microsoft Active Directory or other LDAP service. The NAC-enabled vendor products (such as Cisco ISE, Aruba ClearPass, Fortinet FortiNAC, and Forescout) are an exception to this in certain use cases; most specifically these products are commonly used for registration of personal devices (such as in BYOD), registration of IoT devices, and guest registrations. In these cases, authentication for employees is still verified against a directory service, but local accounts may be created for these other use cases when the user or device doesn't have a domain account.

Regardless of the RADIUS product, there are some common tasks that will need to be completed for it to operate as a fully functional authentication server. For the most part, the order of these tasks won't matter, as long as they're all completed properly. The exceptions to that are: before the policy can be created, the server certificate will need to be applied and the RADIUS server should be connected to the directory services.

Configuration of RADIUS Server for Authentication Sources (such as Active Directory or Other LDAP) Upon provisioning the RADIUS server, one step will be to connect it with the directory services such as Active Directory. This is a one-time process, after which you can specify connected directories as an authentication source for one or more policies, and of course use directory group membership as a constraint condition for processing a policy.

Configuration of RADIUS Server for Connection Policies Included Specification of Supported EAP Types The RADIUS server will likely be processing incoming authentication requests for other services such as user SSL-VPN connections, RADIUS, or TACACS+ for device administrative access, and user authentications for other parts of the infrastructure. In larger environments, it's also likely there will be multiple RADIUS servers.

Because of that, there are criteria specified to tell the RADIUS server whether or not to process the request, and if so, which policy to use for the evaluation. Specifics of these policies vary by product and the step-by-step instructions are outside the scope of this book. For context and to help visualize options, included are sample screenshots from Microsoft NPS.

Figure A.2 is a screenshot from Microsoft NPS showing the Connection Request Policies window with multiple inbound processing rules including Wi-Fi authentication, wired authentication, and SSL-VPN authentication. When using the policy wizards, NPS will automatically create corresponding Connection Request Policies, which can usually be ignored and need not be modified.

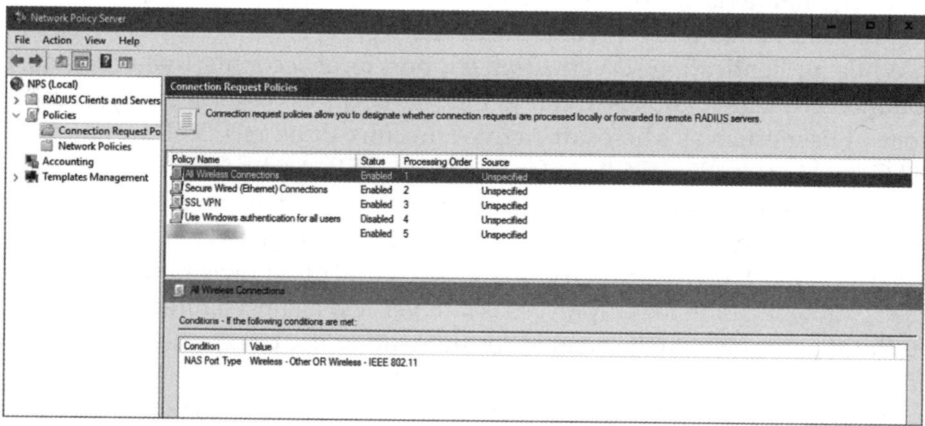

Figure A.2: Microsoft NPS can be instructed to forward or proxy RADIUS requests to other remote RADIUS servers.

Figure A.3 then shows the Network Policies. In this example, five unique policies are enabled, each processing incoming requests for different purposes or from different environments. The first rule is used to process requests coming in from Juniper Mist APs at remote sites and the second rule processes requests from the main Cisco controller. In this example, the Cisco Sunnyvale policy is matching against the Network Access Server (NAS) ID, which is an IP address on the Cisco controller, and access is only allowed if the user is in the SecureWi-FiUsers group in AD.

Within the Network Policy the allowed EAP types are configured. Figure A.4 shows the proper configuration of a policy that allows only PEAP outer tunnel with MSCHAPv2 inner authentication. This part of the configuration also specifies which server certificate to use. The server certificate will need to be installed on and trusted by the endpoints that are connecting to authenticate.

Dynamic VLANs are an easy way to apply segmentation to endpoints, even those on the same SSID. In Figure A.5, the Network Policy settings are configured to return VLAN 99. The Service-Type equal "Framed," Tunnel-Medium-Type of "802," and the Tunnel-Type attribute of "VLAN" are used with Tunnel-Pvt-Group-ID value of "99."

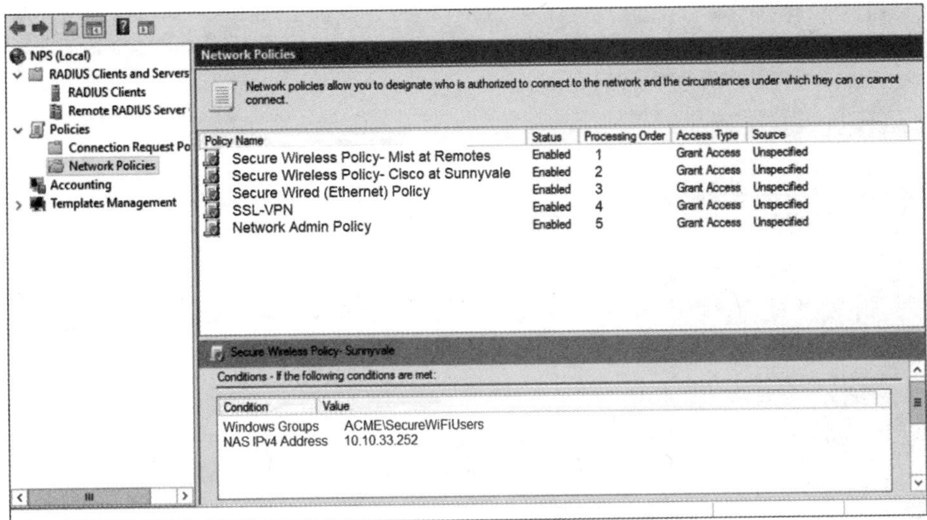

Figure A.3: The Network Policies of Microsoft NPS, which set the criteria for matching incoming requests to the correct policy

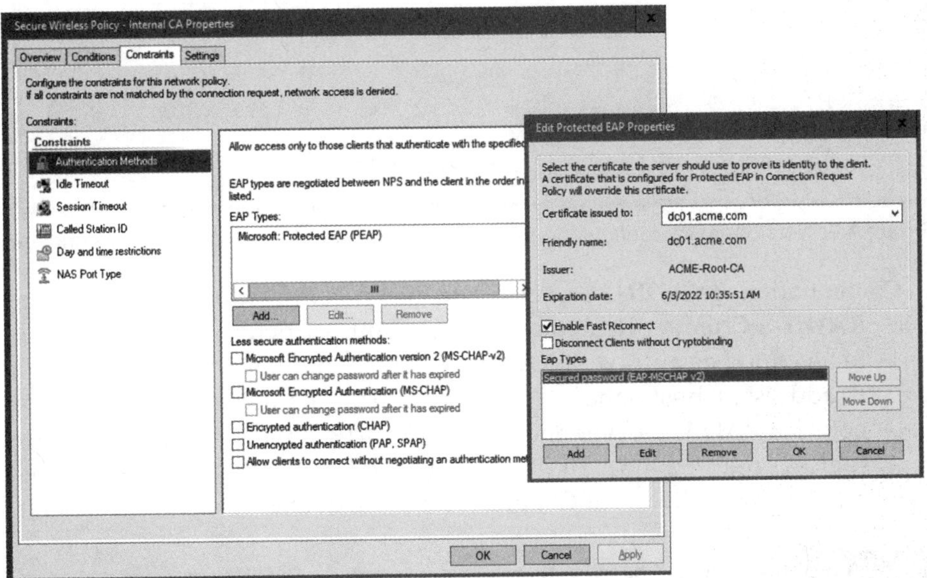

Figure A.4: This view shows the details of a policy specifying Microsoft PEAP as the allowed authentication method. Other options include smart card or device certificates, which could be added to the same policy.

All of these are standard RADIUS attributes that *should be* readable and usable by any enterprise Wi-Fi product.

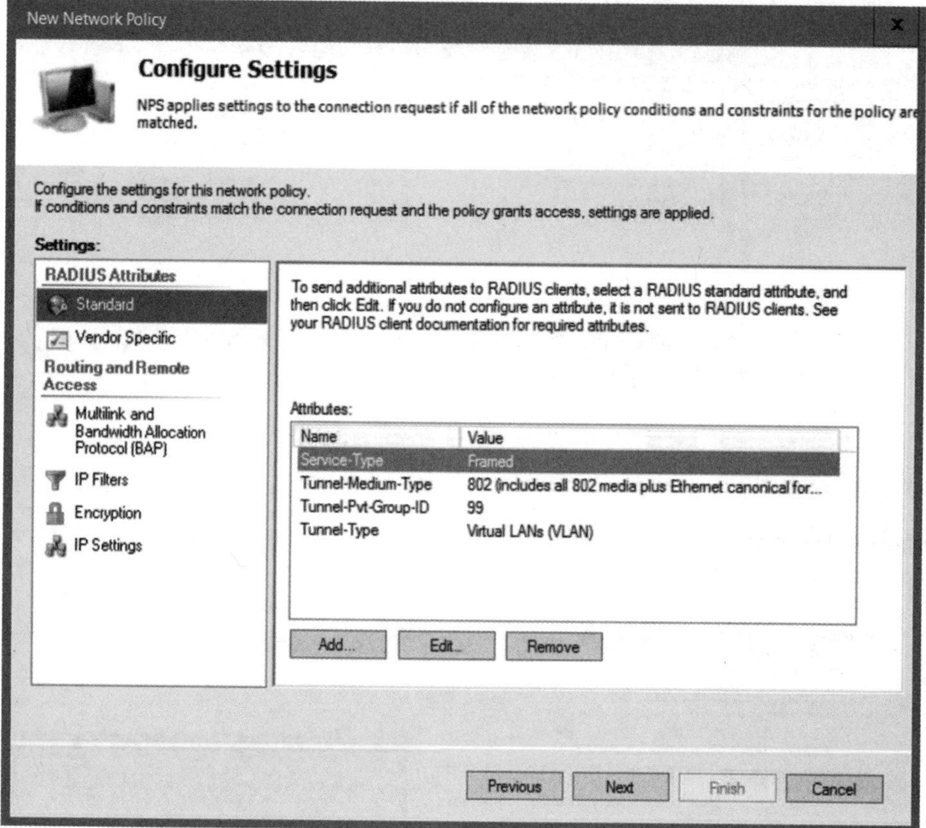

Figure A.5: Network Policy configuration showing the attributes for dynamic VLAN assignment

Configuration of RADIUS Server to Add the Wi-Fi Infrastructure as Approved RADIUS Clients Remember the Wi-Fi infrastructure devices will need to be added as approved RADIUS clients. This can be done by specific IP addresses, hostnames, or IP subnets. The shared secret is entered on both the RADIUS server and Wi-Fi infrastructure but is sent in cleartext (unless protected through RADSEC or another mechanism), so don't use a real password. Revisit Chapter 3 for more details.

NOTE For any 802.1X-based authentication to work, the RADIUS server must have a valid and supported server certificate. Revisit Chapter 2 for more details.

APPENDIX

B

Additional Resources

IETF RFCs

IETF is responsible for most of the Internet protocols in use today. As far as Wi-Fi goes, most standards are either IEEE or IETF RFCs.

Navigating and Reading RFCs

RFCs are a great source of information and are relatively easy-to-read bite-sized documents. There are a few notes when reading and reviewing. It's easy to find documents with a simple keyword search and "+RFC" or "+IETF". You can also start by navigating to the home page at `www.rfc-editor.org`. Some of the documents will be standards, others informational, and some drafts and experimental.

Figure B.1 shows the header from the RADIUS RFC 2865. In it, you'll notice it's been updated by five other RFCs (2868, 3375, etc.), also linked. And it identifies that this current RFC 2865 has superseded or made obsolete RFC 2138. When reading RFCs, it can be helpful to consider them in context. The "updated by" list, if the RFC has not been superseded, usually contains RFCs that add to or extend the RFC.

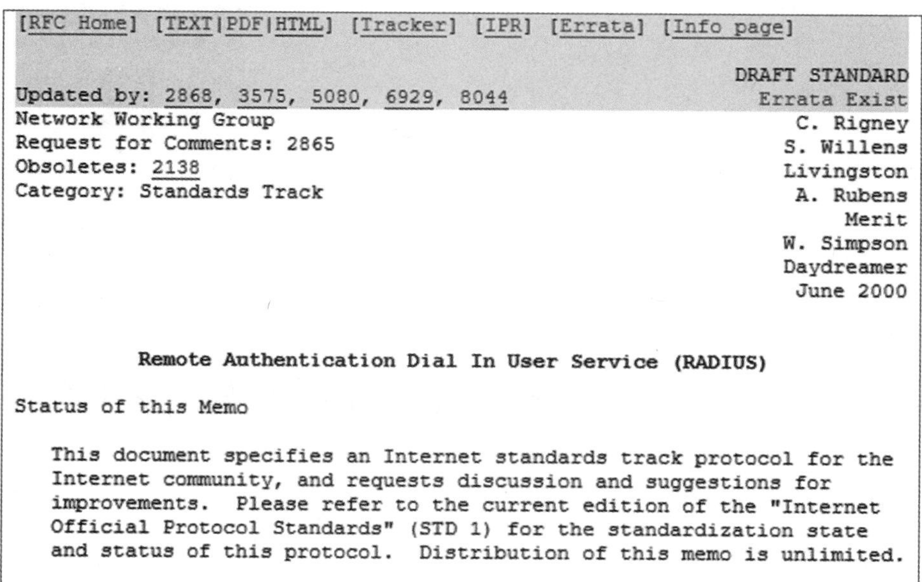
```
[RFC Home] [TEXT|PDF|HTML] [Tracker] [IPR] [Errata] [Info page]

                                                          DRAFT STANDARD
Updated by: 2868, 3575, 5080, 6929, 8044                    Errata Exist
Network Working Group                                          C. Rigney
Request for Comments: 2865                                    S. Willens
Obsoletes: 2138                                               Livingston
Category: Standards Track                                      A. Rubens
                                                                   Merit
                                                              W. Simpson
                                                              Daydreamer
                                                               June 2000


          Remote Authentication Dial In User Service (RADIUS)

Status of this Memo

   This document specifies an Internet standards track protocol for the
   Internet community, and requests discussion and suggestions for
   improvements.  Please refer to the current edition of the "Internet
   Official Protocol Standards" (STD 1) for the standardization state
   and status of this protocol.  Distribution of this memo is unlimited.
```

Figure B.1: Sample IETF RFC header

For example, RFC 6929 referenced in Figure B.1 adds protocol extensions to RADIUS, and two of the linked RFCs offer implementation guidance.

Almost every RFC will include security considerations toward the end, and this is a great source of information.

Helpful RFCs and Links

- RADIUS Types: `https://www.iana.org/assignments/radius-types/radius-types.xhtml`
- Multicast DNS: `https://datatracker.ietf.org/doc/html/rfc6762`
- Extensible Authentication Protocol (EAP) Mutual Cryptographic Binding: `https://datatracker.ietf.org/doc/html/rfc7029`
- Opportunistic Wireless Encryption: `https://datatracker.ietf.org/doc/html/rfc8110`
- RADIUS Attributes for IEEE 802 Networks: `https://datatracker.ietf.org/doc/html/rfc7268`

IEEE Standards and Documents

Most of the technologies covered in this book rely heavily on IEEE standards, aside from Bluetooth and a few other exceptions.

Navigating and Reading IEEE Standards

Wi-Fi is based on the 802.11 standard. You've also seen mention of amendments that added features, such as 802.11w Protected Management Frames, and 802.11r Fast Transition roaming.

Once these become adopted as part of the standard, they're rolled up into the greater 802.11 standard and not available as separate documents the way IETF RFCs are. Similar to the IETF RFCs, though, the IEEE standard home page viewer will show recent amendments to the standard as well as prior standards versions and amendments that have been superseded. When amendments are rolled into the standard, the standard will be revised, and the added amendments noted in the opening. For example, when 802.11-2016 added 802.11ah, 802.11ai, 802.11ak, etc., it became 802.11-2020.

From a naming convention, IEEE standards follow a simple format. For example, 802.11 is part 11 (WLANs) of the 802 (LAN/MAN) standards committee. 802.1 (note, dot-one not eleven) is the Higher Layer LAN Protocols Working Group whose security task group is responsible for our beloved 8021.X standard for Port-Based Network Access Control, a standard used across many network mediums and types, including 802.11 (WLANs) and 802.3 (Ethernet).

The capital letters in IEEE naming conventions denote a standalone standard, such as 802.1X, while lowercase letters indicate an amendment to an existing standard. 802.1AE (MACsec) and 802.1AR (Secure Device IDs) are both full standards, whereas 802.11w (PMF) and 802.11ax are amendments to the 802.11 standard.

Helpful Links

Some of the standards including 802.11 WLANs can be accessed free through the IEEE GET Program, found at `https://ieeexplore.ieee.org/browse/standards/get-program/page`.

The 802 suite of standards includes 802.11 WLANs along with 802.15 PANs and others and can be found at `https://ieeexplore.ieee.org/browse/standards/get-program/page/series?id=68`.

The 802.11 timeline of projects and standards can be found at `https://www.ieee802.org/11/Reports/802.11_Timelines.htm`.

IEEE 802.11 Standard

The cipher suites and AKM suites related to WPA3 can be found in the IEEE 802.11-2020 standard under section "9.4.2.24 RSNE".

Protected Management Frames (formerly 802.11w-2009) was rolled into the 802.11-2012 revision and now part of the 802.11-2020 standard. The PMF components have been absorbed into several disparate areas of the 4,000-page document. The management key was incorporated into the key hierarchy. The description of how management frames are encrypted with pairwise keys and the specification of BIP can be found throughout the document.

In section "9.4.2.24.2 Cipher suites" in the last paragraph of page 1051, a statement of cipher suite use is defined when PMF is negotiated.

Wi-Fi Alliance

The Wi-Fi Alliance is responsible for compliance and now also conformance testing and certifies vendor products (infrastructure and endpoints) against the IEEE and other standards.

Certification by the Wi-Fi Alliance helps guarantee predictable behavior, conformance to standards, and interoperability between products.

Among many others, the Wi-Fi Alliance tests and certifies with these programs:

Wi-Fi CERTIFIED 6™

Wi-Fi QoS Management™

Wi-Fi Easy Connect™

Wi-Fi CERTIFIED WPA3™

WPA3 is a mandatory certification for all Wi-Fi CERTIFIED devices, including Wi-Fi 6. Enterprise architects should note that to be WPA3 certified only requires WPA3-Personal and testing for WPA3-Enterprise is purely optional.

Since WPA3 will be continuously updated and improved, readers are encouraged to stay up-to-date by following information at `https://www.wi-fi.org/security`.

The Wi-Fi Alliance standards documents and test plans are available only to its members and not to the general public.

Blog, Consulting, and Book Materials

As promised, additional resources will be made available at my blog `www.securityuncorked.com`. Materials from this book including tables, templates, and images can be found at `https://www.securityuncorked.com/books`.

Through my company, Viszen Security, I offer additional resources including whitepapers, workshops, and consulting to help IT decision makers with network security, wireless, digital transformation projects, and advanced network access control strategies including zero trust architectures. For technology manufacturers and VARs, I also offer advisory services for technical marketing strategy, technical sales facilitation, and speaking engagements. Find more at `www.viszensecurity.com`.

Compliance and Mappings

Compliance requirements are updated regularly, although often not nearly as quickly as the risk climate changes.

NIST SP 800-53 and ISO 27001

Throughout this book, several references to NIST SP 800-53 standards have been made. While NIST is recognized internationally by many governments and standards organizations, some architects may need or desire to map security controls for secure wireless against ISO standards.

A free NIST to ISO Mapping created and maintained by NIST can be found at `https://csrc.nist.gov/publications/detail/sp/800-53/rev-5/final`. The document is in the right toolbar and titled "Mapping: Rev. 5 to ISO/IEC 27001." The SP 800-53 standard document in PDF and the control catalog spreadsheet in XLS are also helpful.

Table B.1 shows the 20 security and privacy control families from NIST 800-53. Secure wireless architecture includes components from most of these domains.

Table B.1: NIST 800-53 control families

ID	FAMILY	ID	FAMILY
AC	Access Control	PE	Physical and Environmental Protection
AT	Awareness and Training	PL	Planning
AU	Audit and Accountability	PM	Program Management
CA	Assessment, Authorization, and Monitoring	PS	Personnel Security
CM	Configuration Management	PT	PII Processing and Transparency
CP	Contingency Planning	RA	Risk Assessment
IA	Identification and Authorization	SA	System and Services Acquisition
IR	Incident Response	SC	System and Communications Protection
MA	Maintenance	SI	System and Information Integrity
MP	Media Protection	SR	Supply Chain Risk Management

Guidance in this book can be mapped to several areas of the NIST 800-53, including:

AC-2(5) Inactivity timeout

AC-2(7) Privileged user accounts

AC-4(21) Physical or logical separation of information flows

AC-6(3) Network access to privileged commands

AC-6(6) Privileged access by non-organizational users

AC-7 Unsuccessful logon attempts

AC-10 Concurrent session control

AC-17 Remote access

AC-18 Wireless Access

AC-19 Mobile Devices

CA-7 Continuous Monitoring

CA-8 Pen testing

CA-9 Internal system connections

IA-2(1) MFA to privileged accounts

IA-3 Device Identification and Authentication

IA-5 Authenticator Management

IA-5(18) Password managers

IA-12 Identity Proofing

RA-5 Vulnerability Monitoring and Scanning

SC-5 DoS Protection

SC-7 Boundary Protection (7) split tunnel remote

SC-7 (14) Protect against unauthorized physical connections

SC-7 (15) Networked Privileged Access - dedicated managed interface (AC-2, AC3, AU2, I-4)

SC-7 (19) Block comm from non-organizationally configured hosts

SC-7 (20) Dynamic isolation and segregation

SC-7 (21) Isolation of system components

SC-7 (22) Separate subnets for connecting to different security domains

SC-7 (29) Separate subnets to isolate functions

SC-8 (1) Cryptographic protection for confidentiality and integrity

SA-8 (9) Trusted Components

SC-12 Cryptographic key establishment and management

SC-17 PKI Certificates

SC-40 Wireless Link Protection

SC-41 Port and I/O Device access

SC-45 System Time Synchronization

SI- System monitoring, logging, alerting, analysis and correlation, anomaly detection, security inspection

SI-4 System Monitoring

SI-4 (14) Wireless intrusion detection

SI-4 (15) Wireless to wireline communication

SI-4 (22) Unauthorized network services

SI-7 Software, firmware, and information integrity

A few sample rows from NIST's mapping of SP 800-53 to ISO 27001 is found in Table B.2. Note that an asterisk (*) indicates that the ISO/IEC control does not fully satisfy the intent of the NIST control.

Table B.2: Sample excerpt rows from NIST to ISO mapping

NIST SP 800-53 CONTROLS		ISO/IEC 27001 CONTROLS
AC-2	Account Management	A.9.2.1, A.9.2.2, A.9.2.3, A.9.2.5, A.9.2.6
AC-19	Access Control for Mobile Devices	A.6.2.1, A.11.1.5, A.11.2.6, A.13.2.1
CA-7	Continuous Monitoring	9.1, 9.2, A.18.2.2, A.18.2.3*
IA-2	Identification and Authentication (Organizational Users)	A.9.2.1
SC-7	Boundary Protection	A.13.1.1, A.13.1.3, A.13.2.1, A.14.1.3
SC-12	Cryptographic Key Establishment and Management	A.10.1.2
SC-17	Public Key Infrastructure Certificates	A.10.1.2

PCI Data Security Standards

The most current PCI DSS standards can be found at `https://www.pcisecuritystandards.org/pci_security/maintaining_payment_security`. Additional documents including the full PCI DSS Requirements and Security Assessment Procedures are available at `https://www.pcisecuritystandards.org/document_library`.

Among other things, PCI DSS specifies:

- Network segmentation
- Wireless security and monitoring
- Encryption of cardholder data across open or public networks
- Use of firewalls to protect cardholder data
- Physical and logical access control
- Logging and auditing of access to network resources
- Regular testing of security systems and processes

A few of the PCI DSS requirements that impact wireless architecture and security include the following. Testing procedures and guidance are included in the PCI DSS document referenced earlier.

1.2.3 Install perimeter firewalls between all wireless networks and the cardholder data environment, and configure these firewalls to deny or, if traffic is necessary for business purposes, permit only authorized traffic between the wireless environment and the cardholder data environment.

2.1.1 For wireless environments connected to the cardholder data environment or transmitting cardholder data, change ALL wireless vendor defaults at installation, including but not limited to default wireless encryption keys, passwords, and SNMP community strings.

4.1.1 Ensure wireless networks transmitting cardholder data or connected to the cardholder data environment, use industry best practices to implement strong encryption for authentication and transmission.

Cyber Insurance and Network Security

Cyber insurance is a type of insurance an organization can have to cover losses relating to damage to IT systems, networks, and data. This coverage has become especially critical for organizations looking for financial protection from ransomware and other attacks. When an organization is covered by a cyber insurance policy the agency will assist with incident response and recovery, among other things.

However, for an organization to qualify, it must meet stringent guidelines for internal security practices, or the claim will be denied.

The wireless security posture may also impact the organization's ability to qualify for cyber insurance. Some common requirements for insurance related to network security, including Wi-Fi, include:

- Patching
- Backup processes and offline data backups
- Pen testing
- Network and endpoint monitoring
- Multi-factor authentication especially for privileged system access
- Biometric security for certain applications or access
- Password maintenance
- Data encryption
- Development of an incident response (IR) plan
- Monitoring including 24x7

APPENDIX

C

Sample Architectures

The following sample architectures are organized by network use case type (employee, guest, etc.) with each use case then covering the range of security architecture from most to least secure. The content is arranged this way with the understanding that some organizations have higher security requirements for some networks than others. Healthcare is again a great example; it's common to have varying degrees of secure internal networks for staff and biomedical devices, but a completely open guest network to meet business objectives for patient needs.

A few use cases may span both internal and guest network models. For example, BYOD with access to internal resources is covered in the first group, and BYOD with only Internet access is covered under guest networks. The same will be true for certain uses cases of third parties and contractors as well as students. In K-12/primary education students are most often accessing the network with school-managed devices, whereas in higher education and university systems, the students are using personal devices. For that reason, primary education scenarios are covered under the category of "Managed User and Managed Device," and university scenarios are included under the "BYOD/Personal Device" headings.

The two main areas of architecture are:

Architectures for Internal Access Networks "Architectures for Internal Access Networks" will address security for all use cases where a user or device has access to any of the organization's protected assets, such as internal networks.

Architectures for Guest/Internet-only Networks "Architectures for Guest/ Internet-only Networks" will address securing networks for users and devices needing only Internet access. While the backend architecture supporting these guest networks may require access to internal resources (such as DNS or DHCP servers), the model is defined by the end user or endpoint access.

The architectures are described as relative: high, medium, and low. Where appropriate, additional notes for more security are included along with elements to avoid. As you'll see in the coming content, it's appropriate to mix-and-match based on the network in scope—meaning an organization may have a self-described high-security implementation for two internal SSIDs, and a lower security architecture for a guest/Internet-only network.

With security architecture, decisions are complex, non-linear, and interdependent. The beauty of a well-orchestrated security architecture is that there are layers of complementary and compensating controls.

With all of that said, it's simply not possible to outline recommendations for every possible scenario. High, medium, and low are generalizations used to describe the combination of needs as defined in risk and compliance (Chapter 1, "Introduction to Concepts and Relationships"), with available authentication and authorization mechanisms (Chapter 3, "Understanding Authentication and Authorization"), and scoping (Chapter 5, "Planning and Design for Secure Wireless").

The safe bit of guidance I can offer is, based on common practices (not necessarily best practices), many educational institutions opt for low-medium relative security, whereas government and regulated industries will tend toward higher security. That statement is still an incomplete and inaccurate generalization.

Architectures for Internal Access Networks

Internal access networks offer some level of access to an organization's resources other than just Internet. The access could be a connection to an internal network, or some level of privileged access to one or more assets.

What's excluded from this section are use cases to allow guest users access to one or more specific resources, such as allowing access to a single printer or screen casting device; those scenarios are covered in the "Architectures for Guest/Internet-only Networks."

In this section, secure architectures are described for the following access scenarios:

- Managed User and Managed Device
- Non-User-Based Devices
- Contractors and Third Parties
- BYOD/Personal Devices with Internal Access

Managed User with Managed Device

Access scenarios based on a managed user on a managed device is the most common scenario for traditional enterprise models for access by employees and staff. This model also applies to students accessing resources from a school-managed device.

In these use cases, the user is known and managed in a directory such as Active Directory. The device is also known and managed and may be part of the domain infrastructure (also in AD) or managed by an MDM tool or NAC product. Although I'm using Microsoft as an example for domain management, the same thought process follows with any platform including G-Suite and Google Workspace, Google's Managed Microsoft AD, Microsoft Azure AD, or any other directory. Similarly, MDM products range from Microsoft Intune to Google Endpoint Management, Citrix, SOTI, AirWatch, ManageEngine, and others.

In all cases, the managed devices covered here will have the capability to connect to an 802.1X-secured network and the architectures here are based on Enterprise class networks of WPA2-Enterprise or WPA3-Enterprise.

Organizations are encouraged to separate SSIDs, and that may include more than one 802.1X-secured network. In fact, for many environments, that's exactly what the following recommendations prescribe. This means there may be multiple 802.1X-secured networks, each mapped to a different security level of high, medium, or low, and each with a different use case and security domain. See the upcoming topic of "Guidance on When to Separate SSIDs" for more.

Security Considerations

Considerations for the security architecture of managed users on managed devices include:

Classification of Network or Data Sensitivity The overall classification of the network or data is the most influential factor in planning that network's security. Networks that host assets with data labels associated with compliance (including PCI DSS and CMMC as examples) will demand certain levels of authentication, encryption, and segmentation. Beyond the minimum requisite designs, additional controls may be recommended, and different classifications of data (with varying security requirements) should be further segmented.

Segmentation Requirements Network and endpoint segmentation can include techniques like over-the-air inter-station blocking (peer-to-peer blocking) as well as L2/L3 controls enabled on the wired and wireless networks.

Authentication Requirements The networks defined in these architectures are all based on 802.1X/Enterprise-class networks. As such, all users and/or devices will be authenticated via 802.1X, with expectations that higher security requirements prescribe layering or chaining of both user and device authentication.

Group Key Exploits in 802.1X As you've seen, SSIDs are effectively a broadcast domain, by default with shared group keys (even if endpoints are in different VLANs). In 2010, researchers announced the "Hole 196" vulnerability, which specified a suite of possible attacks against Wi-Fi networks, primarily based on exploiting the inherent trust properties of group keys.

This is an insider attack, and a rare one at this point in time. It can only be executed against endpoints connected to the same SSID, on the same AP, but should remain a consideration for highly sensitive environments. In these instances, recommendation for further separating SSIDs or using per-endpoint group keys is advised. Most organizations need not worry about this since an attacker will have several other easier paths into to the network.

Industry-Specific Requirements Specific industries will have additional recommendations and requirements. Please use these sample architectures as a starting point and adjust as appropriate for your use case.

High-Security Architecture

High security is the relative descriptor for networks and environments that require the most protection from attacks, both malicious and unintended insider attacks.

Organizations in industries of financial, healthcare, federal governments, and those with high-value intellectual property should consider high-security networks for any access to internal resources. Highly secured networks can be supplemented with less secure architectures described later, with the proviso that the less secured networks will not afford the same level of access as the highly secured network.

High-Security Architecture Requirements

For highly secured networks being accessed by managed users from managed devices, the architecture will specify:

- 802.1X-secured SSID.
- WPA3-Enterprise Only with a minimum of CCMP-128 or use of the 192-bit suite (with 256-bit keys) for the highest-security environments.
- Authentication of both the endpoint and the user, which can be accomplished with EAP chaining, vendor-specific implementations that add device authentication or directory lookup along with user authentication, or use of a NAC product or advanced authentication server that supports both user and device authentication.
- Only 802.1X authentications will be allowed, and use of MAB or any MAC-based authentication along with 802.1X is not indicated for high-security networks.
- 802.1X allowed EAP methods should specify an encrypted outer tunnel and allow only the requisite methods (this may include PEAP, EAP-TTLS, EAP-TEAP, and EAP-FAST); the policy should not allow lower-security protocols such as un-tunneled MSCHAPv2 or deprecated methods of MD5, LEAP, or CHAP. Use the latest version of TLS and specify cipher suites of ECC vs RSA (see `https://datatracker.ietf.org/doc/rfc4492/` for more).
- Interstation blocking (peer blocking) should be enabled.
- Multicast, mDNS, Bonjour, and any zeroconf networking protocols will be blocked/filtered (allowing for special considerations with IPv6).

Notes on User and Device Authentication in High Security

Although certificates are the most cryptographically secure authentication mechanism, both user and device authentication should be enforced on highly secured networks.

Additional controls for exceptionally sensitive environments may use multifactor authentication and/or a VPN for Wi-Fi connections.

As summarized the preceding bullets, authenticating both the user and device can be accomplished in many ways, including:

- Through EAP chaining

- With vendor-specific implementations that add device authentication or directory lookup along with user authentication
- Use of a NAC product or advanced authentication server that supports both user and device authentication (not a feature supported in Microsoft NPS)

Further, since MAC addresses are an assertion of identity and not authentication, any authentication based on MAC addresses should not be mixed with fully secured 802.1X authentication. This includes MAC Authentication Bypass (MAB) among others. Endpoints that can't participate in 802.1X are addressed later and will use a different SSID, not a shared 802.1X network.

MAC randomization will also impact the ability to rely on MAC addresses for identity.

Medium-Security Architecture

Environments in which the organization is prepared to trade off certain security controls in favor of usability and streamlined operations are described with the relative label of "medium security." Organizations that are security-conscious but not targeted to the degree of highly sensitive environments should consider medium-security architecture.

Medium-Security Architecture Requirements

Internal networks for managed users and devices, based on medium-security architecture should be planned with the following:

- 802.1X-secured SSID
- WPA3-Enterprise Only with a minimum of CCMP-128
- Authentication of the endpoint and/or the user
- 802.1X-secured SSIDs may support limited MAB endpoints, but only if the MAB endpoints are segmented from full 802.1X-authenticated endpoints
- 802.1X-allowed EAP methods should specify an encrypted outer tunnel and allow only the requisite methods (this may include PEAP, EAP-TTLS, EAP-TEAP, and EAP-FAST); the policy should not allow lower security protocols such as un-tunneled MSCHAPv2 or deprecated methods of MD5, LEAP, or CHAP
- Interstation blocking (peer blocking) should be enabled except when allowed by an admin-defined policy
- Multicast, mDNS, Bonjour, and any zeroconf networking protocols should be blocked/filtered in business environments and used in limited areas to support residential use cases such as university residence halls and long-term care facilities

Medium-security architecture differs from high security in bullets 2–4, 6, and 7. For medium-security designs, the WPA3-Enterprise SSID need not support 192-bit mode. In these networks it's also appropriate to authenticate the user or the device—although still advisable to authenticate both when possible.

Supporting MAB with 802.1X in Medium-Security Networks

Another significant variation is that in medium-security environments we introduce the flexibility to support MAB on the same SSIDs as 802.1X-authenticated endpoints. The caveat here is that the MAB devices and the 802.1X devices should be in separate networks, which can be accomplished in many ways including:

- Via static or dynamic VLANs from RADIUS
- Using downloadable ACLs from RADIUS
- Other static or dynamic segmentation method such as NFVs
- With a vendor-specific implementation based on a role or label
- From a NAC product or advanced authentication server with granular policy support and/or endpoint profiling

Remember that even if endpoints are in different VLANs, if they're on the same SSID, they're in the same wireless broadcast domain, creating opportunity for Hole 196 or other insider attacks.

Supporting mDNS and Zeroconf in Medium-Security Networks

In Chapter 6, "Hardening the Wireless Infrastructure," I covered a million and one reasons why zeroconf protocols shouldn't be allowed on secured networks. And I maintain that if a network *does* allow these protocols, by definition, it's not a secured network.

However, networks aren't valuable unless they support the defined business use cases, and some use cases will demand these protocols.

Where possible, I encourage organizations to fully weigh the pros and cons of allowing these protocols. Without rehashing Chapter 6, I'll just issue the reminder that these protocols were designed for home use, not enterprise networks, and exploiting them is one of the first methods used by attackers and pen testers to breach a network.

If you must support a business case and enable some or all protocols, the recommendation is to separate the networks, create a different SSID and VLAN (or other NFV segmentation) for endpoints and networks using these protocols.

For enterprise use cases such as network printing and screen casting, there are products and configuration options that don't rely on vulnerable zeroconf protocols. It may take a bit of extra effort up front, but your network will be exponentially more secure if you put in that work. As an example, you can configure screen casting (such as an AirTame device) to display an IP address to connect to; configure an allow policy through the Wi-Fi infrastructure, and then allow users to connect to and cast to the device without the use of mDNS.

Supporting mDNS and zeroconf protocols recommendations can be summarized as follows:

- Only allow it if absolutely necessary
- If enabled, segment the endpoints and networks using zeroconf with separate SSID and VLAN (or NFV)
- Add mitigating controls and/or additional security monitoring as appropriate, specifically for managed endpoints

Low-Security Architecture

Organizations may have use cases for one or more networks that are appropriate for a lower level of security architecture. As hinted earlier, a lower-security network may be used in addition to higher-security architectures to address different risk postures and use cases.

Low-Security Architecture Requirements

Internal networks for managed users and devices, based on a lower security need could be planned with the following:

- 802.1X-secured SSID
- WPA3-Enterprise Only with a minimum of CCMP-128
- Authentication of the endpoint and/or the user
- 802.1X-secured SSIDs may support limited MAB endpoints
- 802.1X-allowed EAP methods should specify an encrypted outer tunnel and allow only the requisite methods (this may include PEAP, EAP-TTLS, EAP-TEAP, and EAP-FAST); the policy should not allow lower security protocols such as un-tunneled MSCHAPv2 or deprecated methods of MD5, LEAP, or CHAP
- Interstation blocking (peer blocking) is optional
- Multicast, mDNS, Bonjour, and any zeroconf networking protocols may be allowed

Moving from medium- to lower-security requirements impacts the guidance in bullets 4, 6, and 7. With the laxer design, an organization may choose to comingle MAB devices with 802.1X-authenticated endpoints. Interstation-blocking (peer blocking) and the zeroconf protocols may be enabled. Figure C.1 provides a summary of all three security architecture levels.

	Feature	High-Security Architecture	Medium-Security Architecture	Low-Security Architecture
WPA-Enterprise Networks	Mode	WPA3-Enterprise Only with a min. CCMP-128 or use of the 192-bit suite	WPA3-Enterprise Only with CCMP-128	
	Authenticated Entity	Endpoint and user	Endpoint and/or user	
	MAB	No MAB or MAC-based auth	Allowed only if MAB endpoints are segmented from 802.1X endpoints	Allowed
	EAP	Encrypted outer tunnel and locked to secured EAP methods		
	Inter-station communication	Blocked	Blocked except when allowed by an admin-defined policy	Optional
	Zeroconf	Blocked	Blocked in business environments and used in limited areas to support use cases	Optional

Figure C.1: Architecture summary for internal access by managed users with managed devices

Headless/Non-User-Based Devices

Non-user-based devices include what are often referred to as "headless" devices such as printers, VoIP phones, and standard network-connected business endpoints. In addition, this category also covers LAN-based IoT devices including Wi-Fi connected facilities management and monitoring such as HVAC controls, temperature sensors, and lighting control systems.

NOTE **Technically "headless" originally referred to devices that lacked a configuration interface (a printer's front panel would be an example of an interface), however common use today is for "headless" to reference non-user-based devices, even if there is an interface to configure it.**

Headless devices are defined by the lack of a human user attached to them. Depending on their capabilities, these devices may connect to the network using full 802.1X, using MAB, or with a passphrase, meaning Enterprise or Personal class networks.

Security Considerations

Considerations for the security architecture of enterprise-owned devices that are non-user-based are the same as enterprise-managed devices with managed users described previously, including:

- Classification of network or data sensitivity
- Segmentation requirements
- Authentication requirements
- Group key exploits
- Industry-specific requirements

In addition, certain headless and IoT-type devices may also have unique roaming and availability requirements that may impact the security architecture, adding two more considerations:

- Roaming requirements and latency tolerance, which may dictate whether 802.1X (with or without MAB) can be used with FT, or if a passphrase-secured connection is required
- Availability and uptime requirements, which may dictate whether Wi-Fi or another connectivity technology such as private cellular is best suited for availability needs

Here, the authentication requirements will be less stringent. There will be a subset of endpoints that support a full 802.1X authentication using domain credentials or device certificates, but the majority of endpoints in this category will not.

It's worth noting that even if a non-user-based device supports 802.1X authentication, it is not recommended to connect them to the same SSID as traditional OS endpoints covered earlier. See "Guidance on When to Separate SSIDs" later in this section for more.

High-Security Architecture

A high-security architecture for headless devices with internal access may be based on one or more of the following network types:

- 802.1X-secured SSID
- MAB on an 802.1X-secured SSID
- WPA3-Personal Only SSID
- Cellular LAN

Headless Devices on 802.1X-Secured SSID For endpoints such as most network printers and VoIP systems that support 802.1X authentication, device certificates or domain credentials can be configured on the device manually or provisioned through a management platform. As much as possible, the network should be configured similarly to the methods described earlier in "Managed User with Managed Device," which also use 802.1X-secured networks.

In higher-security environments, headless devices, even when authenticating with 802.1X should be on a separate dedicated SSID, not on a shared SSID with internal users on managed devices.

Headless Devices with MAB on an 802.1X-Secured SSID MAB may be used to authenticate endpoints with known and fixed MAC addresses. Please review the caveats and design best practices detailed in the section titled "MAC-Based Authentications" in Chapter 3.

Headless Devices on WPA3-Personal Only SSID The technical editors and I all agree that any network based on a Personal-class SSID is not considered "highly secure." However, this is relative scoring, and some environments will demand passphrase-based networks. For endpoints that may not have a fixed MAC address and/or have roaming requirements that can't be met with FT, use of WPA3-Personal Only networks is appropriate.

Note that higher security requirements should use WPA3-Personal Only and should not use WPA3-Personal Transition Mode for this purpose (which effectively downgrades security to WPA2-Personal levels).

Headless Devices on Cellular LAN Private cellular (aka cellular LANs) offers a secure alternative to Wi-Fi connected headless devices. Cellular LANs are covered in Chapter 8, "Emergent Trends and Non-Wi-Fi Wireless."

These networks include an intentional and known provisioning process, device identification, and mutual authentication to the network, offering a higher degree of security than the other prior two options for headless devices on Wi-Fi. In addition, cellular LANs operate using transmission scheduling and modified modulation and encoding that make it more suitable for hostile RF noise environments, enable connectivity over longer distances, and offer a higher guaranteed QoS for latency-sensitive applications.

Additional Configurations for Headless Devices

In addition to the type of network security profile, the following should be included in the architecture:

- SSID segmentation of headless devices from fully capable and managed endpoints
- SSID segmentation of headless devices of varying security classifications; for example, a biomedical device that could cause injury or death if tampered with should be separated from digital signage boards
- Other segmentation as required; for headless devices of the same security classification, but with different access requirements, use dynamic VLANs (or similar control) to segment endpoints connected to the same SSID
- Granular policies for authorization; the headless devices should have policies applied in the wireless or wired infrastructure allowing communication to and from only the resources required; some may only require Internet access; others will require access to one or more servers or devices
- Access based on the degree of confidence of what a device is before allowing it network access; especially in sensitive networks, the admin and system owner should know without a doubt what an endpoint is and have a way to re-validate on connection if using MAC addresses to identify the endpoint
- In sensitive networks, there should be a mechanism to profile endpoints actively and/or passively and alert when an endpoint profile changes; this is applicable for all Wi-Fi connected headless devices, regardless of how it's connected; this prevents MAC spoofing and endpoints using passphrases without authorization

CROSS-REFERENCE Also see "Other Considerations for Secure IoT Architecture" in Chapter 8.

Medium-Security Architecture

A medium-security architecture for headless devices with internal access may be based on one or more of the following network types:

- MAB on an 802.1X-secured SSID
- WPA3-Personal Only SSID
- WPA3-Personal Only SSID + WPA2-Personal SSID
- Other MAC-based authentication

The first two bullets are described in the preceding section. The third bullet adds the option to support WPA2-Personal and WPA3-Personal SSIDs, and the fourth bullet adds the flexibility for other (non-MAB) MAC-based authentication.

WPA3-Personal Only SSID + WPA2-Personal SSID For medium-level security networks, WPA2-Personal and WPA3-Personal can be supported in parallel. The configuration should include two SSIDs as follows:

- SSID for WPA2-Personal
- SSID for WPA3-Personal Only

The WPA2- and WPA3- networks should not share the same passphrases nor the same VLANs (or other segments). WPA3-Personal Transition Mode is not appropriate for medium security requirements and should be avoided. To recap the additional security requirements:

- Do not re-use any passphrases between WPA2- and WPA3-Personal networks
- Do not use WPA3-Personal Transition Mode
- Place WPA2- and WPA3- secured endpoints on different VLANs
- As with all headless devices, allow only the specific access required for internal resources

Headless Devices with Other MAC-based Authentication MAC authentication can be achieved several ways. Aside from MAB, medium-security networks may use a NAC product or a Wi-Fi vendor's profiling and IoT engine to identify and then authorize a headless device based on its MAC address. This method does not rely on 802.1X or MAB.

As with other aforementioned methods, any authorization that isn't based on strong device identity and authentication should add mitigations such as endpoint profiling to prevent MAC spoofing.

Low-Security Architecture

A lower-security network for headless devices with internal access may be based on one or more of the following network types:

- WPA3-Personal Only SSID
- WPA3-Personal Only SSID + WPA2-Personal SSID
- WPA3-Personal Transition Mode SSID

For networks and headless devices that don't require medium or high security, an additional option is presented.

The WPA3-Personal Transition Mode SSID supports both WPA2 and WPA3 endpoints on the same SSID. This mode is not recommended since it effectively downgrades the overall security to that of WPA2-Personal including all the vulnerabilities for offline dictionary attacks.

With Transition Mode, there are some mitigations for the WPA3-connected endpoints including forward secrecy and PMF, but continued use and sharing of passphrases between WPA2 and WPA3 networks introduces additional vulnerabilities that impact the entire wireless network.

If WPA3-Personal Transition Mode is used, additional recommendations include:

- Use different passphrases for WPA2 and WPA3 connected devices if possible
- Segment WPA2 and WPA3 connected endpoints on different VLANs using dynamic VLAN assignment or other method

Figure C.2 provides a summary of all three security architecture levels for headless devices accessing internal network resources.

	Feature	High-Security Architecture	Medium-Security Architecture	Low-Security Architecture
Classes of Network	WPA-Enterprise	Same as Managed User on Managed Device	N/A	N/A
	MAB/MAC	Allowed only if MAB endpoints are segmented from 802.1X endpoints	MAB and other MAC-based authenticiation allowed	Allowed
	WPA-Personal	WPA3-Personal Only	WPA3-Personal Only and separate WPA2-Personal with caveats	WPA3-Personal Transition Mode also allowed
	Other Wireless	Cellular LAN	Cellular LAN and other IoT	BLE and other IoT

Figure C.2: Architecture summary for internal access by headless devices

Contractors and Third Parties

Contractors and third parties requiring access to a subset of internal resources often fall just between the scenarios of employee access from a managed device, and BYOD access from a personal device. With contractors and third parties granted privileged access, the organization may invoke additional security controls to protect its assets and data.

Scenarios covered here include an unmanaged user on an unmanaged or lightly managed device. It's not only reasonable but it's common for organizations to require contractors to use a NAC agent or privileged access management agent to allow the organization to inspect the endpoint for security posture.

Depending on the contractual relationship and length of engagement, organizations may issue third parties a managed device and treat them just as they would an employee, in which case the security architecture will be in line with the "Managed User with Managed Device" scenario described previously. On the other end of the spectrum, again depending on the relationship and access needs, the organization may treat contractors and third parties similarly to BYOD access scenarios.

For scenarios covered here, it's assumed the third party needs access to internal resources, is onsite or remote, and is accessing the enterprise resources from an unmanaged device. This may include integrators that require management access to some or all of the wireless infrastructure, vendor systems engineers, or external security monitoring services. It also encompasses non-IT users requiring internal access including contractors providing business operations, accounting, and auditing services, among others.

Security Considerations

Securing access for contractors and other third parties should take into account these considerations:

- Source of access, whether the contractor accessing resources from on-prem or remotely
- Level of access, describing the assets or resources the contractor requires access to
- Duration of access, which may be one time, recurring on a cadence, or for a specific length of time
- Security classification of the resource(s) being accessed; an entity accessing a printer for remote maintenance should be handled differently than a contractor accessing the domain servers or security tools
- Nature of the relationship, including whether the person is employed by a vendor partner with a formalized contractual relationship versus a freelance consultant, plus the posture or security certification of the vendor partner if applicable

High-Security Architecture

For contractors or third parties accessing sensitive networks or data, one or more of these methods should be considered:

Formal Vetting and Onboarding Just as with an Employee This is common and appropriate for long-term contractors effectively working as an employee or extension of the organization; these users are often issued a managed device and are then treated just as those in the "Managed User with Managed Device" architectures covered earlier.

Granting Access Through a Privileged Access Management Platform Contractors that are engaged less often or for a shorter time but are accessing highly sensitive data, networks, or are granted privileged access such as admin access to the Wi-Fi infrastructure should be managed through a privileged access management platform.

These users and their access should be well documented; if appropriate, their endpoints should be scanned for security posture before being allowed access; access should be restricted to only what's required. For users that have access to management of the infrastructure, the connection and every command should be recorded for auditing.

Medium-Security Architecture

In medium-security architectures supporting contractor access, the organization may opt to loosen security a bit and add the additional option for VPN- or NAC-based access.

VPN Access for Contractors VPN access can be used to secure and control connections both on-prem and remote but for the purpose of contractors, it's most often for remote access. In wireless architecture, our purview is for contractors, engineers, and TAC that may need management access to the wireless infrastructure for configuration or troubleshooting.

Ideally VPN access for contractors will leverage a product that supports a zero trust strategy, and therefore allows very granular access policies. A contractor shouldn't have full access to the entire network environment.

NAC Access for Contractors NAC products can enforce granular authorization policies including for contractors. Policies can be enforced via wired connections, wireless, and/or remote VPN, depending on the deployment.

With both scenarios, vendor management processes and privileged access management rules should be followed, if defined by the organization. For third parties accessing key infrastructure devices, they should be vetted and have insurance verified, etc.

It is also perfectly reasonable to require third parties that are accessing internal resources to follow your organization's requirements for security scanning and posturing—meaning whether agent or agentless, it's appropriate and common to require interaction with a NAC or MDM.

Low-Security Architecture

In organizations that have less mature processes, lack the resources to implement granular control and auditing, or simply have a higher risk tolerance, contractor and third-party access may be handled in one of the following ways:

Granted Full Network Access Without Inspection or NAC Although not ideal, it's not uncommon for risk-tolerant organizations to simply grant access to certain trusted third parties with no formal vetting, privileged access management, or auditing. These users may be given a domain account and allowed to connect with their own device, which is not scanned for security posture. Others may be granted through a passphrase-secured network with access to the production network.

Granted Use of a Managed Device with Logged-on User Another access scenario for contractors in lower-security environments is to allow the third party to use a managed endpoint that a user has logged on to. This may be the lesser of two evils but isn't always feasible and offers no auditable attribution to actions taken by the contractor; all events will be documented as the logged-on user.

BYOD/Personal Devices with Internal Access

As defined throughout the book, BYOD scenarios may include access to internal resources or Internet-only access. This section covers managed users such as employees or students that are accessing internal resources from a personal device.

Security Considerations

For a refresher on security considerations related to BYOD with internal access, visit "Bring Your Own Device" in Chapter 8. Here are a few of the highlights from that content on considerations:

- Management and ownership model (BYOD, COPE, CYOD, etc.)
- Legal considerations
- Technical considerations

As with all content in "Sample Architectures," the high, medium, and low requirements are defined in part by the organization's risk management and by the sensitivity of the data or resources being accessed. For each use case, it's reasonable to design multiple networks if needed to meet differing privilege use cases.

High-Security Architecture

High security requirements indicate networks with sensitive data or resources that require a high degree of protection. Personal devices accessing assets with high security requirements should plan for:

- BYOD, CYOD, COPE, or COBO models, with additional security vetting required for BYOD
- Access via 802.1X-secured networks only, specifically:
  - WPA3-Enterprise Only SSID
- Use of BYOD onboarding platform or MDM to manage server and device certificates for 802.1X
- Use of MDM platform for posture scanning and monitoring of the endpoint
- Granular access policies allowing only access to and from the resource required by business needs
- Networks with personally managed smartphones, tablets, and laptops should be segmented from enterprise-managed endpoints with a dedicated SSID

Medium-Security Architecture

Personal devices accessing internal network resources with medium security requirements should plan for:

- BYOD, CYOD, COPE, or COBO models, with additional security vetting required for BYOD
- Access via 802.1X- or Personal-secured networks, specifically:
  - WPA3-Enterprise Only SSID, or
  - WPA3-Personal Only SSID
- Optional use of BYOD onboarding platform or MDM to manage server and device certificates for 802.1X
- Use of MDM platform for posture scanning and monitoring of the endpoint
- Access policies allowing only access to and from the resource required by business needs

Low-Security Architecture

Personal devices accessing internal network resources with low security requirements can plan for:

- BYOD, CYOD, or COPE models, with additional security vetting required for BYOD
- Personal device self-registration through portal
- Access via 802.1X- or Personal-secured networks, which may include:
  - WPA2-Enterprise
  - WPA3-Enterprise Only
  - WPA3-Enterprise Transition Mode, or
  - WPA2-Personal
  - WPA3-Personal Only, and
  - WPA3-Personal Transition Mode
- Optional use of BYOD onboarding platform or MDM to manage server and device certificates for 802.1X

Guidance on WPA2-Enterprise and WPA3-Enterprise

Even though WPA3 has just recently been added to Wi-Fi products, any modern endpoint with a standard operating system should be WPA3-capable. WPA3 is required for Wi-Fi 6 and newer devices. Most standard endpoints updated capabilities by the end of 2020, and if the option isn't available, endpoints may simply require a firmware upgrade.

Migrating from WPA2-Enterprise to WPA3-Enterprise

Organizations can quickly and easily migrate existing WPA2-Enterprise networks to WPA3-Enterprise and should not require reconfiguration of any endpoints or RADIUS policies, unless upgrading to the 192-bit security mode.

If there are concerns about the endpoint ability, the intermediary step of moving the SSID from WPA2-Enterprise to WPA3-Enterprise Transition Mode will allow the network admins to verify endpoint connectivity; any Wi-Fi product should show whether a client is connected with WPA2 or WPA3 security, and/or reports can be run.

For small organizations, the intermediary configuration can be short—a week or two. For large organizations additional time may be required but as a guideline, should not exceed six weeks unless there are extenuating circumstances. At the end of the transition time when client support has been proven, the SSID should be updated again from WPA3-Enterprise Transition Mode to WPA3-Enterprise Only mode.

TIP **The defining factor of whether a connection is WPA2 or WPA3 is whether or not PMF is enabled. If PMF is enabled on a WPA2 SSID, by definition it becomes a WPA3 Transition Mode network. Similarly, if an endpoint attaches to a WPA2 SSID with PMF enabled, it is classified as a WPA3 endpoint.**

Supporting WPA2-Enterprise with WPA3-Enterprise

If WPA2-Enterprise support is required in a medium- to high-security environment for an extended period of time, the architect should plan for two SSIDs—one being a WPA2-Enterprise secured network, and the other the WPA3-Enterprise Only network. It's left to the organization to determine whether the WPA2 or WPA3 network should be the newly created SSID.

If most endpoints support WPA3, then it will be more efficient to create a new WPA2-Enterprise SSID, and simply modify the existing WPA2-Enterprise SSID and change it to WPA3-Enterprise Only. The endpoints that don't support WPA3 will need to be modified manually or via group policy to connect to the newly created WPA2 network with a different SSID name.

For highly secured networks, Transition Mode should not be used other than for very short-term migration periods as described previously.

Guidance on when to Separate SSIDs

The one overarching theme of this appendix, and perhaps this entire book, is that today's security considerations are driving architectures toward separating SSIDs—a direct opposition to the decade-old guidance to collapse them.

Why? SSIDs are just an over-the-air broadcast domain. There are numerous known and well-documented vulnerabilities related to misuse of group keys in Wi-Fi. Hole 196 from 2010 is just one of many and it's still a vulnerability and viable attack today, not only for WPA2-secured networks but also WPA3. These attacks are insider attacks and are simply an abuse of the standard as it's written. As such, the Wi-Fi industry (thus far) has elected to not address these issues, and therefore the security architect must design mitigations for these known risks. These attacks are so specific that in most environments there are many easier ways for an attacker to gain entry, but it's worth mentioning.

Even if you've created an SSID with dynamic VLANs, all devices on the same SSID are in the same Wi-Fi broadcast domain and using the same group keys. Attacks include everything from denial of service to installation of malware.

The recommendations dating back to 2008 and 2009 to collapse SSIDs was based on the desire to preserve airtime by reducing the overhead of beacons. Fast forward 10 years, and the Wi-Fi technology has evolved considerably, as has the speeds of 802.11 WLANs. The 802.11 of 10 years ago started to introduce airtime issues in the range of 3–6 SSIDs. Today's technologies can support a greater number of SSIDs without negatively impacting user experience. That doesn't mean we have carte blanche for an unlimited number of SSIDs; the design still requires a delicate balance, but we need not (and should not) continue attempting to collapse our wireless to three SSIDs.

Architectures for Guest/Internet-only Networks

Architectures for users and endpoints that only require Internet access differ from internal access requirements described in the prior section. Internet-only connections don't absolve the organization from all liability but offer greater flexibility and relaxed security requirements.

In this section, secure architectures are described for the following access scenarios:

- Guest Networks
- BYOD/Personal Devices with Internet-only Access

Guest Networks

Guest networks as defined here are Internet-only and are designed for transient guest users. Guest use cases for personal devices are covered under the two BYOD headings, and use cases for contractors and third parties requiring internal access were covered earlier.

Security Considerations

When it comes to guest Internet-only access, the mandates for certain security controls are lessened. The relative labels of high-, medium-, and low-security architectures described in the coming text really maps to how much visibility and control the organization wants over the guest user.

CROSS-REFERENCE **The following addresses SSID-based security. For segmentation best practices and recommendations, visit Chapter 2, "Understanding Technical Elements."**

Some organizations don't allow guest access at all; others do with a formal registration and approval, while some use open networks with no self-registration and not even a captive portal to present acceptable use terms.

As a reminder, the six models for guest access include the following. They're joined by an open network option with no portal at all for a total of seven scenarios.

- Guest self-registration without verification
- Guest self-registration with verification
- Guest sponsored registration
- Guest pre-approved registration
- Guest bulk registration
- Captive portals for acceptable use policies
- Open network with no portal

Revisit the "Captive Portal Security" area of Chapter 3 for details of these models and for more on planning captive portals for guest access.

High-Security Architecture

Organizations that allow guests but wish to have full visibility and auditing of guest access will use one of these methods:

- Guest self-registration with verification
- Guest sponsored registration

Both methods ensure the guest requesting access is a real person and has not falsified data to gain access, such as used a fake email address during a self-registration process. These two methods also ensure someone within the organization knows and expects the guest user and is authorizing them at the time of access (versus pre-provisioning access).

Medium-Security Architecture

Organizations that prefer to have some visibility and auditing of guest access but with a less stringent process may select one or more of these methods:

- Guest pre-approved registration
- Guest self-registration without verification
- Guest bulk registration

Pre-approved registrations are similar to the approval-based registrations mentioned in high-security methods but is included here under medium security since there is an opportunity to abuse an account that was pre-provisioned.

Low-Security Architecture

Organizations that are not concerned with knowing who the guest users are may opt for these methods:

- Captive portals for acceptable use policies
- Open network with no portal

Even if an organization isn't concerned with security for guest networks, it makes sense to enable Enhanced Open, which is encrypted.

BYOD/Personal Devices with Internet-Only Access

As mentioned earlier, BYOD scenarios may include access to internal resources or Internet-only access. This section covers managed users such as employees or students that require only Internet access from a personal device.

As covered in Chapter 8, BYOD users vary from guest users in that most organizations want to remove the friction of forcing employees to repeat a traditional guest captive portal experience daily.

For that reason, additional networks and authentication may be provisioned to allow personal devices on an Internet-only network without enforcing repeating the captive portal.

Security Considerations

For BYOD and personal devices that require Internet-only, the security considerations look a bit different than other use cases and include additional goals such as:

- Protecting domain credentials
- Keeping personal devices off production networks
- Keeping managed endpoints off guest networks

If a personal device is accessing internal resources, it should (in any level of security) be required to have an MDM or other enterprise-managed agent. However, if a personal device is being used for Internet-only access, there's really no reason to go through the hassle and ongoing maintenance involved with an MDM. Fundamentally these are two very different access models requiring unique security architecture.

Because BYOD is a personal device in use by someone with domain credentials, one goal in managing BYOD is to protect the user's enterprise credentials by not exposing them in an unmanaged device. This is especially relevant for personal devices without an enterprise MDM such as those with Internet-only access.

WARNING **Under no circumstance should a managed user enter their credentials on a personal device to (for example) join it to a secured 802.1X network that's enabled for username/password-based authentication. Instead, options for supporting Internet-only personal devices include**

- **A device registration portal**
- **Passphrase-secured SSIDs**

Each of these two options are available in a few configurations that result in varying degrees of security.

High-Security Architecture

Just as with guest registrations, if an organization prefers to know what device belongs to whom, and/or requires some level of attribution, they may choose to implement a device registration portal. These portals are offered natively in some Wi-Fi products and are available in every NAC product, as well as some authentication servers.

To date, registration has always been based on the MAC address. The process can vary but generally follows this flow:

1. Employee/user connects to the BYOD portal from their personal device.
2. Employee/user enters their domain credentials into the secure BYOD portal and registers the personal device (by MAC address).
3. Wi-Fi and/or BYOD registration system registers the device's MAC address as a known (but unmanaged) device.
4. The personal device is granted access for the specified time without revisiting the portal (durations vary but are frequently 90 days to 1 year).
5. Optionally, some portals allow the organization to specify a maximum number of devices per-user.
6. Optionally, some portals offer the option for users to self-manage their devices (this is especially common in higher-education deployments with NAC products).

In today's architectures, any open networks will use Enhanced Open to offer encryption over the air.

Use of passphrase-secured networks is also an option for Internet-only BYOD. Note that passphrase-secured networks for high security requirements should meet these criteria:

- WPA3-Personal Only (do not allow Transition Mode)
- The SSID should have Internet-only access
- The SSID should not be shared with any enterprise managed endpoints including IoT

Medium-Security Architecture

For organizations that don't need to know the identity of Internet-only BYOD devices, they may simply provision a passphrase-secured network for BYOD. Passphrase-secured networks for medium security requirements should meet these criteria:

- WPA3-Personal Only or WPA3-Personal Transition Mode
- The SSID should have Internet-only access
- The SSID should not be shared with any enterprise managed endpoints including IoT

Low-Security Architecture

For organizations that aren't concerned with user friction and aren't concerned with the security of the personal devices, BYOD devices may connect with open networks with no encryption, including other guest networks.

Determining Length of a WPA3-Personal Passphrase

Ah, we've arrived at the million-dollar question: How long should WPA3-Personal passphrases be?

Getting the answer to this was quite the journey. I asked notable cryptographers. I asked pen testers. I asked friends and family. (All of my friends and family work in tech, so that seems reasonable.)

Why Passphrase Length Matters

WPA2-Personal networks have a minimum length of 8 characters. They are susceptible to offline dictionary attacks, and the key was directly derived from the passphrase, hence the term "pre-shared key."

WPA3-Personal networks use a new algorithm, Simultaneous Authentication of Equals, or SAE. The key isn't directly derived from the passphrase, and so the encryption key length is not dependent on the passphrase length; meaning, if a WPA3-Personal passphrase was only three characters long, the cryptographic key would be the same length as if the passphrase were 20 characters long.

That begged the question: Why then do we care about WPA3-Personal passphrase lengths? If an attacker can brute force the WPA3 passphrase, they can join the network and have access to some set of network resources and have some level of access to other peer endpoints on the same SSID. Remember there are insider attacks and an SSID is a broadcast domain.

That means in order to protect passphrase-secured networks, we need to have an appropriate passphrase length.

CROSS-REFERENCE **Visit Chapter 2 for more on WPA2 versus WPA3 and the security enhancements of WPA3.**

Considerations for Passphrase Length

In Chapter 7, "Monitoring and Maintenance of Wireless Networks," I said pen testing offered a relative score of security. How long, with what tools, and how much effort did it take to compromise a target? Security controls don't give us 100 percent protection. They buy us time, and if our architecture is correct, we're alerted before an attack is successful. It's not a matter of "if" it can be breached, but "when" or "how long did it take?"

Cracking passphrases works the same way. The question is: How long does a WPA3-Personal passphrase need to be to offer reasonable resilience against today's attack tools?

The answer? It depends. WPA3 is new in the wild, and there's not a lot of data to analyze and base decisions on. Here's what we do know.

Specifically, the resilience of the passphrase to attacks depends on:

- Capabilities of the AP (a beefier AP can handle more connections per second and therefore more attempts)
- Quantity of APs in the environment (each AP could be targeted for attempts, multiplying the effort)
- Capabilities of the attacker (number of attack devices)

An attacker with a single device, targeting a single AP can operate at speed Y. An attack launched against 100 APs can increase the attack to 100Y and reduce the time to crack to 1/100th the time.

The goal of passphrase security should be considered as well. It's common to use passphrase-secured networks for:

Adding Encryption to What Would Otherwise Be an Open or Guest Network If this is the use case, consider switching to Enhanced Open, and otherwise the passphrase length is not significant as long as the network is Internet-only.

Creating a Barrier to Entry to Avoid Public Use of a Network If the goal is simply to keep passersby off a network, recommendations depend on the nature of that network. For internal access, it should be 802.1X-secured. For guest or Internet-only, passphrase is acceptable and shorter passphrases are sufficient.

Connecting IoT and Headless Devices In this use case, the most secure option is to move to 802.1X if possible. If not, the passphrase length recommendations vary depending on the nature of the IoT device, other endpoints on the same SSID, and the resources they have access to.

Connecting Non-Domain Endpoints with Internal Access This use case may be for contractors or a subset of BYOD models. If the user has internal access, it is recommended to move to an 802.1X-secured network for managed endpoints or use privileged access management (PAM) solutions for third-party access. If there's an unavoidable situation where a user must access an internal network with a passphrase, then the passphrase should be long.

Recommendations for Passphrase Lengths

The following guidance is based on certain assumptions of architecture and without the benefit of historical data (since WPA3 is new).

In the preceding lists, I referenced relative short and long passphrase lengths.

The general consensus is: if you want to protect against a brute force attack, a length of 8 characters is entirely too short, and something close to 20 characters is more appropriate.

With all of the caveats, exceptions, and fine print, I offer the lengths in Table C.1 as rough recommendations to consider.

Table C.1: WPA3-Personal passphrase length recommendations

USE CASE	RESOURCE ACCESS	PASSPHRASE LENGTH	RESILIENCE TO BRUTE FORCE
Adding encryption to what would otherwise be an open or guest network	Internet-only, no critical or secured systems on the same SSID	8–12	Low
Creating a barrier to entry to avoid public use of a network	Internet-only, no critical or secured systems on the same SSID	8–12	Low
Connecting IoT and headless devices	Internet-only, no critical or secured systems on the same SSID	14–20	Medium
Connecting IoT and headless devices	Internal or Internet-only but other critical systems on the same SSID	18–22	High
Connecting non-domain endpoints	Internal	18–22	High

APPENDIX

D

Parting Thoughts and Call to Action

The Future of Cellular and Wi-Fi

Cellular 5G is bringing a new era of connectivity and new opportunity, but Wi-Fi and cellular will continue to be complementary, not competing, technologies.

This section focuses on these two topics and provides some context and comparison for organizations making decisions about cellular augmentation moving forward:

- Cellular Carrier Use of Unlicensed Spectrum
- Cellular Neutral Host Networks

Cellular Carrier Use of Unlicensed Spectrum

One of the questions I'm often asked is: "Will LAA interfere with Wi-Fi?" The short answer is: YES.

For years, mobile network operators (MNOs, aka cellular carriers) have been seeking ways to augment their connectivity and capacity for remote and high-density cellular areas. They do this by using portions of unlicensed spectrum (the same bands we use for Wi-Fi).

Licensed Assisted Access (LAA) is one technology designed for this need. LAA was introduced in 4G/LTE and approved by the FCC in the U.S. in 2016.

It uses spectrum in the unlicensed 5 GHz band which does, in fact, overlap with Wi-Fi. LAA uses cellular (LTE) signaling (not Wi-Fi). Although in most markets (including the U.S., India, and South Korea) LAA uses a Listen Before Talk (LBT) mechanism to attempt to avoid interfering with Wi-Fi, if there is no free airtime, LAA will "share one of the occupied channels equally and fairly with Wi-Fi."

Now, I don't know who gets to define what "equally and fairly" looks like, but in an already crowded Wi-Fi spectrum, LAA will absolutely impact Wi-Fi. The sensing threshold for LAA's LBT is –72 dBm, which is moderate but not necessarily adequate even in an undertaxed Wi-Fi environment. Most environments also suffer severe degradation on Wi-Fi with hidden node problems, which occur almost everywhere at some point.

In 2021, researchers at the University of Chicago in the U.S. studied the impact of LAA on Wi-Fi performance.

"Wi-Fi throughput can degrade as much as 97 percent when coexisting with LAA whereas LAA throughput only degrades 35 percent," the researchers wrote. A summary of the findings and a link to the research can be found at `https://www.cs.uchicago.edu/news/article/laa-Wi-Fi/`. The full body of research can be accessed via `https://arxiv.org/abs/2103.15591`.

The other technology a carrier may propose (instead of LAA) is LTE-U (LTE over Unlicensed), which also uses Wi-Fi's precious 5 GHz spectrum. LTE-U was a precursor to LAA in terms of RF sharing technology and it doesn't employ the same LBT mechanism as LAA, meaning it introduces further opportunity for interference.

Because cellular uses scheduled and coordinated slots and Wi-Fi uses contention-based mechanisms, there's unlikely to be a "fair" and "equal" sharing. Even if there was, the overhead added by the LTE network (such as with LAA LBT) will query and reserve airtime for LTE and impact the performance of both the Wi-Fi and the unlicensed cellular transmissions.

In the LAA and LTE-U models, the MNO (carrier) deploys its equipment within your organization and manages the network, which will use cellular technology over the 5 GHz spectrum. One alternative to the MNO's augmentation of its networks is use of neutral host networks, covered next.

Cellular Neutral Host Networks

Neutral host networks (NHNs) also provide a method to supplement cellular carrier coverage, but they do so in a way that allows the enterprise to retain better control over the RF environment. NHNs are made possible by the recent advances in cellular technology and its use of NFV and SDN. With these two abstractions, cellular communications aren't tied to carrier hardware and software as in prior generations.

In NHNs, the carrier's networks are advertised through the enterprise's cellular infrastructure. That statement may not make sense if you haven't read the "Private Cellular and Cellular LANs" section of Chapter 8.

For organizations already taking advantage of private cellular or cellular LANs, adding NHN capabilities will only require a few knobs and settings to connect to carriers on the backend.

I imagine most technologists will have an immediate "heck, no!" response to the notion of the enterprise serving up carrier networks, but over time, the benefits will greatly outweigh the concerns. Security is not even a consideration in NHN because the authentication and IPSec are handled completely by the respective carrier—the enterprise is not involved in the device identity, authentication, encryption, or security at all. It's only serving as a conduit to pass the traffic to the carrier's network.

In fact, NHN operates on the same principle as single sign-on (SSO) or payment processing. Think of how many online sites you've ordered from and were allowed to pay with your Amazon, Google Pay, PayPal, or Apple Pay account. You've probably also had the option to create an account with a new service or website or simply log on using Google, Facebook, or other social site credentials. It's the same theory with NHN.

In addition, 5G networks will further protect public users including those on NHNs with privacy controls. Cellular devices have a permanent fixed identity—a Subscription Permanent Identifier (SUPI), which is protected through the use of a Subscription Concealed Identifier (SUCI). The closest network analogy is that of a MAC address on a device, and the use of MAC randomization for user privacy. However, in my opinion, the cellular implementation of SUPI/SUCI is a bit more elegant.

NHNs offer enterprises the opportunity to increase cellular connectivity in underserviced areas without handing over control of the RF to the carrier.

If the LAA, LTE-U, and NHN are confusing, Table D.1 provides a comparison.

Table D.1: Overview of cellular, Wi-Fi, and MNO augmentation

TECHNOLOGY	MANAGED BY	USERS/ DEVICES	NETWORK ACCESSED	SPECTRUM USED
LAA/LTE-U	Carrier/MNO	Public Users	Carrier Network	5 GHz
Neutral Host Network	Enterprise	Public Users	Carrier Network	Private Cellular/CBRS
Cellular LAN	Enterprise	Enterprise Devices	Enterprise Network	Private Cellular/CBRS
802.11 Wi-Fi	Enterprise	Enterprise Devices	Enterprise Network	2.4 GHz, 5 GHz, 6 GHz

MAC Randomization

I've made every attempt throughout this book to not go on a rant about MAC randomization, but I can't contain my disdain for it any longer. This topic will remain brief, but I'll touch on:

- The Purpose of MAC Randomization
- How MAC Randomization Works
- The Future of Networking with MAC Randomization

The Purpose of MAC Randomization

It's all about data privacy and consent. MAC randomization was designed to afford users on public networks a level of privacy by obscuring the device identity. The thought was that users could disable app-level location tracking of cellular phones via GPS, but there wasn't a way for a user to prevent a Wi-Fi network operator from tracking the device's movement by MAC address.

What it was not designed to do (necessarily) was modify behavior for enterprise-managed devices on enterprise networks, but MAC randomization is enabled by default, and the impact on enterprise networks is just considered massive collateral damage.

How MAC Randomization Works

Where's my "it depends" stamp? The creation of the random MAC addresses is specified by an IEEE standard and is covered in Chapter 7—see the subsection titled "Client with Invalid (Fake) MAC Address" under "Attacks WIPS Can Detect and Prevent."

How it actually works is dependent on the vendor implementation and the default behavior has varied wildly since its inception, including:

- Unique MAC address on each connection
- Unique MAC address per SSID (unless/until the device is told to forget the network)
- Unique MAC address rotated after XX hours or days
- Randomized MAC addresses for active scanning (when not associated)
- Randomized MAC addresses for Access Network Query Protocol (ANQP) exchanges (when not associated)

Most platforms allow some level of customization and the option for users to enable or disable a randomized MAC address per network. Whether we like it or not, MAC addresses are used for unique device identification for countless (and critical) purposes, such as:

- Captive portal registration including pay-per-use networks
- Portals for device registration including BYOD
- Identifying and securing enterprise-owned headless devices
- MAC-based NAC functions for registration, authentication, and authorization
- Enterprise device authorization with MAC Authentication Bypass (MAB) for endpoints and infrastructure devices
- Security monitoring and event correlation including WIPS
- Network services such as DHCP leases
- Troubleshooting and end-user support
- Network analytics

The impact of randomized MAC addresses on enterprise networks manifests in many ways, including:

- Forcing guest users through captive portal and pay gates multiple times
- Forcing BYOD users through registration portals multiple times (and consuming "seats")
- Driving up costs associated with per-device licensing, which is all based on unique MAC addresses in the system
- Inability to uniquely identify enterprise assets for inventory and security
- Inability to identify endpoints for troubleshooting and trending analysis that informs network configuration modifications
- Inability to properly correlate security events

In addition to impacting the enterprise, the user experience may be negatively impacted, as hinted in the preceding bullets The impact of MAC randomization on captive portals is further detailed in Chapter 3.

The Future of Networking with MAC Randomization

Despite numerous organizations pushing back, IEEE's response to the issue is that (I'm paraphrasing)—MAC addresses should not be used to uniquely identify devices on the network, and they should only be used for layer 2 communication.

Alternatively, IEEE suggests that users and devices should be identified by some other means. The only other option to identify a device is to force some form of authentication, which (in my opinion) only increases the user friction, further invades the user's privacy, and often is not possible for headless IoT devices.

My frustration with the application of MAC randomization is that the industry has tried to solve a political compliance problem with a technical control that's not appropriate for enterprise applications. If the issue is that network operators are using MAC addresses for marketing purposes and financial gain, then just as with all other PII, a governing body should implement a policy protecting the user's privacy and forbid use of that data in that way.

So far, the appeals from numerous organizations to IEEE for MAC randomization are falling on deaf ears. And not only is it unlikely the industry will roll back from MAC randomization, but some vendors are even pushing for further expansion of its use—advocating to require MAC randomization for each and every connected device. If that happens, the IT and infosec communities are going to have a serious problem.

Despite the dismissive rebuttal by IEEE that "MAC addresses shouldn't be used for device identity," the truth is the great majority of current IoT-type endpoints don't support any other option. As I've mentioned several times throughout this book, using a MAC address is (at best) identity and not authentication. But, for now it's all we have to work with in some cases. It will be years before the endpoint industry makes use of unique hardware-based device certificates, and even then, that identity will be just as telling as a MAC address—even more so.

If MAC randomization drives the industry in a better direction for unique device IDs, that will be a positive side effect. My soapbox on the topic isn't that we should keep relying on MAC addresses, but that randomizing MAC addresses in the enterprise doesn't add value and has numerous negative consequences.

In the meantime, it's widely accepted for enterprises to disallow MAC randomization for endpoints connected to their networks. Just like a business can put up a "shirt and shoes required" sign or a facility can say "face coverings required," so, too, can the enterprise stipulate the conditions for using its resources. This is especially true for managed endpoints and those in BYOD programs.

Security, Industry, and The Great Compromise

Throughout this book, one thing that's become abundantly clear is that some technology vendors are far more concerned with their own interests and success than with securing our connected world. Further, the ecosystem of vendors in the space of networking and wireless aren't limited to the enterprise vendors mentioned throughout this book. In fact, many of the decisions that impact our business technology systems are driven by providers in the consumer and residential markets.

In thinking and talking through the state of the technology industry and its influences on critical components of security, one thing becomes abundantly clear: *Security is all about compromise.*

As with anything that comes out of multiple industry organizations such as the IEEE, IETF, and the Wi-Fi Alliance—security, like life, is about compromise. It's a compromise that becomes a tug-of-war between vendors, manufacturers, implementors, and time. Why is security a compromise? Because security is not easy.

Behind the scenes, a delicate dance plays out—a pas de deux of software and hardware, where one's capabilities outstrip its partners'. The entire scene takes place on a backdrop of legacy and less-capable client devices, and all the while, spectators call out for "backward compatibility"—security's greatest nemesis.

Compromise is to be expected, but it's clear the technology industry has work to do, to better examine its decisions, to get input on deployment models, hear customer use cases, and establish reasonable timelines for the adoption of major changes. In talking to colleagues, it's evident some edicts are issued from the ivory tower of standards bodies, without regard for the impact on enterprises, network operators, and even end users.

Their actions, while never ill-intended, bring about unintended consequences.

MAC randomization was one example of a poorly executed industry decree. A choice was made in the name of consumer privacy, but that decision rippled through enterprise environments. It left a wake of chaos as organizations and universities were left unarmed against a barrage of failed security tools, crippled user-support workflows, and routers and wireless controllers that collapsed under the load of bloated ARP and MAC tables.

Security Transition Modes have been another folly of compromise. With the constant advances in security, "backward compatibility" only ever means one thing: relegating security to accommodate the lowest common denominator. Transition Modes give a false sense of increased security—teasing us with the belief that it's additive to what we have or had. In reality Transition Modes downgrade security of the more capable to match that of the less capable.

The compromises of Transition Modes have plagued Wi-Fi for decades. Attackers wait at the ready, prepared to slice through the latest security protocols by exploiting the downgraded security that accompanies Transition Modes. The defenseless end user proves a perfect target. Unlike browser controls, in Wi-Fi the end user isn't even aware of the security level, nor of a downgrade. It's a perfect storm, and one that will capsize even the most capable security architecture.

What's one to do with this information? Well, if you're an architect, be prepared for a world of standards and users that will favor compromise. Balance security requirements with business requirements and remain steadfast in the face of opposition when required. We wish you luck.

For the industry: our hope is that standards bodies, consumer vendors, and the professionals participating in these areas will create space for input into areas that have profound impact on enterprise security architecture. There's an opportunity to accommodate all use cases, and all users, and on behalf of the enterprise IT community, we hope to have a seat at the table.

Index

A

D

N

O

S

X

Z